Springer Geography

The Springer Geography series seeks to publish a broad portfolio of scientific books, aiming at researchers, students, and everyone interested in geographical research.

The series includes peer-reviewed monographs, edited volumes, textbooks, and conference proceedings. It covers the major topics in geography and geographical sciences including, but not limited to; Economic Geography, Landscape and Urban Planning, Urban Geography, Physical Geography and Environmental Geography.

Springer Geography—now indexed in Scopus

More information about this series at http://www.springer.com/series/10180

Xiao Wu · Lingjin Wang

Settlement Spaces: Urban Survival Prospects of China's Special Communities

Empirical Study of Four Types of Representative Community Samples

Springer

Xiao Wu
School of Architecture
Southeast University
Nanjing, Jiangsu, China

Lingjin Wang
School of Architecture
Southeast University
Nanjing, Jiangsu, China

ISSN 2194-315X ISSN 2194-3168 (electronic)
Springer Geography
ISBN 978-981-16-4891-5 ISBN 978-981-16-4892-2 (eBook)
https://doi.org/10.1007/978-981-16-4892-2

This Springer imprint is published by the registered company Springer Nature Singapore Pte Ltd.
The registered company address is: 152 Beach Road, #21-01/04 Gateway East, Singapore 189721, Singapore

National Natural Science Foundation of China (51878142, 51778125, 51678130)

"13th Five-Year Plan of China" National Key Research and Development Program Project (2019YFD1100800)

Postgraduate Research and Practice Innovation Program of Jiangsu Province (KYCX20_0143)

Foreword

Prof. Wu Xiao, a key young pedagogue at our academy, submitted a new book to the press and asked me to furnish a preface. My first reading of the manuscript elicited some of my own thoughts and experiences in contemporary urban planning and social transformation.

Generally, this work in many ways presents the author's own valuable attempts, investigations, and judgments. Two points, however, make the deepest impression: The first is the use of the interdisciplinary perspective of "sociology" to examine and advance our current and increasingly important work in urban planning. The second is the focus on research at the scale of "communities," employing diverse methods to research city systems in the context of urbanization.

Why are we set on establishing a sociological perspective for urban planning? Why are we focusing on residential communities of all types of groups (especially marginalized groups), a question that may be expanded to understand the value of community research? These are the two significant questions that that this book tries to answer.

The following two contextual aspects can strengthen our understanding of the establishment of a sociological perspective for urban planning.

First, we must consider the cutting-edge demand for contemporary international urban planning development. In the last few decades, urban planning has shared an increasingly obvious tendency for specialized delegation with other related disciplines, producing commonly acknowledged interdisciplinary research fields. One of the foremost of these is the attention paid to social, cultural, economic, and policy matters in regional and urban–rural spaces. This also means that effective research on issues of space needs to be considered more comprehensively in the context of socially engineered conditions. The interplay and cross-fertilization between sociology and urban planning have undoubtedly become an internal driving factor and an inevitable direction for contemporary urban planning at the frontier of international practice.

Second, there is a demand for the transformation of current domestic urban planning work. For a long time after reform and opening, China's urban planning department underwent a gradual transformation from being a technical department under the government guidance of a planned economy to becoming a department that assumed more economic roles. The emphasis of its work has shifted from resource allocation and social redistribution to stimulating economic growth and solving employment and income problems. The latest government has proposed its "three 100 million" policy, setting a roadmap for the implementation of a new form of urbanization. Regardless of whether it is the question of the urban assimilation of agricultural migrants or the renovation and securitization of old communities, this work presents ideas and arguments from different angles and at different levels.

In this case, what is the value of community research? Ever since Ferdinand Tönnies (1855–1936) coined the core concept of "community" over a 100 years ago, many scholars have continued to probe this question, building a body of theoretical results and a methodological framework nourished by perspectives from different disciplines while ripening and becoming gradually more comprehensive. Although at the beginning of the twentieth century, Chinese sociologists such as Wu Wunzao and Fei Xiaotong accomplished unique and broadly influential community research, the academic field of urban planning in China is still in a long process of transition from "residential area" research to "community" research. In the eyes of many, "community" is often understood as a general abstraction, a "society" writ small. However, the most obvious feature and greatest value of community research in terms of its methodology and scope are its insight of being able to extrapolate the development of a trend at an early stage. In other words, "communities" provide us insight about "society." Research at a "microscopic scale" can illustrate the "macroscopic structure." As a result, in this work, we can appreciate the rare perseverance and industry of the author and his research group. Working principally from the scale and perspective of "communities," they connected and organized over a decade's worth of research results, promoting and analyzing diverse "special community spaces" that are representative of China's shifting urban and social landscape. In doing so, they touch on and reveal several universal patterns in contemporary Chinese urban social spaces. Such an investigation not only requires keenness of perception, a stalwart attitude, and enduring courage but also calls for interdisciplinary study to cover and master the complex content of "community research," the distinction of types, and a diverse range of techniques.

This book continues Prof. Wu Xiao and his research group's long-term concern for disadvantaged groups, which has become a hallmark of their research. Through empirical research across different types of urban–rural representative community samples, they have encompassed and analyzed international student settlements, protected residential areas, settlements with a fluctuating population, and ethnic Hui settlements, among other diverse special communities and their settlement spaces. This research topic has provided a key window and an important roadmap to analyze urban social spaces in the context of the shifting and complicated social landscape of China's cities. Taking the long-term research and results of the author and his research group as a foundation, this achievement represents their own proactive investigations, thoughts, and answers in the important areas of China's construction of a harmonious society, the discussion (in theory and practice) of a transition period in urban planning, and social values of justice and humanitarianism.

I have known Prof. Wu Xiao for many years. He is an industrious and meticulous scholar who has made equal contributions to pedagogy and research. Prof. Wu has won the recognition and financial support of foster programs such as the Ministry of Education's "New Century Outstanding Talent Support Scheme" and Jiangsu Province's "333 High Profile Talent Foster Program." In terms of research, my colleague has long been engaged in projects that bridge urban sociology and community development, spaces for disadvantaged groups and the means for their protection, and urban design theory and methodology, achieving a body of innovative results. His academic perspective, which straddles urban planning and urban sociology, constitutes a systematization of knowledge that fosters unique and field leading results.

Evidently, interdisciplinary research that integrates perspectives from sociology will continue to thrive. Ideas from research on "communities" are also boundless. I hope that Prof. Wu stays true to his initial ideals on the academic road ahead and achieves even greater success in his research.

Jianguo Wang
Academician of Chinese Academy of Engineering
Professor at School of Architecture
Southeast University
Doctoral Supervisor
Nanjing, China

Preface

This book derives its origin from two enduring academic interests that have been the focus of my research group for many years:

First, how should we approach community research in this new century?

In terms of community research, China's planning sector is currently undergoing a paradigm shift and a practical sea change from a "work unit-oriented nation" to a "community-oriented nation." During the era of the planned economy, China's community research was characterized by its "work unit" and institutional qualities. In terms of theory and practice, the regular mold of "residential areas" proved difficult to break through, meaning there was no way to achieve real community research results. Since opening and reform, the active introduction of Western community research results and experience has significantly raised the scope and understanding of theoretical knowledge in Chinese community research. By contrast, the active practice of community research has yet to emerge from the shadow of "residential areas" and remains content to simply incorporate more market elements while lacking the necessary humanistic and social considerations. After the turn of the century, changes took place in the values and environments of urban construction. A scarcity of resources led to improvement in content, which replaced the sprawling urban expansion of the past. This has become the "new norm" in Chinese urban development. In another respect, the idea of a "people-orientated" society strongly advocated by the central government since the 18th National People's Congress considers the management of people's work and social sphere an important task of the times, hinting at a major change in direction in China's developmental model and direction. In this context, "communities" constitute a basic component and important tool for the transformation from an "incremental model" to an "inventory model," enabling innovative social management and improving the lives of the people in this new era. In terms of theory and practice, "community research" will fulfill an increasingly important social role, opening up a range of opportunities and challenges, thus realizing the substantive transformation from "residential research" to "community research."

The other concern is, as a representative social structure in the urban survival of Chinese special communities, how can settlement spaces survive and evolve?

From ethnic cultures to minority settlements emerging along traditional lines of homogenous congregation to the migrant worker settlements that emerge as a result of rural–urban migration in the wake of opening and reforms; from substantial settlements of international migrants in the context of globalization to policy initiatives for the construction of resettlement areas for land-lost peasants; and from this series of different special communities and representative settlements that commonly exist in every corner of Chinese cities, we can observe the tolerance and resistance that greets the migration of special communities into urban spaces and the complex processes of assimilation and isolation. In another respect, we can use the window of "settlement spaces" to gain insight into the silent changes of urban social spaces. In considering the survival and evolution of "settlement spaces," we can observe the continual process of "integration response" with the relevant nation/city. The selection of our four types of representative settlement samples can be viewed in the following terms: logic of "integration—response" is apparent in the different fields of "international migrants—moving into the country," "landless farmers—requisitioning land in the city," "ethnic minorities—modernization," and "migrant workers—moving into the city." On this basis, we can try to build an integration response theoretical framework equipped for interpretation across a range of contexts. With the help of four types of empirical research, namely "international migrants—moving into the country," "landless farmers—requisitioning land in the city," "ethnic minorities—modernization," and "migrant workers—moving into the city," we can summarize and probe the embedded mechanisms of the urban survival and evolution of Chinese special communities and their settlement spaces and use this to touch on and reveal some universal patterns in contemporary Chinese urban social spaces.

Considering that interest in "community research" has served as the inspiration for this study's focus on "community" as the scale for samples analysis, we can also say that interest in the urban survival patterns of special communities has determined the core topic of our primary material organization: integration response. As a result, our next task and subject of interest is whether we can use the scale of "communities" (settlements) and the topic of integration response to link and integrate different kinds of "settlement spaces for special communities " with representative characteristics and significant samples in the context of shifts in the landscape of China's urban societies.

Based on this logic, the book comprises six chapters that attempt to use an interdisciplinary perspective of "urban planning and sociology" to take a theoretical and empirically integrated, typified research approach to China's "settlement spaces for special communities." In particular, the empirical research section will analyze the four special communities of international students, land-lost peasants, ethnic minorities, and migrant workers from several angles, as well as their settlement spaces to cover different types of representative samples, including samples of international and domestic migration, voluntary settlements and forced settlements, urban populations, and agricultural migrants. The theoretical overview will improve upon the unidirectional integration logic of classical theory and build a bidirectional "integration response" theoretical framework to interpret and summarize the survival and evolutionary patterns of special community settlement areas in Chinese cities.

In terms of its general organization, this study embraces a static analysis of social spaces and a dynamic understanding of community development. It contains systematic research of the present circumstances of communities and in-depth discussions of specific aspects (such as public services). This study contains qualitative and theoretical interpretation and incorporates mathematical and quantitative attempts. Although the empirical section of our results focuses on four representative communities in Nanjing, we hope to offer universally valid ideas in terms of theory and methodology that can represent and piece together a more complete urban survival map for China's special communities and furnish insight into the macroscopic picture of contemporary China's urban social spaces.

This study uses the microscopic "window" of special community settlement spaces in concert with interdisciplinary perspectives to illustrate and examine the survival prospects of special communities in contemporary Chinese cities, attaining a macroscopic view of social space. This conforms to the momentous changes in developmental ideas, models, and trajectories since China's 18th National People's Congress has obvious research value and practical significance for spatial planning control in Chinese cities and cares for marginalized communities and construction of a harmonious, prosperous society. It is therefore of inherent interest for pedagogical, research, and administrative purposes in the fields of urban planning, geography, and sociology.

In closing, I would like to add that, from its initial conception to the final manuscript, this book has been through nearly a decade of exploration, selection, and polishing before finally settling into its present form, bringing the topic of "integration response" to bear at the scale of "communities" and integrating theory and empirical research within a coherent system. The data collection, foundational research, and composition of each chapter were completed by members of my research group, especially Liu Jia (Chap. 3), Liu Xihui (Chap. 4) and Wang Xiao (Chap. 5), who made core contributions to the corresponding chapters. This study also enjoys the substantial support as a National Natural Science Foundation of China (51878142, 51778125, 51678130), "13th Five-Year Plan of China" National Key Research and Development Program Project (2019YFD1100800) and Postgraduate Research and Practice Innovation Program of Jiangsu Province (KYCX20_0143).

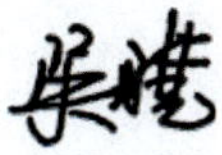

Nanjing, China Xiao Wu

Contents

Authors and Contributors

About the Authors

Xiao Wu holds a doctorate in engineering and is currently Professor at School of Architecture at Southeast University; Doctoral Student Supervisor; Chief Secretary of the Urban Renewal Academic Committee of the Chinese Urban Planning Academic Society; Member of the Urban Image Academic Committee; Board Member of the Chinese Architecture Academic Society Urban Design Sub-Committee; Ministry of Education "New Century Outstanding Talent Support Scheme" Fund Recipient; Jiangsu Province "333 high profile talent fostering program" Awardee; Jiangsu Province "Six Key Talent Summit" funded project Awardee; Jiangsu Tertiary Education "Blue Program" Exemplary Youth Core Pedagogical Training Awardee; and KTH Royal Institute of Technology Visiting Scholar.

Wu Xiao has long been engaged in research in the fields of urban social studies and community development, community spaces for disadvantaged groups and provisions for their protection, urban design theory and methodology; he has held and been chiefly responsible for six National Natural Science Foundation projects, three National Technology Supported (or Key Research and Development) Program projects, one World Bank Project, one provincial-level key research and development program, over ten ministerial and provincial-level research, and HR and curricula reform projects; he has published over 130 papers in core journals and published and translated over 16 books and textbooks. In scientific research, he has earned two second-class Ministry of Education Science and Technology Progress Awards, 1 first-class Huaxia Construction Science Technology Award, and 1 Jiangsu Institute of Higher Education Philosophy and Social Sciences Research Exemplary Achievement Award. In his engineering practice, he has won two National level Exemplary Urban–Rural Planning Design Awards, 18 ministerial and provincial-level Exemplary Survey Design and Exemplary Engineering Design Awards. In terms of teaching, he has won four provincial-level Pedagogical Achievement Awards. Graduate students under Wu Xiao's supervision have earned one provincial-level Outstanding Master's Thesis Award, and his undergraduate students have won two provincial-level Outstanding Graduation Design Team Awards. Undergraduates under his supervision have also attended over 30 events of all types at the National Tertiary Education Urban Planning Competition. (E-mail: seuwxiao9999@163.com)

Lingjin Wang earned a doctorate from the urban planning department of School of Architecture, Southeast University, in 2019. The main focus of her research is special community spaces, in particular those of urban dwelling farmers and international migrants. To date, she has published ten articles in core journals and presented at domestic and international conferences. She has taken part in over ten practical engineering projects and has won two Jiangsu province Outstanding Urban–Rural Planning Awards. (E-mail: zhenbh0201@163.com)

Contributors

Jia Liu Hunan Architectural Design Institute, Changsha, China (contributor to Chap. 3)

Xihui Liu Shanghai Metrodata Technology Co., Ltd., Shanghai, China (contributor to Chap. 4)

Xiao Wang Urban Investment Operation of China Construction Third Engineering Bureau Co., Ltd., Wuhan, China (contributor to Chap. 5)

Chapter 1
Introduction

Abstract "Settlement spaces" are a representative life structure supporting the urban survival of Chinese special communities. They also constitute a special type of community that instigates urban change and responds to dramatic social shifts. The attributes of their marginalization and universality have already become one of the key samples for analyzing and influencing the core variables of Chinese urban social spaces. In light of this, this study adopts an interdisciplinary perspective that applies "urban planning and sociology" at the scale of "communities." Addressing the topic of "integration-response," we integrate empirical and theoretical methods to investigate China's special communities and the evolution and survival of their settlement spaces. This chapter establishes the significance of our research in the context of previous studies, identifies empirical samples, content frameworks, and technical methodologies. Our selection of representative community samples covers different types and sources, including international and domestic migration, voluntary and forced settlements, urban population, and displaced farmer population.

Keywords Research significance · Identifies empirical sample · Content framework · Technical methodology

1.1 Research Background

1.1.1 Diversified Transformations: Emergence of and Changes in Special Groups in Urban Areas

Since the founding of the People's Republic of China, the first half of its development saw forceful interventions of the planned economy by the government and cumulative impact of successive political movements. Under such influences, communities at this stage essentially functioned as a highly administrative extension of the units they were affiliated with. Research on communities was also characterized by distinct unit attributes and institutional features. Community structure and urban social space, as

X. Wu and L. Wang, *Settlement Spaces: Urban Survival Prospects of China's Special Communities*, Springer Geography,
https://doi.org/10.1007/978-981-16-4892-2_1

a spatial mapping of the grand social structure that was static, closed, and homogeneous in nature, was stagnant or even annihilated for a long time (e.g., minority settlements). Along with the reform and opening up, the second half of the development saw a golden age of rapid development in China and a critical period for its socioeconomic transformation and reconstruction of policies and systems. The urban society underwent varying degrees of changes in population structure, industrial economy, social culture, and lifestyle, among other aspects. During this stage, different main lines of change came up consecutively or overlapped with each other. Such transformations not only created a batch of emerging urban special groups in China but also altered the characteristics and developmental directions of a bundle of traditional special groups.

Consider the rapid growth of urbanization after the reform and opening up as an example. With the development of industrial allocations, several emerging small and medium-sized cities have popped up along coasts or rivers and at locations with convenient transportation or abundant resources. Alongside the reconstruction and expansion of urban space, cities have grown steadily in number and population size, with the level of urbanization breaking through 50% in 2011. This indicates that urbanization in China officially entered the second half of the development and would, in turn, bring about profound social changes. Against this macro backdrop of urbanization, China introduced reforms in the rural system, transformation of township and village enterprises, and changes in farmers' perceptions. Following these revolutions, the rural surplus labor force that had long been regulated and suppressed by the urban–rural divide policy was liberated. Due to the gap in socioeconomic conditions between urban and rural areas as well as the ensuing push and pull effect, a mega-scale "migrant worker boom" is noted in China. Billions of migrant workers transfer from rural to urban areas and from inland to coastal regions. Conversely, the rapid expansion of urban space and the "wave of enclosures," which involved expropriation of rural collective land, have spawned a massive number of land-lost peasants who have lost their land and, hence, urgently need to settle in cities. This new type of passive urbanization group shows striking and inherent differences from the aforementioned active urbanization of migrant workers (see Table 1.1).

Another illustrative example is the transnational population migration under the influence of globalization. As a migrant-exporting country, China has seen three large-scale "waves of migration" since the twentieth century. Over these years, the country enhanced its comprehensive national power, elevated its international status, revolutionized information technology, modernized transportation systems, and established increasingly close connections between domestic enterprises and international capital and markets. On these bases, China transitioned to the role of an emerging migrant-importing country and began accepting an increasing number of international immigrants. According to the statistics published by *The Annual Report on Chinese International Migration* (2018), as of 2017, China had approximately one million foreign immigrants, of which international students accounted for the highest percentage (443,000 people), followed by two increasingly expanding groups: overseas high-level talent and professional skilled workers.

Table 1.1 Overview of evolution of land-lost peasants in the process of urbanization in China

Time	Stage of urbanization	Development of land-lost peasants	General situation of land-lost peasants
1950s	Initial stage	First appearance	Land-lost peasants fundamentally transitioned from agricultural to industrial workers and truly transformed from rural residents to urban residents
1958–1978	Stage of sharp fluctuations	Slow growth	The large number of steel industry projects launched during the Great Leap Forward period attracted a massive number of rural laborers seeking employment in urban areas. This marked a quasi-initial stage for their change in identity features
1979–1992	Stage of recovery and development	Faster growth	The land-lost peasants under discussion belong to the passive urbanization group generated under economic construction and urban expansion (different from previous land-lost peasants). However, after losing their land, they started encountering various problems such as declining living standards, difficulty obtaining employment, and inability to enjoy urban social security
1993–	Stage of accelerated development	Large-scale appearance	This passive urbanization group has grown exponentially alongside the rapid development of the urban economy. However, the land-lost peasants have also faced mass incidents associated with land acquisition and increasingly grave resettlement issues

Source Yang Tao and Shi Guo-qing. Review on the Problem of Landless Peasants in China [J]. *Social Sciences in Nanjing*, 2006, 7(10): 10–16

Another case in point pertains to the identity reconstruction of ethnic minorities in the process of urban space production and reproduction. Under the impact of modernization, the unique cultures of ethnic minorities, which serve as an integral composition of traditional Chinese cultures, have inevitably confronted a new crisis of survival. Due to the implementation of large-scale urban renewal and revitalization projects, the disintegrated social structure of ethnic minority communities have gradually shifted from single-ethnicity agglomerations to ethnically mixed settlements. The intermixing of settlement members has also accelerated the penetration of new cultural factors and ultimately added a diversified dimension to family relations, religious beliefs, lifestyles and customs, social organizations, etc. With refined division of labor in urban professions and improved education level, the original members of the settlements are offered a wider range of lifestyle and employment choices that differ from their previous modes and transcend ethnic constraints. All of these led to changes and challenges different from the past to the traditionally special group of ethnic minorities.

1.1.2 Housing Selection and Agglomeration: Emergence and Reconstruction of Settlement Space for Special Groups

Amid such intricate social changes in China, various emerging special groups opt for different housing modes and residential locations through active or passive selection methods on the basis of manifold subjective and objective factors, thereby forming dispersed or clustered residential patterns. This has engendered the emergence of new-type settlement space for special communities. Conversely, the original settlement space of special communities has undergone major transformation and multidimensional reconstruction in social, spatial, and other attributes due to the diverse changes occurring in urban society and the subsequent ongoing impact (e.g., socioeconomic transformations and modernization waves). In a nutshell, as the "settlement space for special communities" has come to serve as a typical social structure for special groups subsisting in Chinese cities, its marginalization attribute and universal existence have become the key samples used to analyze and influence the social space of Chinese cities.

Consider the massive number of migrant workers who transferred into cities in the context of rapid urbanization as an example. This type of floating population often has diversified choices for housing modes and residential locations (see Table 1.2). Of them, "rental housing," perceived as a dominant form of housing selection, can be further divided into two models in terms of spatial distribution. In urban areas, despite the higher density of migrant workers, dispersed rentals and settlements constitute the primary spatial pattern due to relatively limited settlement scales and scattered housing distributions. By contrast, a clustered settlement space is more likely to appear in urban–rural fringe areas, where migrant workers, companies, and housing sources display relatively concentrated distributions. Accordingly, a series of settlements have been formed through active or passive selection in these areas. The "migrant worker settlements" focused on herein refer to a type of settlement spontaneously formed in the rural–urban fringe by migrant workers through active selection and agglomeration based on a comprehensive consideration of multiple factors, including housing, rent, commuting, and management.

Another example involves ethnic minorities under the waves of modernization. Having remained in a stable state of closure for a prolonged period of time, their traditional settlement space has ultimately shifted from a single-ethnicity pattern to an ethnically mixed mode due to the rapid flow of people, logistics, and information in modern society. While exposing practical problems such as cramped housing, land shortage, transportation difficulties, and environmental pollution, the change in their settlement pattern has also triggered transformation and reconstruction in their

Table 1.2 Comparison of various types of migrant worker settlement space

Primary housing mode		Spatial pattern	Primary employment	Spatial distribution on the community/unit dimension	Note
Active selection	Hotels	Dispersed settlement	Mainly business service and cottage industries	Primarily suburban and secondarily urban areas	
	Places of relatives or friends	Dispersed settlement		Directly related to the place of residence of relatives and friends	Displaying a certain degree of randomness in location distribution
	Rental housing	Active settlement & Dispersed settlement		Primarily suburban and secondarily urban areas	Suburbs: Residing in clustered settlements across broad expanses in a dispersed manner Urban areas: Primarily dispersed settlements
Passive selection	Collective dormitories	Passive settlement	Manufacturing industry	Suburban areas	Related to the degree of suburbanization of enterprises and the location of dormitories provided by the government/enterprises
	Construction sites	Passive settlement	Construction industry	Directly related to the location of construction projects	Associated with urban incremental construction and stock renewal activity areas

Source Wu Xiao. *Metropolitan Floating Population of Employment Space Analysis: An Empirical Study for Migrant Workers* [M]. Nanjing: Southeast University Press, 2010:1445

own social attributes (e.g., demographic composition, family relations, community organization, lifestyle, customs) and spatial attributes (e.g., community functions, time and space of residence, time and space of employment, supporting facilities).

In addition, with the continuous influx of international students, two typical types of international student settlements have been formed depending on their modes of housing selection, namely, passively selected settlements (on-campus apartments for international students) and actively selected settlements (special communities formed around colleges and universities through spontaneous rentals). Using such settlements as their support and foundation, this group exhibits rich and unique features in its everyday living space and spatiotemporal trajectory in terms of residence, learning, shopping, and recreation, among others. Meanwhile, the formation and evolution of their settlement space have also brought multi-faceted variations in and comprehensive impact on the surrounding communities.

1.1.3 *Integration and Response: Survival and Evolution of Settlement Space for Special Communities from the Perspective of Community*

The "settlement space for special communities" is a form of typical space that emerges ubiquitously or is currently undergoing transformation in the context of the pluralistically changing Chinese urban society. Such space is also one of the basic variables that bear a mutually influencing relationship with the social structure and spatial pattern of the city wherein it is located. The survival and evolution of such settlement space are essentially a process of continuous "integration-response" between the said space and the place of population inflow. During this process, the settlement not only responds to and encapsulates the drastic changes of society but also brings manifold variations to its surrounding communities and even cities. On this account, treating the exploration of the topic "integration-response" as a major focus and challenge in community research may help shed light on the survival and evolution mechanism of the "settlement space for special communities" in Chinese urban society. Moreover, the study of "integration-response" is also one of the key areas in the discussion on the theory and practice of urban planning amid various social changes in China, reflection of humanistic implications, and care for social marginal groups.

Consider migrant worker settlements as an example. Due to the long-term constraint of the urban–rural dual structure and dearth of policy guidance and efficient management, migrant workers have been one of the disadvantaged groups in urban areas. They exist in enormous numbers yet enjoy limited resources; they contribute to urban economy yet encounter exclusion and alienation in social and cultural life. Despite their long residence in and familiarity with cities, they have profound difficulty surpassing the multiple thresholds for citizenization. These issues have resulted in the disorderly and unruly side of migrant worker settlements during their savage growth.[1] [...] In a sense, these problems precisely reflect the negative impact and integration barriers that this type of settlement space engenders in the process of becoming "embedded" in the place of migration. They also serve as the driving force behind the promotion of the "citizenization of migrant agricultural population" against the backdrop of the new-type urbanization in China.

Another case in point pertains to international student settlements. Alternating between integration and response, international students still manage to construct their own uniquely imprinted everyday living space and spatiotemporal trajectory

[1] Common problems for migrant worker settlements include disordered functional arrangement, inadequate public infrastructure, disorganized space and landscape, glaring deficiencies in territorial management, frequent occurrence of social problems and crimes, lack of social security for residents, and low levels of cultural education. Consider the migrant workers settlement in the Hongshan area of Nanjing as an example. Here, 79.24% of the migrant workers have an educational level below middle school, 49.41% have not received relevant occupational training, 53.26% work more than eight hours a day, merely 1.97% and 4.92% enjoy work-related injury insurance and medical care, respectively, and only 10.49% report satisfaction with the current living environment.

and have imperceptibly changed and influenced surrounding communities to varying degrees (see Table 1.3). Such changes and influences include increasingly complicated demographic composition, open and international spatial arrangement, unique business atmosphere, community economy, and abundant international community activities, among others. All these changes have arisen from the derivation of a new type of "transnational space" and various responses from surrounding communities. This not only foreshadows the implantation and penetration of globalization directly facing localities and communities today but also highlights the significant role of local governance and community construction in modern times.

Another example involves land-lost peasant resettlements. During the process of urbanization, from agricultural civilization to industrial civilization and from the identity of farmers to that of citizens, land-lost peasants are also confronted with various latent problems and potential conflicts in terms of spatial resettlement, which has, in turn, impeded their smooth "integration" into cities. Such problems and conflicts include exclusion under the influence of the urban–rural dual system (absence of education, medical care, and social security systems), social discrimination and prejudice in the places they migrated to, and predicaments for personal transformation (substantial obstacles to transformation in their ideology, employability, lifestyle, etc.). Conversely, such slow integration processes have also brought urban areas negative responses such as undesirable public order and unbalanced public services.

In conclusion, this book outlines the urban survival of special communities in China on the basis of the "**community**" scale, using the interdisciplinary perspective that combines "**urban planning and sociology**" through empirical and theoretical research. In terms of empirical research, the discussion centers on four types of special groups, i.e., international students, land-lost peasants, ethnic minorities, and migrant workers and analyzes their settlement space from different perspectives. The analysis includes and examines typical samples of different settlement types: international migrants and domestic migrants, active selection of residence and passive selection of residence, and urban population and migrant agricultural population. As for the theoretical summarization, this research provides a multi-scenario interpretation and summary of the survival and evolution patterns displayed by the urban settlements of special communities in China through the construction of a two-way linkage theoretical framework of "integration-response."

1.2 Significance of the Study

Since the 18th National Congress of the Chinese Communist Party, the central government initiated a series of discussions on various social issues such as enhancing social construction, innovating social governance, harmonizing social relations, and improving livelihood conditions. While emphatically stressing the main line of "humanism," the discussions also presage the prospect that China is about to make major, different-than-before adjustments to the concept, model, and orientation of

Table 1.3 Various Types of Daily Activities and Spatial Patterns for International Students

Living space		Learning space	Shopping space	Leisure space
Off-campus community	On-campus apartment			

**Source* Own elaboration

its development. Further, they indicate that urban and rural planning will assume more onerous social functions and professional challenges than previously. Against this background, the "settlement space for special communities" built on the basis of the "community" scale has become one of the key areas and strategic directions for promoting the overall coordination of urban and rural areas, improving the welfare of the general population, and deriving social justice in current urban–rural planning schemes.

(1) **Giving Attention to the Requirements of Special Groups**

As an essential component of the urban social structure, various special groups are often differentiated and derived along with the changes and transformations of our urban society. This study selects special groups such as international students, land-lost peasants, ethnic minorities, and migrant workers as typical representatives. Based on these special groups, this study analyzes the vastly disparate static characteristics and dynamic laws of their settlement spaces from different perspectives and explores their unique and differentiated social, economic, and spatial requirements. The results may facilitate a better understanding of the overall appearance of Chinese urban society while disclosing multiple issues that various groups (marginal groups in particular) urgently need to resolve. Moreover, what is examined in this study is precisely a marginal field that must be faced squarely in the establishment of a humanistic and harmonious society and construction of a moderately prosperous society today.

(2) **Strengthening Community Research Systems**

In terms of "integration-response," this research connects and analyzes settlement spaces for multiple types of special communities using a "community" perspective. Its primary objective is to thoroughly examine our "society" through "communities" and outline the "macro structure" via a "micro scale." This study seeks to provide a key window and unique path to analyze the intricate social changes in Chinese cities. Moreover, the direction offered herein serves as a continuation of the attention the project team has long dedicated to the space of marginal groups. Through typified and interdisciplinary research that combines social and spatial attributes, this study enriches the existing community research systems and realizes the academic transformation from "residential area research" to "community research."

(3) **Attaching Emphasis to the Combination of Theoretical and Empirical Research**

Based on the samples of international student settlements, land-lost peasant resettlements, minority settlements, and migrant worker settlements, this research employs technical approaches such as time geography, the SPSS statistical software, and structural equation modeling (SEM) to investigate and demonstrate the evolution patterns, current status and features, and integration strategies of these settlements. Conversely, this research is built on the foundation of optimizing the logic of one-way integration in classical theories. Through the tentative construction of a two-way linkage theoretical framework of "integration-response" between "the settlement space for special communities and the place of population inflow," the study further provides a multi-scenario interpretation and summary of the mechanisms adopted by special communities for their survival and evolution in Chinese cities. While offering an in-depth exploration of the spatial linkages of special community settlements, this study may effectively make up for the inadequacies and limitations in empirical sample types and theoretical interpretation systems observed in previous studies on similar topics.

1.3 Research Progress

The samples of typical settlements selected in this study involve sundry types of special groups, including international students, land-lost peasants, ethnic minorities, and migrant workers. For some social groups, completely equivalent and matching concepts may be missing from the international arena. However, given the similarities and comparability in social, economic, and spatial attributes displayed by overseas marginal groups, e.g., the floating population (or migrants) and racial (or ethnic) groups, the progress of international research on settlements for special groups abroad still bears solid reference value and significance for China under the new-type urbanization strategy.

1.3.1 Progress of Related Research Abroad

Academic interests in the settlements of the floating population (or migrants) abroad began mainly with the continued research and protracted debates in sociology and philosophy, centering on the classic proposition of "tradition/modernity." In response to the population migration arising against the backdrop of rapid urbanization, a number of disparate or even contradictory cognitive models have been successively constructed, including the "opposition/assimilation" model (e.g., Thomas, Zanaiecki; Chicago School of Sociology; Kearney (1986)), the "coexisting" model, "dependency" theory, and "articulation" model (e.g., Frank 1967; Meillassoux 1981; Castell 1975; Weist), and the "embeddedness" theoretical model (e.g., Portes 1998). Under the influence of different theoretical and cognitive systems, various floating population research systems have been established in the field of sociology, including the formation mechanism of social space (e.g., Stark 1984; Park 1925; Faist 2000), migrant communities (e.g., Levitt 2007; Smith 2001), and social integration (e.g., Guarnizo 2001). In the sphere of urban planning, more practice-based, long-term remediation, and exploration have been carried out to address various urban problems associated with population migration and settlement (e.g., slums and inner-city recession). Through repeated adjustments and policy revisions, a wealth of experiences, lessons, and related results have been accumulated, whereby the academic community has achieved a deepened understanding of how the floating population and its settlements are correlated to urban areas.

Regarding the study on racial (or ethnic) settlements, foreign scholars have also established a series of unique theoretical systems through long-term explorations. The theory of regional stratification holds that the spatial characteristics of settlements and family units serve as determining factors in selecting the location of residences (e.g., Logan and Molotch 1987). The theory of regional context argues that more serious discrimination, stricter land policies, or larger independent minority settlements exist in large cities, resulting in intimate correlation between the overall scale of the city and residential isolation (e.g., Farley and Frey 1994; Logan). In terms of spatial assimilation, two sets of theories are proposed. One argues that spatial assimilation is a transitional step from cultural adaptation to social structural assimilation (e.g., Massey and Mullan 1984), while the other holds that spatial assimilation stems from socioeconomic variability (e.g., Massey & Denton (1985)) and that ethnic minorities display lower levels of correlation between socioeconomic status and isolation patterns (e.g., Denton and Massey 1988; Iceland et al. 2005; Wilkes 2006; Logan et al. 2004; Massey and Fischer 1999). In addition, others scholars have conducted empirical research on employment opportunities and job-housing spatial match in black communities (e.g., Kain 1968; Holzer 1991; Ellwood 1986). Still other scholars have analyzed the spatial characteristics, evolution patterns, and motivations of ethnic minorities from a more macro level of regional and urban dimensions (e.g., Goodman 1978; Emília 2010; Eric et al. 2005).

Notably, the formation of these two types of settlements stated above mostly originates from the active selection and spontaneous derivation of the floating population (or migrants) and racial (or ethnic) groups with inherent inevitability and plausibility. Aside from the two types, a multitude of scholars have focused on the products created by marginal groups through passive housing selection, i.e., public housing or resettlement areas (which may serve as a comparison of resettlement areas for land-lost peasants in China). These scholars conducted comprehensive discussions on macro policies (primarily taking reference from international cases) and from the perspective of meso- and micro-level planning and design. For example, Whitehead (2007) and Schwartz (2010) introduced and interpreted housing security policies in typical countries (e.g., the United States, the United Kingdom, France, and Singapore) regarding public housing, rental vouchers, and low-income housing tax subsidies, among others. In addition, Barlow (1994) and Harloe (1995) compared the practices of different countries in terms of social systems, economic operation modes, housing security types, and various other aspects. For example, the HOPE VI program implemented in the U.S. proposed the idea of developing communities with complex functions and mixed hierarchies; the Netherlands also emphasized in *Differentiated Cities* the necessity to attract different target groups through the diversification of housing types, thereby alleviating social isolation and decline of low-income settlements.

Finally, based on the progress of foreign research, several judgments are formed as follows:

(1) In terms of the products created by the floating population (or migrants) and racial (or ethnic) groups through active housing selection and gregarious settlement, foreign scholars started their research early and formed abundant and complete theoretical systems and technical approaches on the basis of the community scale. Many scholars also conducted multi-scale comprehensive discussions on public housing or resettlement areas wherein marginal groups reside through passive selection.

(2) There is room for improvement and further exploration in terms of the results of foreign relevant research on subgroups (specific research on "international students" has not yet established a mature system[4]), local directional expansion (in addition to studies on job-housing spatial match and social integration, other research directions include everyday living space, public service level, etc.), and integration of research ideas (combination of theoretical interpretation and empirical analysis), among others.

(3) For land-lost peasants and migrant workers, completely corresponding concepts do not exist in international academia. Regarding international students and ethnic minorities, while such concepts are present in foreign research, they are subject to inherent differences in structural systems, historical evolution, cultural backgrounds, and various other aspects between China and overseas countries. As such, when studying the settlement spaces of the special groups described above, this book should ensure reference from and continuation of foreign experience while profoundly reflecting China's own differentiation and characteristics. This has reserved greater space and posed tougher challenges for this research.

1.3.2 Progress of Related Research in China

In this section, counterpoint focus and classification overview methods are adopted to filter out extensions and less relevant details. The discussion focuses on special groups, which include international students, land-lost peasants, ethnic minorities, migrant workers, and their settlement spaces (on the community level). The results of directly related Chinese research are summarized as follows:

(1) **International Student Settlements**
Chinese research on international migrants is still developing. Judging from the scale of research, relatively limited results have been accumulated at the overall level. Current studies mainly involve the condition of migrants worldwide and in international regions (e.g., Li 2005), characteristics of different types of migrants (e.g., Han 1994), and dynamics of international migrants in China (e.g., Liu 2014). By contrast, relatively abundant research has been conducted on international migrants at the community level. Research topics primarily include (1) case studies exploring the life, adaptation, and conflict of foreigners based on typical cities such as Shanghai, Beijing, Guangzhou, and Yiwu (e.g., He 2008; Wu 2004); (2) empirical research that analyzes social integration among international migrants on the foundation of sociology and geography

fields (e.g., Niu 2011); and (3) theoretical rationale on the formation and governance model of international migrant communities (e.g., Qian et al. 2013). However, similar to foreign research, an extremely limited number of Chinese studies specifically focus on the subgroup of "international students," thus leaving room for further analysis of their everyday living space and community response in this study.

(2) **Lost-land Peasant Resettlements**

Chinese scholars started relatively late in their research on resettlement of land-lost peasants. Current studies mainly focus on the following three aspects. (1) Research on related policies: Such research primarily reviews the public housing policies of various countries or regions (e.g., Sweden, the United Kingdom, the United States, Singapore, Hong Kong, etc.), summarizing and drawing lessons from their development mechanism, scope of coverage, funding sources, and subsidization model, among others (e.g., Tian 1994; Fan 2012; Hao 2007; Cao 2006; Li 2012; Zhao 2010); (2) Research on planning and construction: Such research mainly puts forward strategies and suggestions from a macro-level dimension of location selection, a meso-level dimension of resettlement planning, and a micro-level dimension of housing design. In particular, Ma (2007) suggested that affordable housing should be located in urban fringe areas with convenient transportation based on calculations conducted as per David Ricardo's rent model; Jiao Yi-xue (2007) provides standards for mixed construction of housing allocation and diversified household design in affordable residential areas; and Huang (2007) proposes outdoor environment design strategies for resettlement housing of land-lost peasants in light of the lifestyles and customs of relocated farmers. (3) Research on village reconstruction: Such research describes and assesses a large number of village reconstruction cases related to land-lost peasants. Representative examples include research on reconstruction of urban villages in Guangzhou, Shenzhen, Xi'an, and other places (Peng 2001; Li 2002) and research on the phenomenon of hollow villages in Jiangsu, Henan, and other places (Li, 2001; Zhao 1998). However, compared with the discussion on the spatial attributes of resettlement areas for land-lost peasants, Chinese scholars have dedicated far less attention to their social attributes (particularly social integration). Research has been focused more on other groups (e.g., migrant workers, new immigrants), with hypotheses proposed on the basis of the results of related foreign research (e.g., Tian 1995; Wang et al. 2008; Ma 2008) or index systems constructed for integration evaluation (Zhu 2002; Wang and Luo 2007; Huang 2010). In terms of measurement and statistical techniques, there is also room for further improvement.

(3) **Minority Settlements**

Chinese researchers largely focus on typical cases of minority settlements for Hui, Uyghur, and Bai groups, among others. Multidimensional research has been conducted on the basis of diverse disciplines, which include the following. (1) In the field of sociology, the cultural characteristics, social structure, and transformational patterns of minority settlements are explained (e.g., Bai 2005;

Zhang 2004; Zhou and Ma 2004); (2) In the realm of urban planning, research has been conducted mainly from a spatial perspective, focusing specifically on spatial structure, renovation and reconstruction, cultural preservation, land use control, evolution of social space, etc. (e.g., Gao 2005; Chen 2000; Dong 1996, 2000)); Wei et al. 2004; Zhang and Hou 2006; Wu et al. 2008; Song 2012); (3) In the area of architecture, research on the micro level focuses on the regional architectural characteristics and spatial succession patterns of minority settlements. In addition, the development direction and design concept of regional ethnic architecture were investigated against the background of globalization and urbanization (e.g., Li 2009; Yang 2009; Wang 2009); (4) In the sphere of geography, research has been conducted on the scale of regional geography to shed light on the characteristic of spatial agglomeration of ethnic minorities and the influence of multiple factors (e.g., Zhou et al. 2002; Yi et al. 2013). Judging from the research results, studies in this respect mostly involve case-based empirical research, which attaches priority to "point"-oriented in-depth exploration rather than "plane"-oriented exposition; accordingly, further improvement is required in the combination of theory and empirical research.

(4) **Migrant Worker Settlements**

Regarding interdisciplinary research directly aimed at "migrant worker settlements" by combining the studies of "floating population" and "residential space," an incomplete and unbalanced state of development is observed in the current Chinese academia. (1) Majority of the more systematic and influential research results are found in the fields of sociology and economics. Masses of scholars focused on typical cases such as "Zhejiang Village" (e.g., Xiang 2000; Wang 1995; Liu 2002; Zhou 1998). Scholarly attention has also been paid to locality issues associated with social integration and population management within the settlements; researchers specializing in this area include Yu (2002), Liu (2000), Li et al. (1991), Hu and Zhao (1997). In other fields (notably urban planning), most research results related to migrant worker settlements were obtained after the 1990s. Aside from the phased progress (2001–2010) (Wu 2005,2001,2010) attained by our project team, disintegrated and limited research results have been produced. Representative research includes feature studies by Yuan 2014, Rao 1999, Zhang 2000, Yang 2003, Zhang 2006). However, these studies mainly offer analyses and qualitative descriptions of the current status of settlements. Certain blind spots and deficiencies are present in research system construction, digital technology application, and regional directional expansion (e.g., the level of public services in the settlements), among others.

1.3.3 General Review

Due to the inherent differences in structural systems, historical evolution, cultural background, etc., between China and other countries, some glaring disparities also exist in conceptual significance, characteristic influences, and developmental stages of special groups and their settlement space. This has, in a sense, determined that the reference and continuation of Western experience should be screened and selected based on actual local conditions.

In general, foreign scholars started their research on related theoretical research and practical exploration relatively early. Over the years, considerable systematic and mature academic achievements, technical methodology, and practical experience have been accumulated. However, there is room for improvement in terms of in-depth discussion on subgroups, expansion of regional directions, and integration of research ideas. Conversely, Chinese research has been conducted primarily on the basis of different disciplines. However, aside from systematic and influential results achieved in the field of sociology, existing research progress is sporadically found in a few regional case studies and empirical analyses, with emphasis on current situation analyses, qualitative descriptions, and experience summaries based on social practice surveys. In the future, new breakthroughs and progress may be achieved in the extension of typical samples and subgroups, application of digital technology, expansion of regional directions, and combination of theory and empirical research.

On this ground, this research starts with a micro-level perspective of "communities" to examine the typical spaces that are ubiquitously emerging or undergoing drastic changes against the grand backdrop of urban social transformations in China. Special communities investigated herein include international students, land-lost peasants, ethnic minorities, and migrant workers. Interdisciplinary research is carried out with a combination of theoretic and empirical investigations, qualitative and quantitative approaches, overall observation, regional in-depth discussion, static analysis, and dynamic tracking. Further, a macro overview of contemporary Chinese urban social space is outlined and explored in this study.

1.4 Object and Scope of Study

During the current critical period of socioeconomic transformation and policy system reconstruction, the "settlement space for special communities" has become one of the core variables and key windows for analyzing the urban social spatial structure in China. In light of this, this research prudently selects special groups from various social groups, which, along with their various settlement spaces, are used as the objects of analysis. The selected groups include international students, land-lost peasants, ethnic minorities, and migrant workers. Based on the micro perspective of "communities," this study attempts to construct a relatively complete picture depicting the survival of special communities in urban areas.

The fundamental reason lies in the conspicuous commonalities shared by these four types of settlement spaces. For example, all these communities display the characteristics of marginality and universal existence (what "marginality" signifies does not simply pertain to "spatial locations" in its narrow sense, but rather "the relative marginalization of the social status of the communities and the heterogeneity of their social compositions"). In addition, these communities continue to realize and advance their own survival and evolution through the process of "integration-response" that occurs between themselves and the place of population inflow. Conversely, these settlements are also vastly different from each other. They not only have strikingly distinct formation backgrounds and evolution patterns but also manifest differentiated and diversified individual characteristics and social attributes. Moreover, they exhibit quite dissimilar spatial attributes and categorical models in terms of location distribution, spatial arrangement, job-housing patterns and commuting modes, daily activities, public services, etc. Due to the coexistence of these numerous similarities and differences, these four types of "settlement spaces for special communities" selected in this study acquire a prerequisite foundation of comparability as well as unique sample significance and representative characteristics that are missing from other groups and communities.

In light of this, we may define the selection of the four typical community samples as **specific manifestations and reflections of the common logic of "integration-response" in differentiated fields, namely, "international migrants-country of population immigration," "land-lost peasants-city of land acquisition," "ethnic minorities-modernization," and "migrant workers-city of population inflow."** Specifically, in terms of the representative cases selected for the four types of samples, the empirical exploration of this study is mainly carried out within the confines of Nanjing City. Of them, the study of international student settlements is based on the sample of Nanxiu Village, covering international students in the Gulou campus of Nanjing University, Suiyuan Campus of Nanjing Normal University, and Nanxiu Village community; the study of land-lost peasant resettlements is based on the samples of Mufu Villa, Tianjingshan Community, Yaolin Xianju, and Xishan Huayuan, covering different locations, stages, scales, forms, and characteristics; the study of minority settlements is based on the well-known case of the Qijiawan Hui community in Qinhuai District, which also serves as one of the critical conflict points between modern development and traditional culture preservation in the old town of Nanjing; for migrant worker settlements, the urban village of Taiping Community in Yanziji Street, Hongshan Village in Hongshan Street, and Hongmiao Quarter in Xinjiekou Street are selected as research samples. Despite the limited spatial scope of the empirical samples, we hope to provide enlightenment and reference that bear universal significance in theory, methodology, and practice and further outline and examine the macro overview of contemporary Chinese urban social space from a point-to-plane perspective.

1.5 Research Content and Methodology

1.5.1 Research Content

This research connects and analyzes the four types of typical settlement spaces for special communities using a "community" perspective, with each chapter allocated for one specific type of special community and its settlement space. The main structure of this research is illustrated below.

(1) Introduction

This chapter primarily introduces and reviews the relevant backgrounds of this research and related research progress at home and abroad; defines the object and scope of this research; and explicates the content, methodology, innovation, and characteristics of this research.

(2) Community Sample I—Living Space and Community Response: Research on International Student Settlements.

This chapter considers the international student settlement of Nanxiu Village, Nanjing City, as an example. With focus on the two major issues of everyday living space and community response, parallel examinations are conducted on the daily activities, life trajectories, and ensuing influences and effects of "international students," a typical international migrant group, on a community scale. Furthermore, in response to the existing problems confronting international student settlements, comprehensive strategies are proposed to optimize the external environment and reduce their integration barriers. In essence, the selection and inspection of this type of community sample involve a specialized discussion on the field of "international migrant-country of population immigration" under the topic of "integration-response."

(3) Community Sample II—Spatial Resettlement and Social Integration: Research on Land-lost Peasant Resettlements

This chapter takes land-lost peasant resettlements as the object of study. First, the social attributes, spatial attributes, and contributory causes of spatial resettlement for this type of passive urbanization group are compared and analyzed. Second, the characteristic patterns and influence mechanisms for the social integration of land-lost peasants are investigated and explored. Further, the strategies and measures to improve the current spatial resettlement and social integration of land-lost peasants are discussed. In essence, the selection and inspection of this type of community sample involves a specialized discussion on the field of "land-lost peasants-City of land acquisition" under the topic of "integration-response."

(4) Community Sample III—Waves of Modernization and Root Climbers: Research on Transformation of Hui Minority Settlements.

This chapter considers the quintessential case of the Qijiawan Hui settlement in Nanjing City as the object of study. With particular emphasis on the specific social and cultural background and distinctive ethnic characteristics of this group, the study offers an in-depth analysis and examination of the transformation patterns displayed by minority settlement space in China. In addition, the existing problems with the evolution of the Hui settlements are disclosed, and a series of strategies and suggestions for the renewal and improvement of their social space are put forward. In essence, the selection and inspection of this type of community sample involves a specialized discussion on the field of "ethnic minorities-modernization" under the topic of "integration-response."

(5) Community Sample IV—Service Supply and Facility Utilization: Research on the Level of Public Services in Migrant Worker Settlements

This chapter conducts a specialized study on the level of public services from the two angles of "objective supply and subjective utilization" based on first-hand collected data of migrant worker settlements. First, this study assesses and compares the level of supply of public service facilities in migrant worker settlements and explores the underlying reasons behind the characteristics of facility supply. Next, this study investigates and analyzes the characteristics of the utilization of public service facilities among migrant workers in these settlements and reveals the laws governing group differentiation in the use of facilities. Finally, it presents the match patterns of facility "supply-utilization" and discusses the approaches to enhancing the level of public services in migrant worker settlements based upon different match types. In essence, the selection and inspection of this type of community sample involve a specialized discussion on the field of "migrant agricultural population-city of population inflow" under the topic of "integration-response."

(6) Conclusion: Integration-Response–Survival and Evolution of Settlement Space for Special Communities.

On the basis of the empirical research on typical communities stated above, this research optimizes the logic of one-way integration in classical theories and construct the core "integration-response" theoretical framework. In addition, this study provides a multi-scenario interpretation and summary of the survival and evolution patterns displayed by the urban settlements of special communities in China from a diversity of perspectives, including "international migrants-country of population immigration," "land-lost peasants-city of land acquisition," "ethnic minorities-modernization," and "migrant workers-city of population inflow."

1.5.2 Research Methodology

Based on the disciplines of urban planning and urban sociology, this research focuses on the topic of integration-response. With the studies of "special social groups" and "residential space" combined, typified and interdisciplinary community research that covers wide breath and profound depth is conducted using a comprehensive range of research techniques. Specific methods are shown as follows (see Table 1.4):

Table 1.4 Application of basic methodology in this research

Research methodology		Primary target	Objective
Stage of data collection	Field observation	All types of settlement space for special communities	To obtain spatial attribute data associated with the various types of settlement space for special communities (e.g., location distribution, spatial arrangement, job-residence commuting, public services, and daily activities), primarily micro data at the community level
	Questionnaire statistics	Special groups including international students, land-lost peasants, the Hui minority, and migrant workers	To obtain data on the social, economic, and spatial attributes associated with the various types of special communities (e.g., employment, income, education level, residential area, kitchen and toilet layout, and daily activities), primarily individual sampling and first-hand data at the community level
	Thematic interview	Functional competent departments at all levels (statistics bureau, three housing offices, public security bureaus, sub-district offices, neighborhood/village committees, etc.)	To obtain general overviews, background information, and statistical data associated with the various types of special communities (e.g., the urban distribution, residential and employment status, community size, and formation and evolution of various special communities)
	Literature review	Monographs, newspapers, internet, etc	To introduce and reference community research theories and methods from previous research with a similar background of social transformation (including globalization, urbanization, and modernization) and similar research subjects (various special communities and their settlement spaces)
Stage of research analysis	Classification analysis	All types of special communities and their settlement spaces	To classify and analyze the evolution patterns, characteristics of the current status, and integration strategies of different special communities and their settlement spaces

(continued)

Table 1.4 (continued)

Research methodology		Primary target	Objective
	Statistical analysis	All types of special communities and their settlement spaces	To conduct quantitative analysis of different special communities and their settlement spaces regarding job-residence commuting, spatiotemporal characteristics of daily life, social integration mechanisms, public service levels, etc
	Comparative analysis	All types of settlement space for special communities	To provide a multi-scenario interpretation and comparison of the motivation and mechanism for the survival and evolution of the various special community settlements in the cities of China under the theoretical framework of "integration-response."

1.6 Innovation and Characteristics of Research

This research considers the micro-level "community" as the spatial subject of study and "special communities" as the social subject of study, mainly in consideration of their sample value and uniqueness that sharply distinguish themselves from other communities and settlements. This study also involves one of the essential fields in China that explore the theory and practice of urban planning under the background of social changes and reflect social justice and humanistic significance. In the process, this research focuses on the following innovations and endeavors:

(1) Construct a Multi-scenario and Two-way Linkage Theoretical Interpretation System for Settlement Space of Special Communities.

This research focuses on four types of typical communities that are emerging or undergoing changes in the context of urban social transformations in China, seeking to break the logic of one-way integration and single-scenario divide in classic theories, and further construct a two-way linkage framework of "integration-response" between the "settlement space of special communities and the place of population inflow." The research combines the background of globalization and domestic context of new-type urbanization. Further, it provides a multi-scenario interpretation and summarization of the survival and evolution mechanism for the "settlement space of special communities" that bear typical sample significance. While offering an in-depth revelation on the spatial mechanism of such settlement space, the study may also effectively make up for the deficiencies and limitations of previous similar studies in theoretical interpretation.

(2) Establish a Multi-mainline and Multi-dimensional Framework for Empirical Analysis of Settlement Space of Special Communities.

Based on the samples of international student settlements, land-lost peasant resettlements, minority settlements, and migrant worker settlements, this research conducts investigations and empirical studies from different perspectives. On one hand, new lines of reference are introduced to construct an empirical framework intertwined with multiple mainlines, e.g., exploration of community characteristics and intra-group differentiation that examines "social and spatial attributes," group job-housing migration, and daily activity trajectory simulation under the "space–time" coupling. On the other hand, this study conducts a comprehensive analysis of special communities and their settlement spaces from different angles and dimensions, e.g., combination of static analysis and dynamic tracking and combination of supply and demand sides. Compared with the single-line structure of previous studies, this research may provide more effective substantiation and support for theoretical interpretation.

(3) Improve on and Provide Diverse Digital Technology Methods for Research on Settlement Space of Special Communities.

In addition to traditional data collection methods such as field observation, questionnaire statistics, and thematic interviews, this research further enhances and expands the fields and paths of digital technology application, seeking new possibilities for technical optimization and expansion. In particular, diverse methods have been integrated, improved, combined, and applied for quantitative evaluation, statistical analysis, and spatial illustration to examine the job-housing commuting patterns, the spatiotemporal characteristics of everyday life, social integration mechanism, and public service levels, among others. The technical methods applied include time geography, SPSS statistical software, structural equation modeling (SEM), and analytic hierarchy process (AHP). Through the application of these research methods, more reasonable and persuasive scientific conclusions may be generated with verification and amendments introduced to traditional analysis and qualitative research approaches.

References

Bai Y. Root Climbers—The Hui Settlement against the Background of Urban Modernization.Yinchuan: Ningxia People's Publishing House, 2005.

Cao Z. British housing system and its inspiration for China. China Real Estate. 2006;27(10):78–78.

Castell M. Immigrant workers and class struggle in advanced capitalism: the west European experience. Politocal Soc. 1975;(5).

Chen Z. The characteristics and changes of the spatial structure of the Hui settlements in Ningxia. Hum Geogr. 2000;5:39–42.

Chen Y, Lv B, Zhang C, Tan X. Social space features and causes of the Hui settlement area in Hebalin District, Lhasa City. Resources Environ Yangtze Basin. 2013;22(1):33–9.

Dong W. Urban reform driven by free market economy—a preliminary study on the self-construction and renewal of Xi'an Hui district. Urban Planning Rev. 1996;5:42–5.

Ellwood DT. The spatial mismatch hypothesis: are there teenage jobs missing in the Ghetto. In The Black Youth Employment Crisis, ed, 1986.

Emília MR. A methodology to approach immigrants' land use in metropolitan areas. Cities. 2010;(27):137–53.
Eric F, Chiu L, Emi O. Spatial distribution of suburban ethnic businesses. Soc Sci Res. 2005;(34):215–35.
Faist T. The volume and dynamic of international migration and transnational social spaces. Oxford: Oxford University Press. 2000.
Fan X. "Drawing on" the U.S. affordable housing system. Urban Development. 2012;31(5): 32–3.
Frank AG. Capitalism and underdevelopment in Latin American. New York: Monthly Review Press; 1967.
Gao Y. Urbanization and Community Research of Minority Areas in Northwest China. Beijing: Publishing House of Minority Nationalities, 2005.
Goodman JL. Urban residential mobility: places, people and policy. Washington, DC: The Urban Institute; 1978.
Guarnizo LE. The economic of transnational migration. California: University of California; 2001.
Guo P. Research on the Reconstruction Strategy for Urban Villages in Nanshan District, Shenzhen. [Master's Thesis]. Harbin: Harbin Institute of Technology, 2001.
Han G, Yang J. Overview of international population migration. Geogr Territorial Res. 1994;4:30–4.
Hao J. Development of low-cost public housing in the British housing supply system. Urban Planning Int. 2007;22(1):67–71.
B, He. On the characteristics and integration of Korean inhabitation area of Beijing: taking Korean village in Wangjing for example. Urban Problems. 2008;159(10):59–64.
S, Hu, Zhao M. Research on the community service-oriented management model of floating population. Chin J Population Sci. 1997;4:22–9.
Huang T. Research on outdoor space environment design of jiangsu rural relocated resettlement community. Nanchang: Nanchang University; 2007.
Kain JF. Housing segregation, negro employment and metropolitan decentralization. Quarterly J. Econ. 1968;(82).
Kearney M. From the invisible hand to visible feet: anthropological studies of migration and development. Ann Rev Anthropol. 1986;(15).
Levitt P, Jaworsky BN. Transnational migration studies: past developments and future trends. Ann Rev Sociol. 2007;33(1):129–56.
Li X. Study on the Inner-decaying village and the Countermeasures with Jiangsu Province as the case. Urban Planning Rev. 2001;25(6):8–13.
Li Y. Research on the reconstruction mode of "Urban Village" in the City in Xi'an. [Master's Thesis]. Xi'an: Xi'an University of Architecture and Technology, 2002.
M, Li. Research on international migration: scope, framework and significance. J Xiamen Univ. 2005;16(3):44–50.
Li W. Research on the Hui Architecture in Ningxia. [Master's Thesis]. Tianjin: Tianjin University, 2009.
M, Li, et al. The influence of floating population on the development of large cities and countermeasures. Beijing: Economic Daily Press (First Edition); 1991.
Li J, Li W, Li Z, et al. Research on the policy of affordable housing in Singapore and its inspirations. Urban Planning Int. 2012;27(8):36–42.
Logan JR, Molotch HL. Urban fortunes: the political economy of place. Berkeley, CA: University of California Press; 1987.
Ma G. Research of Social Security Commodity Housing [D]. [Master's Thesis]. Shanghai: Tongji University. 2007.
Massey DS. Mullan BP. Processes of hispanic and black spatial assimilation. Am J Sociol. 1988;(89):836–73.
Meillassoux C. Maidens, meal and money: capitalism and the domestic community. Cambridge University Press; 1981.
Niu Z. A study on the construction of international communities in Beijing from a cultural perspective: taking the development of Maizidian and Wangjing communities as examples. In Urban

Internationalization Forum 2011: Metropolitan Governance in the Process of Globalization; 2011. Pp. 168–76.
Park RE, Burgess EW, McKenzie RD. The City Chicago. Chicago: University of Chicago Press; 1925.
Qian Q, Feng Z, Bo W. The social spatial characteristics and formation mechanism of Nanjing international community: based on the survey of the international community around Mufuyuan Street. Urban Planning Int. 2013;28(3):98–105.
Rao X. Peripheral vision: a new dimension for the exploration of human settlement environment. J Archit. 1999;6:48–51.
Smith P. Transnational urbanism. Malden: Blackwell; 2001.
Stark O. Migration decision making: a review article. J. Development Econ. 1984;(14):251–259.
Tian D. Housing policy: lessons from international experiences and China's practical choices. Beijing: Tsinghua University Press; 1994.
Tian K. Investigation, analysis and reflection on the urban adaptability of migrant workers. Soc Sci Res. 1995;19(5):90–5.
Wang C. Social Mobility and Social Structure: Research on Beijing's Zhejiang Village. Hangzhou: Zhejiang People's Publishing House (first edition), 1995.
Wang S. Research on the Evolution and Renewal of the Dwellings of the Bai Minority in Dali. [Master's Thesis]. Kunming: Kunming University of Science and Technology, 2009.
Wang G, Shen J, Liu J. Citizenization of peasant migrants during urbanization in China—a case study of Shanghai. Population Development. 2008;29(1):3–23.
Wu X. The "Rural Communities" in cities: a study on the current status and integration of floating population settlements. Urban Planning Rev. 2001;12:25–41.
Wu FL, Webber K. The rise of "foreign gated communities" in Beijing: between economic globalization and local institutions. City. 2004;21(3):203–13.
Wu M, et al. Research on the settlement of floating population under the background of urbanization in China: a case study of Jiangsu Province. Nanjing: Southeast University Press; 2005.
Wu X, Wu J, Wang H, Chen H, Xia M. Research on the changes of minority settlements amidst the waves of modernization—a case study of the Qijiawan Hui settlement in Nanjing. Planners. 2008;9:15–21.
Wu X, et al. Metropolitan floating population of employment space analysis: an empirical study for migrant workers. Nanjing: Southeast University Press; 2010.
Xiang B. Cross-border Communities. Beijing: Life·Reading·Xinzhi Sanlian Life Bookstore Co., Ltd (first edition), 2000.
Yang C. An investigation of the social conditions in the floating population settlements in Lianban. Urban Planning Rev. 2003;11:65–9.
Yang C. Study on the Evolution and Inheritance of Landscape Architecture of the Zhuang Minority. [Master's Thesis]. Harbin: Northeast Agricultural University, 2009.
Yang W, Zeng J, Li Z. Protection and development—research on the current status and development countermeasures of Muslim communities in Mainland China. J Tianjin Univ (social Sciences). 2004;1:71–4.
Yuan Y. research on the spatial differentiation of urban poverty in China. Beijing: Science Press; 2014.
Yue J. The overview of Hong Kong public housing and its enlightenment. Urban Planning Int. 2010;25(3):97–104.
Zhang G. Poor population settlement and its countermeasures: a case study of Beijing. Urban Planning Rev. 2006;(1): 40–46, 54.
Zhang Z. On the theory of vacant village in Hebei Province. J Hebei Normal Univ (natural Science Edition). 1998;22(4):573–6.
Zhang D, Hou X. Urban development of minority settlements and preservation of ethnic cultures: a case study of Guzhang County, Hunan Province. Planners. 2006;8:29–32.

Zhang M, Shi A, Sun M, Yi X. The settlement and distribution of floating population in the periphery of large cities in the Pearl River Delta: a case study of Pinghu Town, Shenzhen. Urban Planning Rev. 2002;5:63–5.

Zhou C, Ma X. A discussion on the paradigm of the social structure of urban Hui people—a case study of the structural changes of the Hui settlement in Beijing. J Hui Muslim Minority Study. 2004;4:35–7.

Zhou S, Zhu L, Wang W, Fan Z. The impact of the development of urban traffic arteries on the evolution of minority communities: a case study of Madian Hui settlement in Beijing. Soc Sci Beijing. 2002;4:34–50.

Zhu L. On the urban adaptation of peasant workers. Jiang Hai Acad J. 2002;45(6):82–8.

Chapter 2
Community Sample I: Life Space and Community Response: A Study of International Student Settlements

Abstract Since opening up and reform, the constant strengthening of comprehensive national power and continual promotion of exchanges with the outside world has led domestic institutions of higher education to embark on the "send students out and welcome students in" road of internationalization and globalization. More and more Chinese institutions of higher education have received international recognition, while ever more foreigners are choosing to come to China to study and work. The daily growing community of international students often forms different types of student settlements around institutions of higher education. The daily activities and life space of these settlements are not only unique in obvious and overlooked ways but also—by virtue of long-term interaction—instigate change at different levels and different degrees in the surrounding community. This chapter takes the international student settlements of Nanxiucun, Nanjing (including forced settlements of collective dorms as well as flats and voluntary settlements in the surrounding community), as its case study. Using the two major topics of everyday life space and community response as our point of entry, we investigate the daily activities, life habits, and influence of "international students" as a representative international migrant community at the community scale. In addition, concerning current problems in international student settlements, we suggest a comprehensive strategy for the improvement of the external environment and removal of obstacles to their integration. The selection and survey of this type of community sample substantively constitutes a dedicated examination of the "integration-response" topic in the area of "international students moving into the country."

Keywords International student settlement · Everyday life space · Community response · Nanxiucun community · Optimization strategy

X. Wu and L. Wang, *Settlement Spaces: Urban Survival Prospects of China's Special Communities*, Springer Geography,
https://doi.org/10.1007/978-981-16-4892-2_2

2.1 Definition of the Research Object

2.1.1 Definition of International Students

2.1.1.1 About International Migrants

In 1922, the fourth session of the International Labor Conference made the first proposal regarding how the countries should define "international migrants," creating a unitary standard of evaluation. The conference furnished clear definitions for basic concepts relating to international migrants, such as "emigration" and "immigration." In 1953, the UN Department of Economic and Social Affairs Statistics Division made a proposal for the standardization of "international migrant statistics," mainly addressing the issue of how to define "long-term migrants who are not of local origin (including those who have gained and those who are yet to gain citizenship). This was the first time that a standard, established under the auspices of the UN, asserted that those residing abroad "for more than a year" are "international migrants."(Rajan 2006) In 1976, the UN Department of Economic and Social Affairs Statistics Division amended the regulations listed above. A new statute stipulated that "international migrants" include "those who aim to reside long term, have been in the host country for over a year, and continue to reside in that country, which also extends to those who intend to immigrate to a country for the long term but have not yet completed a full year of continuous residence, or those who have lived for a year or more abroad but are not currently living in that country." In 1998, the Recommendations on Statistics of International Migration (hereafter 1998 Recommendations) were published, finally providing a range of different definitions, both exhaustive and concise, for "international migration." Among the concise definitions is the following: An international migrant is anyone who changes their usual country of residence. Those who travel abroad in the short term for the purposes of recreation, holidays, business, medical treatment, or religious pilgrimages are not included (DESASD (Department of Economic and Social Affairs Statistics Division, UN) 1998). The International Organization for Migration has also furnished a definition for "international migration": generally, international migration refers to people who cross borders to reside in another country, including temporary residential populations. Those who travel for tourism or short-term business trips are not ordinarily included in the purview of migrants.[1]

Comparing the definitions of these authoritative organizations, we can observe that there are three key points to all definitions of "international migration": 1. crossing the borders of a sovereign nation; 2. continuously residing in a foreign country; and 3. the purpose of migration. Simultaneously, we need to note that in the current world migratory populations are highly diverse. In the wake of new developments

[1] Cf. "International Migration and Development" on the "International Organization for Migration" website: http://www.iom.int/jahia/Jahia/pid/271.

Table 2.1 International migration categorization

Categorization basis	Categories
Scale of Migration	Individual migration, migration in small groups, large-scale migration
Distance of Migration	Short-distance migration, long-distance migration; migration across oceans, migration across continents
Motivation for Migration	Migration for survival, migration for development; voluntary migration, forced migration
Legality	Legal migration, illegal migration; regular migration, irregular migration
Duration of Migration	Short-term migration, long-term migration; temporary migration, permanent migration
Migrant Identity	Independent migration, dependent migration; migrant workers, migration to be reunited with families, asylum migration, migrant students
Purpose of Migration	Employment, family reunification, education, business investment, recreation, and asylum-seeking

Source Li Minghuan. Definition and Types of International Migrants—with Special Reference to Chinese Migrants [J]. Overseas Chinese History Studies, 2009(2): 1–10

not only do the aims and reasons for cross-border migration change prolifically but international migration also falls into multiple categories (see Table 2.1).

2.1.1.2 On International Students

From the categorization of the aforementioned table we may gather that international students constitute a significant aspect of international migration. Their motivation for migration is developmental or voluntary; from a legal standpoint, their migration is legal and regular; in terms of duration, their migration is short term or temporary. However, although studying abroad is accompanied by working in the host country, or immigration, this can turn into long-term or permanent migration. In terms of migration identity and purpose, international students may be classed as study migrants. We must also add that there is currently no way to draw distinctions in terms of scale and distance among international students.

2.1.1.3 International Students in This Study

The "international students" discussed in this chapter are individuals residing in China voluntarily and temporarily for the purpose of study. They are non-Chinese citizens with a high school diploma or a similar academic qualification that have registered to study or are engaged in research at a Chinese institute of higher education, or research facility. They constitute a special sub-category of regular international migration.

2.1.2 *Definition of International Student Settlements*

2.1.2.1 On International Student Settlements

International migrant communities are geographic spaces for communal survival and development of large numbers of international migrants of different nationalities, languages, economic potentials, social statuses, cultural backgrounds, and customs. These communities can provide residents with a sense of spiritual identification, belonging, and safety (Ma 1998; Smith 2001; Portes 1987). In line with the progress of globalization, Chinese cities have greeted new opportunities for transformation. Urban construction and development is progressively becoming more culturally diverse and international. The friction and integration that characterize cultural diversity have driven the thriving development of international communities in China's larger cities. For example, Wangjing district in Beijing has become home to China's largest settlement of Koreans. Settlements for mainly white collar European and American workers have appeared in Pudong and Gubei Hongqiao districts in Shanghai. The Xiaobei road area in Yuexiu district, Guangzhou, is home to a sizable settlement of Africans. Korean settlements are currently forming around Xita street in Shenyang, Chengyang district, in Qingdao and Chengdong district in Nanjing (Wu 2004; Zhigang et al. 2008; Kim 2003). The international student settlements researched in this chapter form a representative branch of international migrant communities.

2.1.2.2 The International Student Settlements in This Study

On the basis of the relevant definitions listed above, the international student settlements discussed in this study refer to international migrant settlements formed by individual or multiple communities in the form of university dorms or privately rented accommodation either attached to a campus or in the surrounding area.

It is important to note that this study has expanded the scope of international student settlements as a community to include university-organized dormitories for international students. International student settlements may therefore be divided into two categories: the first is forced settlement communities, that is, accommodation (or dormitories) organized and managed by the university; the second is voluntary settlement communities, that is, spontaneously formed communities of international students renting accommodation in the vicinity of the university.

2.1.3 *Definition of Everyday Life Space*

2.1.3.1 Concerning Everyday Life

To define the scope of "everyday life space," we must first clarify what is meant by the concept of "daily life." This concept first arose in philosophy. The Czech philosopher Kosik (1979) once vividly described the essential characteristics of daily life as being that which is natural and comfortable. He indicated that in everyday life, activities, and lifestyles become instinctive, subconscious, and unconscious without requiring thought. These mechanisms, objects, people, sports, work, the environment, and the world as part of lifestyles are not comprehended in terms of their innovativeness and reality but are accepted as component parts of an already familiar world. Simultaneously, everyday life is full of multifarious dimensions and possibilities, which can come under the control of individuals using their talents and wits. Drawing from Kosik's viewpoints, it should be acknowledged that the everyday lives would differ for individuals and the era they live in. Undeniably, these differential everyday lifestyles share a common trait: they are all directly related to one's survival and reproduction. Heller (2010), a core member of the Budapest school, defined everyday life as "the aggregate of those individual reproduction factors which, *pari passu*, make social reproduction possible" in *Everyday Life*. Generally speaking, everyday life is an amalgamation of an individual's direct environment, home, and nature as well as a collective term for various activities that is meant to sustain an individual's survival and reproduction.

To humans, the world of everyday life is the most fundamental of disciplines. That said, the biggest challenge in studying everyday life is defining its boundaries, to which we can reference some common criteria proposed by scholars of everyday life. Everyday life, as the object of study, is a daily routine and repetitive lifestyle. It is universal and excludes random occurrence of events. Chinese scholar Xuewei (2007) summarized everyday life as activities that are common, secular, and less ritualistic in nature. These include conscious activities such as work, residential and entertainment activities as well as incidental activities that are secondary to the main activity or occur randomly and can take place alongside or after the primary event. Examples of incidental activities include routine interactions, conversations, strolling on the streets, and accidental encounters, among others.

2.1.3.2 Regarding Everyday Life Space

Initially, scholars of everyday life space were mostly drawn from the humanities and geography disciplines; thus, their studies on city spaces mostly concentrated on macro-perspectives and abstract ideas by focusing on the structures and models of social spaces. However, examining everyday life and its urban spaces from the "human" perspective remains overlooked. The study of such everyday urban spaces also hardly incorporates relevant theories of urban sociology. In recent years, and alongside the growth of urban studies, urban sociology has become an important research area in the discipline of sociology. Gradually, correlation studies between structural changes to spaces in urban society and everyday lifestyles came to the attention of researchers (Gang 2004).

Presently, the more common understanding of everyday spaces is that spaces are utilized by people for activities in their everyday lives. Amid their social lives, people often need to move between their homes, workplaces, consumer spaces, and non-consumer public spaces. As described by time geographers, existing objects take up space. The spaces taken up by activities of everyday life are therefore everyday life space (Xuewei 2007).

As for the classification of everyday life space, it depends somewhat on the rich diversity of everyday activities. Some scholars believe that urban everyday lives can be distilled into four categories: work or school, residential, shopping, and leisure (Wang 2000; Xingzhong 2000). There are also scholars who posit that everyday live is dominated by family, work, consumer spaces, and non-consumer public spaces; therefore, the spaces of everyday life of contemporary urbanites often include their workspace, consumer spaces, interaction and leisure spaces, and residential spaces, among others (Chengbi 2004). That said, these perspectives are similar in that the occurrences of those activities are within a specific scope, and therefore the spaces of everyday life can be further segmented as shown below (see Table 2.2).

2.1.3.3 Everyday Life Space in This Study

The everyday life space examined in this chapter mainly comprises four typical lifestyle spaces arising from international student settlements (based on the International Student Settlements in Nanxiu Village, Nanjing City): **residential spaces, learning spaces, shopping spaces,** and **leisure spaces**. Such distinctions are relatively comprehensive in covering the common aspects of the residents' everyday life without being weighed down by details. This will be helpful in tackling the core of the issue holistically. While the residential spaces largely correspond to the geographical boundaries of the international student settlements, leisure and shopping spaces often expand to much wider horizons depending on the settlement zone and university (from the community to city blocks and even to Tier 1 cities).

The residential spaces in this study are not limited only to indoor spaces that people live in but also include city blocks, neighborhoods, or specific plots of land that involve human activities, which can be private or public in nature. The most fundamental activities of everyday life, of sustaining the individual's life, as well as a portion of family interactions are completed in residential spaces. Learning spaces primarily refer to spaces where international student communities engage in learning activities (e.g., classes, self-study, seminars, etc.), such as the classroom and library. Leisure spaces primarily refer to public urban spaces that satisfy their needs for leisure and entertainment and can host five types of activities: culture and entertainment; interactions, sports, games, and hobbies. Concurrently, this study zooms in on the narrow definition of consumer spaces, that is, shopping spaces as

Table 2.2 Different classifications of spaces of everyday life

Space Classification	Sub-classifications
Residential Spaces	Outdoors (community, city block, villages, etc.) and indoors (family)
Unit Spaces	Workspace, learning spaces, etc
Consumer Spaces	Broad definition: spaces for everyday shopping, spaces that correspond to consumer behavior in leisure and entertainment activities Narrow definition: spaces for everyday shopping (small shops in city blocks, large departmental stores in the city center, chain supermarkets, general merchandise stores, convenience stores, etc.)
Leisure Spaces	Categorized by functions: mental stimulation (including card games, reading, watching movies, etc.), relaxation (including raising pets, craftswork, calligraphy arts, collecting objects, visiting markets, etc.), fitness (including ball sports, strolls, specialized fitness training, etc.), interaction (including visiting friends, organizing gatherings, chatting with family, etc.), and public service (including community management, specialized services, etc.)
	Categorized by distance: community, city block, land section, and city
	Categorized by space characteristics: closed-door leisure spaces (e.g., cafes, bars, cinemas, theaters, discoes, theme parks, etc.) and public urban spaces (e.g., bustling streets and quiet parks)
	Categorized by markets: leisure and entertainment spaces with public service value (e.g., community squares), sub-market leisure, entertainment spaces (e.g., parks), commercial leisure, and entertainment spaces (e.g., bars)

Source Yu Xiangyang and Wang Xingzhong. Spatial Structure of Commercial Entertainment Venues in Urban Communities [J]. *Human Geography*, 2003, 18(2):30–36
Li Quanbao. An Analysis of space–time Characteristics of Elderly Everyday Activities in Nanjing City, with a Focus on Shopping and Leisure Activities [D]. Nanjing: Southeast University, 2015: 20–21
Zhang Xuewei. Everyday Life Space [D]. Shanghai: Tongji University, 2007: 42–43

venues where the residents shop in their everyday lives as well as the corresponding spatial nodes (excluding online shopping methods).

Because of the complexity of spaces of everyday life, this study has made the following distinctions and classifications of the four types of spaces and their corresponding activities (see Table 2.3).

2.1.4 Defining Community Response

2.1.4.1 Concerning Community Response

The author realized that in the process of reviewing and consolidating the community response concept, academia has yet to reach a unified and clear understanding.

Table 2.3 Sub-categories of spaces and activities in this study

	Basis	Sub-category	Activity type
Residential spaces	—	Outdoors (community) and indoors (family)	Various family activities; some interactions
Learning spaces	Function	Classroom, library, self-study room, seminar room, etc	Classes, self-study, reading, discussions, etc
Shopping spaces	Tier	Commercial spots in the community (within 1 km)	Shopping for daily amenities (e.g., food, soap, laundry detergent, etc.)
		Regional commercial centers (1–2 km)	Shopping for products based on choices, such as fashion, cosmetics, and books
		Urban commercial centers (2–3 km and beyond)	Shopping for special gadgets, such as household appliances and electronics[2]
Leisure spaces	Distance	Within the community (within 1 km)	Cultural and entertainment types: music, dancing, movies, online (games), radio broadcasting, etc Integration types: chatting, visiting friends, meeting for meals, gatherings, etc Fitness types: ball sports, strolls, specialized fitness training, etc Holiday types: visiting parks, hiking, short trips, etc Hobbies types: music, chess, calligraphy, painting, photography, collecting objects, writing, inventing new ideas/objects, etc. (Chongqi 2006)
		City blocks (1–2 km)	
		Within the main city area (2–3 km and beyond)	

Source Own elaboration

Nonetheless, existing studies generally agree that societal integration and local response are two contrasting concepts.

Amid theories on immigrant cultural identification and integration, there are two perspectives, that of the "melting pot" and "pluralism" (Bo and Haiying 2010). Similar studies have shown that social identification and integration are multi-dimensional, and the expansion of the social identification concept is among one of the development trends. For instance, Samuel P. Huntington offers the following

[2] The author has referenced the classifications of consumer products in Yu Jianyuan's "Classification of Consumer Products according to Purchasing Habits and Its Marketing Significance: daily amenities, products based on choices, special products, and non-seeking products. This study has classified the shopping types based on what the consumer would purchase: daily amenities, products based on choices, and special products.

classifications of the dimensions and types of social identification: associative (e.g., age, gender, blood relations, and race), cultural (e.g., ethnicity, tribe, nation, religion, and civilization), local (e.g., city blocks, villages, and countries), political (e.g., organizations, factions, parties, and nations), economic (e.g., job duties, industry, and class), and social (e.g., clubs, societies, and social statuses). These have provided a standardized scope for social identification studies (Huntington 2013). As for local response, it is more manifest as the response or understanding toward a group of people as well as the influence or changes brought about to the surrounding community. Related studies have mostly been based on real-world case studies. For instance, Li Zhigang conducted a survey of local residents, business operators, local media, and online public opinion in Guangzhou's Xiaobei neighborhood, where Africans reside. Using questionnaires and interviews, he described the negative African neighborhood response that was mutually constructed, and further uncovered the complex influence that the international migrants had continually brought to the locals. The addition of a new "non-national space" undoubtedly compounds the difficulty of urban management and highlights the importance of managing localization and community development in our time. This also suggests that locality and community need to face the arrival and permeation of globalization head on (Zhigang et al. 2009).

2.1.4.2 Community Response in This Study

The community influenced examined in this chapter primary refers to the overall influence and changes brought to the surrounding community features, spatial delineation, industrial economy, interaction, and identification amid the students' everyday life and development of everyday life space during the formation and evolution processes of international student settlements.

2.2 Description of International Students and Their Everyday Lifestyles

2.2.1 *An Overview of International Students in Nanjing*

2.2.1.1 Population

Since the economic reforms in China, various higher educational institutes in Nanjing have started to recruit international students from abroad, the scale of which has continued to grow dramatically in the last 10 years. Currently, there are 17 schools in Nanjing that accept students into their full-time undergraduate programs. Relevant data (see Fig. 2.1) shows that the total number of international students in Nanjing has grown from 5500 in the year 2000 to approximately 15,000 in the year 2017—a near threefold increase. The number of international students in each school has grown

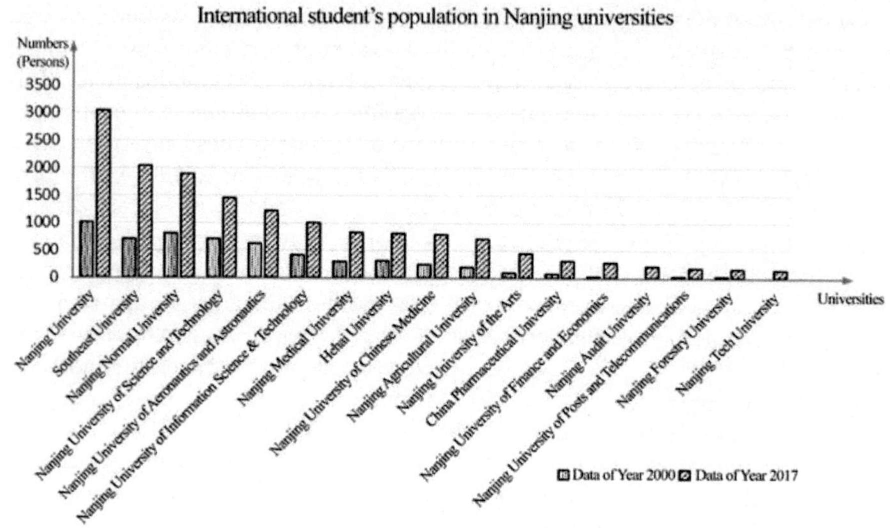

Fig. 2.1 Bar chart depicting international student population sizes in Nanjing Higher Educational Institutes in 2014 and 2017. *Source* Own elaboration based on data from the official websites of the various higher educational institutes in 2000 and 2017

as well, with the largest increase found in Nanjing University, Hohai University, and Nanjing Medical University. The international student population size is the greatest in Nanjing University, Southeast University, and Nanjing Normal University. Nearly half of all international students in Nanjing are studying in one of these three institutions.

2.2.1.2 Basic Attributes

Differences in the nationality distribution, specializations, and degree types of international students arise as different schools have their own teaching characteristics and academic focuses. This study has tabulated relevant data from Nanjing University, Southeast University, and Nanjing Normal University as these three schools have the largest international student population size in Nanjing and can serve as a useful reference for the overall situation. The basic information is as follows (see Table 2.4).

In terms of national distribution, European and North American students dominate the international student population in Nanjing University, followed by Japanese and Korean students, and then Southeast Asian students. The ratios are also fairly balanced across different nationalities. For Southeast University, students from Southeast Asia constitute the largest percentage of international students, followed by Japanese and Korean students, and African students. The student population is mostly dominated by students from developing countries. For Nanjing Normal

Table 2.4 An overview of international students in Nanjing University, Southeast University, and Nanjing Normal University

	National distribution	Majors	Type of studies
Nanjing University	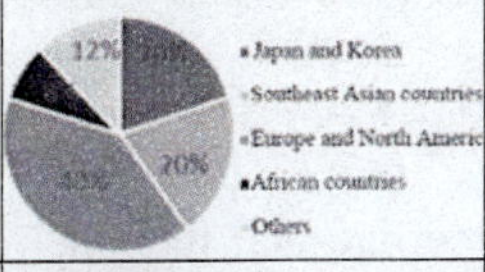	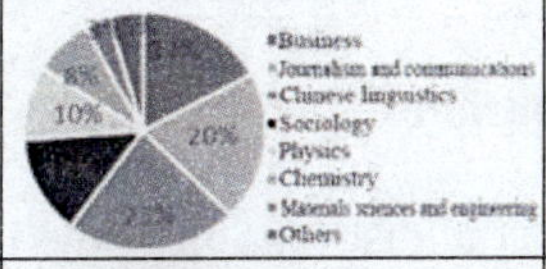	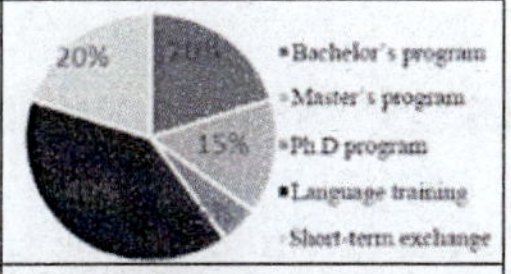
Southeast University	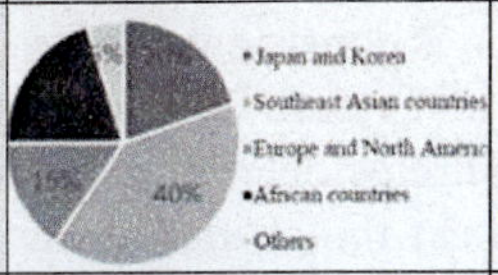	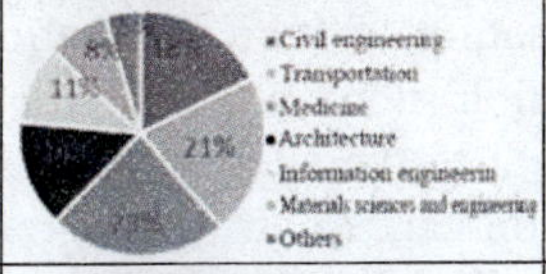	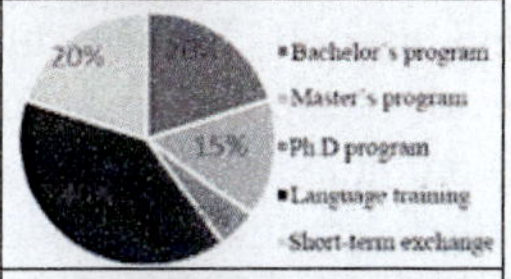
Nanjing Normal University	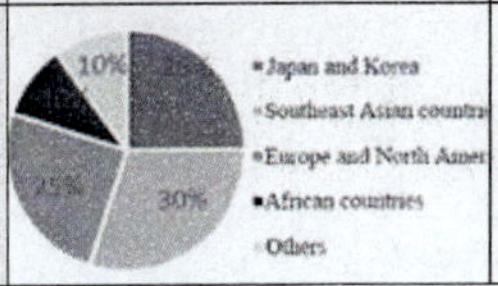	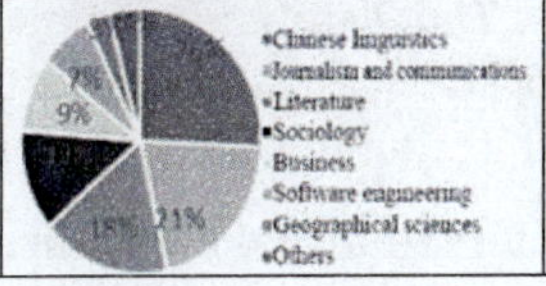	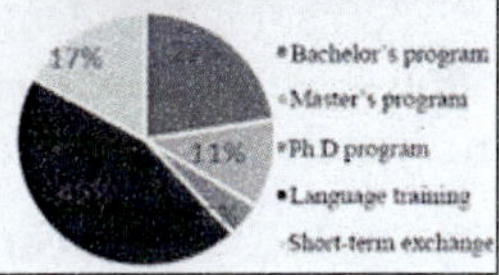

Source Nanjing University's Office of International Cooperation and Exchanges (https://wb.nju.edu.cn/main.htm); Southeast University College of International Students (http://cis.seu.edu.cn/); Nanjing Normal University's Office of international Cooperation and Exchanges (http://gjc.njnu.edu.cn/)

University, Southeast Asian students also make up the highest proportion of international students, followed by European, North American, Japanese, and Korean students. As the earliest higher education institute to accept international students in Nanjing, Nanjing Normal University has a more diverse international student population.

In terms of course specializations, the international students at Nanjing University are enrolled in more varied courses. Among liberal arts subjects, business, journalism and communications, and sociology disciplines attract the greatest number of students. Among the sciences and technical fields, physics, chemistry, material sciences, and modern engineering are popular choices. For Southeast University, the international students tend to choose technical specializations, such as civil engineering, transportation, architecture, and information engineering, as well as practical and application-based majors, such as medicine. As for Nanjing Normal University, the internal students are more concentrated in humanities courses, such as linguistics, cultural studies, and sociology.

In terms of the degree types, the international students at Nanjing University also pursue more varied pathways, among which students in language immersion programs and degree programs each comprise 40% of the international student population. The remaining 20% is made up of students on short-term exchanges. For Southeast University, international students take up 76%, the highest weightage compared with the other two universities, while the rest are students on short-term exchanges. Nanjing Normal University's situation is fairly similar to that of Southeast

University, in that non-degree international students far outnumber students pursuing degrees.

2.2.1.3 Spatial Distribution

At the urban level, the spatial distribution of the international students' community and intra-city higher education institute distribution is consistent. At the community level, the spatial distribution shows clustering to an extent. Therefore, the international students' attachment to the educational institutes determines how the institutions' spatial distribution can, to an extent, reflect the spatial characteristics of the international students. The distribution characteristics of international students within the city can be mapped out as a whole after tabulating the data on international students from educational institutions in Nanjing (see Fig. 2.2).

City level: External clusters, internal concentration

The spatial distribution of international students within Nanjing City is consistent with the distribution of the higher education institutions, demonstrating the characteristics of "external clusters and internal concentration." On the one hand, the higher educational institutions within Nanjing City form a cluster distribution in the old city area: Xianlin University City, Pukou University Town, and Jiangning University Town. On the other hand, the institutions inside the main city area are mostly concentrated in the old city zone, and the number of institutions gradually falters beyond the old city. One main reason for such a spatial formation can be attributed to how the city typically expands outwards, from the old city area to the main city area. Therefore, the educational resources concentrated in the old city area are very rich as the higher educational institutions and their initial (old) school zones tend to crowd this area. Another reason for the distribution of higher educational institutions beyond the main city area is because of the government's organization and planning of the three university towns in a bid to centralize the resources of higher educational institutions' resources, which are gradually moving outward. Therefore, clusters at the three university towns were formed in the external region.

Community level: Mixed co-living population at the macro level, more homogenous clustering at the micro level

At the community level, residential spaces distribution for international students show characteristics of mixed co-living at the macro level but more homogenous clustering at the micro level. On the one hand, be it actively choosing to rent outside school premises or passively staying within the school compound, the international students mostly live with other races and nationalities at the macro community level. On the other hand, they would also prefer to live with international students from the same nation or race and therefore live within a smaller proximity; they will stay with their kind, especially for those who actively choose to stay outside of the school. There are two primary reasons for this spatial formation. First, the international students come from all around the world, and such mixed co-living patterns are inevitable

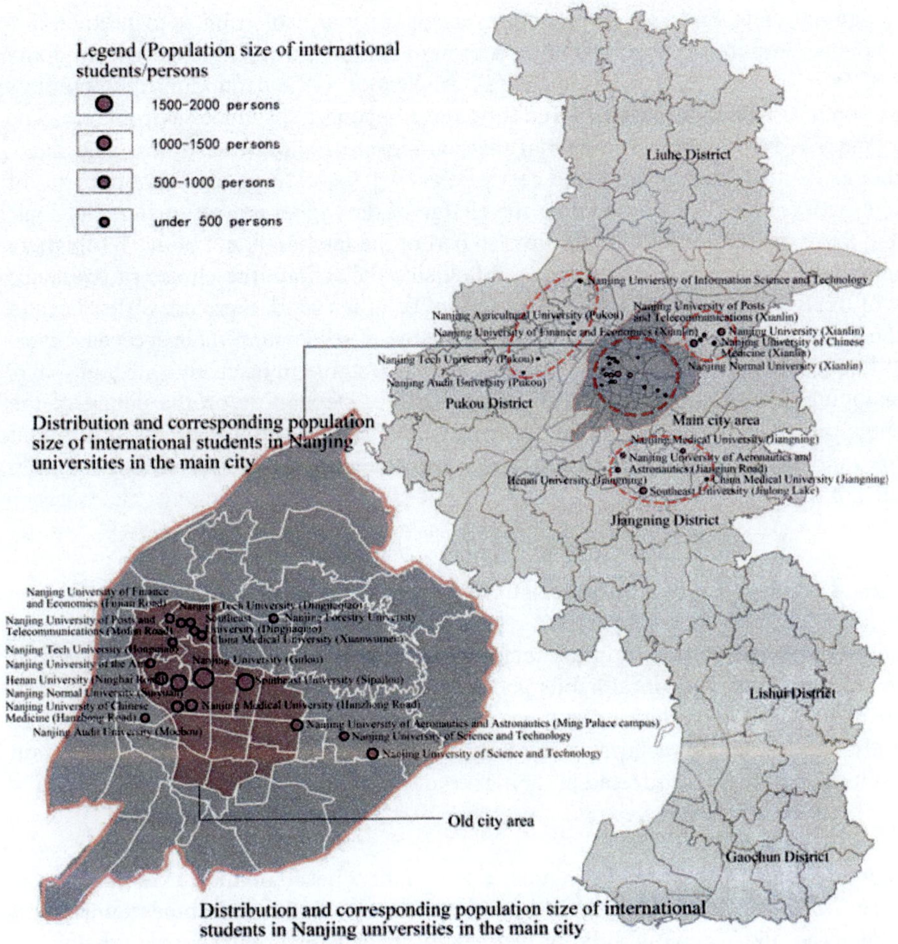

Fig. 2.2 Spatial distribution of international students in Nanjing (Left: Main city area; Right: City area). *Source* Own elaboration based on the data from the official websites of the universities

be it in school-arranged dormitories or renting outside of the school to live with the locals. Second, international students from the same nation or race have more similar cultural backgrounds and traditional roots, which help them to better engage and interact with one another. As such, they are likely to stay closer to each other within a smaller geographical scope.

School level: Tightly knit connections, external expansions

In terms of international students' management and everyday life at the school level, the spatial distribution is that of tightly knit connections and external expansions. On the one hand, the students' residential and learning lifestyles are closely connected to the school, for instance, their studies are primarily concentrated within school compound and their living arrangements are also primarily distributed in international students' dormitories or surrounding communities. On the other hand, their shopping and leisure activities are based around the school and surrounding community's settlements, which allow them to easily move beyond the school environment. The primary reasons for such a spatial distribution are as follows. On the one hand, the residential arrangements and services, as well as learning and development, of international students come under the charge of the higher educational institutions, and these international students are also part of the larger student body. While there may be living arrangements within and outside the school, the choice of residence will likely be closer to the school zone. On the other hand, there are differences in the activities and spatial needs of the international students in their everyday lives. While residential and learning activities are more stable in terms of time and space, shopping and leisure activities are more flexible. Depending on the nature of the shopping and leisure activity, the students will engage in activity spaces of different tiers and distances, and it is possible for them to venture around the school and even toward the city.

2.2.1.4 Community Characteristics

Given their cross-border migrant identities, the international students tend to differ in areas such as behavioral habits and cultural identification. Simultaneously, they are also subject to unified and routine management and training as part of the student body. These dual characteristics not only affect their everyday activities to a certain extent but are also manifested in their everyday activities.

First, the uniqueness of individual identity

Being foreigners, they are distinguished from others based on their external appearances. This results in them being a special group within the Chinese university compound and receiving varying degrees of special treatment or care whether in society or within the school. In the interviews, the international students mentioned that, as foreigners, they could feel that the Chinese nationals are friendly toward them and therefore enjoy different treatment from the Chinese students. Conversely, being foreign does not help the international students to expect equal treatment in everyday life and may cause them to demand special treatment everywhere they go.

Second, the youthfulness of the students

Most of the international students who come to China are youths and therefore have the common traits of the typical young person. They are generally mature or completely mature in their physical and psychological development and possess a certain scientific and cultural understanding. They also have their own thoughts and pursuits. When interacting with the international students, one can discover the following three characteristics: first, they are confident in their abilities, intelligence, and skills and strongly express their need for independence; second, they hope to be respected, especially by their teachers or fellow international students; and third, they feel that time is abundant and they hope to have more cultural, sports, or entertainment activities. Therefore, this community hopes that while the school is able to satisfy their basic lifestyle needs, it is also able to take their learning and spiritual needs into consideration at the same time.

Third, diverse cultures and customs

The international students' cultural customs are primarily manifested as overall multi-elemental inclusivity and living with others despite their differences. For international students as a whole, the multicultural structures and consequent differences in lifestyles and customs make it such that any one nation or cultural type would not dominate. When interviewed, although most of the international students are willing to try foods from other countries, they are not able to adapt in the long run to foods from other countries, including China. Instead, they will cook foods from their own hometowns. Likewise for festive occasions, while the international students will be interested in the festivities and customs different from their own, few are able to truly participate and assimilate. Instead, international students with the same nationality and similar cultural backgrounds will often gather and celebrate their own festivities. As for religious customs, which can be highly diverse, international students tend to show more inclusivity and non-interference toward those of different faiths. International students who share the same faith will find it easier to engage in deep interactions on religion and culture.

Evidently, while all international students are passionate toward the traditions and culture of their own nation, ethnic group, or religion, they do not discriminate against other cultural expressions. International students from the same or similar backgrounds often interact and contact each other because of their similar faiths, hobbies, and lifestyle habits, which help to bridge and foster connections. Gradually, a common international student culture develops a sub-culture that is different from the values, perspectives, thought processes, and behavioral norms from mainstream campus culture.

Fourth, limitations in interactions and integrations

International students' languages, thought processes, values, cognitive structures, and behavioral habits are influenced by their homeland culture; therefore, it is inevitable that they will encounter challenges in interacting and assimilating into the local foreign culture. Mu Dan discovered from interviews with international students that they generally believe that interacting with other international students in schools is easier than interacting with Chinese students. They also share more common languages and interact more deeply with other international students than Chinese students.[3] Through questionnaires and interviews with international students, Chen Xiaoming et al. found that the majority responded that they interact with classmates or friends of the same country (42.0%) and with classmates or friends from different

[3] Gansukh Nyamdolgor. *Research on the Problems in the Life and Study of International Students and Their Countermeasures* [D]. Northwest A&F University, 2014.

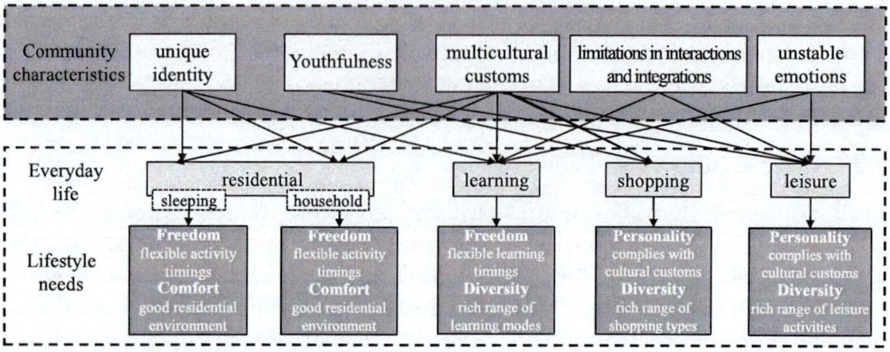

Fig. 2.3 The correlation between international students' group characteristics, everyday life, and needs. *Source* Own elaboration

countries (41.9%). Only 16.1% interact with Chinese classmates or friends.[4] This shows that there are relatively larger differences between international and Chinese students in terms of behavioral habits, entailing certain challenges and limitations for them in assimilating into the Chinese university environment.

Fifth, unstable emotions

When moving from a familiar society, culture, and environment into one that is foreign to them, international students are likely to encounter cultural clashes due to differences in cognitive structures, thought processes, values, and behavioral modes. Once these differences have been registered in their minds as unexplainable and irreconcilable with the cultural values they are familiar with, the international students may develop feelings of loss, loneliness, anxiety, fear, and even disgust. Sociologists and anthropologists have termed this impact and shock to one's psychological state due to sudden changes in inner cultural worlds and external cultural influences as "culture shock."[5] Those affected often appear to be prone to anger, depression, and self-isolation because deep within them, they experience a lack of security as well as loneliness and helplessness.

In summary, the international students' unique identity, youthfulness, multicultural customs, and limitations in interactions and integrations as well as unstable emotions will be directly reflected in the residential, learning, shopping, and leisure aspects of their everyday lives. These will then manifest into various needs, such as flexible activity timings, comfortable and personalized activity spaces, and diverse activities (see Fig. 2.3).

[4] Chen Xiaoming, Ou Qing, Wang Zulei. Investigation and Analysis of the Adaptability of International Students in Beijing [J].*International Chinese Language Education*, 2015(1): 19–32 + 198.

[5] *Zhou Xiaohong. Modern Social Psychology* [M]. Shanghai: Shanghai Renmin Press, 1997: 531.

2.2.2 *Overview of the Everyday Life of International Students in Nanjing*

As a special group, international students lead different everyday lives from other communities in terms of the type of activity, time, space, and approaches. Therefore, when studying the everyday life spatial attributes of international students, we also need to describe what their everyday lives are like.

2.2.2.1 Overview of Everyday Life Activities of International Students

International students engage in diverse everyday life activities. For instance, Zhou Yuan (2010) surveyed international students from South China University of Technology on their everyday activities and lifestyle habits. His informal interviews with the students discovered that their everyday activities are dominated by leisure and entertainment activities, such as fitness, shopping, drinking, and gatherings. These differ significantly from the lives led by Chinese students. As an example, while international students will go to the gym or shopping in the afternoon, Chinese students will either be revising school work in the library or working part-time. The lifestyle habits also greatly differ. While international students are more into entertainment activities, Chinese students focus more on their studies. International students lead exciting lives outside of their classes and are motivated to enjoy life. In contrast, Chinese students are more conservative and show more interest in their academics and part-time work. Chen Xiaoming et al. (2010) also conducted written interviews on how international students adapt to their new lives in China and discovered that international students are far more attracted to leisure and entertainment activities, such as shopping, dining, and gatherings, than learning activities.

This study also used sample surveys to collate data on the everyday activities of international students in Nanjing universities and validated and refined the aforementioned conclusion to a certain extent. In other words, besides mandatory residential activities, such as sleeping and household chores, international students prefer leisure activities, and learning only takes up a quarter of their time. Besides leisure activities, the international students also enjoy cultural and entertainment events, interacting with others, and sports activities (Fig. 2.4).

2.2.2.2 Time Attributes of International Students' Everyday Life

As shown by the sample survey data on the everyday life space of students in Nanjing's International Student Settlements (see Fig. 2.5), the international student community engage in extremely varied everyday life activities. Within the same time frame, there are many different possible activities that they can participate. In terms of time, there is also an evident difference between a working day and a rest day, in which there is an overall hybrid attribute of routine and spontaneity.

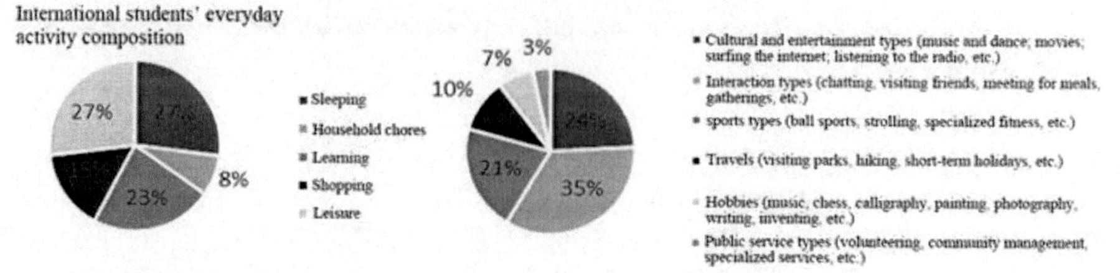

Fig. 2.4 International students' everyday activity composition (the Right Image Shows the Composition of Leisure Activities). *Source* Data from the research group's sample surveys on the everyday life space of International Student Settlements in Nanjing (2017)

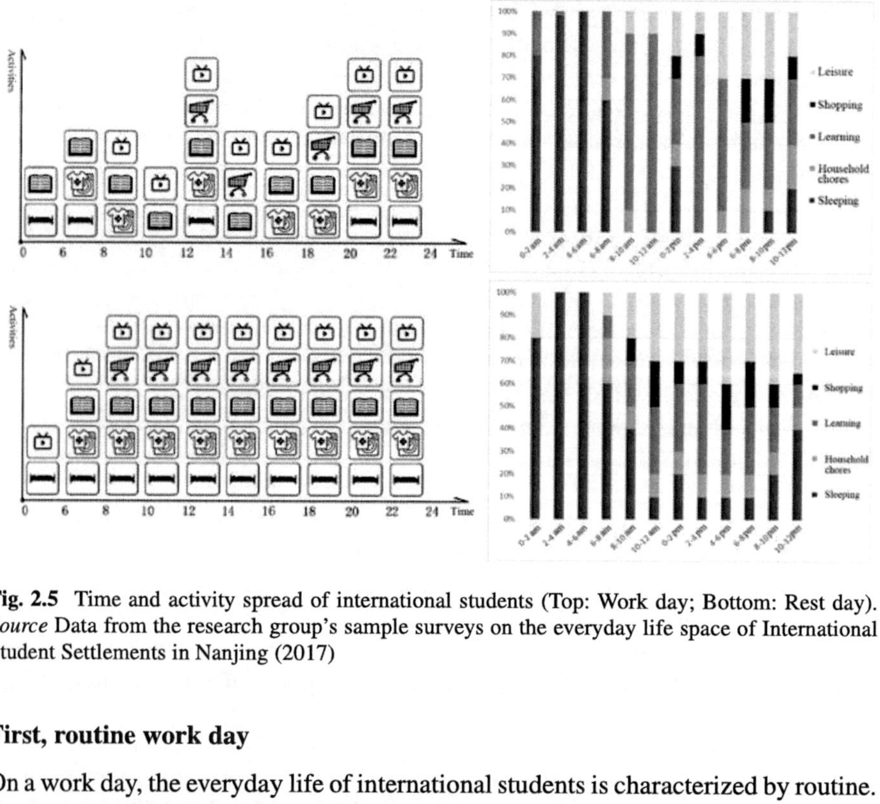

Fig. 2.5 Time and activity spread of international students (Top: Work day; Bottom: Rest day). *Source* Data from the research group's sample surveys on the everyday life space of International Student Settlements in Nanjing (2017)

First, routine work day

On a work day, the everyday life of international students is characterized by routine. On the one hand, as part of the student body, the international students also face stress from coursework, thus learning activities take up the largest proportion of their time on a work day. In fact, these activities almost run through the day. On the other hand, everyday life revolves around learning activities. Other than sleeping time, all other activities are mostly arranged outside of learning time. They take up very little time and are also very scattered around the students' schedule.

Second, spontaneous rest days

The everyday lives of international students on rest days manifest attributes of spontaneity. On the one hand, shopping and leisure activities take up a higher portion of their time compared to work days filled with lesson time and course stress. The time distribution is also wider. On the other hand, their time on rest days is also very flexible and free, such that any activity is possible at any given time. Many activities and behavior undertaken are not subject to time restrictions but are spontaneous and can take place in parallel.

2.2.2.3 Spatial Attributes of International Students' Everyday Life

First, limited independence

International students' everyday life space have limited independence. On the one hand, because of their unique identity and different lifestyle habits and cultural customs, their residential spaces are relatively independent. There tends to be designated housing or dormitories for international students within the school compound, which are managed separately from the other students. As for international students who choose to rent housing outside of school, they also tend to choose to stay with international students with similar lifestyle habits, cultural backgrounds, or race. Additionally, their leisure activity spaces are also more independent and more geared toward organization and participation within the international student body. On the other hand, the everyday life space of international students are independent on the relative scale, as being within the Chinese society and the school environment, it is inevitable for them to be in contact and interact with the local environment. Their learning and shopping activities also take place in shared spaces. It can therefore be said that the everyday life space of international students have limited independence, which reflects the balancing of assimilative and repelling forces experienced by this special group of international students that arise when interacting in the local environment.

Second, frequency of activity falls as the students move beyond their usual circles

The everyday life space of international student show a diminishing pattern in terms of frequency as they move beyond their usual circles. This is manifest in how their own attachment patterns and restrictions in school management, among other factors, result in their everyday life space being concentrated around the premises of the school (especially for residential and learning spaces). The residential location and school form two overlapping and symmetrical centers, and as the distance and type of activities increase, the frequency of participation in these activities falls.

Field studies show that the everyday lives of international students are mostly concentrated inside the school and the space surrounding the residential area as well as supermarkets, squares, and parks near the residential area. The spatial scope of their activities also expands because of shopping and leisure activities. Therefore, the scope of activities can be classified accordingly as basic lifestyle circle (within

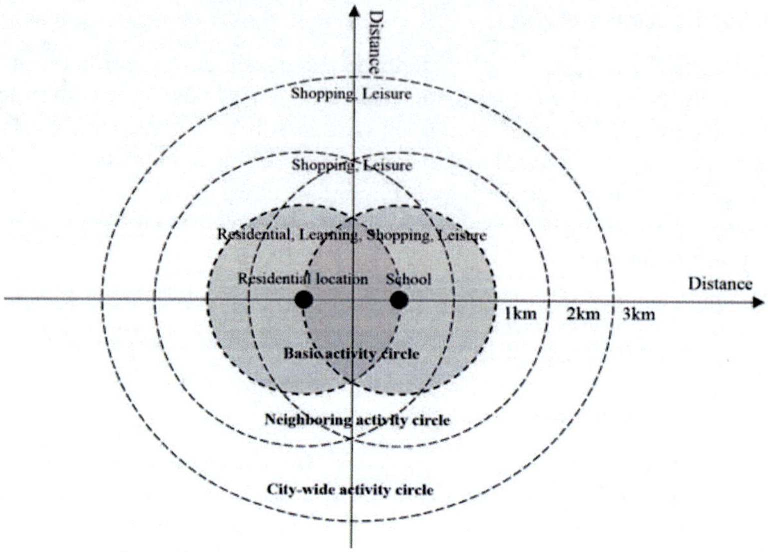

Fig. 2.6 Shopping activities diminishing pattern. *Source* Zhang Xuewei. *Everyday Life Space* [D]. Shanghai: Tongji University, 2007

1 km), neighborhood activity circle (1–2 km), and city-wide activity circle (2–3 km and beyond). The residential and learning activities are mostly or all concentrated within the basic lifestyle circle, while shopping and leisure activities are distributed across the basic lifestyle circle, neighborhood activity circle, and even city-wide activity circle depending on the shopping objective and type of activity as well as in a diminishing pattern (see Fig. 2.6).

2.2.3 Determining the Overall Study Trajectory

2.2.3.1 Study Scope

Based on the above overview analysis on international students in Nanjing, and in tandem with the study trajectory drawn from urban planning and community perspectives, as well as the characteristics of everyday life and feasibility of field research studies, the scope of spaces and groups in this study is defined as follows (see Fig. 2.7, Table 2.5).

In terms of the scope of spaces, the study first made an assessment regarding the international student recruitment situation in the various Nanjing universities. The international student population size was the largest in Nanjing University (Gulou) and Nanjing Normal University (Suiyuan). They are also geographical neighbors; thus, a rather large International Student Settlement was formed within and

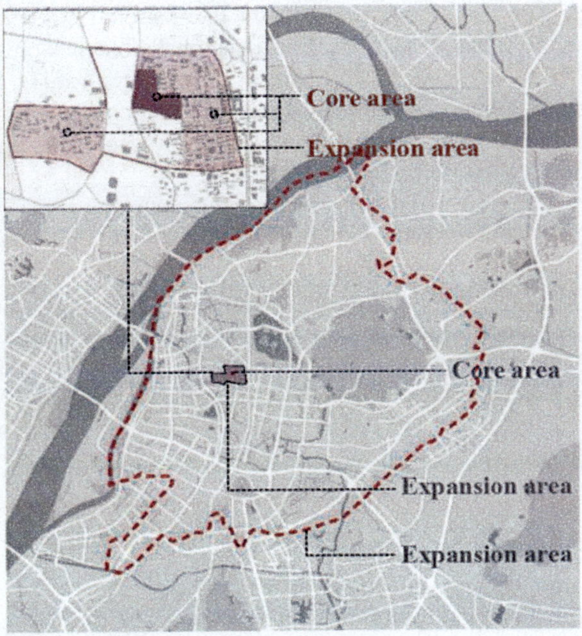

Fig. 2.7 Determining the study scope. *Source* Own elaboration

Table 2.5 Basic information of Nanjing University Gulou Campus, Nanjing Normal University Suiyuan Campus, and Nanxiucun Community

Tier	Scope of spaces	Corresponding activity circle	Object of study
Core spaces	Nanjing University Gulou campus, Nanjing Normal University Suiyuan campus, and Nanxiucun community	Basic activity circle (within 1 km)	Residential and learning spaces
Expanded spaces	Within the school compound, Nanxiucun and surrounding community (including the Nanxiucun community, Qingdaolu community at Hunanlu Street, the Hujuguan community at Huaqiaolu Street, and Suzhoulu community at Ninghailu Street)	Basic activity circle (within 1 km)	Shopping and leisure spaces; community response
Extended spaces	Nanjing main city area	Neighboring activity circle (1–2 km) and city-wide activity circle (2–3 km and beyond)	Shopping and leisure spaces

Source Own elaboration

outside the school—Nanxiucun International Student Settlement.[6] Therefore, the study selected Nanxiucun International Student Settlement, which can serve as a typical example, as the object of this empirical study. Simultaneously, based on the study criteria of everyday life space and community response, as well as considerations for field research, the scope of spaces was further distinguished into three tiers.

There are two primary types of the groups studied. The first group refers to international students who are currently living in Nanjing University Gulou campus, Nanjing Normal University Suiyuan campus, and Nanxiucun community. The second group refers to the people in the surrounding community, or locals living or working in the school, Nanxiucun community, and the surrounding community, with a focus on Ninghai Street, including areas under it, such as Shanghailu New Village, Taogu New Village, Nanchegua Small District, Huaxinxiang Small District, and Nanyinyang Small District. The second group also includes business operators and public service personnel.

2.2.3.2 Study Subjects

The survey subjects of everyday life space (based on the statistical approach for questionnaires) were drawn from the Nanxiucun International Student Settlements, that is, the core spaces of the study have been identified as a significant source of samples for the sample survey and specifically comprise international students from Nanjing University Gulou campus, Nanjing Normal University Syuyuan campus, and Nanxiucun community (see Table 2.6).

The surrounding community response (via interviews) targeted individuals drawn from the area bordering Ninghai Road on the west, Qingdao Road on the east, Beijingxi Road on the north, and Guangzhou Road on the south (see Fig. 2.8). The total area is 0.6 km^2, encompassing international students' settlement in Nanxiucun and the nearby Hujuguan, Qingdao Road, and Suzhou Road community, among others. There are many universities in this vicinity, accounting for the rich cultural atmosphere. There are also many international students and expats living in this area. There are three main groups who were interviewed for this study: local residents (including resident population within the area scoped out in the study), store operators (including foreign dining establishments, fashion stores, bars, beauty and hair salons, and casinos) and public service personnel (including employees of the various residential committees, international students management departments of the universities, US–China cultural center, among others).

[6] The Nanxiucun International Student Settlements studied in this book encompasses the international student settlements in Nanxiucun community, Nanjing University Gulou campus, and Nanjng Normal University Suiyuan campus.

Table 2.6 Basic information of Nanjing University Gulou Campus, Nanjing Normal University Suiyuan Campus, and Nanxiucun Community

Name	International Student Settlements within the school		International Student Settlements outside the school
	Nanjing University Gulou campus	Nanjing Normal University Suiyuan campus	Nanxiucun community
Geographical Location	Situated in the Gulou zone of Nanjing City. Using the Hankou Road as a gauge, the campus is divided into the Nanyuan and Beiyuan. Nanyuan is the living area while Beiyuan serves as the area for learning, research, and development	Situated in the Gulou zone of Nanjing City. The southern side of the campus is next to the Nanjing Brain Hospital, the eastern side leads to Ninghai Road, the northern side is next to Hankouxi Road, while the western side is next to Hujuguan. The campus is divided into the eastern and western zones, with the former designated for learning and the latter for recreation	Situated in Gulou zone of Nanjing City. The western side of the community is next to Shanghai Road, the northern side is next to Beijingxi Road, and the southern side is next to Hankou Road. It is also within 10 min' walk to Nanjing University and Nanjing Normal University
Area (km^2)	0.5	0.3	0.15
Population size	14,000 students, of which approximately 1900 are international students	11,000 students, of which approximately 1000 are international students	7250 permanent residents and 576 foreigners (including approximately 400 international students at Nanjing University and Nanjing Normal University, and nearly 50 students from Hohai University and Nanjing Medical University)
Spatial formation		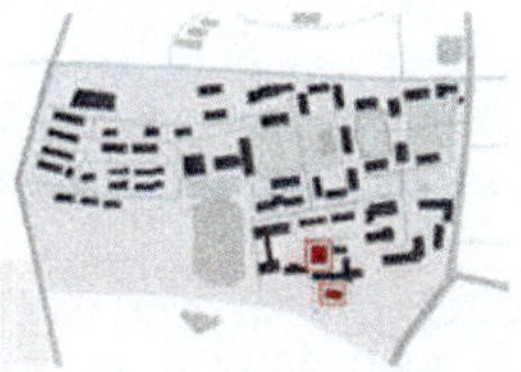	

Source Own elaboration

2.2.3.3 Study Content

The questionnaire on "everyday life space" is centered around the everyday life space of the Nanxiucun International Student settlements (based on static analysis). It has

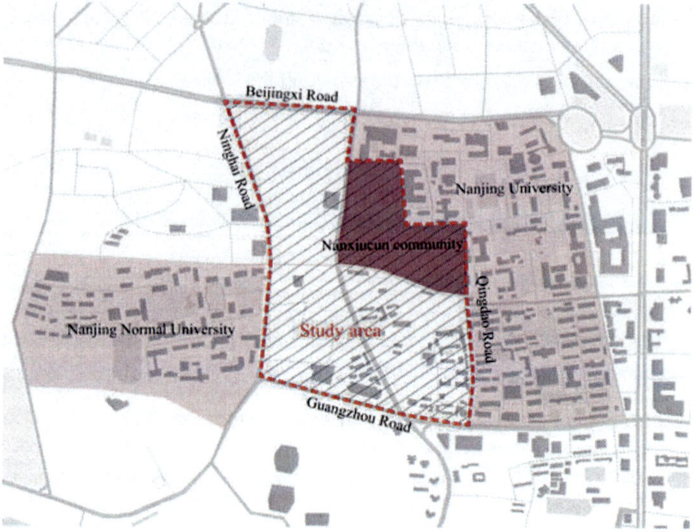

Fig. 2.8 Scope of study. *Source* Own elaboration

three main parts: the first is related to basic personal attributes, with a focus on understanding basic information of the interviewees, including their gender, nationality, educational level, income level, and residential locations; the second part collates data about the everyday life space of the international students, with a focus on the interviewees' residential, learning, shopping, and leisure activities as well as the spaces utilized for these activities; and the third part is about the time and space related to the activities, with a focus on the interviewees' time structure in their everyday lives and the corresponding spaces. Based on this information, the characteristics of the various activities, differences in this community, and their space–time trajectory are then further segmented and delineated. The questionnaire design and sampling criteria are set as follows.

Everyday life is viewed in the time span of 24 h (00:00 to 24:00), broken up into 12 blocks of 2 h each for the purposes of observation and data tabulation of the activities undertaken.

There are six distance categories for everyday life: residential location, ≤ 500 m, 500–1000 m, 1000–2000 m, 2000–3000 m, and >3000 m. The modes of mobility are on foot, bicycles, motorcycle/electric bicycles, buses, trains, taxis, and private cars.

Everyday life activities are segmented into four types: residential (sleeping or household chores), learning, shopping, and leisure.

The consumer items[7] of everyday life's shopping activities can be further classified based on consumers' shopping behavior and habits: convenience items (also known as daily amenities, which refer to the commodities that consumers require and will purchase repeatedly in their everyday life, such as food, beverages, soap, and powder detergent); goods of choice (while these are more expensive than convenience items, consumers are more willing to spend time comparing them across different stores before deciding on which item to buy, such as clothes and household appliances); and special items (consumers are more willing to spend time buying these items because of certain preferences, such as television, refrigerator, and make-up).

Everyday leisure activities (Chongqi et al. 2006) can be further segmented as culture and entertainment types (music and dance, movies, surfing the internet, radio programs, etc.), interactive types (chatting, visiting friends, meeting for meals, gatherings, etc.), fitness (ball sports, strolls, specialized fitness programs, etc.), travels (visiting parks, hiking, short-term travels, etc.), and hobbies (music, chess, calligraphy, painting, photography, writing, inventions, etc.).

The interviews on "surrounding community response" focus on the community response surrounding the Nanxiucun International Student Settlements (based on dynamic analysis). This comprises three parts. First, interviews are conducted with public service personnel, such as the committees and the universities' international student offices, to understand the residential structure and nationality composition of the international student body—as well as historical data on living habits—and public facilities for foreign personnel. Second, the researcher sought to understand the social attributes of local residents, review the historical circumstances of the area's functional uses and ancillary facilities, and consolidate the attitudes toward international students and their actions with them. Third, store operators were interviewed to gain an understanding on the local e-commerce scale, state of business, and store layouts, among others; special consideration is given to the development of unique industries related to international students, and their attitudes toward the international students and their actions with them are consolidated.

2.2.3.4 Data Collation

Questionnaires on "everyday life space" targeted international students at Nanjing University Gulou campus, Nanjing Normal University Suiyuan campus, and Nanxiucun community. Considering that the students either passively stay within school premises or actively pursue accommodations outside, the following principles were abided by random sampling, typicality of the study area, and comprehensive coverage of all residential types. Based on these principles, 210 sets of questionnaires were sent out, and 200 responses were received (90 from those living on campus, and 110

[7] This author has referenced the classifications of consumer products in Yu Jianyuan's "Classification of Consumer Products according to Purchasing Habits and Its Marketing Significance": daily amenities, products based on choices, special products, and non-seeking products. This study has classified the shopping types based on what the consumer would purchase: daily amenities, products based on choices, and special products.

from those living outside campus). The response rate is 95% and complies with the standard sampling rate (more than 5%).

Interviews on "surrounding community response" targeted the local residents, store operators, and public service personnel in Nanxiucun and the surrounding community, thereby covering all types of personnel in the surrounding community. Based on this principle, in-depth interviews were conducted with 54 people, comprising 40 local residents, 10 store operators, and 4 public service personnel.

2.3 Everyday Life Space in International Student Settlements

This chapter has selected a typical example of International Student Settlements in Nanjing for the case study. A comprehensive analysis is conducted on the individual attributes and everyday life space of those from the Nanxiucun International Student Settlements. The analytical focus is on the everyday activity patterns and spatial characteristics in the settlements (based on the residential, learning, leisure, and shopping activities), which in turn present space–time trajectory of the typical everyday life of the international students body.

This chapter has scoped out the study period as one day (either work or rest day) in the international students' life. The earlier determined scope of core, expanded and extended spaces are respectively matched to the residential, learning, leisure, and shopping spaces to be studied (see Fig. 2.9). Various study methods were used, including field observation, questionnaires, statistical analyses, and comparative analyses, among others.

2.3.1 Overview of Individuals in the International Student Settlements

Overall, the international students studied are basically evenly divided between the two genders. In terms of age, most of them are between 20 and 30 years of age, with the highest proportion being those aged 22 and 25 years. The following sections present further analyses on the basic information of the international students in Nanxiucun.

2.3.1.1 Nationality and Religion

Students in Nanxiucun International Student Settlements come from over 70 countries and have diverse religious beliefs. Based on the national distribution, religious

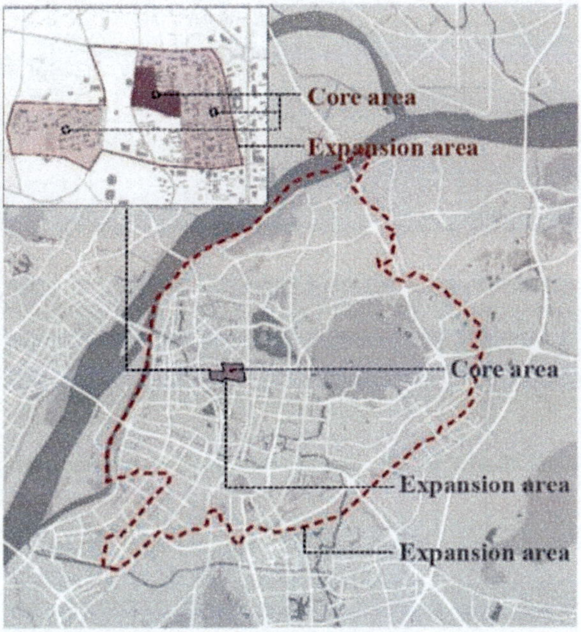

Fig. 2.9 Scope of space in this study. *Source* Own elaboration

beliefs, and cross-analyses between the two, the following can be observed (see Fig. 2.10).

First, in terms of national distribution, nearly 40% of the international students in Nanxiucu International Student Settlements come from South Asian and Southeast Asian countries, making up the largest proportion. There is a more even spread across European and North American, Japanese and Korean, and African nationalities, each at 15%. Latin American and Australian students make up the minority remainder. Overall, the international student body in Nanxiucun International Student Settlements shows diversity and broad distribution across nationalities.

Second, in terms of religious beliefs, most of the international students in Nanxiucun International Student Settlements identify with religious groups (82%), which are diverse and among which Islam forms the largest group, followed by Christianity, Buddhism, and Catholicism. There are also other religious groups that take up a significant proportion (such as traditional African beliefs and state religions in Southeast Asia).[8] Additionally, almost 20% of international students are free thinkers. That said, the international students were not active participants of religious activities, and over 60% of them said they had never participated in any. Around 30% participated in one activity each month on average, while fewer than 10% participate on a weekly basis. While such behavior may appear paradoxical given how majority of the international students have religious beliefs, the primary reason is because there is a lack of public spaces for religious activities in China as well as no public

[8] China recognizes Islam, Christianity, Catholicism, Daoism, and Buddhism. Religious groups of these faiths are permitted by the government to engage in legal religious activities.

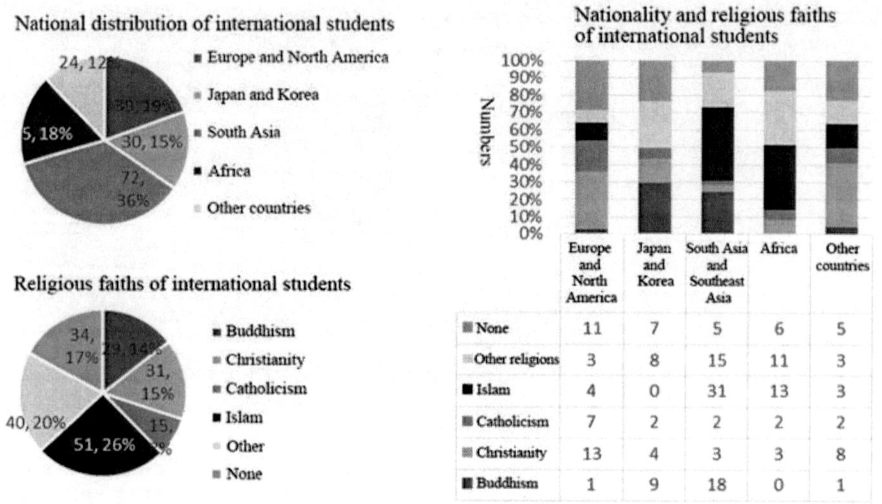

Fig. 2.10 International students' nationalities and religious beliefs. *Source* Data from the research group's sample surveys on the everyday life space of International Student Settlements in Nanjing (2017)

organization of religious activities. In addition to these, there are also other reasons such as language barriers. Therefore, many international students choose to engage in religious activities in private venues (most of which are residential locations), mostly alone or in small groups. Overall, there are many religious international students in Nanxiucun International Student Settlements, among whom there is a rich diversity, but the religious activities lean more toward solo activities, with few large-scale and public religious activities for participation.

Third, in terms of the relation between nationality and religion, there are large religious differences among international students from different countries and regions. These are largely manifest, on the one hand, in the proportion of international students with religious beliefs differing across countries, with students from South Asian and Southeast Asian countries being more religious (over 90%), while 30% of those from Europe and North America are free thinkers. On the other hand, there are also significant differences in terms of religious groups across different countries, with those from Europe and North America being mostly Christians or Catholics, while Japanese and Korean students are mostly Buddhists or Catholics. The religious beliefs among South Asian and Southeast Asian students are more diverse, with Islam and Buddhism being more dominant, alongside other beliefs such as Islam, Judaism, and other small religious groups. Overall, the international students' nationalities and religious beliefs are often related to the religious context and political orientations in their respective home countries and show rich diversity.

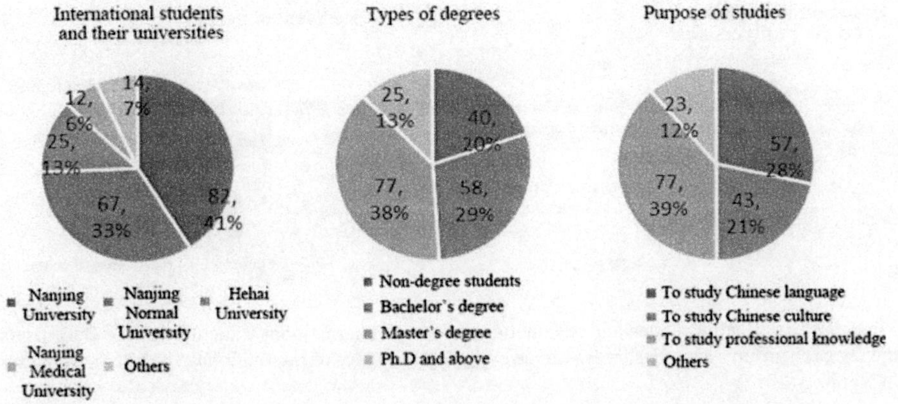

Fig. 2.11 Types of degrees pursued by international students. *Source* Data from the research group's sample surveys on the everyday life space of International Student Settlements in Nanjing (2017)

2.3.1.2 Types of Degrees Pursued by the International Students

The international students in Nanxiucun International Student Settlements show the following characteristics in the coursework they are pursuing (see Fig. 2.11). First, because of the geographical locations and school management, the international student body in this settlement primarily comprises students from Nanjing University and Nanjing Normal University and a smaller group of students from Hohai University and Nanjing Medical University. Second, affected by the universities' international students recruitment policies, almost 40% of the international students in this settlement are pursuing master's degrees, forming the majority, followed by those pursuing bachelor's degrees (almost 30%). As Nanjing University and Nanjing Normal University also accept non-degree students for exchanges and language immersion, 20% of the international students are also on short-term exchanges or immersion programs. Third, the objectives of students in this settlement are varied, with 40% of the international students targeting to equip themselves with professional knowledge, nearly 30% who want to study the Chinese language, and around 20% who are interested in Chinese culture and therefore chose to study in China.

2.3.1.3 Residential Location

There are two residential modes for international students in Nanxiucun International Settlements: passively staying inside the school and pro-actively renting outside, among which 55% of international students stay outside the school at rented spaces within the Nanxiucun community, while 45% stay at apartments (or dormitories) offered by the universities. In terms of residential period, international students are mainly temporary residents, with those having stayed in Nanjing for 2–4 years making up the largest percentage (nearly 40%), while new arrivals take up around

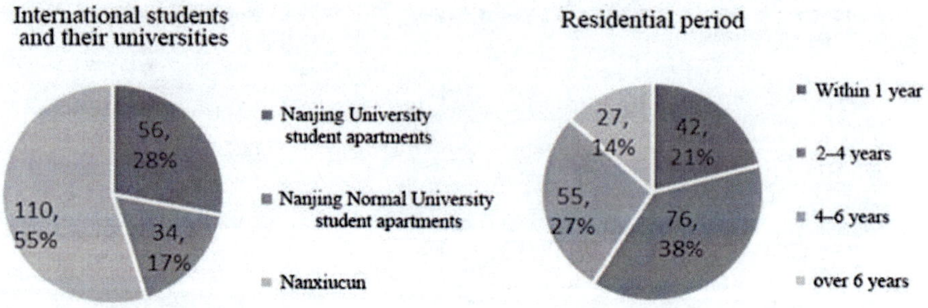

Fig. 2.12 Residential modes and residential periods of international students. *Source* Data from the research group's sample surveys on the everyday life space of International Student Settlements in Nanjing (2017)

20%, and those who have stayed in Nanjing for over 6 years make up 14%. Overall, the international students in Nanxiucun International Student Settlements have relative autonomy over their residences, and the period of stay is largely temporary (see Fig. 2.12).

2.3.1.4 Consumption Levels

The consumption level of international students at Nanxiucun International Student Settlement is generally high. This author tried to compare the monthly expenses between an international and a Chinese university student (expenses beyond tuition and accommodation fees) (see Fig. 2.13). The *2016 Report on Chinese University Students' Perspectives on Consumption* revealed that the Chinese student's median

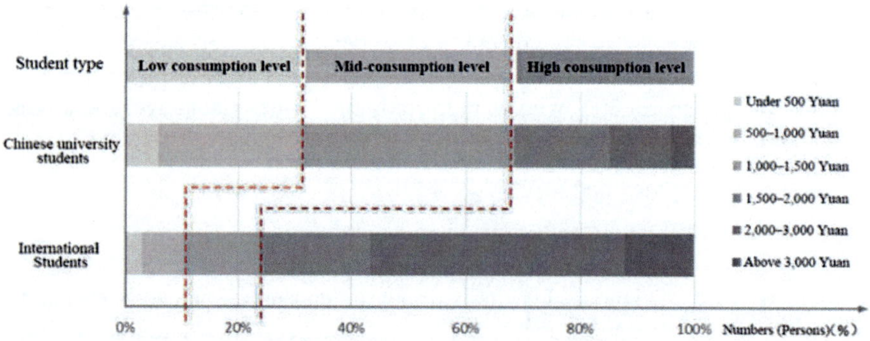

Fig. 2.13 Comparing monthly expenses between international and chinese students. *Source* Data from the research group's sample surveys on the everyday life space of International Student Settlements in Nanjing (2017); The *2016 Report on Chinese University Students' Perspectives on Consumption* published on Tencent Questionnaires platform

monthly expenses were around 1212 Yuan. Taking this as a basis, student consumption can be viewed as low (under 1000 Yuan), mid-range (1000–1500 Yuan), and high (above 1500 Yuan). Applying this to the international students, it seems that only 23% of international students fall under the low and mid-range consumption level, and nearly 80% of international students spend as much as Chinese students in the high-level category, among which international students who spend 2000–3000 Yuan each month make up the highest percentage (over 40%). This shows that the overall consumption level of international students in Nanxiucun International Student Settlement is higher than that of Chinese university students. This is closely related to the overall economic level of the international student's home country or individual family conditions.

2.3.1.5 Level of Integration

Related studies (Longyu 2003) by Chinese scholars on international students adapting to the Chinese culture were referenced. Three indicators were selected to represent the level of integration of international students: the international students' level of understanding toward Chinese culture, composition of people they interact with, and frequency of interactions with locals. In conjunction with the questionnaire results on Nanxiucun International Student Settlements, the varying levels of the different items were classified and given values accordingly. The findings were as follows (see Fig. 2.14).

The level of integration of international students in Nanxiucun International Student Settlement was classified as high (19–21), mid (13–17) and low (3–11). It was discovered that 60% of international students belong to the mid-level category,

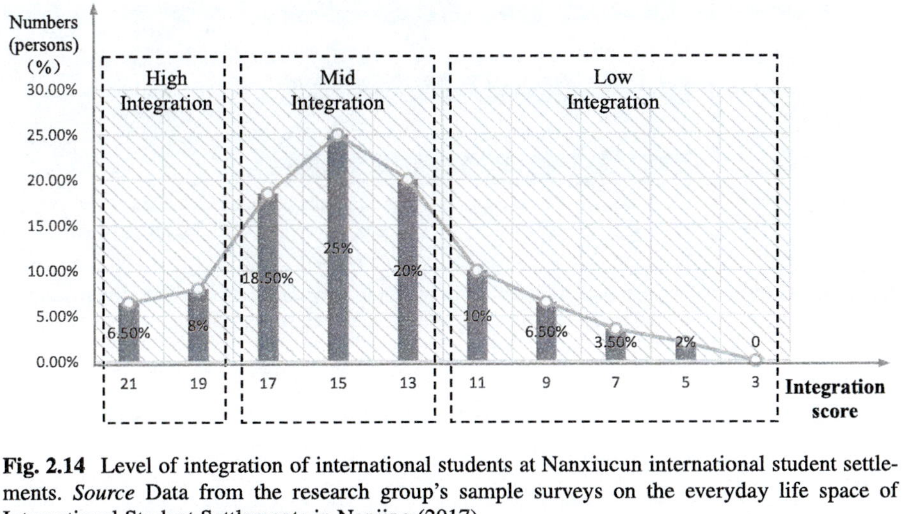

Fig. 2.14 Level of integration of international students at Nanxiucun international student settlements. *Source* Data from the research group's sample surveys on the everyday life space of International Student Settlements in Nanjing (2017)

signifying that in terms of interaction and integration in a foreign country, international students are adapting relatively well. They are able to interact with people from other countries and regions comfortably and can also relatively adapt well to life in a foreign land. Simultaneously, nearly 15% of international students belong to the high-level category, signifying that they are able to completely assimilate into the local environment. It is noteworthy that there remain over 20% of international students who belong in the low-level category; they mostly only interact with people from similar countries or backgrounds and are unable to assimilate well into the local lifestyle. Overall, while international students are able to assimilate into the local environment as a whole, there remain differences between individuals, and there are varying levels of integration.

In terms of the willingness to stay, overall, the intention is relatively weak for international students in Nanxiucun International Student settlements. Over 70% of international students choose to leave Nanjing after graduating, for reasons including further studies in other countries, longing for home, difficulties in adapting in China, and family factors, among others. For those who choose to stay in Nanjing, most do so because they like Nanjing or believe that there are better prospects in China, etc. This also validate the integration level analysis presented earlier (see Fig. 2.15).Generally speaking, among those who come under the mid-level category, the proportion of those willing to remain in Nanjing are relatively higher.

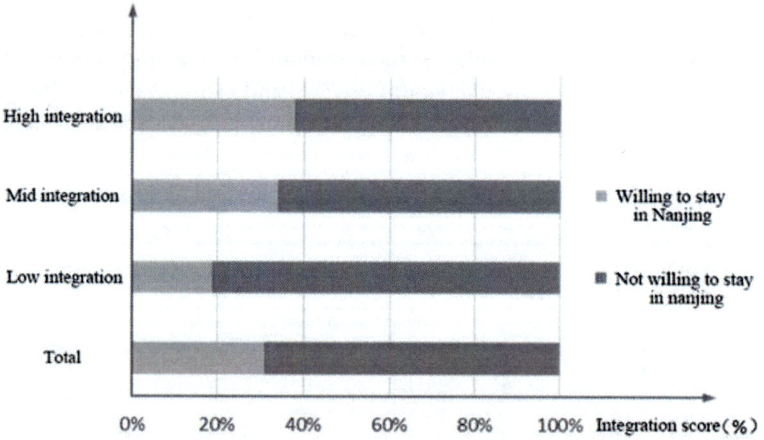

Fig. 2.15 Analysis of international students' willingness to stay. *Source* Data from the research group's sample surveys on the everyday life space of International Student Settlements in Nanjing (2017)

2.3.1.6 Correlation Between International Students' Individual Attributes and Their Everyday Life Space

To deepen the scope of study on the four types of everyday life space of International Student Settlements, this study conducted a logistic regression analysis on the relation between the international students' individual attributes and their space–time choices in everyday life. The aim is to select more targeted standard attributes to analyze the differences in everyday lives in the international student body.

For instance, to analyze the differences in residential and activity spaces of international students with different individual attributes, the data was processed by calculating the weighted average of residential and activity spaces, which were classified accordingly as "within school" and "outside school." Each respectively took up 52 and 48%. Further, logistic regression analyses were conducted using Statistical Product and Service Solutions (SPSS) (for results, refer to Table 2.1). It was discovered that nationality indicator ($p = 0.032$), religion indicator ($p = 0.011$), and level of integration indicator ($p = 0.020$) had a significant influence on the choice of residential spaces (see Table 2.7).

The individual attributes of international students amid the four types of everyday life timing and spaces were analyzed using logistic regression (as the residential activity timings were relatively fixed, the regression analysis was applied only on its relation with space). The results were consolidated and the various individual attributes were evaluated for significant influences (see Table 2.8). These serve as a basis for the subsequent, more specific analyses, as shown below.

Table 2.7 Logistic regression analysis on the international students' choice of residential spaces based on individual attributes

International students' individual attributes	B	S.E	Sig
Gender	0.395	0.470	0.348
Nationality	0.296**	0.142	0.032
Religion	0.338**	0.130	0.011
Type of degree	– 0.173	0.265	0.512
Residential location	——	——	——
Consumption level	0.124	0.241	0.607
Level of integration	0.321**	0.145	0.020
Constant	– 2.019	1.478	0.170
Model Chi-square (df)	7.684 (5)		
-2Log Likelihood	117.879		

Independent variables: 0 = within school, 1 = outside school; significance: *$p < 0.1$; **$p < 0.05$; ***$p < 0.01$).

Table 2.8 Logistic regression results of international students' individual attributes and everyday life space

Type		Gender	Nationality	Religion	Type of degree	Residential location	Consumption level	Level of integration
Shopping spaces correlation	Space		√	√				√
Learning spaces correlation	Time				√	√		
	Space				√	√		
Shopping space correlation	Time	√						
	Space	√					√	
Leisure space correlation	Time	√						√
	Space	√						√

Source Own elaboration

2.3.2 *Residential Spaces at International Student Settlements*

2.3.2.1 Overall Characteristics of International Students' Residential Activities

I. First, time characteristics of residential activities

The residential activities at International Student Settlements are relatively fixed, primarily comprising sleeping and household chores. These are necessary everyday activities. In terms of activity timings (see Fig. 2.16), sleeping time has an overall regular pattern, while household chores timings are relatively scattered and take up minimal time. Additionally, the residential activity time on rest days is more flexible compared to that of work days.

II. Second, spatial characteristics of residential activities

There are primarily two residential modes in Nanxiucun International Student Settlements: passively choosing to stay in international student apartments within school premises (including Nanjing University international student apartments and Nanjing Normal University international student apartments) and pro-actively choosing to stay in a community outside of school (Nanxiucun community serves as an example) (see Table 2.9). In terms of geographical location, the former is located within school premises, whereas the latter is within a 500 m range from the schools. They are all relatively close to each other. In terms of student capacity, the latter has a higher number of international students, signifying that international students prefer to stay outside of the school. In terms of rental expenses, the monthly expenses were tabulated after accounting for currency conversion, and it was found that the monthly expenses for school and outside school housing were around 1500–2000 Yuan, without much

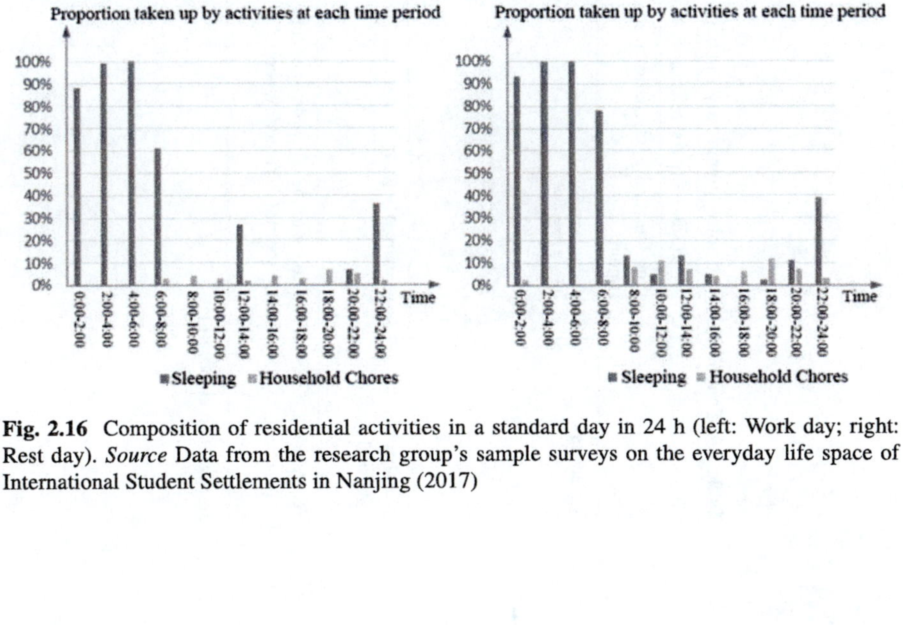

Fig. 2.16 Composition of residential activities in a standard day in 24 h (left: Work day; right: Rest day). *Source* Data from the research group's sample surveys on the everyday life space of International Student Settlements in Nanjing (2017)

Table 2.9 School (Passive Choice) community and outside school (Proactive Choice) community

Type	Living in school (passive choice)	Living outside school (proactive choice)	
	Nanjing University international student apartment	Nanjing Normal University international student apartment	Nanxiucun community
Location	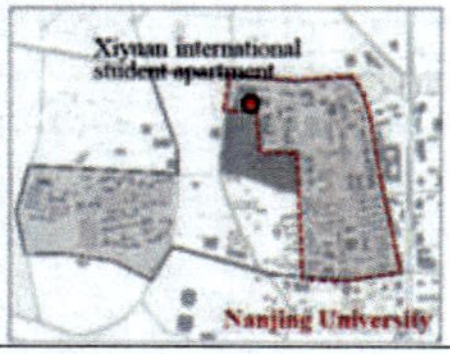	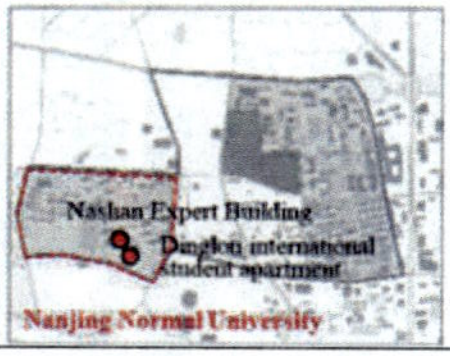	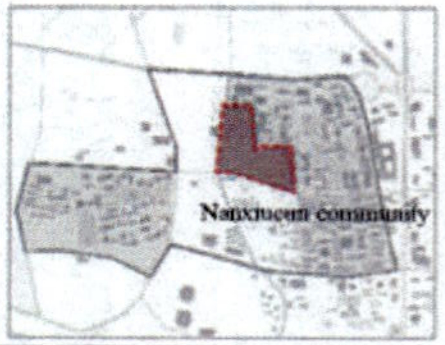
Number of international students	Approximately 200	Approximately 400	Approximately 550
Rental	Double room: 52 Yuan/day/person Single room: 67 Yuan/day/person	Nanshan Expert Building: 4800 Yuan/person/schooling term Dinglou: 5000 Yuan/person/schooling term	1500–2000Yuan/month/person

Source Own elaboration

differences. Therefore, rental cannot be taken as a critical factor that influences the students' choice of housing.

2.3.2.2 Differences in International Student Body's Residential Spaces

This study is grounded in urban planning and community perspectives, further exploring the time and space characteristics of international students' residential activities from "social attributes" to "spatial attributes." Social attributes are derived from the SPSS regression analyses on international students' individual attributes and everyday life space. Nationality, religion, and level of integration were selected for analyses as they are highly related to residential spaces. The influence of spatial attributes was analyzed through field observation and on-site records, primarily involving the residential spaces in the community outside school (proactive residential choice) and within the school (passive residential choice).

Influence of nationality on international students' residential activities (social attribute)

In terms of nationality, students from Europe and North America take up the lowest percentage among those who are living within the school premises (20%). This percentage is also lower than the average among international students. South Asian and Southeast Asian students make up the largest proportion among those who are living within the school premises (55%), while Japanese and Korean students are closer to the average (see Fig. 2.17).

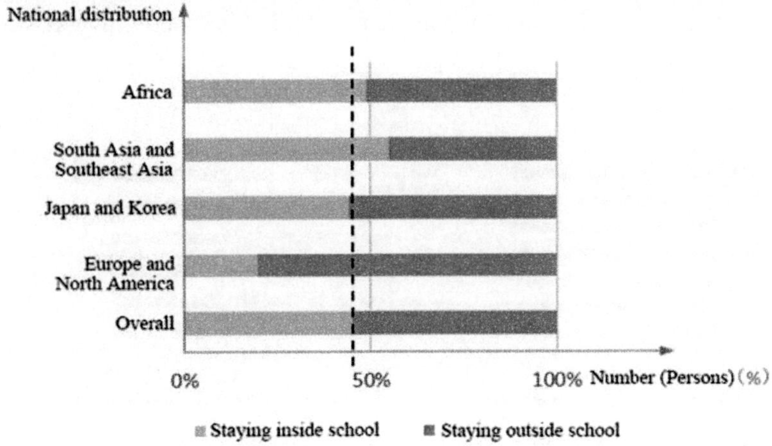

Fig. 2.17 Proportion of international students living in each residential mode. *Source* Data from the research group's sample surveys on the everyday life space of International Student Settlements in Nanjing (2017)

The reason for this can be attributed to differences in the living environment, cultural backgrounds, races, and religions across the different home countries and areas. Relatively speaking, international students from Europe and North America are more inclined toward freedom and personality and place more emphasis on individual and lifestyle privacy. As for students from South Asia, Southeast Asia, and Africa, they adapt more readily to group living, while Japanese and Korean students are somewhere in between.

Influence of religion on international students' residential activities (social attribute)

In terms of religious beliefs, Christian students living within school premises take up the lowest percentage among international students (39%), followed by Islam (42%) and others (40%). These are lower than the average value for international students. Conversely, Buddhist students living within school premises make up the highest proportion (56%), followed by free thinkers (50%). In terms of the frequency in participating in religious activities, the international students are basically found to follow this pattern: the more they participate in religious activities, the more likely they are to be staying outside of school (based on Buddhist and free thinkers) (see Fig. 2.18).

The reason for this pattern can be attributed to different religious values, perspectives, lifestyle, and customs (including diet, festivities, religious activities, etc.). Relatively speaking, students of Christian, Islam, and other faiths may require more private and flexible lifestyle spaces, while Buddhists and free thinkers are better able to adapt to group living. Additionally, international students will mostly opt for private venues or their residences to participate in religious activities, instead of participating in large-scale public religious events. Therefore, the frequency of religious activities and the mode of participation will affect the residential choices to a certain extent.

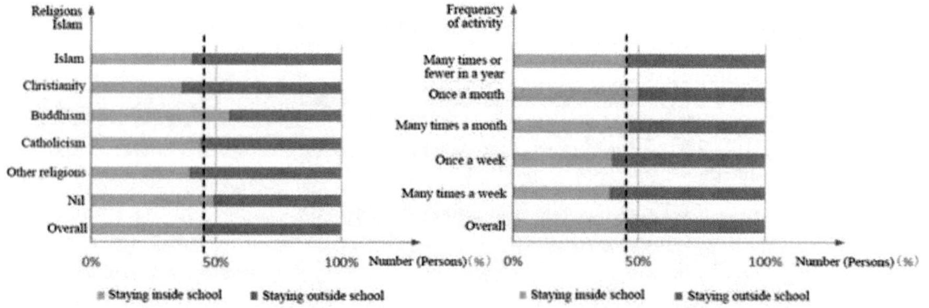

Fig. 2.18 Composition of international students according to their residential modes, religious beliefs, and participation frequency. *Source* Data from the research group's sample surveys on the everyday life space of International Student Settlements in Nanjing (2017)

Influence of integration level on international students' residential spaces (social attribute)

In terms of the level of integration, the proportion of international students who have assimilated very well into the local community and who choose to stay in school apartments or dormitories are slightly more than the average. The lower the level of integration, the more likely for the international students to stay outside of school (see Fig. 2.19).

It can thus be seen that the level of integration and choice of residence are closely related and mutually influencing. Because of reasons such as personality, international students with high levels of integration will find it easier to assimilate into the local community and are more willing to live together in school with students from other countries and races. As the interactive relations and dependence become more tightly knit, the level of integration will in turn be influenced and thus increase. Conversely, students with lower levels of integration are more inclined toward living in the community surrounding the school.

Influence of Housing Conditions on the International Students' Residential Activities (Spatial Attribute)

Based on questionnaires and field observation, the residential modes were classified and compared based on residential area, residential numbers, interior facilities, and community facilities. The factors influencing the selection of residence were then summarized (see Table 2.10 and Fig. 2.20).

In terms of residential area, the size of international student apartments is approximately 20 m^2. As the type of housing and residential numbers differ for housing outside of the school, this study has focused on units for 2–3 persons, with an area of

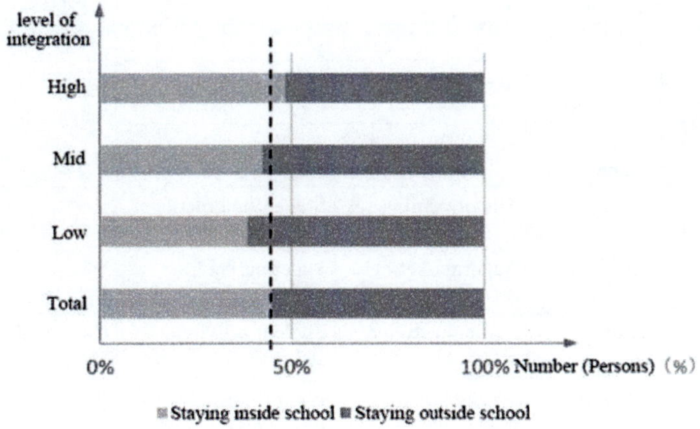

Fig. 2.19 Composition of international students according to residential modes and levels of integration. *Source* Data from the research group's sample surveys on the everyday life space of International Student Settlements in Nanjing (2017)

Table 2.10 Summary of housing conditions of the school community (Passive Residential Choice) and outside community (Proactive Residential Choice)

Type		Living in school (passive choice)		Living outside school (proactive choice)	
		Nanjing University international student apartment	Nanjing Normal University international student apartment	Nanxiucun community	
Residential area		Approximately 20 m^2	Approximately 20 m^2	Under 30 m^2	7%
				30–50 m^2	34%
				50–90 m^2	47%
				Above 90 m^2	9%
Residential numbers		1–2 persons	2 persons	1 persons	5%
				2–3 persons	63%
				4–5 persons	26%
				More than 5 persons	6%
Interior facilities	Kitchen	×	×	√	
	Toilet	√	√	√	
	Tap water	√	√	√	
	Broadband connection	√	√	√	
	Television	×	×	√:71% × :29%	
	Heater	√	√	√	
	Air conditioner	√	√	√	
	Washing machine	△	△	√	
Community facilities	Dining	Student dining	Student/international student dining	×	
	Shopping	Supermarket in school	Supermarket in school	×	
	Leisure activities	Student activity center	Student activity center	Community activity room	
	Learning	Classroom, self-study room, work studio, library	Classroom, self-study room, work studio, library	×	

(continued)

Table 2.10 (continued)

Type		Living in school (passive choice)		Living outside school (proactive choice)
		Nanjing University international student apartment	Nanjing Normal University international student apartment	Nanxiucun community
	Fitness, sports activities	Track, sports center	Track, sports center	Community fitness equipment

Note The interior facilities mentioned in the table can be differentiated as for self-use (r facilities men△), or not existing (×)
Source Data from the research group's sample surveys on the everyday life space of International Student Settlements in Nanjing (2017)

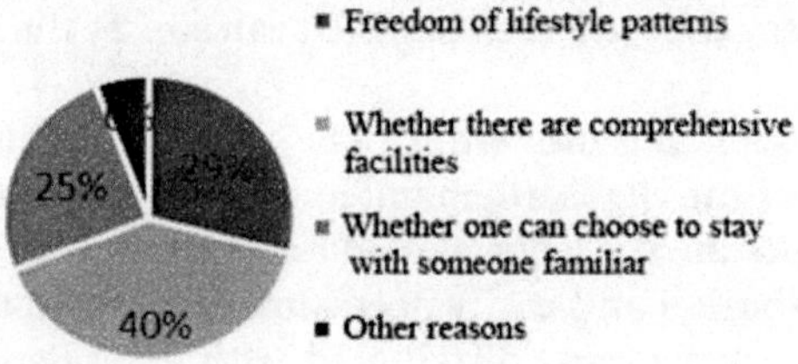

Fig. 2.20 Main factors of consideration for international students when choosing residence. *Source* Data from the research group's sample surveys on the everyday life space of International Student Settlements in Nanjing (2017)

50–90 m^2. Considering the shared spaces for standard housing, the residential area per capital is actually larger than that of school apartments.

In terms of residential numbers, the school's international student apartments offer single and double rooms. Those who select double rooms will need to adjust to their roommate's lifestyle habits, while those who stay outside of the school mostly choose the flatmates they want and generally can have their own room. It can thus be seen that housing outside the schools offer better privacy and more freedom than school apartments.

In terms of interior facilities, the school's international student apartment does not come with individual kitchens and televisions, while washing machines are for shared usage. The facilities for outside housing are more comprehensive, and the standard of living is higher.

In terms of community ancillary facilities, the international student apartment is located within the school compound and thus has better access to various facilities in the school. The necessary learning spaces are available to the international students, and the learning culture is also much stronger. As for the community outside school, there is access only to community activity room(s) and the fitness area; there is no provision of public spaces for learning. Therefore, from the perspective of students, the various facilities offered within the school are better furnished and convenient.

Influence of surrounding facilities on international students' residential spaces (spatial attribute)

The spatial distribution and overall characteristics of the facilities can be classified accordingly after studying the various facilities within 1000 m (basic lifestyle activity circle) of the Nanxiucun International Student Settlements (see Table 2.11). Considering that the three International Student Settlements at Nanxiucun are within close proximity and tightly knit, the study around this area is representative of the shared ancillary facilities and current situation in these three settlements.

Overall, the various facilities near the Nanxiucun International Student Settlements are comprehensive and exhibit the following characteristics:

First, comprehensive and convenient: There are comprehensive medical centers, commercial venues, parks, squares, fitness spots, banks, and transportation facilities near the settlements, which can satisfy the students' everyday needs.

Second, wide-ranging and numerous: The various facilities near the settlements can be found at the city, settlement, and community level. The scale and variety of spaces make the students spoiled for choice.

Third, even distribution and easily accessible: The various facilities nearby the settlements are rather evenly distributed with good density. They are also near the schools and can provide quick and convenient lifestyle services.

2.3.2.3 Space–time Model of International Students' Residential Activities

According to the above analyses on the overall characteristics and group differences of international students' residential spaces, a time model of the international students' residential activities can be derived (see Fig. 2.21). It can be seen that the residential activities of international student are mostly fixed at their residential locations; however, the timing of the activities will differ depending on the individual. Therefore, a time pattern of "single location, lengthy activities" is manifested.

2.3.3 Learning Spaces at International Student Settlements

2.3.3.1 Overall Characteristics of International Students' Learning Activities

The learning activities at Nanxiucun International Student Settlements mainly comprise classes, self-studies, and exchanges and discussions, of which self-studies take up the largest proportion, whereas exchanges and discussions the smallest. While

Table 2.11 Summary of surrounding facilities at Nanxiucun international student settlements

Type	Medical facilities				Parks and squares		
	City level		Settlement level	Community level	City level	Settlement level	Community level
Nature	General	Specialized	General	Community health stop	Park	Park, square	Shared spaces
Name	Drum Tower Hospital, Jiangsu's People Hospital	Nanjing Brain Hospital, Nanjing Chest Hospital, Children's Hospital of Nanjing Medical University	Nanjing Normal University Hospital	Community health centers at Ninghai Road and Tongren Street, Nanjing University Community Health Center	Qingliangshan Park, Drum Tower Park, Wulongtan Park, etc	Drum Tower Square, Dazhongting Park, Defenguan, etc	Green spaces and square within the community
Location of spaces							
Summary of characteristics	• Comprehensive and varied medical facilities nearby, which are evenly distributed. Overall, there is good access to medical facilities • There are basic health service centers within the school and around the city blocks, which can satisfy basic medical needs				• The nearby city- and settlement-level parks and squares are numerous and well varied, which can basically satisfy diverse spatial needs • Shared spaces within the school are abundant and diverse but are relatively lacking for housing outside the school		

(continued)

Table 2.11 (continued)

<table>
<tr><td rowspan="2">Type</td><td colspan="5">Medical facilities</td><td colspan="4">Parks and squares</td></tr>
<tr><td colspan="2">City level</td><td colspan="2">Settlement level</td><td>Community level</td><td>City level</td><td colspan="2">Settlement level</td><td>Community level</td></tr>
<tr><td rowspan="3">Type</td><td colspan="9">Commercial facilities</td></tr>
<tr><td colspan="5">Shopping centers, supermarkets</td><td colspan="2">Dining</td><td colspan="2">Leisure and entertainment</td></tr>
<tr><td>City level</td><td colspan="2">Settlement level</td><td colspan="2">Community level</td><td colspan="2">——</td><td colspan="2">——</td></tr>
<tr><td>Name</td><td>Greenland Center, Golden Eagle International Shopping Center</td><td colspan="2">New World Department Store, Lichee Square, Huanya Katherine Square</td><td colspan="2">Ce-Mart, Suguo Supermarket, Shiji Hualian Supermarket, departmental stores, etc</td><td colspan="2">Nanxiucun bar street, foreign food establishments, restaurants, snack stalls, etc</td><td colspan="2">Internet cafes, game cafes, leisure and health activities, etc</td></tr>
<tr><td>Location of spaces</td><td colspan="5">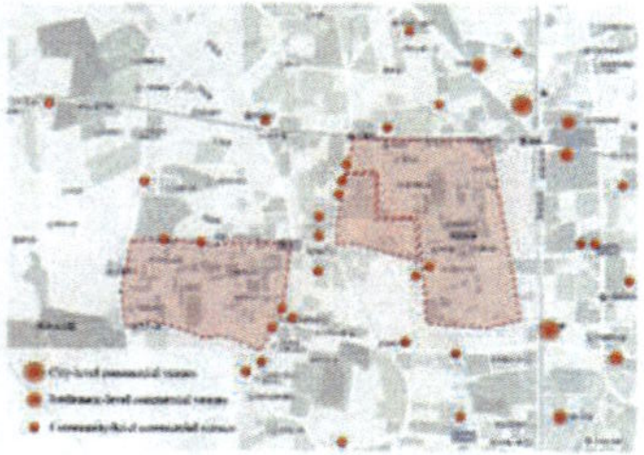</td><td colspan="4">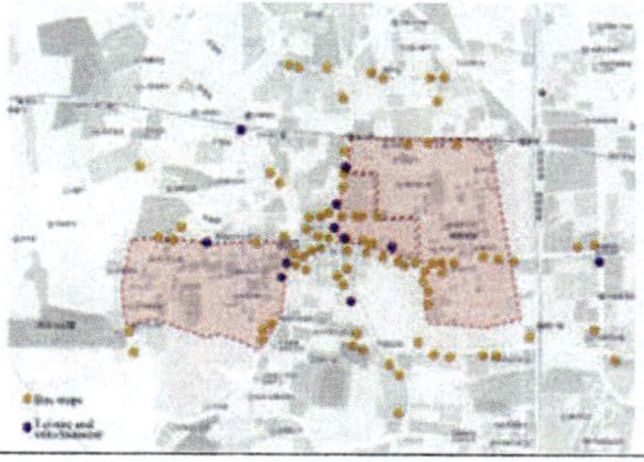</td></tr>
</table>

(continued)

Table 2.11 (continued)

Type	Medical facilities			Parks and squares		
	City level	Settlement level	Community level	City level	Settlement level	Community level
Summary of characteristics	• Extensive and well-varied shopping venues that can satisfy everyone's shopping needs • Rich diversity of nearby dining facilities and food items, including foreign foods • The nearby leisure and entertainment facilities are dominated by internet cafes and board game cafes, which can fulfill the needs of the youthful consumers					

Type	Fitness facilities	Banks	Public transportation
Nature	Stadiums, fitness equipment, gyms	Banks, post office banks, automated teller machines (ATM)	Train stations, bus stops, bike-sharing
Location of spaces	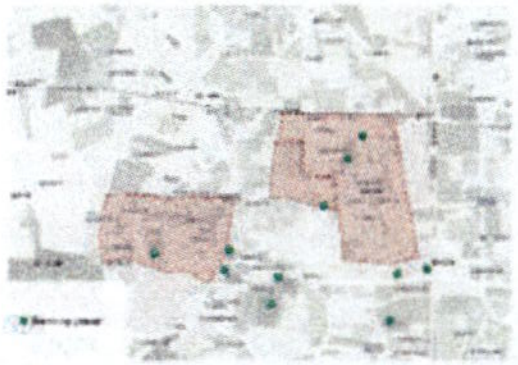	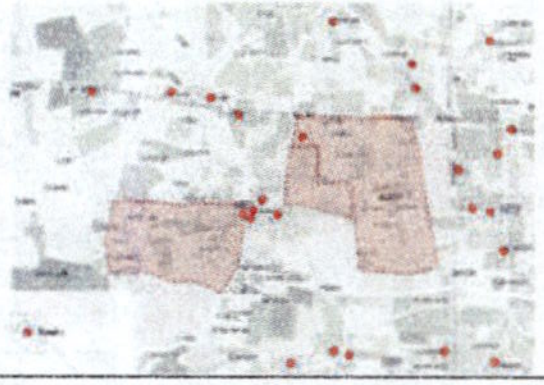	
Summary of characteristics	• There are large stadiums inside the school and nearby • There are numerous well-equipped gyms and outdoor fitness equipment	• The nearby bank operations are comprehensive and numerous, capable of satisfying the needs of the students	• Transport is convenient, with multiple train and bus stops • Bike-sharing services can meet short-distance mobility needs

Source Data from the research group's sample surveys on the everyday life space of International Student Settlements in Nanjing (2017)

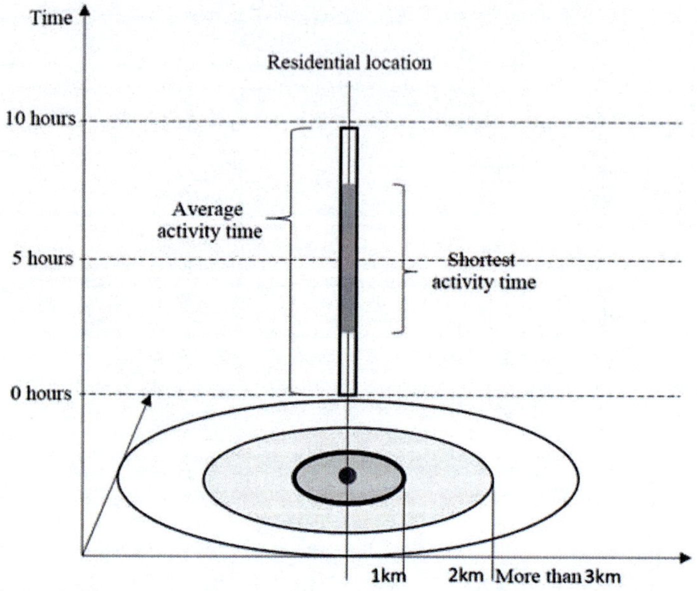

Fig. 2.21 Time model of international students' residential activities. *Source* Own elaboration

classes are mandatory, international students will also likely choose to self-study. In terms of study group sizes, almost half of the international students will choose to study on their own. A significant number will also choose to study in groups of 2–5. Therefore, it can be seen that international students are relatively more independent and passive (see Fig. 2.22). The specific analyses are as follows.

First, time characteristics of learning activities

In terms of the time duration of learning activities (see Fig. 2.23), it can be seen that on the one hand, there are significant differences in the proportion of learning activities in a day between a work day and rest day, with the former having an evidently longer

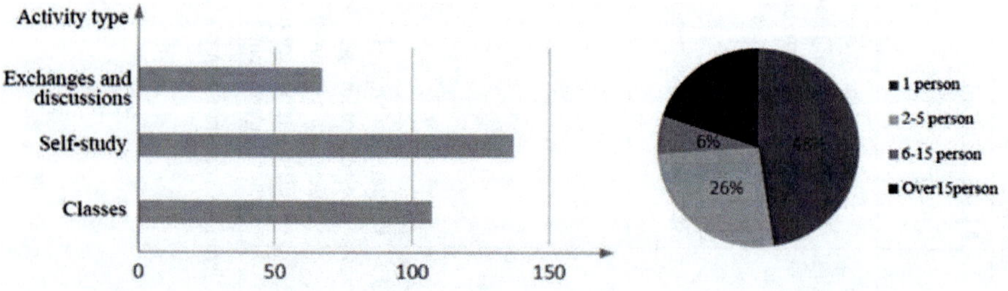

Fig. 2.22 International students' learning modes (Left) and size of study group (Right). *Source* Data from the research group's sample surveys on the everyday life space of International Student Settlements in Nanjing (2017)

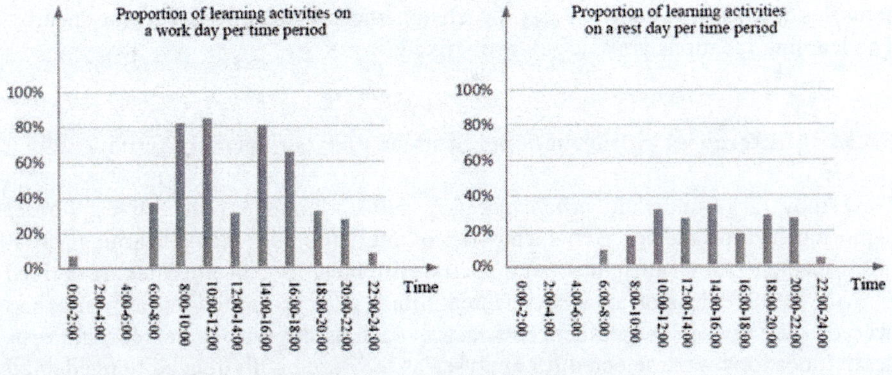

Fig. 2.23 Proportion of learning activities in a standard day (left: Work day; right: Rest day). *Source* Data from the research group's sample surveys on the everyday life space of International Student Settlements in Nanjing (2017)

duration of learning activities than the latter. On the other hand, the time distribution of learning activities on work and rest days differs. The learning time and class time basically correspond on a work day, showing an evident concentration in time period, while on a rest day, learning time distribution is more scattered and flat.

Second, spatial characteristics of learning activities

In terms of the distance traveled for learning activities (see Fig. 2.24), it is easily discovered that, first, international students' learning activities are inclined to take place nearby, mostly concentrated in various spaces within 1000 m proximity (i.e., basic lifestyle activity circle). Second, the learning activities are mostly concentrated within 500 m proximity and mostly at learning facilities within the school. Third, on rest days, there is greater flexibility when it comes to choosing spaces for learning activities, and a significant portion of international students will choose to study at home. Overall, there is a certain difference between the distance traveled for learning

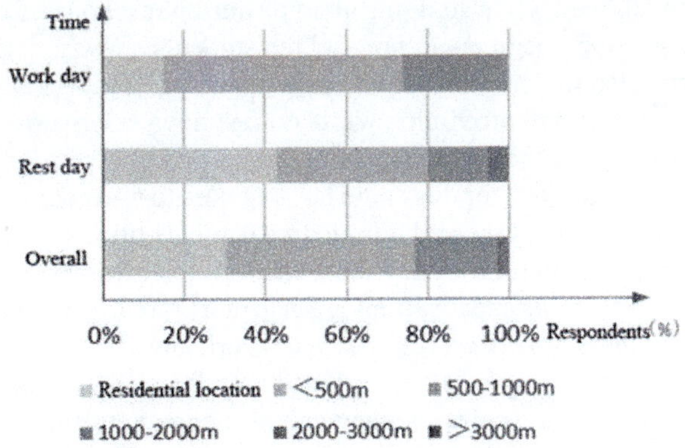

Fig. 2.24 Distance traveled for learning activities on a work day and rest day. *Source* Data from the research group's sample surveys on the everyday life space of International Student Settlements in Nanjing (2017)

activities on a work day and rest day, which nonetheless generally takes place nearby. The learning locations are also relatively fixed.

2.3.3.2 Differences in International Student Body's Learning Activities

This study is grounded in urban planning and community perspectives, further exploring the time and space characteristics of international students' learning activities from their "social attributes" and "spatial attributes." Social attributes are derived from the SPSS regression analyses on international students' individual attributes and everyday life space. The influence of factors such as the type of degrees and residential locations were selected for analyses as they are highly related to residential spaces. The influence of spatial attributes was analyzed through field observation and on-site records, primarily involving the various learning spaces in schools, especially teaching facilities.

Influence of the type of degrees on international students' learning activities (social attribute)

There are significant differences among international students pursuing different types of degrees in terms of their learning timings and spaces (see Table 2.12).

In terms of learning duration, research students (including master's degree students, Ph.D students, and above) not only spend more time studying compared to bachelor's degree students but also far more than that of non-degree students. Compared to research students, bachelor's degree students and non-degree students' learning activities are more concentrated during class time. Their time is more structured, and the differences between a work day and rest day are also greater for them. Specifically, on rest days, they study significantly lesser compared to work days, and the study timings are also more scattered.

In terms of learning spaces, a higher proportion of research students chose to study at home, while bachelor's degree and non-degree students are more inclined to study within 500 m proximity, among which, bachelor's degree students show a more significant time difference in terms of their mobility, such that they travel out for classes, while studying at home during rest hours. Compared to research students, bachelor's degree and non-degree students show greater differences in the distance traveled for learning activities on their work and rest days. On a work day, they study within 500 m proximity, while on rest days, a large proportion of them will study at home or further away (1–2 km).

The reasons for this could be attributed to differential teaching and learning modes. Bachelor's degree and non-degree students primarily undergo lessons, assignments, and examinations. They have fixed timetables and class locations, therefore their learning tasks and locations are also more fixed. Their learning hours are more regular, and there are differences between work and rest days. As for research students, they have fewer fixed classes and mostly will arrange varied learning activities for themselves (besides classes, they can also engage in self-study, experiments, exchanges and discussions, etc.). They require more time for their in-depth research; therefore, their learning time and spaces are more scattered, manifesting characteristics whereby they have long study hours and multiple study locations.

Table 2.12 International students' learning activities and space–time mapping based on the Degrees pursued

		Proportion of learning activities undertaken by international students at different timings	Space–time mapping
Bachelor's degree students	Work day	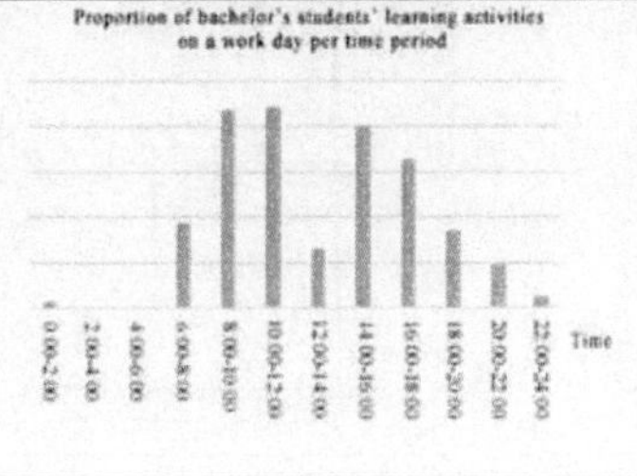	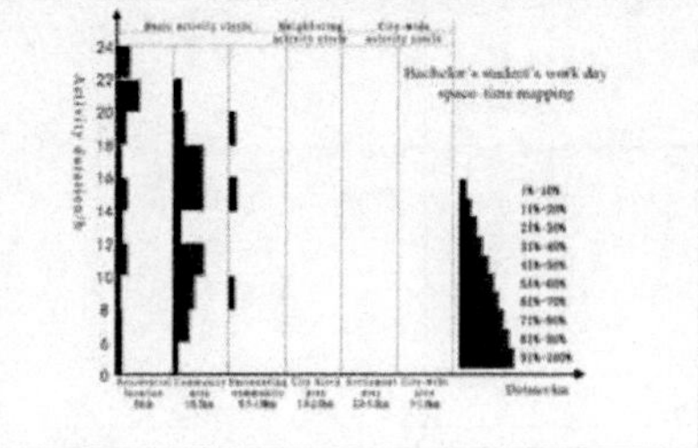
	Rest day	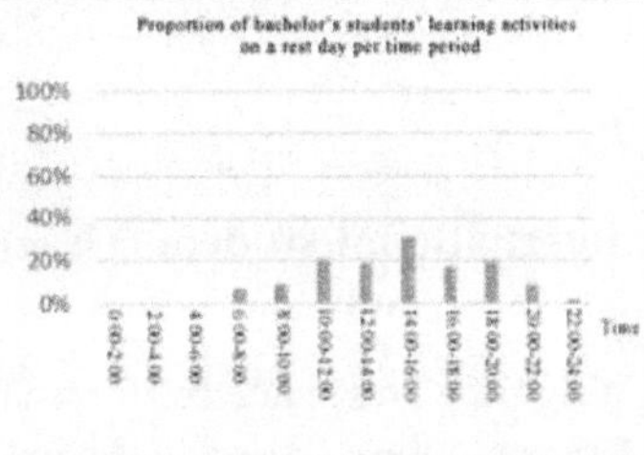	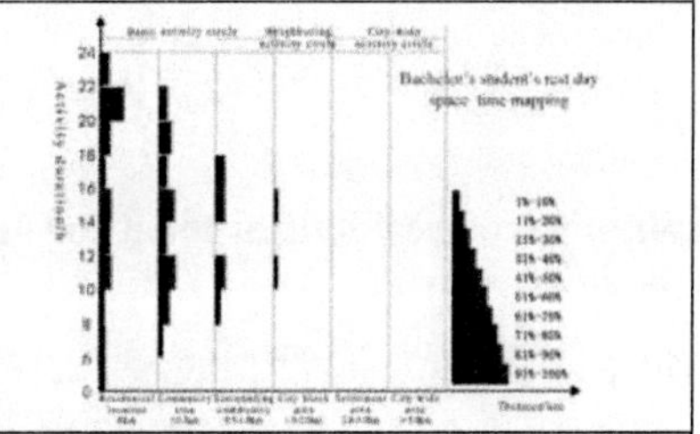
Research students	Work day	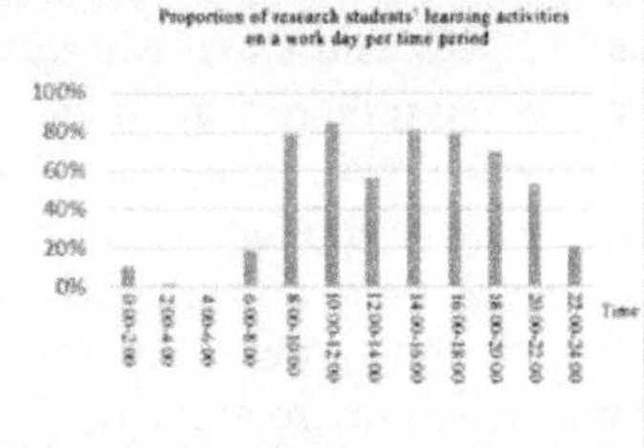	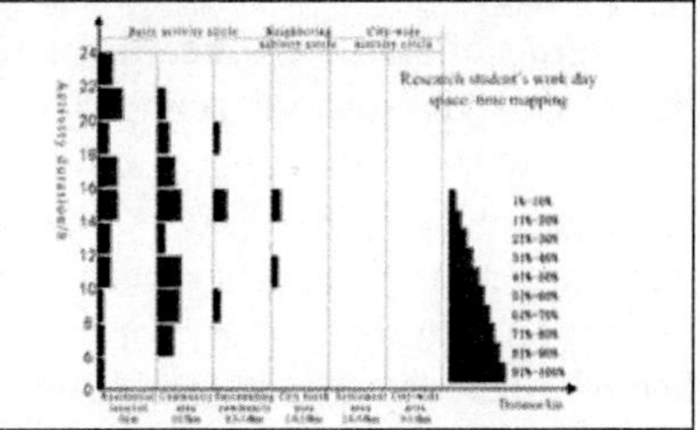
	Rest day	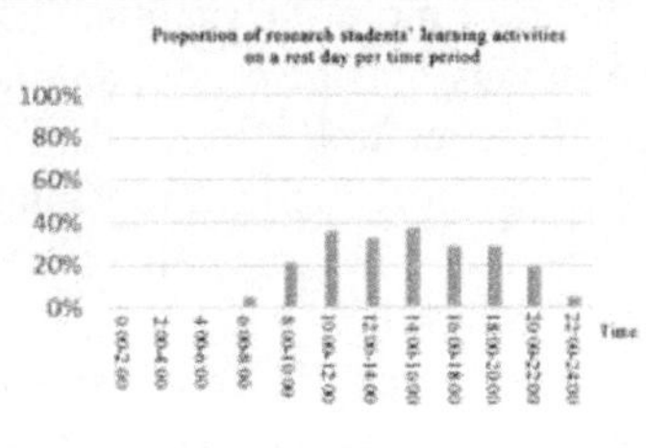	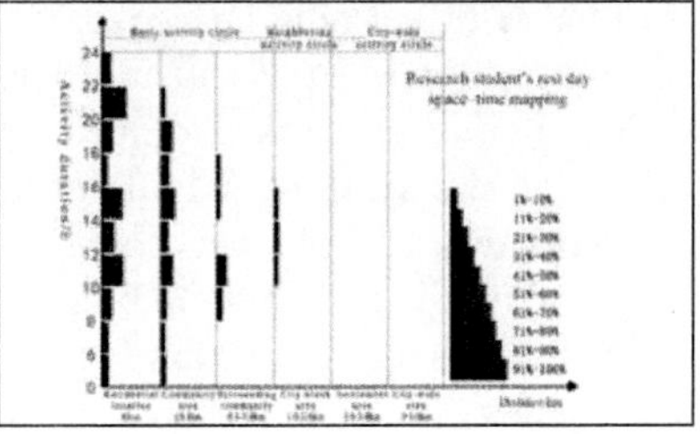

(continued)

Table 2.12 (continued)

Non-degree students	Work day	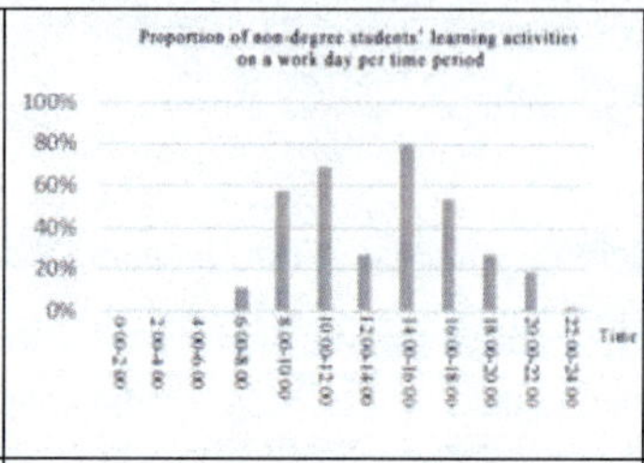	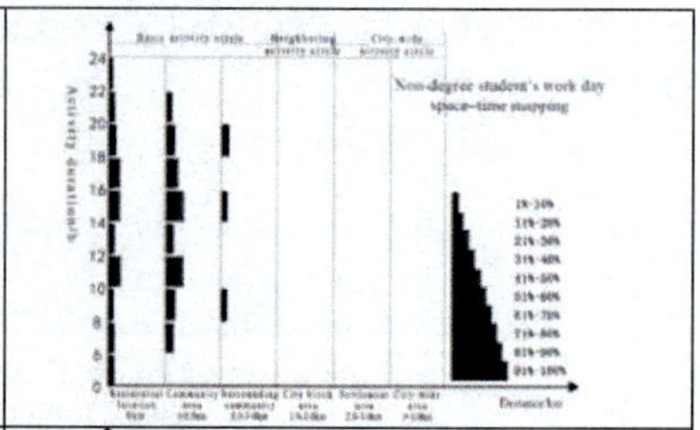
	Rest day	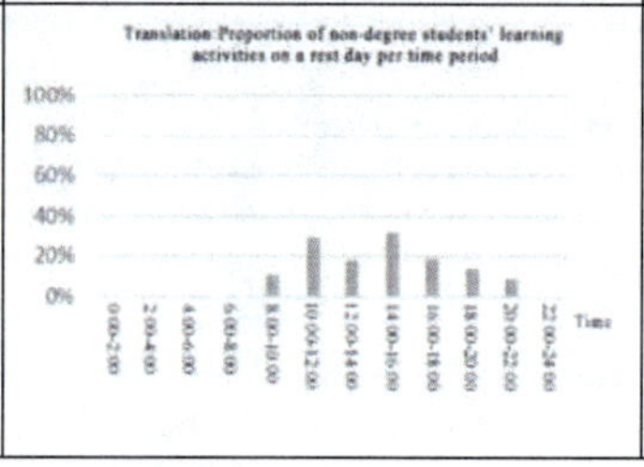	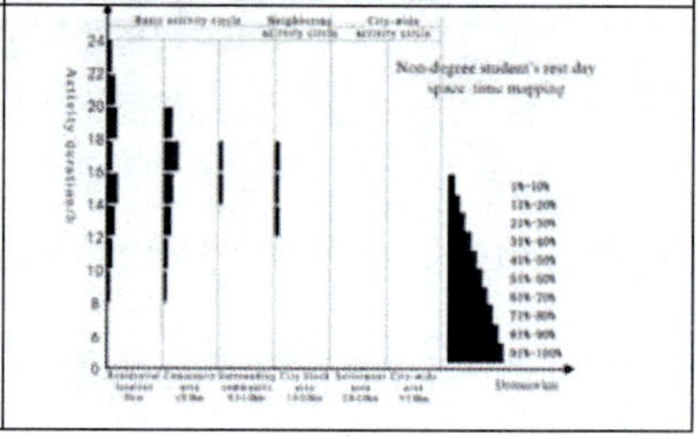

Source Data from the research group's sample surveys on the everyday life space of International Student Settlements in Nanjing (2017)

Influence of residential location on international students' learning activities (social attribute)

There are significant differences in learning time and space for international students staying at different spots (see Table 2.13).

In terms of learning time, the international students staying in school study more than those living outside school. The former has a more regular and strict approach when it comes to learning. As for the latter, they will make learning arrangements based on what they like and require; therefore, it is common for them to study late at night. In terms of the differences in learning on work and rest days, both groups of international students study for a significantly longer period of time on a work day than on a rest day, and the time distribution is also more regular.

In terms of learning space, there were a lot more international students staying in school who will study outside. Those staying outside school are more inclined to study at home. There is also not much difference in the time taken to travel out to study. Conversely, those staying in school are more inclined to venture out to study on a work day, and study at home on a rest day. They may also prefer going out to study in daytime and study at home when night falls.

The reasons for this can firstly be attributed to the differential community management modes. Students in school apartments often share a room with another person, and the strict management entails that the activities undertaken will have a more fixed and inflexible schedule. Therefore, the students will have a more regular approach to utilize time under those restrictions. Those staying outside of the school are independent from school restrictions and thus can enjoy more freedom and flexibility. Second, the lifestyle approaches are also different. International student apartment

Table 2.13 Proportion of learning activities undertaken by international students at different timings and space–time Mapping of these activities

		Proportion of learning activities undertaken by international students at different timings	Space–time mapping
Students living in school	Work day	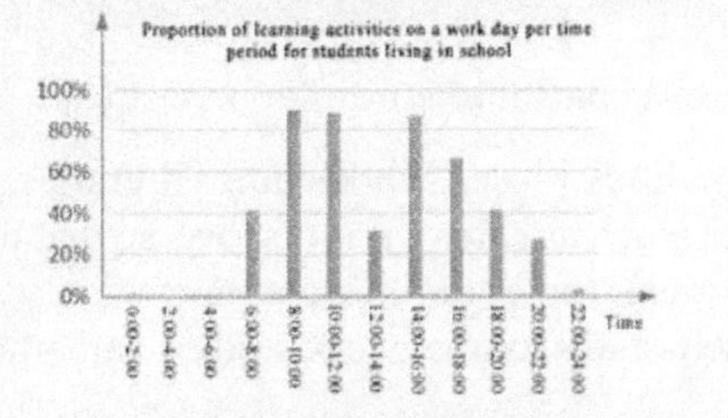	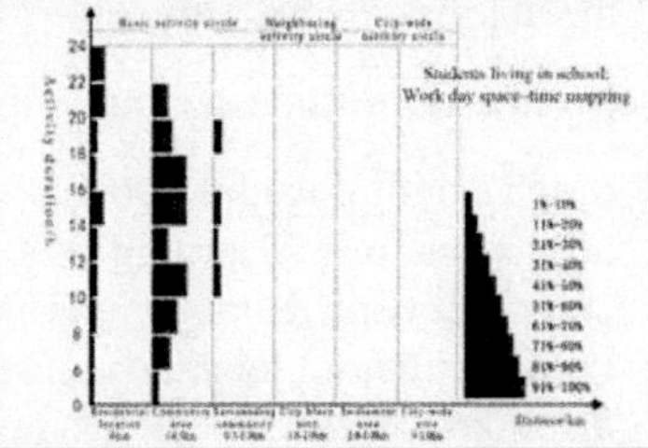
	Rest day	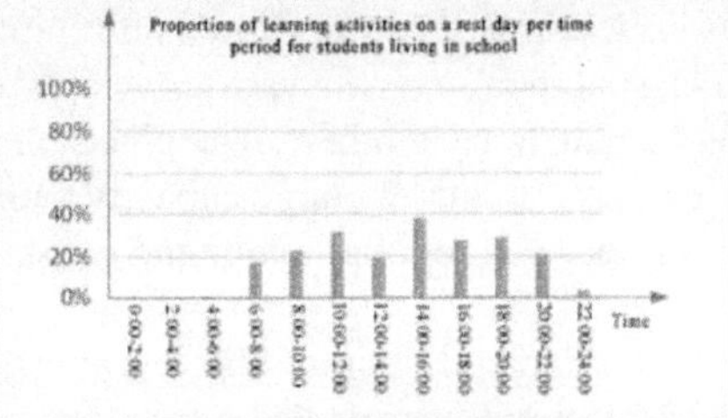	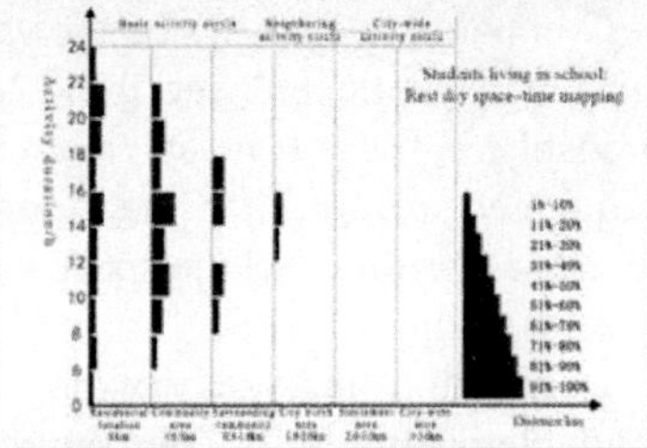
Students living outside school	Work day	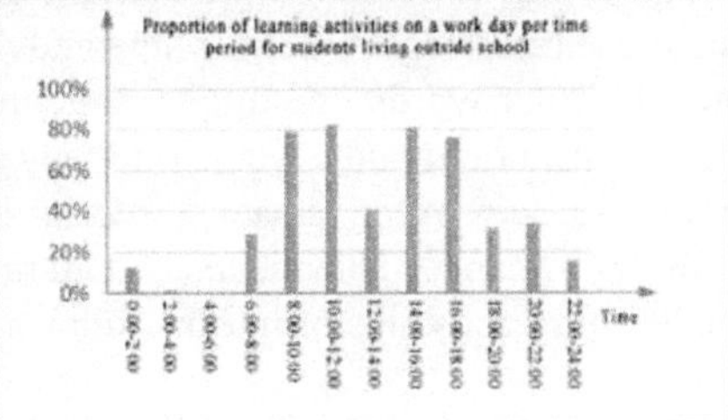	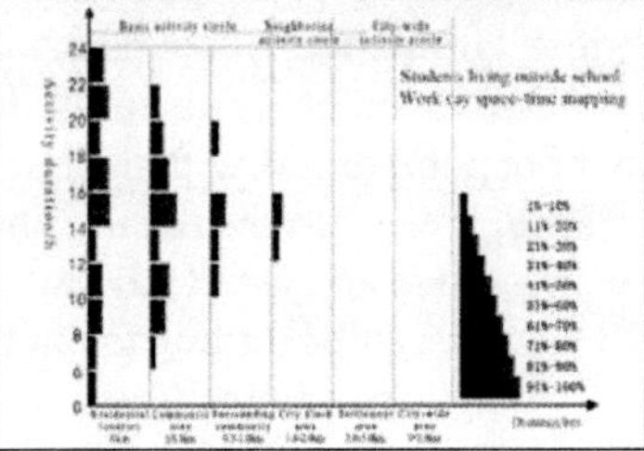
	Rest day	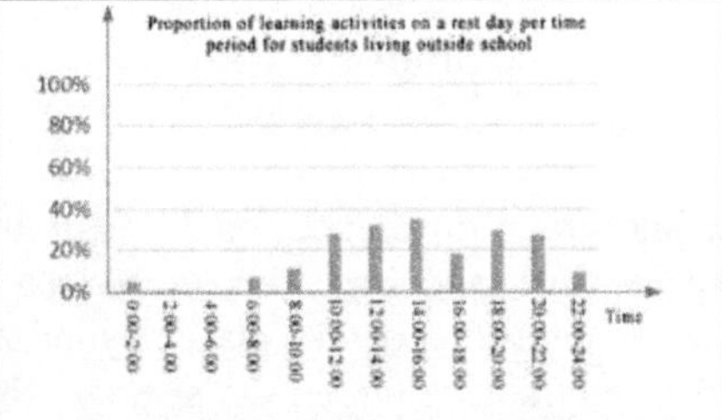	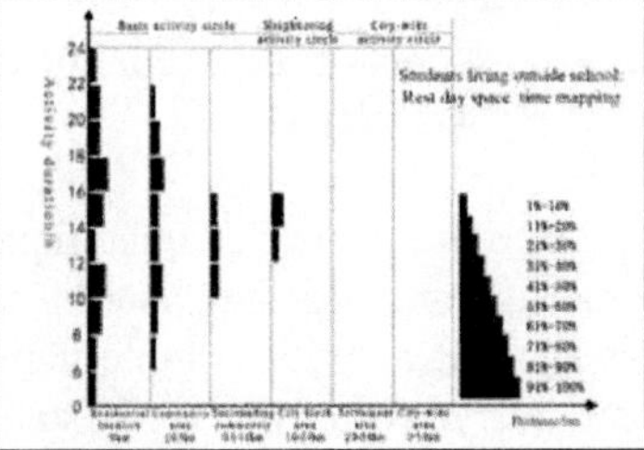

Source Data from the research group's sample surveys on the everyday life space of International Student Settlements in Nanjing (2017)

are group living spaces where residents are mutually influencing. While enjoying the rich academic culture, students are also subject to the lifestyles, habits, and paces of life that may be very different from their own. The individualized learning environment and conditions inevitably face restrictions. Conversely, international students staying outside of school face little external interferences. They enjoy high privacy and freedom and thus a higher proportion of them also choose to study at home. Third, geographical locations also come into play. Compared with those staying outside of school, those staying in the school's international student apartments are nearer to various learning spots. It is thus more convenient for them to venture out of the apartments to study.

Influence of spatial attributes on international students' activities

Through examining classrooms, self-study rooms, work studios, libraries, residential locations, and other learning venues (such as cafes), it is discovered that libraries are the most frequented among international studies for their learning activities, followed by their residential locations and self-study rooms; work studios and other learning spots ranked lower.

Comparative analyses on the various learning locations based on venue characteristics were conducted (see Table 2.14), specifically on the various venues' capacities, atmosphere, formation, and learning format, among others. The places that are more frequented are similar in the following ways: relatively large capacity, quiet environment for learning, NS the space has a linear and regular layout and is used primarily for self-study.

Correlation analyses on the various characteristics of the learning spots were conducted using SPSS (see Table 2.15). Using Spearman's correlation, a significant positive correlation is found between the learning spot's atmosphere and the number of students who chose to study there. In other words, the quieter the environment, the more popular it is with the international student body. Using one-way analysis of variance, it was discovered that the significances of the capacity, interior layout, and learning mode all passed the test at 5% significance. This means that all three factors will have a significant impact on the learning activities of international students.

2.3.3.3 Space–time Model of International Students' Learning Activities

Based on the overall characteristics and group differences analyses on the international students' learning spaces, a space–time model was developed (see Fig. 2.25). The learning activities of international students do not take place beyond 2 km of the basic everyday activity circle and the neighboring circle. On average, the learning duration and distance traveled do not have any significant correlation. The learning activities are also evenly distributed within a specific scope. Therefore, their learning activities display the space–time characteristics of taking place at multiple spots within a limited distance and are evenly distributed across time.

Table 2.14 Comparison of international students' learning spots

	Classroom	Self-study room	Work studio	Library	Residential location	Others (e.g., cafe)
Capacity	30–100 person	30–50 person	Approximately 10	30–80 person	1 person	Approximately 20
Atmosphere	Quiet, restrained	Quiet, restrained	Rather quiet and restrained	Quiet, restrained	Quiet, free	Rather noisy and free
Learning mode	Attending class	Self-study	Self-study; exchanges and discussions	Self-study	Self-study; exchanges and discussions within a small group	Self-study; exchanges and discussions
Interior formation	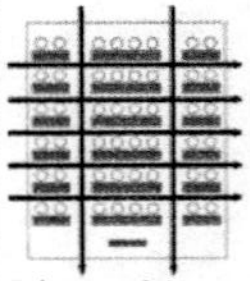 Linear layout	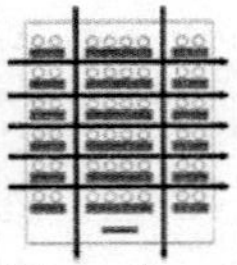 Linear layout	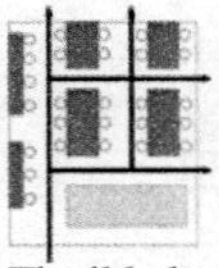 Flexible layout	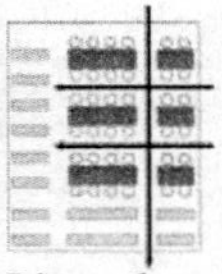 Linear layout	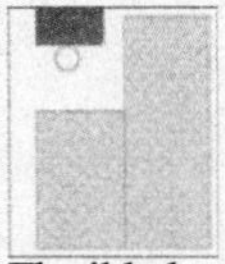 Flexible layout	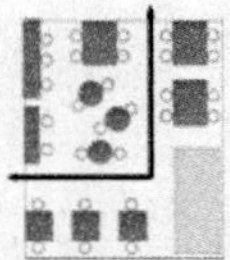 Flexible layout
Number of students who study in this location, as a percentage of total respondents (%)	18	24	11	38	21	7

Source Data from the research group's sample surveys on the everyday life space of International Student Settlements in Nanjing (2017)

Table 2.15 Correlation analyses on learning spots' characteristics

		Capacity	Atmosphere	Interior formation	Learning mode
Spearman correlation analysis	Correlation coefficient	——	0.512	——	——
	Significance (two sided)		*0.031*		
One-way analysis of variance	F	7.288	——	5.941	10.325
	Significant level	*0.020*		*0.040*	*0.009*

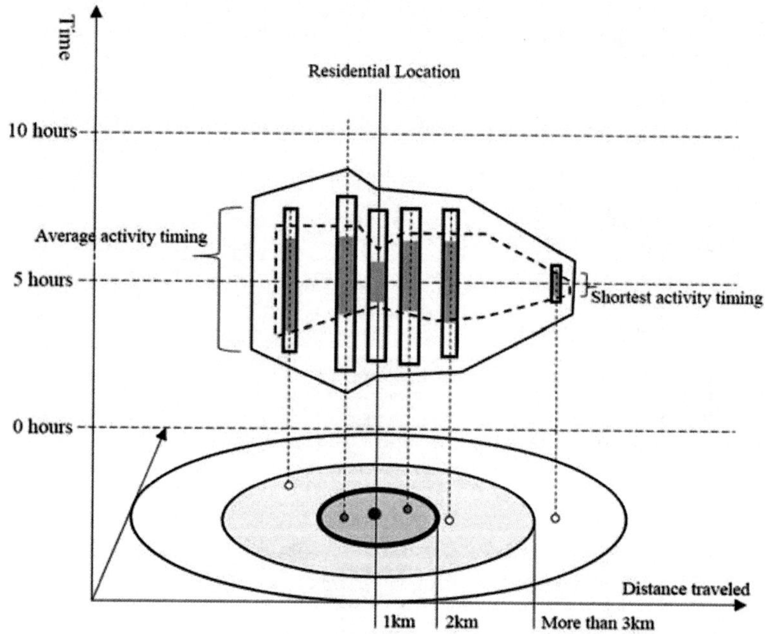

Fig. 2.25 Space–time Mmodel of international students' learning activities. *Source* Own elaboration

2.3.4 *Shopping Spaces at the International Student Settlements*

2.3.4.1 Overall Characteristics of International Students' Shopping Activities

Among the 200 questionnaires returned, there were 77 respondents who engage in shopping activities in their everyday lives. They primarily buy everyday amenities;

goods of choice and special items were bought less often. In terms of traveling for shopping, riding on motorcycles/cars, bicycles, and walking are among the more common modes. Some also choose to take public transportation, such as trains and buses.

First, temporal characteristics of shopping activities

It can be seen that there is a certain regularity in terms of the time taken for shopping and that there are significant differences between a work and rest day. First, the proportion of international students who engage in shopping activities is not high, probably because of limited economic independence and purchasing power. Second, the international students' shopping activities mostly take place in the day and are scattered in terms of time distribution. Third, more shopping is done on a rest day compared to a rest day and takes place freely throughout the day. On a work day, the students can only shop after class (see Fig. 2.26).

Second, time characteristics of shopping activities

Based on the international students' shopping activities, we can consolidate and develop an abstract three-tiered concentric structure of the shopping spaces. Their shopping activities show a declining pattern as they move further away from the core (see Fig. 2.27).

First circle: Shopping is concentrated in this tier (within 1 km of their residential location). This is the main shopping tier of the international students (also their basic everyday activity circle), in which 77% of shopping activities on a work day and 55% of shopping activities on a rest day take place.

Second circle: Expanded shopping tier (within 1–2 km of their residential location). The time spent on shopping activities dropped in this tier, the rate of diminution being 0.81 on a work day and 0.45 on a rest day. This is the secondary shopping tier for international students (also their neighboring activity circle).

Third circle: Marginal shopping area (at 2–3 km and beyond the residential location). The total time spent on shopping activities continues to drop, with the rate of

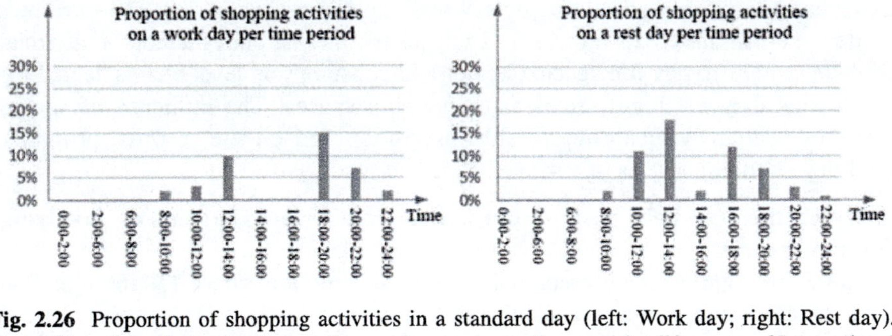

Fig. 2.26 Proportion of shopping activities in a standard day (left: Work day; right: Rest day). *Source* Data from the research group's sample surveys on the everyday life space of International Student Settlements in Nanjing (2017)

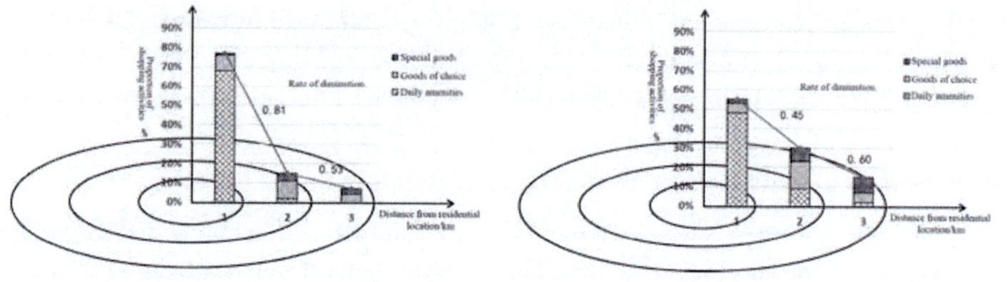

Fig. 2.27 Map of international students' diminishing shopping activity level according to distance (left: Work day; right: Rest day). *Source* Data from the research group's sample surveys on the everyday life space of International Student Settlements in Nanjing (2017)

diminution being 0.53 on a work day and 0.6 on a rest day. This tier also refers to their city-wide activity circle.

The following characteristics of the shopping spaces are evident. First, the international students' shopping activities generally align with the residential location-centric principle, in which the further away the shopping area is from the residential location, the time spent on shopping activities fall correspondingly. Second, the space–time structure of shopping activities differs for work and rest days; the rate of diminution in terms of distance on a work day is higher than that of a rest day, and shopping activities are also much more concentrated within 1 km on a work day than on a rest day. Third, there is a significant difference in the types of shopping done at different distance tiers. Daily amenities are mostly bought within 1 km, while goods of choice and special goods take place more often as the distance broadens.

2.3.4.2 Intra-Group Differentiation Among International Students' Shopping Activities

This study is grounded in urban planning and community perspectives, further exploring the time and space characteristics of international students' shopping activities in terms of "social attributes" and "spatial attributes." Social attributes are derived from the SPSS regression analyses on international students' individual attributes and everyday life space. Gender and consumption level were selected for analyses as they are highly related to shopping spaces. The influence of spatial attributes was analyzed through field observation and on-site records, primarily involving shopping spaces of various levels and types.

Analyzed through field observation and on-site records, primarily involving shopping spaces

There are significant differences in shopping time and space for international students of different genders (see Table 2.16):

In terms of shopping time, female students spend a significantly higher proportion of their time on shopping than male students. Both male and female students

Table 2.16 Proportion of shopping time for international students of Different genders and mapping of rate of Diminution of shopping activities based on distance

<table>
<tr><th></th><th></th><th>Proportion of recreation activities undertaken by international students at different timings</th><th>Map of rate of diminution of shopping activities based on distance</th></tr>
<tr><td rowspan="2">Male students</td><td>Work day</td><td>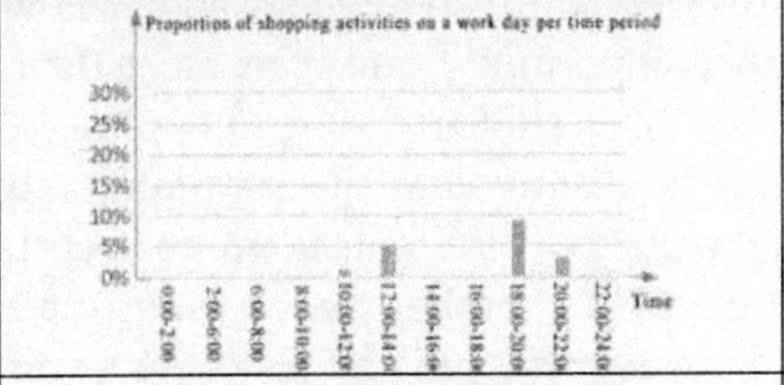</td><td>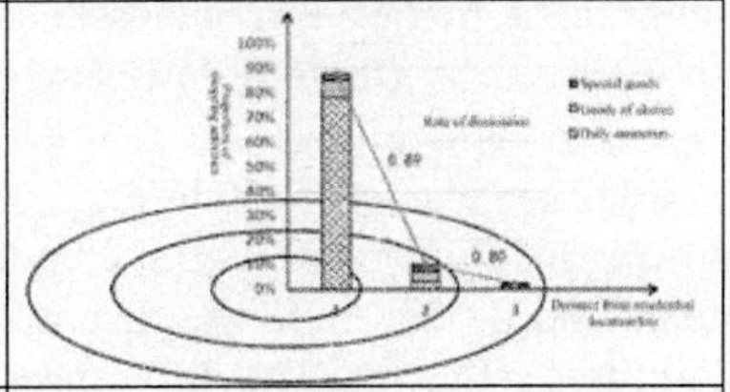</td></tr>
<tr><td>Rest day</td><td>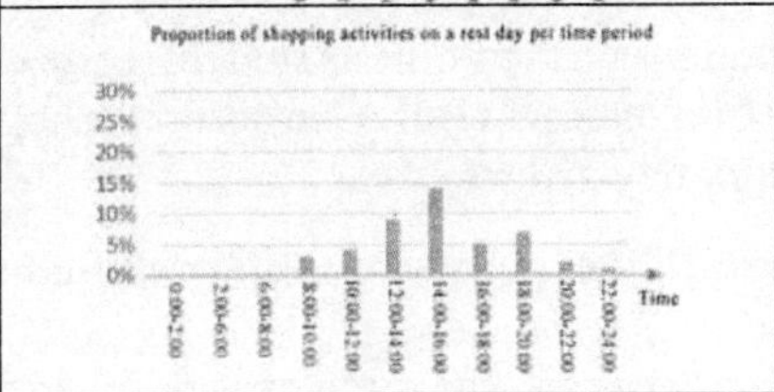</td><td>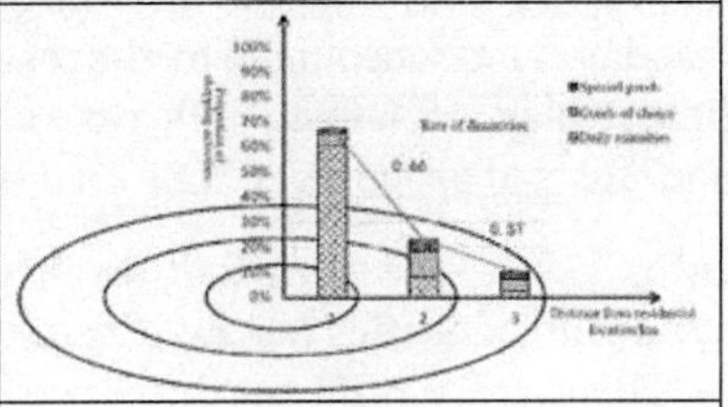</td></tr>
<tr><td rowspan="2">Female students</td><td>Work day</td><td>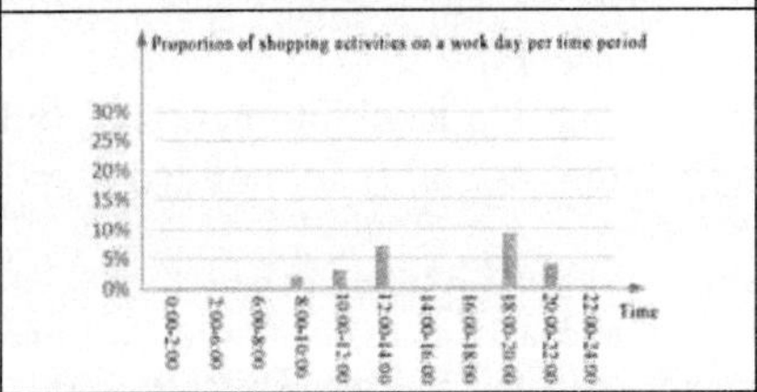</td><td>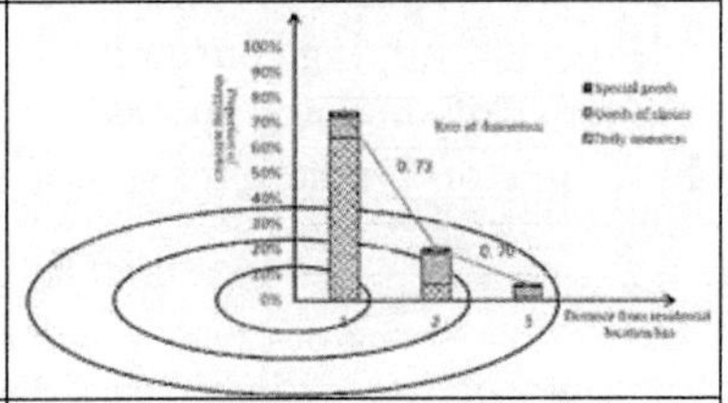</td></tr>
<tr><td>Rest day</td><td>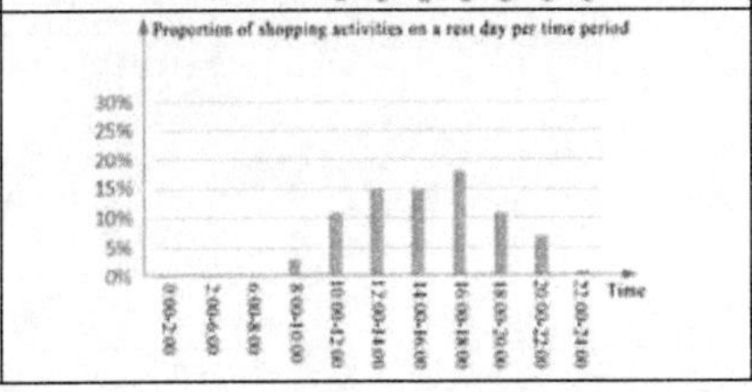</td><td>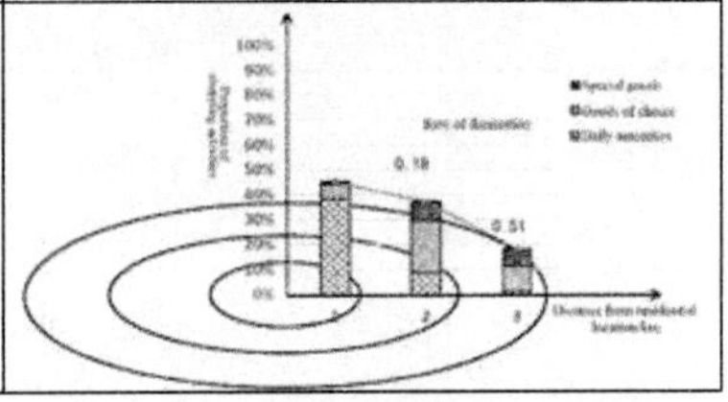</td></tr>
</table>

Source Data from the research group's sample surveys on the everyday life space of International Student Settlements in Nanjing (2017)

demonstrate shopping characteristics of shopping intermittently and on a work day and shopping continuously on a rest day. Of which, the difference between shopping time on a work day and rest day are greater for female students. Additionally, the female students also spend a significantly higher proportion of their time on shopping on a rest day compared to male students.

In terms of shopping spaces, the rate of diminution for male students' shopping activities on work and rest days are higher than that of female students. The male students' shopping activities are also more concentrated within 1 km. As for female students, they engage in significantly more diverse shopping and at different distances; shopping for goods of choice and special items occur more often at further distances. The female students' spatial structures of shopping activities on work and rest days also differ greatly compared to that of male students. Long-distance shopping also constitutes a higher proportion of their time on a rest day.

The reasons for this are related to gender psychology. Male students mostly view shopping activities as a means to satisfy basic lifestyle needs and thus will choose to shop nearby (thereby lowering time opportunity costs) to meet their shopping objectives where possible. However, female students view shopping activities as entertainment or a lifestyle experience and thus hold the shopping and selection process in higher regard. As such, they will go to different shopping spots to compare items and do not mind the distance traveled. Another reason is attributed to their identity as students. The differences between shopping behavior on a work day and rest day can be accounted by the restrictions on their time as students (especially class time, which is relatively inflexible). This therefore resulted in different shopping time patterns and space selections on a work day and rest day.

Influence of consumption level (monthly expenses) on international students' shopping activities (social attribute)

There are significant differences in terms of shopping spaces for international students with varying consumption levels (see Table 2.17).

In terms of shopping spaces, high consumption level international students have a lower rate of diminution for shopping activities and spaces on work and rest days

Table 2.17 Proportion of shopping time for international students of different Consumption Levels and Mapping of Rate of Diminution of Shopping Activities Based on Distance

	Work day	Rest day
High consumption level students	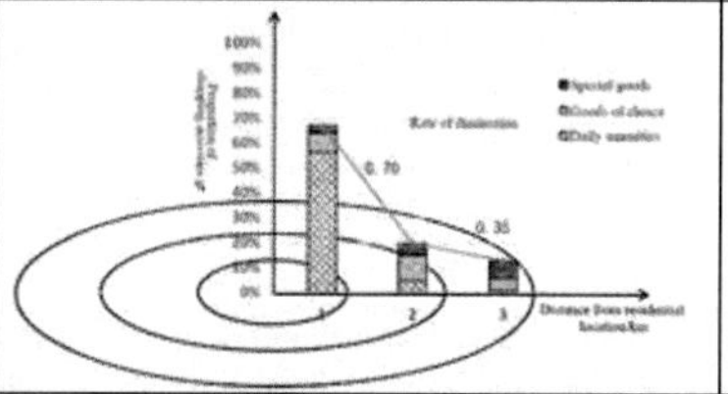	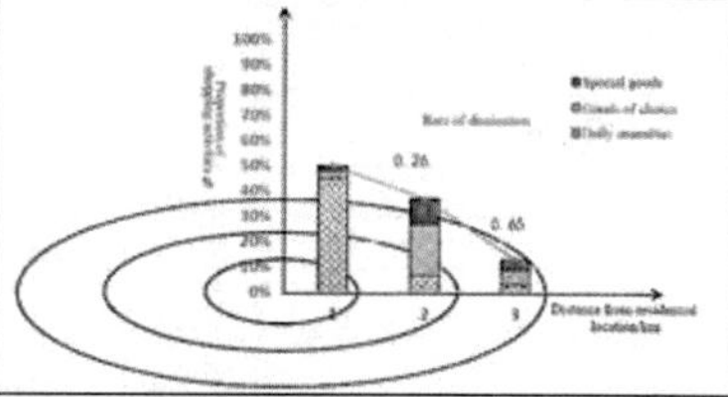
Mid-consumption level students	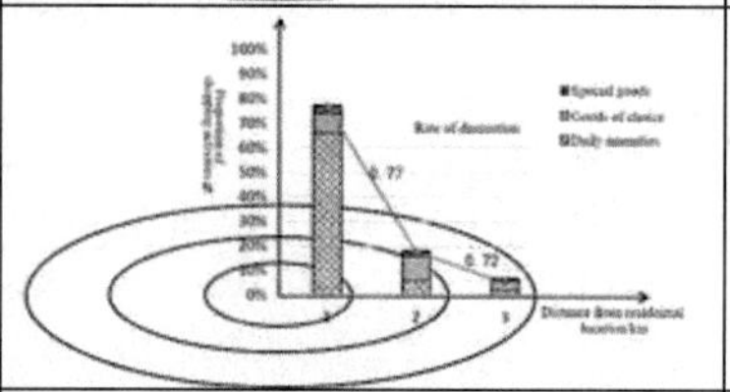	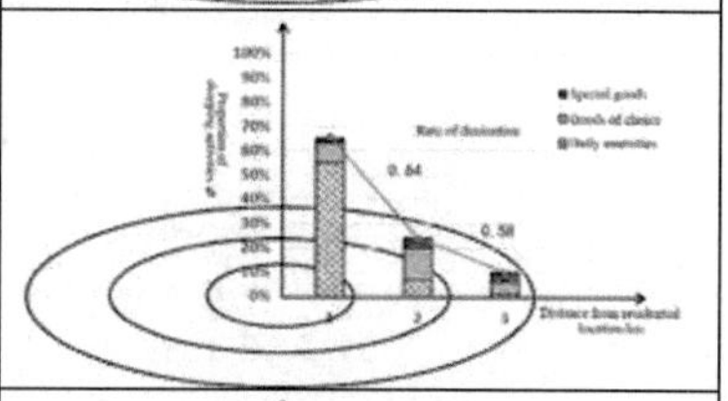
Low consumption level students	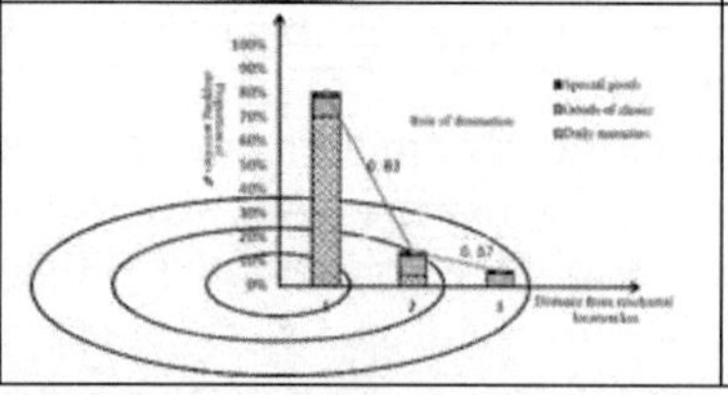	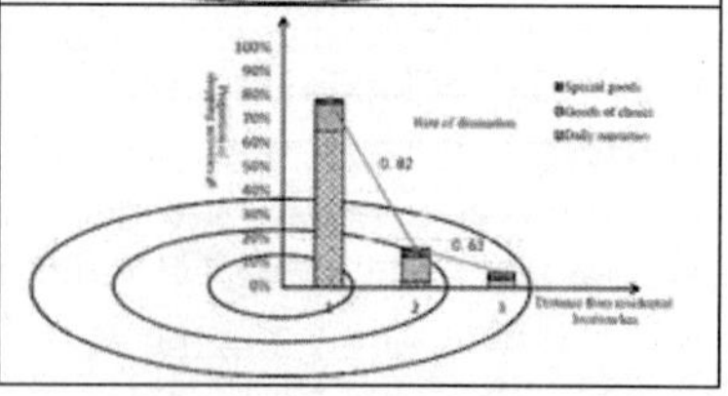

Source Data from the research group's sample surveys on the everyday life space of International Student Settlements in Nanjing (2017)

compared to that of mid- and low-level consumption level students. International students with a high consumption level are more willing to travel further to achieve more diverse shopping objectivities. Compared to the mid and low consumption level students, the high consumption level international students display greater differences in terms of the spatial structure of shopping activities for work and rest days. Long-distance shopping on a rest day also constitutes a higher proportion of their time.

The reason for this is because of differences in consumption perspectives. The differences in consumption levels have brought about inherent differences in the international students' consumption perspectives and shopping variety. Low consumption level students basically purchase daily amenities nearby, and their consumption perspective is that of satisfying basic lifestyle needs, convenience, and low prices. High consumption level students will focus more on the level of consumption and quality of the goods; they will purchase more special items and goods of choice besides daily amenities and are also willing to spend time and travel further for shopping. Mid-consumption level students are somewhere in between the high and low consumption level students.

Influence of spatial attribute on international students' shopping activities

Further investigations and consolidation of data on eight typical shopping venues from city-wide, settlement, and community levels show that the highest proportion of international students chooses to shop at community supermarkets, followed by large supermarkets at the settlement level. City-wide spots, commercial centers, and settlement tier shopping squares constitute the lowest proportion.

Comparison analyses were made on the various shopping spots (see Table 2.18) based on the venue characteristics, including level of shopping spot, type, distance, tier, and type of goods, among others. The shopping venues frequented more often have the following similarities: community-level commercial establishments, located within 1 km from residential location, mid-to-low tier in terms of goods, and mostly selling daily amenities and goods of choice.

Correlation analyses on the various characteristics of the shopping spots were conducted using SPSS (see Table 2.19). Using Spearman's correlation, a significant negative correlation is found between the shopping spot's level and distance from residential location. In other words, the lower the level and the nearer the shopping spot is to the residential location, the more popular it is with the international student body. Using one-way analysis of variance, it was discovered that the significance values of the shopping spot type, tier, and types of goods sold all passed the test at 5% significance. This means that all three factors will have a significant impact on the shopping activities of international students.

2.3.4.3 Space–Time Model of International Students' Shopping Activities

In accordance to the analyses on overall characteristics of international students' shopping spaces and group differences, a space–time model on the international

Table 2.18 Comparison of different types of international students' shopping venues

Name	Xinjiekou	Hunan Road	Zhujiang Road	Huanya Katherine Square	Ce-Mart	Aishang World	Suguo Supermarket	Nanjing University Supermarket
Level	City level		Settlement level			Community level		
Type	Commercial center	Large-scale commercial street	Commercial street	Shopping plaza	Large-scale supermarket	Small-scale commercial street, shopping center	Supermarket Convenience store	Supermarket in school
Distance (km)	3	2	1.5	1.5	1	1.5	0.5	<0.5
Tier	Mid-high tier	Mid-tier	Mid-low tier	Mid-tier	Mid-low tier	Mid-low tier	Mid-low tier	Low tier
Type of products	Goods of choice; Special items	Goods of choice; Special items	Goods of choice; Special items	Daily amenities; Goods of choice	Daily amenities; Goods of choice; Special items	Daily amenities; Goods of choice	Daily amenities	Daily amenities
Number of students who shop in this location, as a percentage of total respondents (%)	13	33	27	18	38	21	51	62

Source Data from the research group's sample surveys on the everyday life space of International Student Settlements in Nanjing (2017)

Table 2.19 Correlation of shopping venues' characteristics

		Level	Type	Distance	Tier	Type of products
Spearman correlation analysis	Correlation coefficient	−0.397	____	−0.536	____	____
	Significance (two sided)	*0.029*		*0.010*		
One-way analysis of variance	F	____	5.227	____	13.061	9.750
	Significant level		0.033		*0.007*	*0.014*

students' shopping activities can be derived (see Fig. 2.28). It can be seen that the shopping spots of international students are distributed throughout within the city-wide activity circle, and overall, the sample shows that the average activity duration shortens as the distance widens. The smallest individual shopping duration will however increase as distance increases. These manifest the space–time characteristics of "broad spatial distribution and diminishing activities as the distance widens, while the time distribution increases as the distance widens."

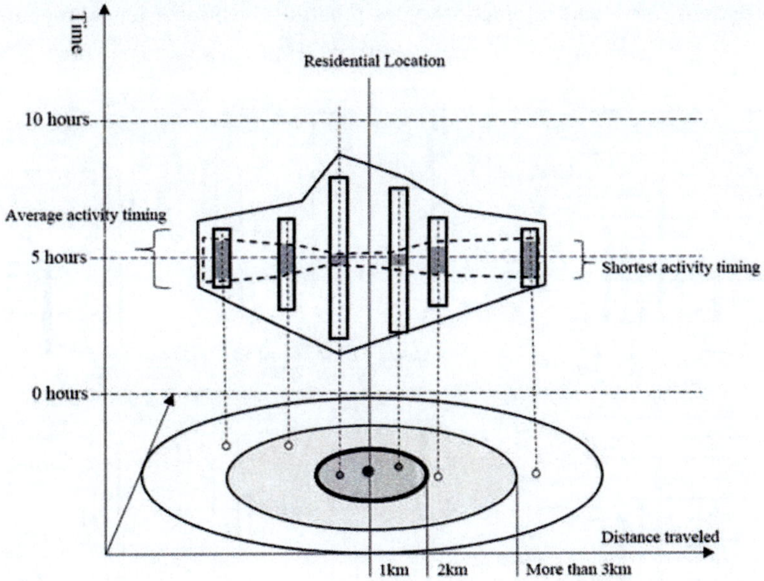

Fig. 2.28 Space–time model of international students' shopping activities. *Source* Own elaboration

2.3.5 *Recreation Spaces at International Student Settlements*

2.3.5.1 Overall Characteristics of International Students' Recreations Activities

This study has classified the internationals students' activities as culture and entertainment types (music, dance, movies, surfing the internet, radio programs, etc.), interaction types (chatting, visiting friends, meeting for meals, gatherings, etc.), fitness types (ball sports, strolls, specialized fitness programs, etc.), trips (visiting parks, hiking, short-term trips, etc.), and hobbies (music, chess, calligraphy, painting, photography, writing, inventions, etc.). The space–time characteristics are summarized below (see Fig. 2.29).

I. First, time characteristics of recreation activates

There are significant differences in terms of the time duration for the various recreation activities. The international students also display significant preferences for some activities over others. These can be seen as described below. First, in terms of the proportion of time spent on the various recreation activities, international students prefer culture and entertainment types (music, dance, movies, surfing the internet, radio programs, etc.) and interaction types (chatting, visiting friends, meeting for

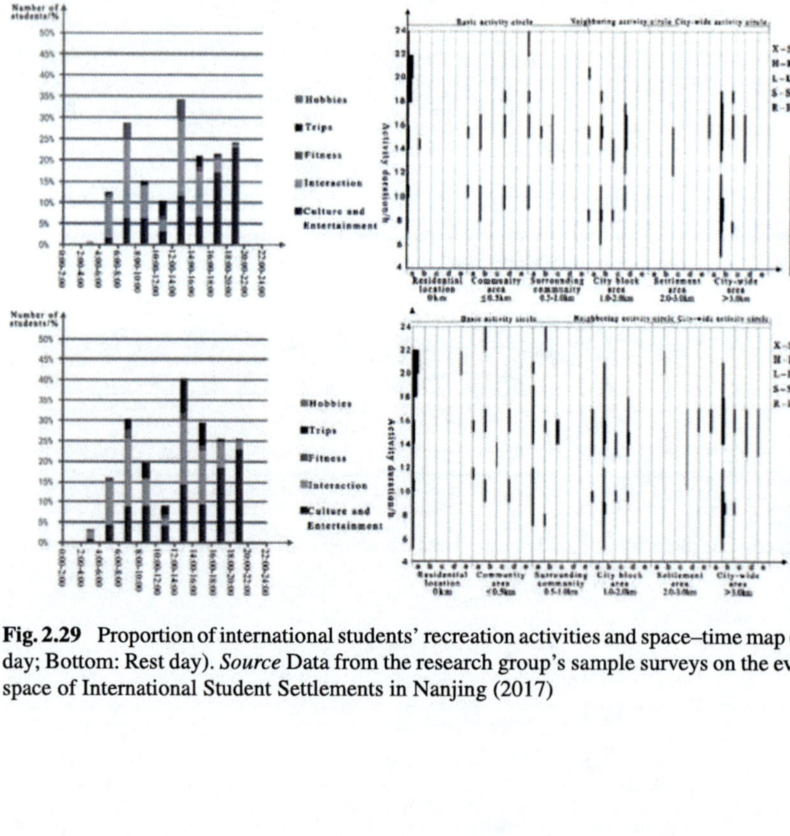

Fig. 2.29 Proportion of international students' recreation activities and space–time map (Top: Work day; Bottom: Rest day). *Source* Data from the research group's sample surveys on the everyday life space of International Student Settlements in Nanjing (2017)

meals, gatherings, etc.). No matter whether it is for a certain time period or throughout the day, these two activities take up a large proportion. These are then followed by fitness types (ball sports, strolls, specialized fitness programs, etc.), which are distributed throughout the day. While these are regular, the time spent on them is shorter. Lastly, trips (visiting parks, hiking, and short-term trips) and hobbies (music, chess, calligraphy, painting, photography, writing, inventions, etc.) take up a smaller proportion, and they occur more frequently and for a longer duration on a rest day than on a work day.

There is a certain difference in the duration and type of recreation activities on a work day and a rest day in the following areas. While international students engage in recreation activities on a work day for a shorter period compared to that on a rest day, there are overall similar patterns in terms of time distribution. There are also differences in the types of recreation activities engaged on a work day and rest day. There are not many differences for culture and entertainment, interactions, and sports type activities, but trips occur more frequently and for a longer duration on a rest day.

II. Second, spatial characteristics of recreation activities

There are diverse venues for recreation activates. While there are rather large differences across the venues in terms of the types of recreation activities, there are also possibilities for cross-exchanges and sharing of spaces. For instance, culture and entertainment, as well as interaction activities, can occur at various places and at varying distances. Fitness activities take place at relatively fixed locations. A myriad of recreation activities can also take place at residential locations.

Additionally, there is a certain correlation between the location and types of recreation activities. Interaction and sports activities are largely concentrated within the city block area, while hobbies as well as culture and entertainment activities are engaged at diverse and spread-out venues.

2.3.5.2 Intra-Group Differentiation Among International Students in Terms of Recreation Activities

This study is grounded in urban planning and community perspectives, further exploring the time and space characteristics of international students' recreation activities in terms of "social attributes" and "spatial attributes." Social attributes are derived from the SPSS regression analyses on international students' individual attributes and everyday life space. Gender and level of integration were selected for analyses as they are highly related to recreation spaces. The influence of spatial attributes was analyzed through field observation and on-site records, primarily involving the various levels and types of recreation spaces.

Influence of gender on international students' recreation activities (social attribute)

There are significant differences in recreation time and space for international students of different genders (see Table 2.20).

Table 2.20 Proportion of recreation activities undertaken by international students of different genders and space–time mapping

		Proportion of recreation activities undertaken by international students at different timings	Space–time mapping
Male students	Work day	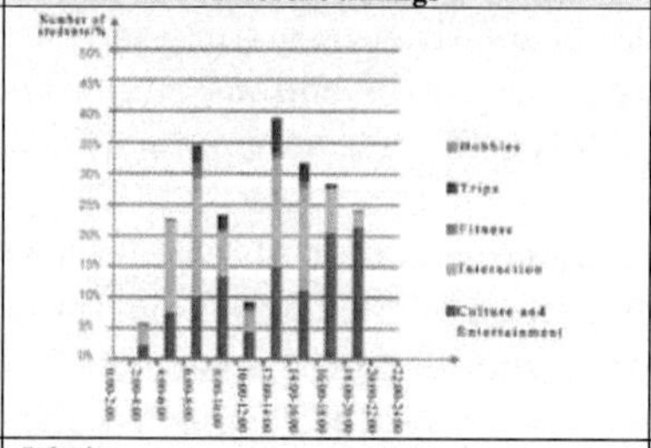	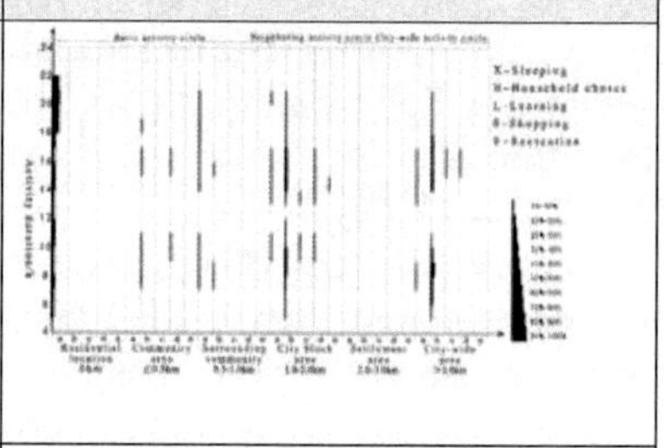
	Rest day	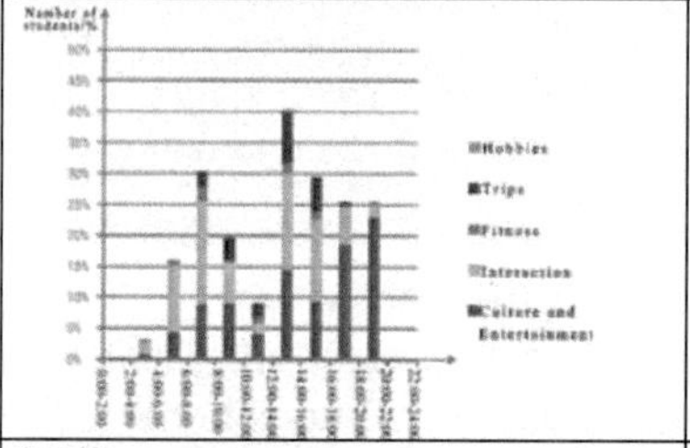	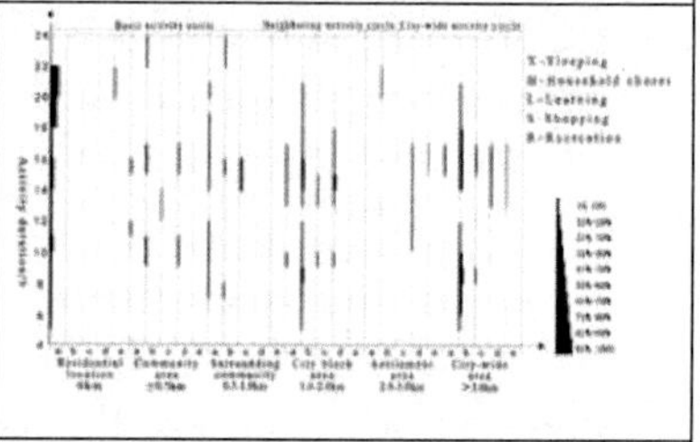
Female students	Work day	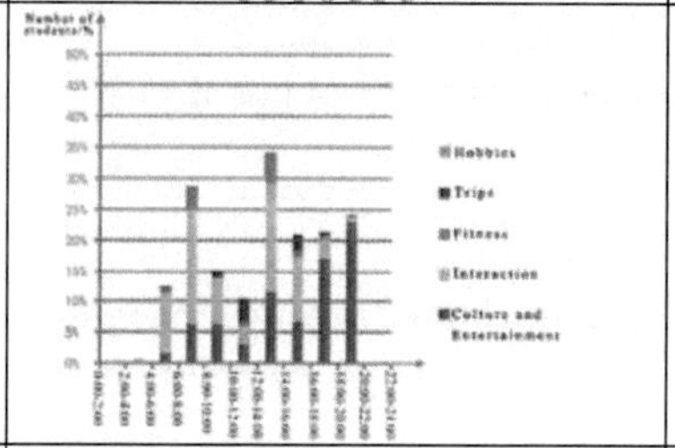	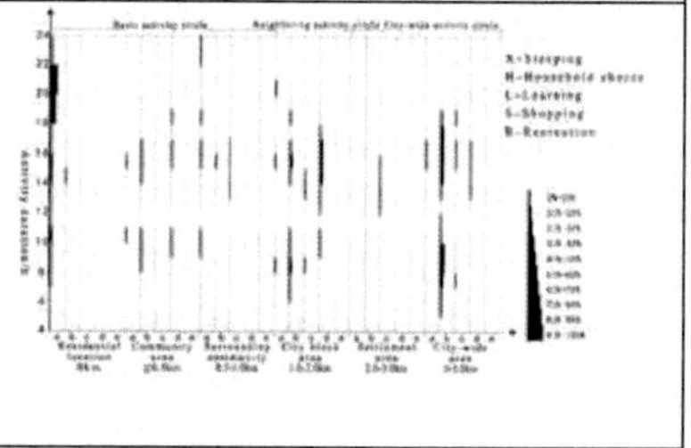
	Rest day	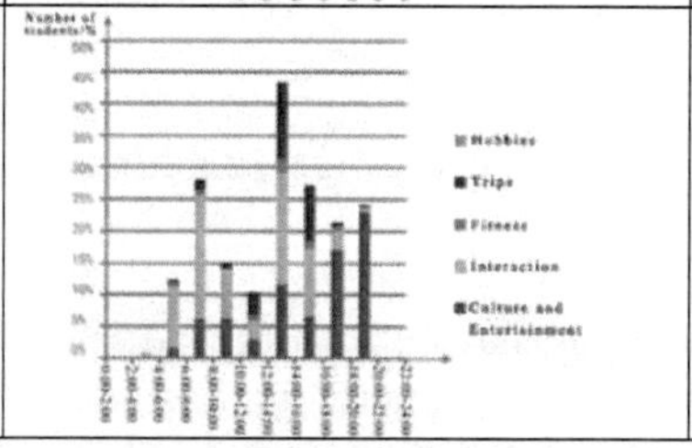	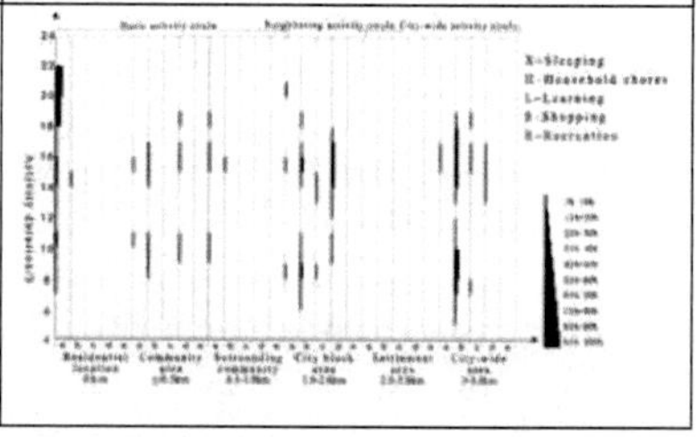

Source Data from the research group's sample surveys on the everyday life space of International Student Settlements in Nanjing (2017)

In terms of recreation time, there are no major differences in the overall duration of the activities, overall time distribution (greater in the day than at night), and the different types of recreation activities for male and female students. Among which, male students are more into recreation activities that are long and continuous, while female students are more into recreation activities that are shorter and in bite-sized amounts. Female students also display a bigger difference in terms of the recreation activities engaged on a work day and rest day, and the proportion of recreation activities on a rest day is also significantly higher compared to that of male students.

In terms of recreation spaces, male students engage in recreation activities within 1 km (basic everyday activity circle) at a higher proportion than female students. Female students tend to engage in recreation activities in which the locations are more diverse in the day and closer to their residential location at night. For male students, their locations can be more varied for both day and night recreation activities; specifically, a higher proportion of them also travel further on a rest day (city-wide activity circle), though such a difference is not evident for female students.

The reasons for these could be attributed to the correlation between gender and activity attributes. Male and female students have different preferences for recreation activities. Naturally, there are differences in their choice of recreation activities, resulting in different activity time and spaces and differential characteristics of "long and consecutive" and "short and bite-sized," respectively. Another reason lies with psychological security. Out of safety considerations, female students will choose to engage less in long-distance recreation activities at night compared to male students.

Influence of level of integration on international students' recreation activities (social attribute)

There are significant differences in terms of recreation time and space for international students with varying levels of integration (see Table 2.21).

In terms of recreation time, high integration level international students spend a significantly higher proportion of their time on recreation activities than mid and low integration level international students. They will also spend more time on extrovert recreation activities (such as culture and entertainment activities), while mid and low integration level international students prefer mono-faceted and introvert recreation activities (such as hobbies). Mid and high integration international students also display larger differences in their recreation time on a work day and rest day compared with low integration level students. This is manifest as short, intermittent activities on a work day and long, continuous activities on a rest day.

In terms of recreation spaces, international students of varying integration levels manifest a similar pattern in terms of distance and time distribution for recreation activities: "long-distance extrovert activities and nearby introvert activities" and "long-distance activities in the day and nearby activities at night." Compared with low integration level students, mid and high integration level students manifest a larger difference in terms of recreation spaces on a work day and rest day, and long-distance activities significantly increase on a rest day.

The reasons for this can be attributed to their preferences. High integration level students are mostly cheerful and extroverted, enjoy socializing with others, and find it easier to assimilate into the community through varied recreation activities that have mass participation. Having closer interactions with different communities over time entails that members in the community become more reliant and connected with one another, which will in turn cause integration level to increase. The converse is true for low integration level students. The level of integration and choice of recreation activities are interlinked and mutually influencing.

Table 2.21 Proportion of recreation activities undertaken by international students of varying integration levels and space–time mapping

		Proportion of recreation activities undertaken by international students at different timings	Space–time mapping
High integration level students	Work day	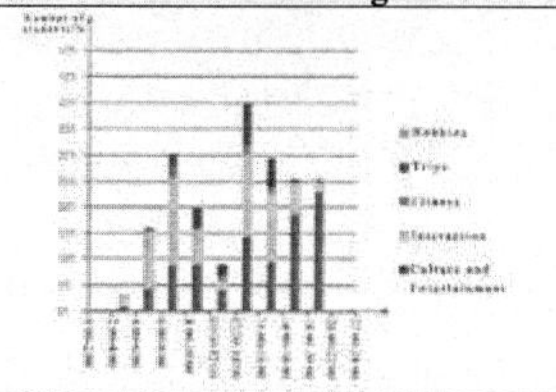	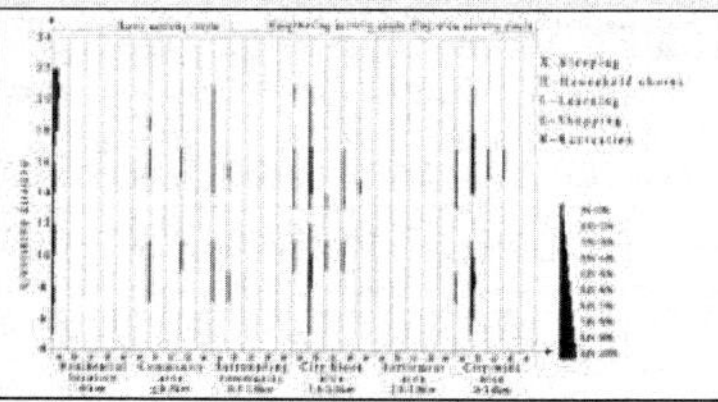
	Rest day	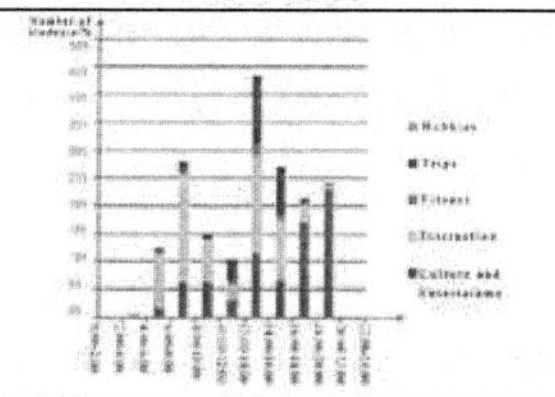	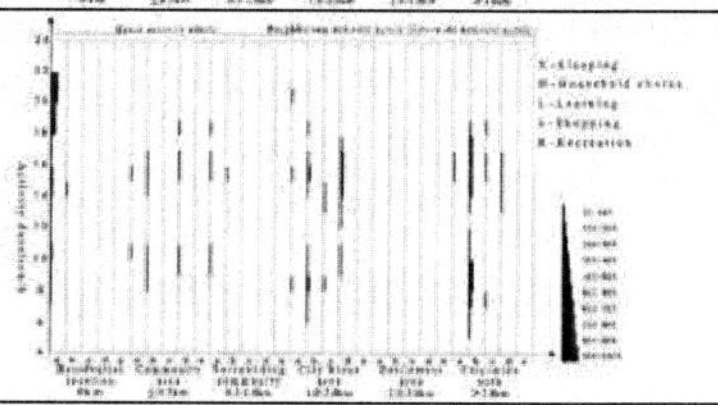
Mid-integration level students	Work day	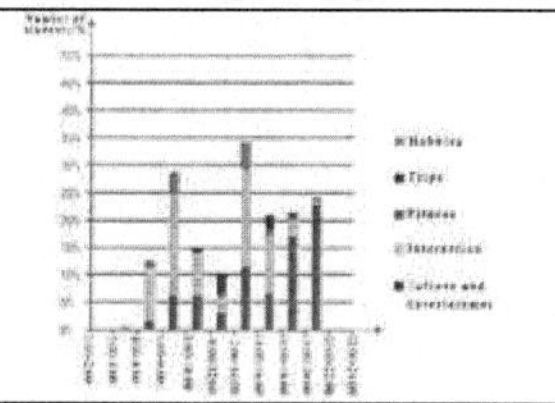	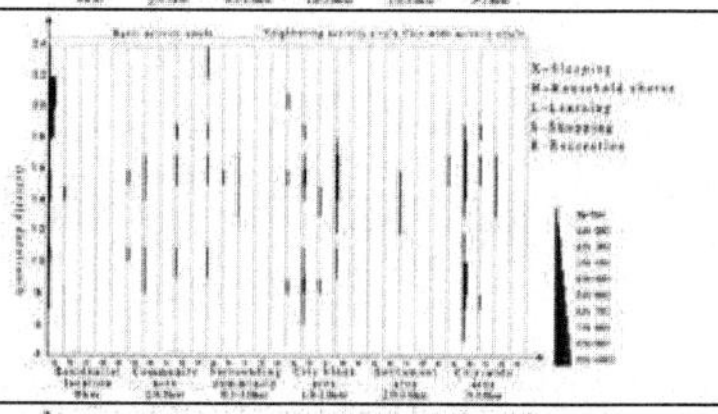
	Rest day	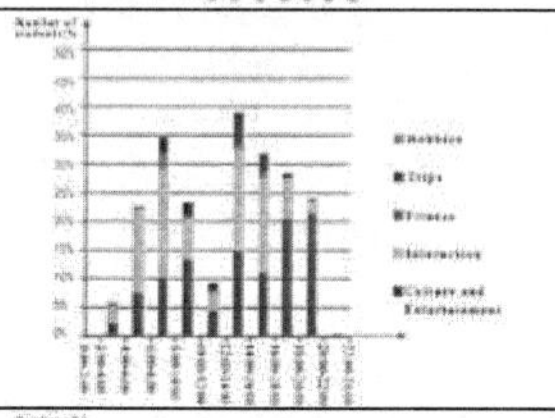	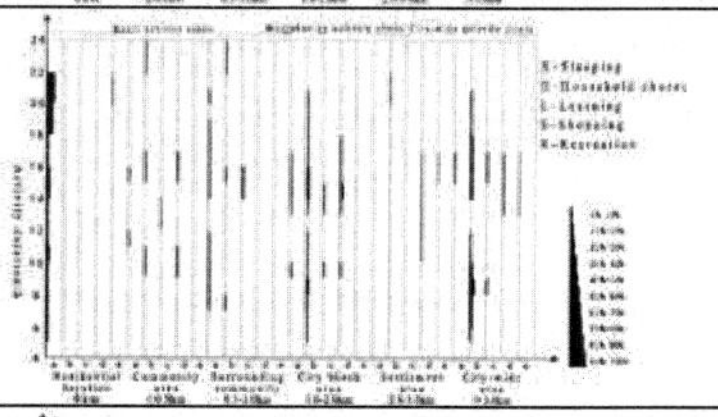
Low integration level students	Work day	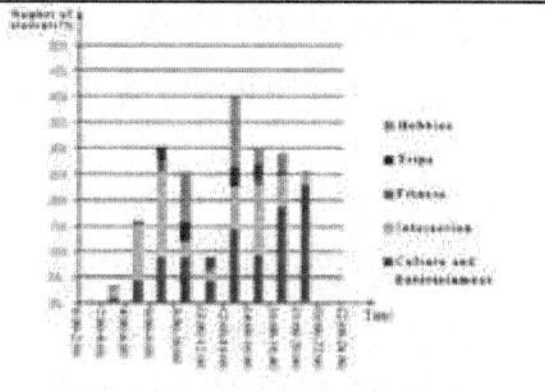	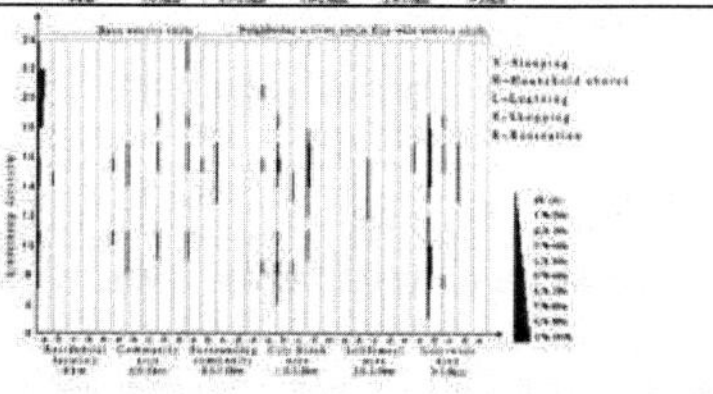
	Rest day	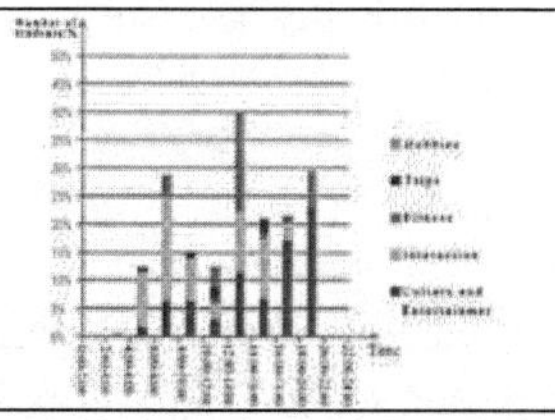	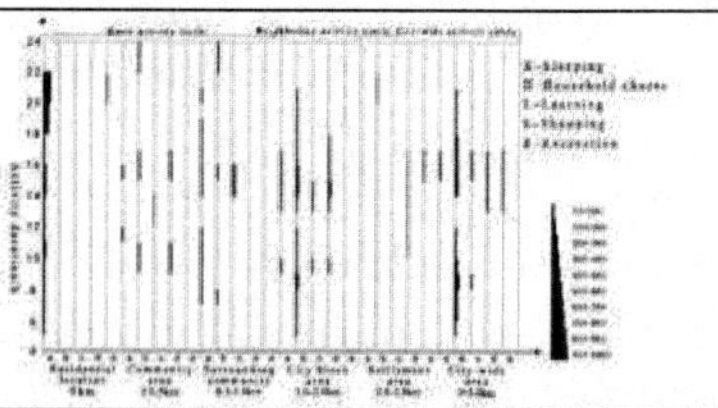

Source Data from the research group's sample surveys on the everyday life space of International Student Settlements in Nanjing (2017)

Influence of spatial attributes on international students' recreation activities

Investigations and data tabulation on culture and entertainment activities (music, dance, movies, surfing the internet, radio programs, etc.), interaction activities (chatting, visiting friends, meeting for meals, gatherings, etc.), fitness activities (ball sports, strolls, specialized fitness programs, etc.), trips (visiting parks, hiking, short-term trips), and nine typical recreation venues show that the greatest proportion of international students choose to visit cafes and bars as their recreation activities. This is followed by gyms, stadiums, and residential locations. Those who choose internet cafes, cinemas, and parks form the lowest proportion.

Comparison analyses on the venue characteristics of various recreation activity spaces were conducted (see Table 2.22), including the types of venues, spatial layout, size, distance, atmosphere, and quality of the space. The recreation venues that are frequented by more people have the following similarities: indoor public shared spaces for interaction activities, with a capacity for 10–30 people, relatively near to residential locations and have a bustling, free atmosphere.

Correlation analyses on the various characteristics of the recreation venues were conducted using SPSS (see Table 2.23). Using Spearman's correlation, a significant positive correlation is found between the distribution of proportion of international students with the recreation venue's quality of space and a negative correlation with the recreation venue's distance from residential location. In other words, the better the quality of the space and the nearer the recreation venue is to the residential location, the more popular it is with the international student body. There is no correlation between the space size and the distribution of the proportion of international students. Using one-way analysis of variance, it was discovered that the significance values of the spatial layout and atmosphere passed the test at 5% significance, while the type of venue does not exhibit any correlation. This means that the spatial layout, distance from residential location, atmosphere, and quality of the space all have a significant impact on the recreation activities of international students.

2.3.5.3 Space–Time Model of International Students' Recreation Activities

Based on the overall characteristics and group differences analyses on the international students' recreation spaces, a space–time model was developed (see Fig. 2.30). It can be seen that the proportion of international students' recreation activities occurring at 1–2 km of the neighboring activity circles is the highest. This area also forms the central point in which activities taking place at a shorter and further distance diminish. There is no pattern that can be found between the shortest duration of individual recreation activities and the distance traveled. Therefore, the space–time characteristics manifested are "diminishing level of activities beyond the central point, and spontaneous activity duration."

Table 2.22 Comparison of international students' recreation venues

Name	Home	Cinema	Internet cafe	Bar	Specialty restaurant	Cafe	Gym	Stadium	Park
Type	Culture and entertainment			Interaction			Fitness		Trips
Spatial layout	Indoor Private	Indoor Semi-private	Indoor Public shared space	Indoor Public shared space	Indoor Semi-public shared space	Indoor Public shared space	Indoor Public shared space	Outdoor Public shared space	Outdoor Public shared space
Size (capacity)	1 person	30–100 persons	30–80 persons	10–30 persons	20–50 persons	10–30 persons	20–50 persons	More than 100 persons	More than 1000 persons
Distance (km)	0	1	0.5	0.5	0.5	0.5	1.5	0.5	2
Atmosphere	Quiet Free	Quiet Restrained	Relatively quiet Relatively free	Bustling Relatively free	Bustling Relatively free	Bustling Relatively free	Relatively bustling Relatively free	Bustling Free	Bustling Free
Quality	Good	Good	Normal	Normal	Normal	Good	Good	Normal	Good
Proportion (%)	24	18	13	52	23	61	33	25	19

Source Data from the research group's sample surveys on the everyday life space of International Student Settlements in Nanjing (2017)

Table 2.23 Correlation of recreation venues' characteristics

		Type	Spatial layout	Distance	Size (capacity)	Atmosphere	Quality of the space
Spearman correlation analysis	Correlation coefficient	——	——	– 0.536	0.167	——	0.231
	Significance (two sided)			*0.010*	0.188		*0.041*
One-way analysis of variance	F	3.440	7.227	——	——	5.397	——
	Significant level	0.077	*0.026*			*0.031*	

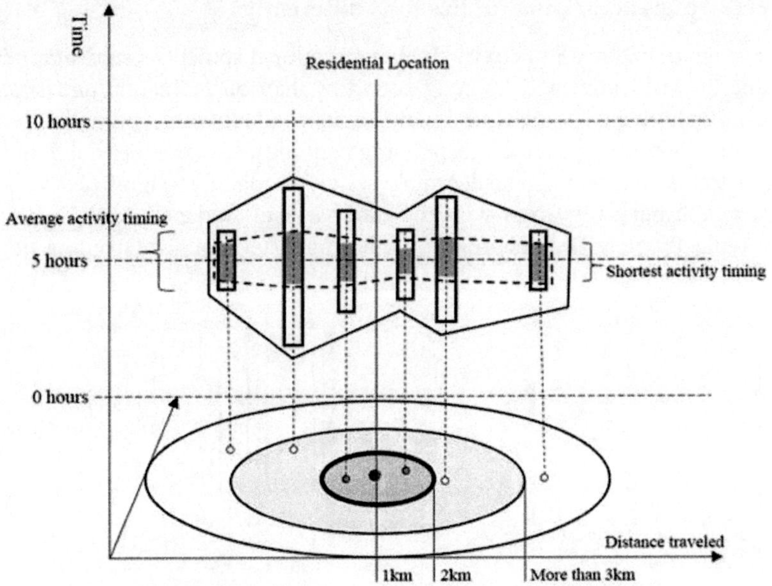

Fig. 2.30 Space–time model of international students' recreation activities. *Source* Own elaboration

2.3.6 *Space–time Trajectory of International Students Everyday Life*

Based on the international students' space–time characteristics of various activities, group differences, and model analyses set out above, this section will analyze the international students' overall everyday life space–time characteristics. Eight typical everyday life scenarios were also selected for further analyses to dive deeper and more directly into the space–time characteristics and patterns of international students.

2.3.6.1 Overall Space–Time Characteristics of Everyday Life

The following characteristics of international students' everyday life in terms of time and space can be seen (Fig. 2.31).

I. First, significant differences between a work and rest day

On the one hand, in terms of activity time, the international students' learning activities on a work day is higher on a work day than on a rest day. Recreation and shopping activities on a rest day would be higher on a rest day than a work day. On the other hand, the activity spaces on a work day are relatively near and fixed on a work day, manifesting regular patterns. On a rest day, the types and distances of activity spaces become more diverse and spontaneous to a certain extent.

II. Second, different activities utilize time differently

On the one hand, in terms of activity time, international students spend significantly more time on residential and learning activities than on recreation and shopping activities; this is related to their designated student identity as well as the level of necessity and flexibility of the corresponding activities. On the other hand, in terms of time patterns, residential and learning activities manifest significantly regular patterns, which basically align with the international students' resting and class timings, typically classified according to morning, afternoon, and evening timings.

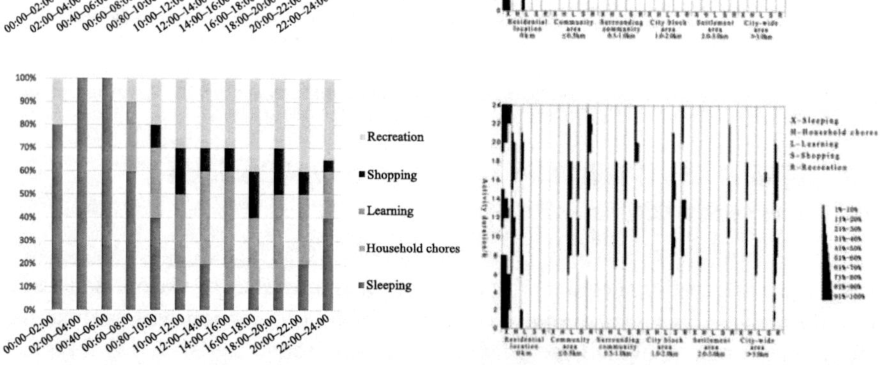

Fig. 2.31 Proportion of international students' everyday life activities and space–time mapping (Top: Work day; Bottom: Rest day). *Source* Data From the research group's sample surveys on the everyday life space of International Student Settlements in Nanjing (2017)

In contrast, recreation and shopping activities have no fixed timings and are largely scattered throughout learning hours.

III. Third, the choice of spaces for different types of activities display different characteristics

On the one hand, the venues for residential and learning activates are relatively fixed, mostly concentrated at residential locations, classrooms, and libraries. These are distributed near the residential locations. On the other hand, there are rather large differences in the recreation venues because of the type of activity. It is similar for shopping spaces, which will often differ depending on the shopping types and duration.

2.3.6.2 Typical Model of Everyday Life Space–Time Characteristics

The overall space–time characteristics of Nanxiucun international students generally can be seen as representative of the everyday life patterns for most international students. However, to come to a more comprehensive understanding of the typical characteristics of the various types of international students, there is a need to further segment and select some typical case studies for further analyses. Therefore, SPSS was used to group the sample's everyday life models for work and rest days based on the space–time distribution and patterns of international students' everyday lives. The classifications are as follows (see Tables 2.24 and 2.25).

First, international student clusters on a work day

Five clusters emerged at the preliminary stage. As the actual sample size of cluster 2 and cluster 5 are relatively small, they will not be considered. Slight adjustments were made to the cases based on the actual samples of the various activities, eventually deriving three typical groups of the international students (see Table 2.26).

Table 2.24 Clustering of everyday life models of the international student population (Work Day)

Cluster center						
Type		1	2	3	4	5
Variable	Household activities	0.3	0.8	0.2	0.6	0.5
	Learning at residential location	*7.2*	2.1	0.7	2.1	0.6
	Learning outside	2.3	1.7	*5.5*	0.9	1.2
	Shopping activities	0.5	0.4	0.2	0.2	0.3
	Recreation activities at residential location	1.6	5.2	0.9	1.3	3.8
	Recreation activities outside	0.9	0.9	1.9	*4.9*	0.4
Number of cases		26	3	57	22	2

Table 2.25 Clusters of everyday life models of the international student population (Rest Day)

Cluster center						
Type		1	2	3	4	5
Variable	Household activities	0.3	0.8	0.2	0.6	0.5
	Learning at residential location	*4.5*	2.1	0.3	0.9	0.2
	Learning outside	1.2	*4.8*	0.5	0.3	0.9
	Shopping activities	0.5	0.4	*5.2*	1.1	2.3
	Recreation activities at residential location	0.9	1.8	1.9	*4.9*	1.1
	Recreation activities outside	1.8	0.9	2.2	2.1	*7.5*
Number of cases		15	12	20	17	36

Table 2.26 Three typical clusters of international students' everyday life cases

Type	Primary characteristics	Sample size
Learning at residential locations	The students studied at their residential locations for 6–8 h daily, ventured out for a short and fixed period of time, and traveled nearby	23 (21.9%)
Learning outside	The students studied for 6–8 h daily, which correspond to their class timings	55 (52.4%)
Engaging in outdoor recreations activities	The students engaged in 5 h of recreation activities daily and traveled within the city block area	20 (19.0%)

These three groups comprise 93.3% of the overall sample size, with neither cross-interactions nor overlaps. These can encompass and represent most everyday life models of the international student body.

9.0% student characteristics of the everyday life of international students who learn at their residential locations

Compared with the general international student's everyday life space–time characteristics, this type of international students' everyday life displays the following characteristics (see Fig. 2.32). The student engages in learning activities for a rather long period of time in a day (around 6–8 h), far higher than other activities besides sleeping. Most of the time, such students will stay at their residential location and will only venture out during meal times and in the evening. Their outdoor activities are mostly dominated by shopping and recreation activities, the duration of which are rather short and which occur mostly in the surrounding community.

The attributes of international students who fall under this category are majoritarily women (the ratio of men to women is 1:2.5); mostly from Japan, Korea, and Southeast Asia; primarily studying for their master's degrees or Ph.Ds; mostly staying outside of campus; varying consumption levels; and relatively low integration levels.

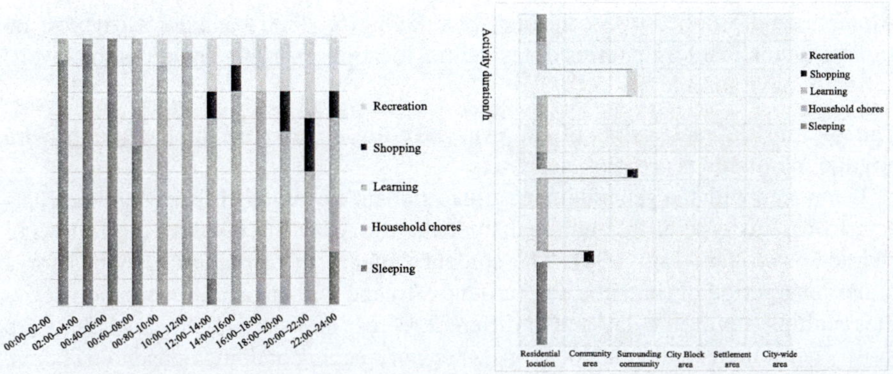

Fig. 2.32 Proportion of activities undertaken by international students who Learn at their residential locations and the trajectory of activities. *Source* Data from the research group's sample surveys on the everyday life space of International Student Settlements in Nanjing (2017)

Space–time characteristics of the everyday life of international students who learn outside

Compared with the general international student's everyday life space–time characteristics, this type of international students' everyday life displays the following characteristics (see Fig. 2.33). The student engages in learning activities for a rather long period of time in a day (around 6–8 h). Such students lead extremely regular lifestyles and allocate appropriate time for learning and resting. Besides sleeping and doing household chores, most of their time is not spent at their residential locations, though the distance traveled tends to be within the city block area.

The attributes of international students who fall under this category are as follows: no significant inclinations toward either gender; most are from Europe, North

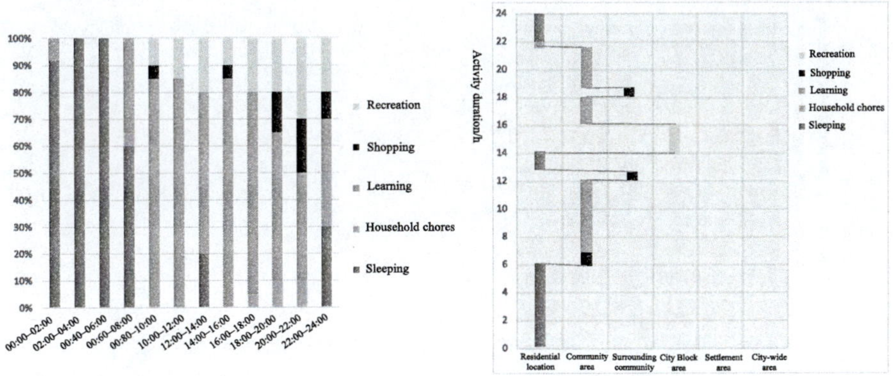

Fig. 2.33 Proportion of activities undertaken by international students who learn outside and the trajectory of activities. *Source* Data from the research group's sample surveys on the everyday life space of International Student Settlements in Nanjing (2017)

America, and Southeast Asia; primarily studying for their bachelor's degrees; no significant inclinations in terms of residential locations; varying consumption levels; and mid-integration levels.

Space–time characteristics of the everyday life of international students who engage in outside recreation activities

Compared with the general international student's everyday life space–time characteristics, this type of international students' everyday life displays the following characteristics (see Fig. 2.34). The student engages in recreation activities for a rather long period of time in a day (around 5 h) and will engage in recreation activities multiple times in a day and at different recreation locations. Such students are mostly not at their residential locations when not sleeping or doing household chores, though they will usually travel within the city block area. Their learning time in the day are also relatively scattered, and they have a habit of learning at night.

The attributes of international students who fall under this category are mostly male (the ratio of males to females is 3.5:1); mostly from Europe, North America, Japan, and Korea; primarily bachelor's degree and non-degree students; majority stay outside of campus; relatively high consumption levels; mostly mid- and high integration levels.

Second, clustering of everyday activities of international student clusters on a rest day

Of the preliminary five clusters, slight adjustments were made to cases based on the actual samples of the various activities, ultimately deriving five typical clusters of international students on a rest day (see Table 2.27). The five typical clusters represent 95.2% of the overall sample size, with neither cross-interactions nor overlaps. These can encompass and represent most of the everyday life models of international students on a rest day.

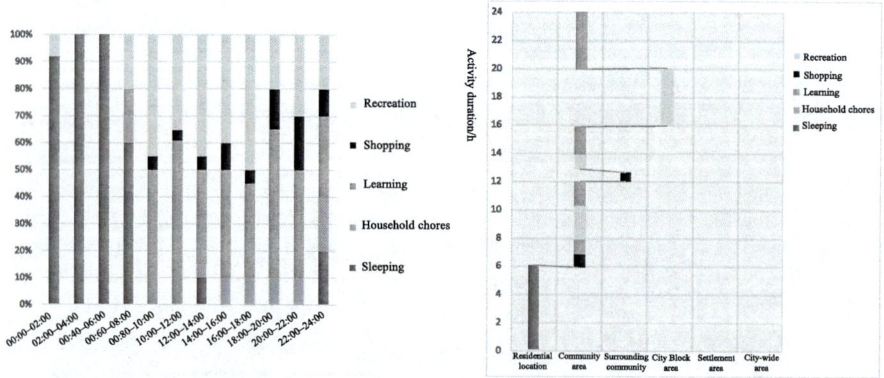

Fig. 2.34 Proportion of activities undertaken by international students who engage in outside recreation activities and the trajectory of activities. *Source* Data from the research group's sample surveys on the everyday life space of International Student Settlements in Nanjing (2017)

Table 2.27 Five typical clusters of international students' everyday lives

Type	Primary characteristics	Sample size
Learning at residential locations	The student learns at home for 5–6 h daily, ventures out for a short, fixed period of time, and travels nearby	15 (14.3%)
Learning outside	The student learns for 5–6 h daily and has relatively freer time compared to a work day	12 (11.4%)
Engaging in shopping activities	The students shops for around 5 h and is inclined toward shopping for a long time and at farther distances	20 (19.0%)
Engaging in recreation activities at residential location	The student engages in recreation activities at home for 4–5 h in a day and spends 95% of his/her time at home	17 (16.2%)
Engaging in outdoor recreations activities	The student engages in recreation activities for around 5 h in a day and travels within the city block area	36 (34.3%)

4.3% students engage in recreation activities for around 5 h a day and travel within the city block locations

Compared to the general international student's everyday life space–time characteristics, this type of international students' everyday life displays the following characteristics (see Fig. 2.35). The student engages in learning activities for a shorter duration compared to a work day but nonetheless is still long overall (around 5 h) and is far longer than other everyday activities besides sleeping. Such students are mostly at their residential locations and will only go out during meal times and in the evening. Their outside activities are also primarily dominated by shopping and

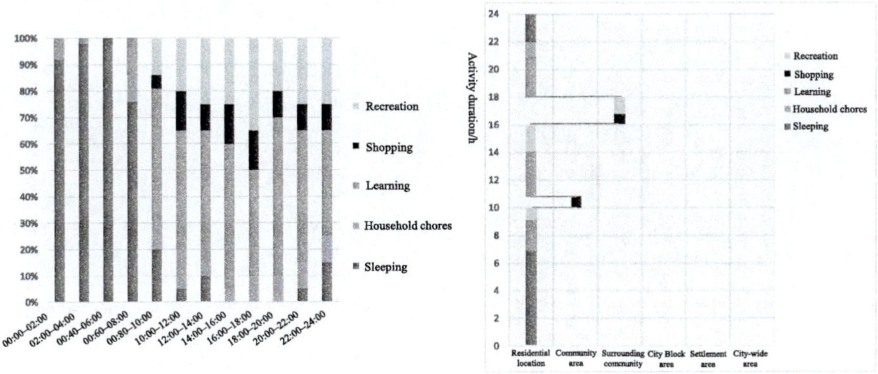

Fig. 2.35 Proportion of activities undertaken by international students who learn at their residential locations and the trajectory of activities. *Source* Data from the research group's sample surveys on the everyday life space of International Student Settlements in Nanjing (2017)

recreational activities, which take up relatively short periods of time and mostly occur within the surrounding community.

The attributes of students who fall under this category are as follows: no significant gender inclinations; mostly from Europe, North America, and Southeast Asia; pursuing the different types of degrees; primarily staying outside of campus; varying consumption levels; and have mid and low integration levels.

Space–time characteristics of the everyday life of international students who learn outside

Compared with the general international student's everyday life space–time characteristics, this type of international students' everyday life displays the following characteristics (see Fig. 2.36). The student engages in learning activities for a shorter duration compared to a work day but nonetheless is still long overall, around 5–6 h in a day. Such students lead extremely regular lifestyles and allocate appropriate time for learning and resting. Besides sleeping and doing household chores, most of their time is not spent at their residential locations. The distances traveled tended to be within the city block area.

The attributes of international students who fall under this category are as follows: no significant inclinations toward either gender; most are from Southeast Asia; primarily studying for their bachelor's or master's degrees; no significant inclinations in terms of residential locations; varying consumption levels; and mid-integration levels.

Space–time characteristics of the everyday life of international students who engage in shopping activities

Compared with the general international student's everyday life space–time characteristics, this type of international students' everyday life displays the following characteristics (see Fig. 2.37). The student engages in shopping activities for a longer

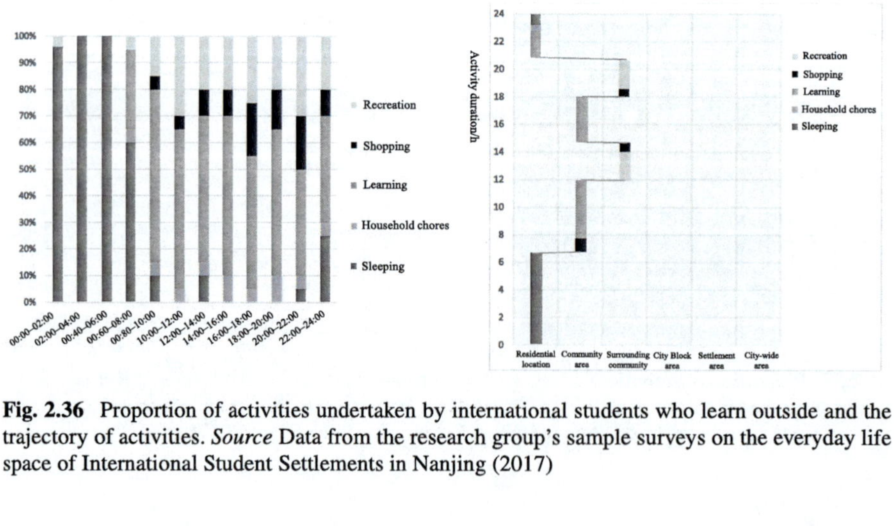

Fig. 2.36 Proportion of activities undertaken by international students who learn outside and the trajectory of activities. *Source* Data from the research group's sample surveys on the everyday life space of International Student Settlements in Nanjing (2017)

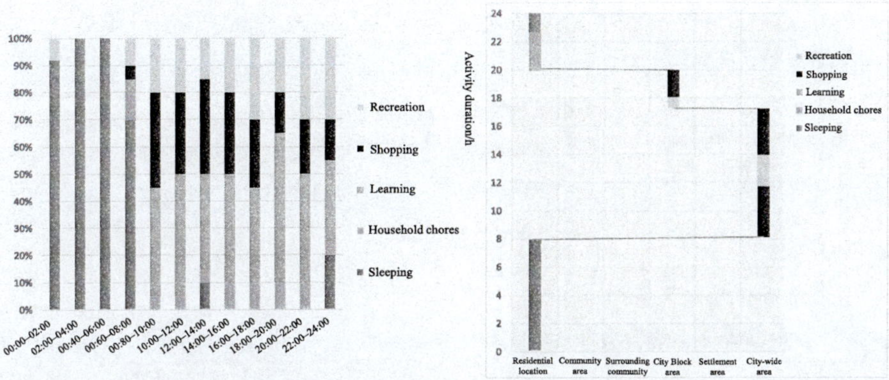

Fig. 2.37 Proportion of activities undertaken by international students who engage in shopping activities and the trajectory of activities. *Source* Data from the research group's sample surveys on the everyday life space of International Student Settlements in Nanjing (2017)

time than other types of students, and the activity is also more concentrated on a rest day than a work day, typically around 4 h in a day (mostly in the day time). Such students are inclined toward shopping at far distances and will usually choose shopping spots outside the community. When not sleeping or doing household chores, they are mostly not at their residential location and have a habit of studying at night.

The attributes of international students who fall under this category are mostly female (the ratio of males to females is 1:4); mostly from Europe, North America, Japan, and Korea; primarily pursuing bachelor's or master's degrees; no significant inclinations toward any residential locations; relatively high consumption levels; and have mid-integration levels.

Space–time characteristics of the everyday life of international students who engage in recreational activities at their residential locations

Compared with the general international student's everyday life space–time characteristics, this type of international students' everyday life displays the following characteristics (see Fig. 2.38). The student engages in recreation activities for a long period of time in a day, around 5 h in a day, and will engage in recreation activities multiple times in a day and mostly at their residential location. Such students engage in most of their activities at their residential locations. Their learning time in the day are relatively scattered, and they have a habit of learning at night.

The attributes of international students who fall under this category are mostly female (the ratio of males to females is 1:3); mostly from Southeast Asian and African countries; primarily pursuing master's or doctorate degrees; majority stay outside of campus; mid-consumption levels; and have low and mid-integration levels.

Space–time characteristics of the everyday life of international students who engage in outside recreation activities

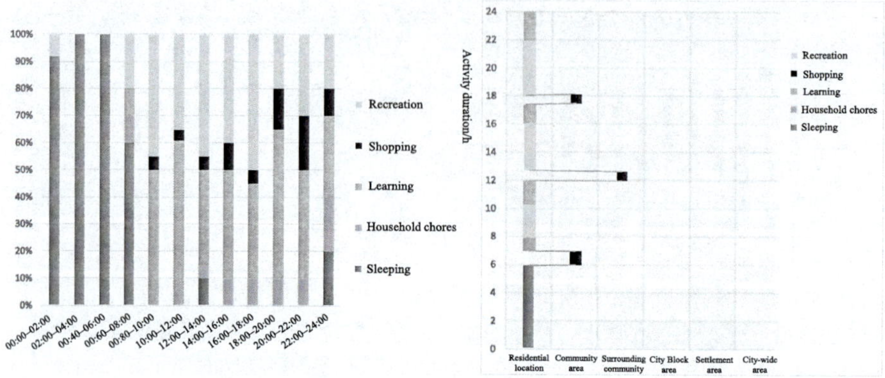

Fig. 2.38 Proportion of activities undertaken by international students who engage in recreation activities at residential locations and the trajectory of activities *Source* Data from the research group's sample surveys on the everyday life space of International Student Settlements in Nanjing (2017)

Compared with the space–time characteristics of ordinary international students' daily life, their daily life has the following characteristics (see Fig. 2.39). They spend a long time in leisure activities (including day and night), and the activity time is relatively concentrated, but the location is different; most of the time except bedtime and housework, they are not in their residence, and the travel distance is relatively long. The study time of this kind of students is relatively scattered in the daytime and less in the night.

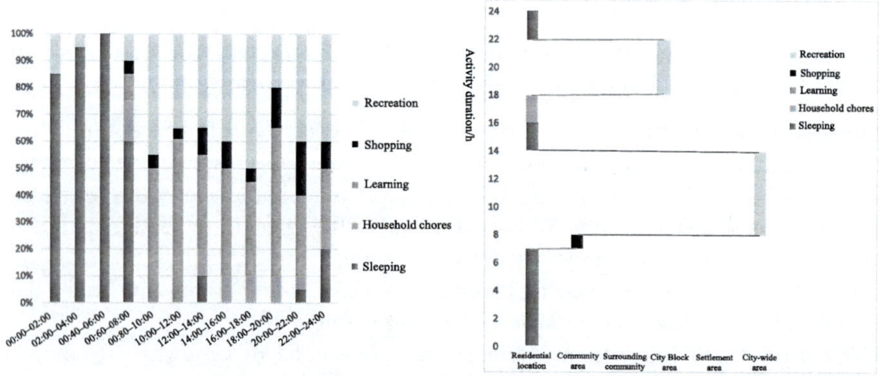

Fig. 2.39 Proportion of activities undertaken by international students who engage in outside recreation activities at different time periods on a rest day and the example of daily activities trajectory. *Source* Data from a sampling survey of daily life space in Nanjing International Student Settlements by the research team (2017)

The corresponding group attributes of international students include the following: no obvious gender tendency, mostly from Europe and the United States, mainly undergraduate and nonacademic students, no obvious tendency of living place, including all kinds of consumption level groups, and the degree of integration is at the medium level.

2.4 Response from the Surrounding Communities to the International Student Settlements

From the perspective of residents, operators and public service personnel of Nanxiu Village and its surrounding communities, this section will analyze the response of its surrounding communities in terms of social attributes, spatial layout, industrial economy, identity, and communication, which are strongly related to international student settlements and can reflect the characteristics of "internationalization." A dynamic analysis of the response of its surrounding communities from the following four aspects is conducted—the evolution of the International Student Settlements in Nanxiu Village is divided into three stages, which is close to Nanjing University and Nanjing Normal University, where the relative concentration of international students. The off-campus community is used as a research space (see Fig. 2.40); comprehensive application of interview method, field observation method, comparative analysis method, statistical analysis method, and other methods reveal the evolution and changes of surrounding communities before and after the formation and

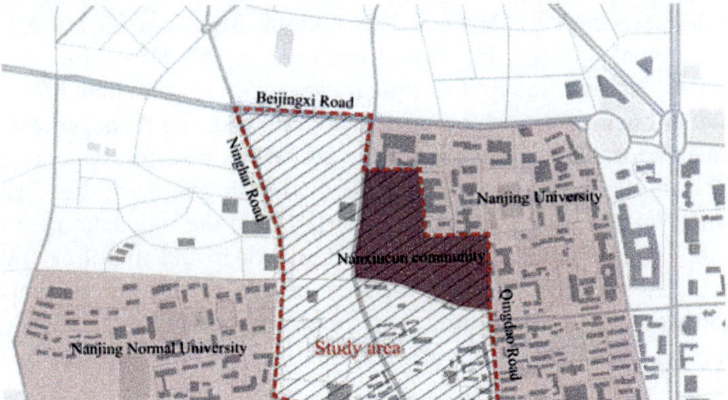

Fig. 2.40 Research space scope. *Source* Own elaboration

development of International Student Settlements through cross-stage longitudinal comparison.

2.4.1 Evolution Background and Stage Division

2.4.1.1 The Formation and Evolution of the International Student Settlements in Nanxiu Village

First, the policy changes of the introduction of international students by the state

The earliest history of China's acceptance of international students can be traced back to the Warring States Period. In the prime of Tang Dynasty, more than 1000 people studied in Chang'an, and the number of international students who came to China during the Ming Dynasty reached as many as 3000.[9] After the founding of the People's Republic of China, it accepted the first batch of 33 students from Eastern European countries in 1950. Now it has experienced more than 60 years of ups and downs. Not only has the program for international students made great progress from policy to management but the scale has also grown from the beginning. Dozens of people have increased to more than 400,000 people now. Its related policies have also undergone five stages of changes.

Initial test stage (1950–1965) My country began to recruit international students in a planned way. At that time, the students were mainly from Eastern Europe, neighboring countries, Asia, Africa, and Latin America, and the number was extremely limited for the purpose of exchanges. The "Interim Measures for the Administration of International Students in China from the People's Democratic Countries" promulgated in 1955 was the first administrative measure formulated for the international student community, which provided a basis for the management of international students for the purpose of friendly exchanges with foreign countries; this was approved in 1962. The first statutory document in the field, the "Trial Regulations on the Work of Foreign Students (Draft)," laid the foundation for the legal management of international students. The international student policy at this stage is basically formulated separately for different countries, and various management tasks are jointly completed by multiple departments, scattered, and lacking in system, and all aspects are still in the exploratory stage.

Complete stagnation stage (1966–1977) The Cultural Revolution completely interrupted the originally imperfect enrollment plan for international students. It was not until the end of the Cultural Revolution that the country slowly resumed the enrollment plan for international students, but no new policies were issued in the short term.

[9] Jin Xiaoda. Introduction to Education for Foreign Students [M]. Beijing: Sinolingua Press, 1998.

Entering the right track stage (1978–1988) After the economic reform and opening up, foreign exchange and cooperation have been developed on an unprecedented scale. A series of policies on the education of international students have been issued successively (such as the "Measures for the Administration of Foreign Students" in 1985), gradually establishing a three-level management system of the central government, provinces, cities, and universities and a management mechanism of coordination and cooperation among government departments. The plan is on the right track, which also provides conditions and possibilities for the expansion of the scale of international students under the new situation.

Comprehensively systematic stage (1989–2009) The state has gradually liberalized the management of the education of foreign students in China, and universities have gradually changed from absolute obedience to government arrangements to self-sponsored learning, which has led to a rapid increase in the number of international students. The "Regulations on Recruitment of Self-financed Foreign Students in China" promulgated in 1989 devolved the enrollment power to universities and liberalized the right of general higher education institutions to recruit self-financed international students. International student education has also begun to shift from international compulsory assistance to self-finance. Primarily, it has greatly promoted the development of the education of foreign students in our country and has started the rapid increase in the number of international students in China; in 1996, the China Scholarship Council was formally established, marking that the education of foreign students in China will be managed by the government from a centralized and unified management to the government's macro management. The transformation of autonomous management of colleges and universities; the "Administrative Rules on the Acceptance of Foreign Students by Colleges and Universities" promulgated in 2000 provides legal protection for the education and management of foreign students in China. At this stage, my country began to systematically formulate various policies based on the overall development of international student education in China, which led to the extraordinary development of international student education in China. From more than 7000 in 1990 to more than 200,000 in 2009 marked an increase of nearly 30 times.

The stage of improvement and promotion (2010 to now) The national plan for international students' enrollment and collection tends to be improved, and the number of international students has steadily increased. In 2010, "the Plan of Studying Abroad in China" proposed that "we should establish a system of studying in China that is suitable for China's international status, education scale, and level." It can be seen that this stage emphasizes further building a good legal environment to complete and improve the system and mechanism for the development of studying abroad in China.

Second, the enrollment of international students in Jiangsu Province

The development of international student education in Jiangsu colleges and universities is roughly synchronized with the development of national international student education, and it also shows different phase characteristics: Jiangsu's international student education started in 1955, but until the economic reform and opening up, the number of international students accepted has been very small. Almost all scholarship students were recruited through government channels. During the Cultural Revolution (1966–1976), the enrollment of foreign students in Jiangsu Province was suspended and resumed in 1976; from 1978 to 1988, the number of foreign students in Jiangsu colleges and universities increased from 200 to 510, an increase of 1.5 times; in the 10 years from 1988 to 1988, the number of international students increased from 510 to 2497, an increase of 3.9 times, and the growth rate during the same period far exceeded that of Japan and the United States. After 2000, the number of international students in universities in Jiangsu Province entered a stage of rapid growth until 2006, and the number of international students in the province has reached nearly 9000. Since 2010, the Jiangsu Provincial Government has further opened up the "Jasmine Jiangsu Government Scholarship for Studying in Jiangsu." The number of international students has increased from 14,142 in 2010 to nearly 26,000 in 2015. Its scale has been ranked among the top four in the country in recent years, second only to Beijing, Shanghai, and Tianjin (Ying 2016) (see Fig. 2.41).

Third, the enrollment of international students in Nanjing University and Nanjing Normal University

Presently, the members of the International Student Settlements in Nanxiu Village are mainly from Nanjing University and Nanjing Normal University. The changes in the enrollment of international students in the two universities are on the one hand affected by my country's economic development and opening up to the outside world. On the other hand, it is closely related to national and local international student policy changes.

Nanjing University has been enrolling international students since 1955 and has so far received nearly 20,000 international students from more than 100 countries. After long-term teaching exploration and practice, it has formed the characteristics

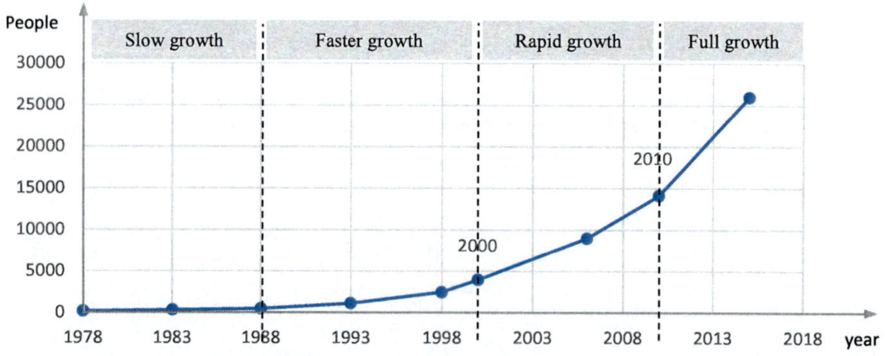

Fig. 2.41 Population growth trend of international students (degree students) in Universities of Jiangsu Province (1978–2015). *Source* Zhang Ying. Investigation and Research on the Current Situation of Foreign Students' Education in Universities of Jiangsu Province [D]. Nanjing: Nanjing Normal University, 2016

and advantages of cultivating international high-level Chinese talents. And a large number of "China Experts" and "Sinologists" trained are active in the political, academic, diplomatic, business, and military circles of various countries; Nanjing Normal University has recruited Vietnamese students since 1965 and has recruited international students from more than 100 countries. There are more than 10,000 international students. Since the 1980s, it has been designated as a supplementary base for southern Chinese, the only Chinese education base in Jiangsu Province, an HSK test center, and a national teaching base for Chinese as a foreign language.

The development of international students at Nanjing University and Nanjing Normal University is basically the same as the overall law of Jiangsu Province. Before 1990, the number of international students was small and the growth was slow; after 1990, the number of international students increased because of the country's relaxation of the policy for international students to come to China. Since the twenty-first century, as the country's policies on the employment and business of foreign students and foreigners in China have become more liberalized and relaxed, the international student population has formed a considerable scale; after 2010, the number and proportion of international students have significantly improved.

Fourth, the evolution of Nanxiu Village and surrounding communities

Nanxiu Village and its surrounding communities were first built around the 1980s, originally belonging to the unit compound of universities, surrounding enterprises, and public institutions or early commercial houses; in 1998, following the "State Council's Proclamation on Further Deepening the Reform of the Urban Housing System and Accelerating Housing Construction," the issuance of the "Notice" marked the withdrawal of the housing distribution system that has been in place for nearly 40 years. It also promoted the unit housing in the area to gradually move to the market. Around 2000, the number of renters in the area increased significantly, among which there were some problems. Not even a few international students from Europe, America, Japan, and South Korea and foreign teachers choose to live here; since 2010, the increasingly open policy of studying in China by the country and colleges and universities has brought more and more international students to this area, accompanied by more and more foreigners operating and opening stores in Nanxiu Village and its nearby communities.

From the analysis above, it can be seen that it is the mutual influence and joint effect of the country's policy pertaining to studying in China, the policy and scale of international students in Jiangsu Province, the scale of international students from Nanjing University and Nanjing Normal University, and the development of Nanxiu Village and surrounding communities that promoted the gradual formation and development of the International Student Settlements in Nanxiu Village.

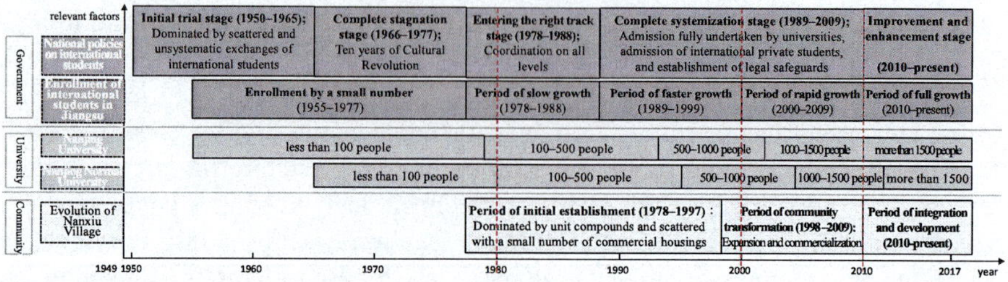

Fig. 2.42 Schematic diagram of the evolutionary background stage of the international student settlements in Nanxiu Village (since 1949). *Source* Own elaboration

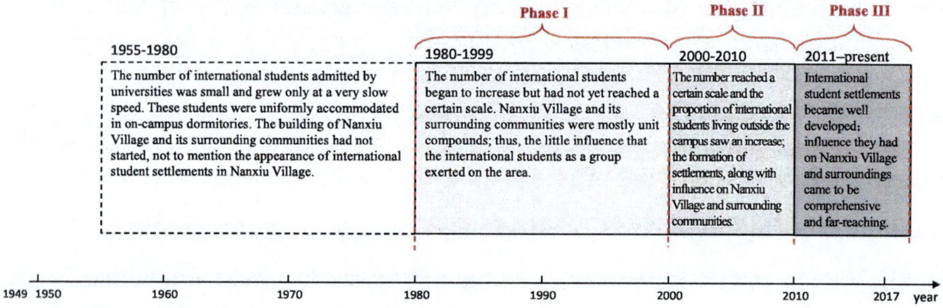

Fig. 2.43 The evolution stage division of the international student settlements in Nanxiu Village. *Source* Own elaboration

2.4.1.2 The Evolutionary Stages of the International Student Settlements in Nanxiu Village

The emergence and growth of International Student Settlements under the wave of globalization is an inevitable product of my country's contemporary urban development and social progress, and its emergence is bound to have multiple impacts on the local area. The author summarizes and analyzes the national international student policy, the recruitment of international students in Jiangsu Province, the scale of international students from Nanjing University and Nanjing Normal University, and the evolution of Nanxiu Village and surrounding communities (see Fig. 2.42). The evolution of the International Student Settlements is divided into three stages[10] (see Fig. 2.43): the unformed stage (Stage I: 1980–1999), development stage (Stage II: 2000–2010), and mature and perfect stage (Stage III: 2011 to present).

Based on the stage division and combined with the first-hand data of the research object—International Student Settlements and its surrounding communities—the author focuses on the "internationalization" characteristics that are closely related to the International Student Settlements. This is undertaken by starting from the four

[10] Before 1980, the number of international students in Nanjing University and Nanjing Normal University did not form a large scale, and NanxiuVillage and its surrounding communities had not yet been established. Therefore, it has not been considered.

aspects of social attributes, space layout, industrial economy, and identity communication, combining qualitative and quantitative historical analysis and dynamic research, aiming to explore the multiple impacts of the emergence of international student groups on the local area, and then providing the next step, which is the reference basis for problem conclusion and strategy formulation.

2.4.2 Community Response to Social Attributes

2.4.2.1 The Response of Nationality Composition

The nationality composition of the International Student Settlements in Nanxiu Village shows the response characteristics of "**a significant increase in the scale of international students, a diverse composition, and community differentiation**" (see Table 2.28, Fig. 2.44).

The response to its nationality composition mainly stems from the following three aspects (see Fig. 2.45).

First, the liberalization of the policy for international students in China After 2010, my country has further opened up and improved the education, employment, and service systems for international students. For example, my country provides various preferential scholarship systems for international students as well as counterpart education support for some underdeveloped countries, which has attracted more and more international students. International students from many countries, especially the countries along the "the Belt and Road" in recent years, have become the main exporting countries for international students in China. All these measures have resulted in a rapid increase in the number of international students.

Second, the impact of the objective conditions of the local community On the one hand, the aboriginal population in Nanxiu Village and its nearby communities is seriously aging and even has reversed population growth. The gathering of international students quickly and effectively changed the demographic composition of the local community; on the other hand, this stems from the location of the community. Depending on the conditions, Nanxiu Village and Hujuguan communities are both located in the first radiation circle of Nanjing University and Nanjing Normal University, with obvious geographical advantages. Not only the number of international students has increased rapidly but the living area has also continued to increase with the increase in international students. Expanding outwards, while the international students in Qingdao Road and Suzhou Road communities in the second radiation circle started late, grew slowly, and were relatively limited in scale, which directly led to the differentiation between neighboring communities.

Third, the increase of China's international influence On the one hand, since the economic reform and opening up, with the continuous development of China's scientific research economy and the continuous improvement of overall national strength, the educational resources of many universities have gradually been recognized in the international education industry, and this has attracted more and more foreigners to come to China. On the other hand, with the continuous improvement of China's cultural exchanges with foreign countries, the broad and profound Chinese culture has generally stimulated great interest in Chinese culture among students from all over the world, thereby promoting the diversification of the nationality composition of international students in China.

Table 2.28 The evolution of the nationality composition of the international student settlements in Nanxiu Village

Time phase	Nationality composition (Overall and partial)	Response characteristics by stage
Phase I: 1980–1999	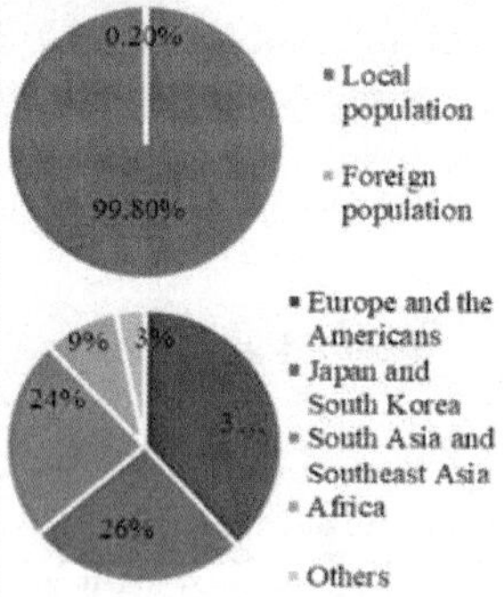	• In the first 10 years of this stage, the number of international students in the community was very small (only a dozen), and the growth rate was slow; however, in the next 10 years, the number of international students and the proportion of the community gradually increased • Mainly foreign students from Europe, America, Japan, Korea, and Southeast Asia live here • There are differences between different communities. Among them, the Nanxiu Village community has the fastest growth rate and the largest scale of international students (reaching 2% of the community population), while the Suzhou Road and Qingdao Road communities have not yet started
Phase II: 2000–2010	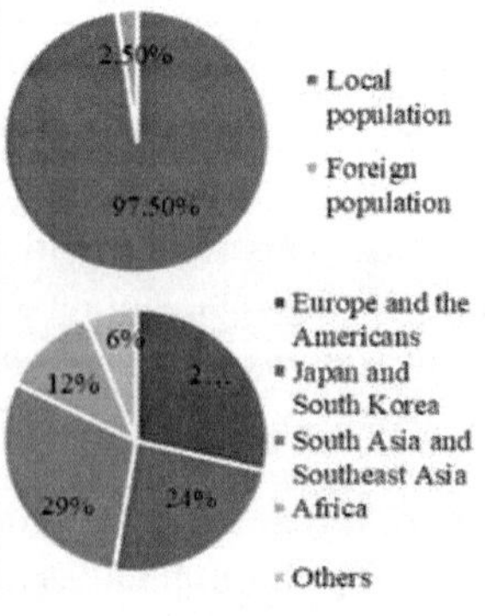	• At this stage, the international students in the community have entered a period of rapid growth, accounting for about 4 to 5% of the total population of the community • The distribution of nationalities is showing an increasingly diversified trend, mainly in Europe, America, Southeast Asia, Japan, and South Korea, and the proportion of Africa and other countries has also increased • The differences between communities have further widened. The proportion of international students in Nanxiu Village community (6%) is significantly higher than that of the other three communities, followed by Hujuguan community (4.5%) and the proportion of Qingdao Road Community and Suzhou Road Community has also increased

(continued)

Table 2.28 (continued)

Time phase	Nationality composition (Overall and partial)	Response characteristics by stage
Phase III: 2011 to now	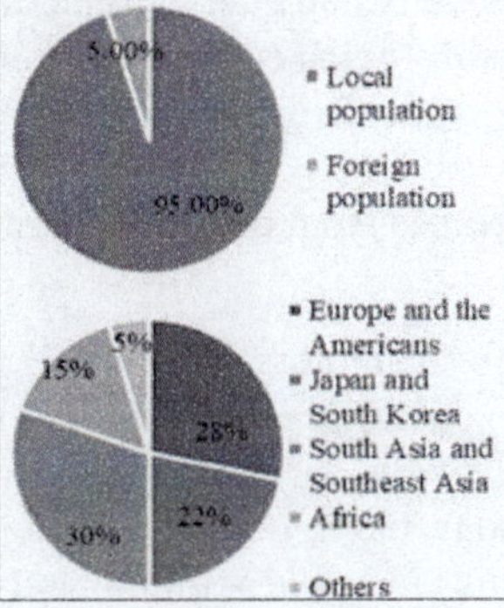	• The scale of international students at this stage is still expanding, but the growth rate is slowing down • The nationality distribution of international students is becoming more balanced and diverse • In terms of community differences, the number of international students in the Nanxiu Village community has reached about 7%, while the other three communities are all between 4 and 5%

Source Sampling survey data of the research team on the International Student Settlements in Nanxiu Village (2017)

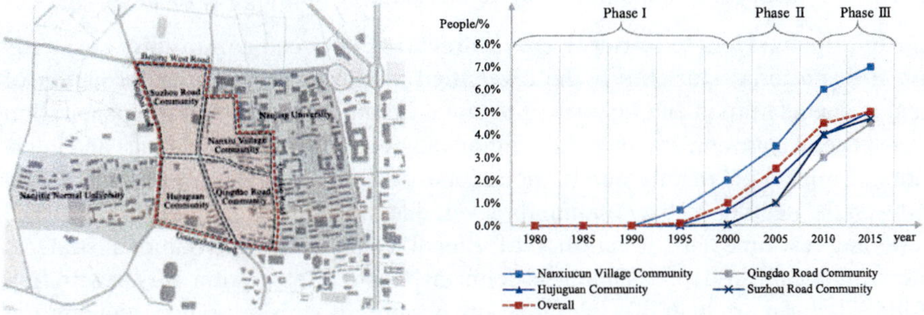

Fig. 2.44 Evolution of the proportion of international students in Nanxiu Village and surrounding communities (since 1980). *Source* Data from a sample survey of everyday life space in Nanjing International Student Settlements by the research team (2017)

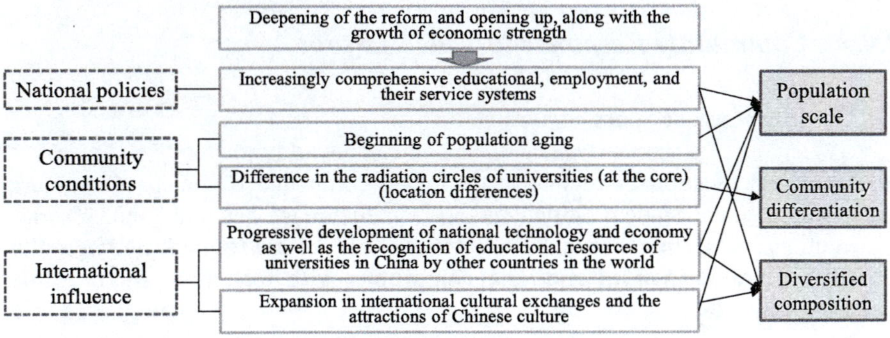

Fig. 2.45 Response analysis of the nationality composition of Nanxiu Village and its surrounding communities. *Source* Own elaboration

2.4.2.2 The Response of Life Customs

Because of the influx and gathering of international students in the International Student Settlements in Nanxiu Village, local residents and international students have shown the response characteristics and "**increasingly rich customs composition and increasing customary participation but religious activities are isolated from each other**" (see Table 2.29).

The response to their life customs mainly comes from the following two aspects (see Fig. 2.46).

First, cultural penetration and integration in the context of globalization In the era of informationization and networking, the exchanges between various nations and countries have become more frequent, which affects our daily life, food, clothing, housing, and transportation all the time. Among them, the most influential is the global spread of Western cultural concepts. This kind of penetration and integration is most obvious in food and festivals, and it changes the existing lifestyles of Chinese people, presenting a complex and unique duality—the positive effect of enriching the cultural composition and the negative influence of diluting the local culture.

Second, the barriers to in-depth communication between community residents and international students If the diversified participation and transformation of local residents in food and festivals is to some extent caused by cultural penetration in the general environment, then the religious segregation of local residents and international students is mainly due to the second—the superficiality of the interaction between the people. At the community level, due to language, cultural background, work, and rest time, etc., it is difficult for local residents and international students to communicate in depth. This also confirms that the integration and penetration of ideas is far more than lifestyle. It is much more difficult to change, and cultural integration also encounters deeper isolation and exclusive inertia than participatory integration and functional integration.

2.4.3 Community Response to Space Layout

2.4.3.1 Response of Space Structure

As the overall functional structure of the city is compounded, the space structure of the International Student Settlements in Nanxiu Village has undergone a series of evolutions, presenting response characteristics of "**from a single function and a closed group to an open international community with dots, lines, and multiple functions**" (see Table 2.30).

The response of the space structure mainly comes from the following four aspects.

First, the connections among communities become stronger Since 2000, Nanjing's urban planning system has gradually improved, and the urban spatial

Table 2.29 The evolution of living customs in the Iinternational student settlements of Nanxiu Village

Phase	Acceptance of eating habits	Participation in festival activities	Understanding of religious activities	Phased response characteristics
Phase I: 1980–1999	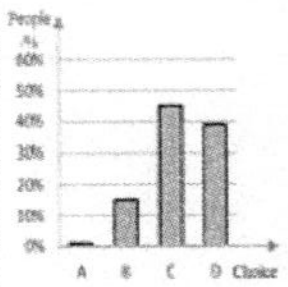	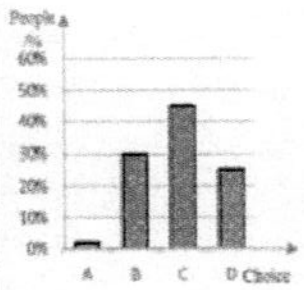	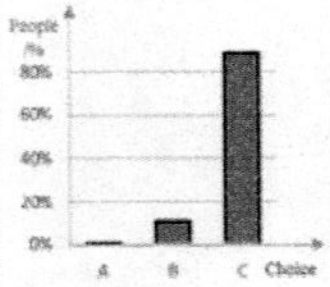	[Diet] Few people choose foreign food, and the frequency is low [Festival] Most people know something about it, but they don't know how to celebrate foreign festivals. A small number of people occasionally try and participate in it [Religion] Few people understand or pay attention to the religious activities of foreign students
Phase II: 2000–2010	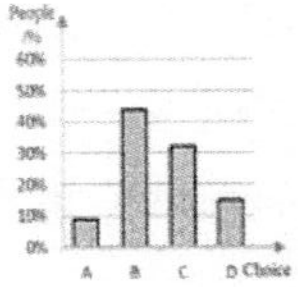	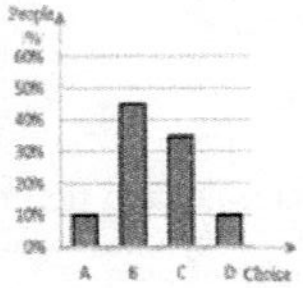	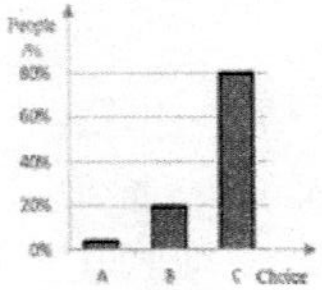	[Diet] Some people began to choose and accept foreign food [Festival] Most people know something about it, and their participation has increased [Religion] Generally speaking, they are still unfamiliar with the religious activities of foreign students, but a few people have begun to pay attention to them
Phase III: 2011 to now	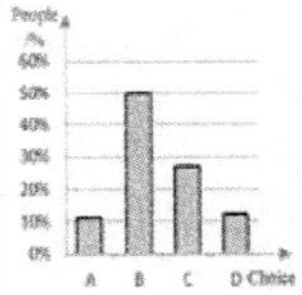	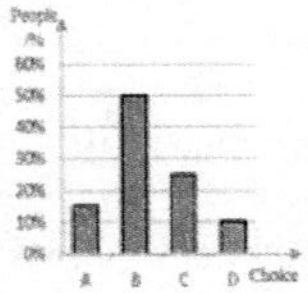	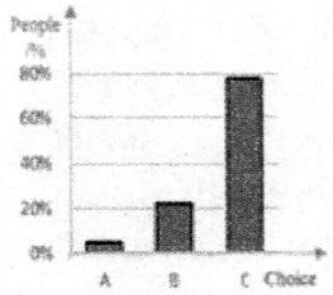	[Diet] The frequency of choosing exotic foods has generally increased [Festival] Further increase in understanding and participation [Religion] Generally speaking, the understanding of the religious activities of international students is still limited, and the level of attention is low

(continued)

Table 2.29 (continued)

Phase	Acceptance of eating habits	Participation in festival activities	Understanding of religious activities	Phased response characteristics
Interpretation	Frequency of foreign food selection: A. Once a week or more B. 1–2 times a month C. Once a few months D. Hardly	Understanding and participating in foreign festivals: A. Well understood and participate B. Understand the part and participate occasionally C. Knows a little, but won't participate D. Don't understand and participate	Do you pay attention to or understand the religious activities of international students? A. Very concerned and understanding B. Interested but don't understand C. Don't care and don't understand	

Source Sampling survey data of the research team on the International Student Settlements in Nanxiu Village (2017)

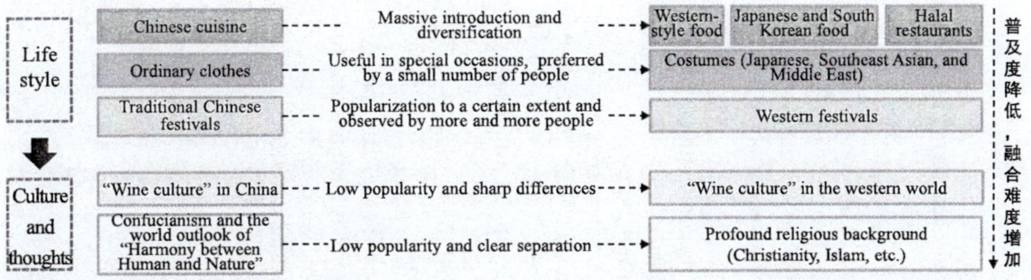

Fig. 2.46 Response analysis to the living customs of Nanxiu Village and its surrounding communities. *Source* Own elaboration

structure has also been optimized day by day. Urban land has begun to be developed in a mixed manner, resulting in many mixed-use land combining residential and supporting service facilities (see Fig. 2.47).The business development axis promoted by this has also strengthened the sharing and connection between communities to a certain extent.

Second, the development of the commercial service industry With the diversification and maturity of the population composition of Nanxiu Village and surrounding communities, the residents' increasingly complex and abundant daily needs have directly stimulated the growth and optimization of the category, scale, and layout of commercial service facilities in this area.

Third, the manifestation of international features With the expansion of the international student population, driven by the catering service industry with international characteristics, a certain scale and unique exotic commercial street has gradually formed in the local area, which has not only further strengthened the commercial agglomeration and attraction of the region but also enriched the functions of land use and closes space connections.

Fourth, the dilemma of old city renewal and development Because of the shortage of land, the increasing number of commercial service facilities in the old city can only use the original commercial stores or transform the bottom floor residence as the bottom business. The banded space layout cannot gather to form a larger scale under the constraints and squeeze of the actual conditions, which is not conducive to the formation of more regional radiation and "international" features of exotic blocks in this area.

2.4.3.2 Response of Public Service Facilities

From the evolution of the layout of public service facilities in the International Student Settlements of Nanxiu Village, it can be seen that the changes in its overall layout show response characteristics "**from scattered distribution to strip extension and from localized commercial service industry outlets to a public service system with international features**" (see Table 2.31).

Table 2.30 The evolution of the space structure of the international student settlements in Nanxiu Village

Phase	Space structure	Land use function	Phased response characteristics
Phase I: 1980–1999	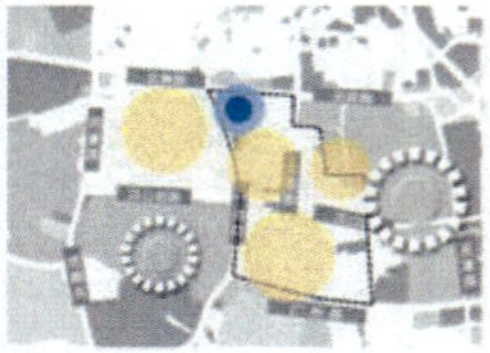	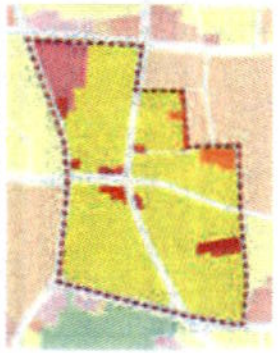	• The initial period of the construction of Nanxiu Village and surrounding communities • Relying on existing colleges and universities, various residential communities have been gradually established, mainly in the courtyards of enterprises and institutions, and some commercial houses are also available • The spatial structure of this period was mainly endogenous in the group, and the connection between the courtyards and groups was weak, but the internal functions of the group were self-contained
Phase II: 2000–2010	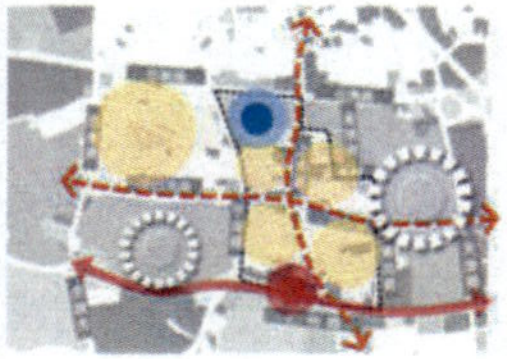		• The mature period of Nanxiu Village and surrounding communities is also the forming stage of International Student Settlements • With the transformation of urban spatial structure, the maturity of communities and the marketization of housing, Guangzhou Road has gradually developed into an urban secondary commercial belt • A business node with Suning Global Center as the core is formed, and two spatial connection axes of Shanghai Road and Hankou Road, which are mainly commercial functions, have been formed in the area, and the linear connection between the groups has been strengthened
Phase III: 2011 to now	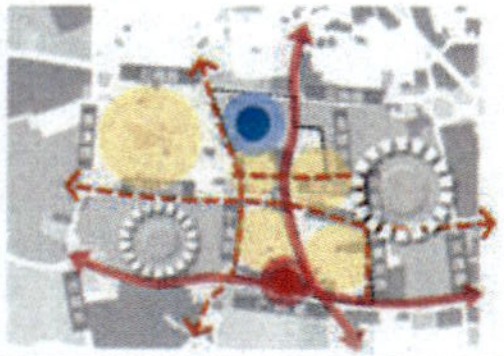		• The maturity of the International Student Settlements in Nanxiu Village • In the space structure built in the previous stage, the Ninghai Road spatial connection axis that strengthens the traffic connection has been formed, and the Nanxiu Village-Pingcang Lane spatial connection axis deep into the group, effectively improving the hierarchy of the district structure • There is still a lack of large-scale core nodes, which is also the focus of future structural optimization

(continued)

Table 2.30 (continued)

Phase	Space structure	Land use function	Phased response characteristics
Legend			

Source Sampling survey data of the research team on the International Student Settlements in Nanxiu Village (2017)

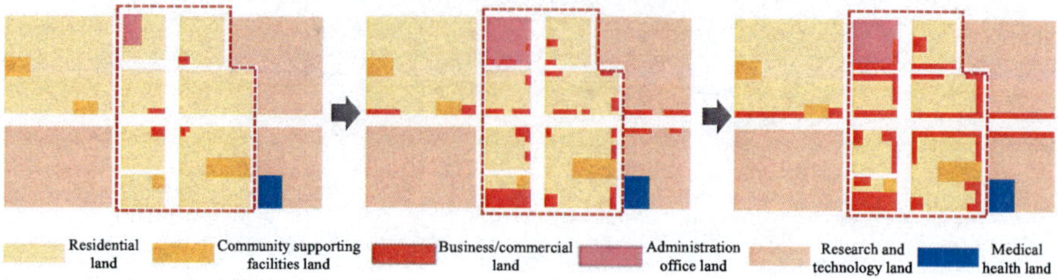

Fig. 2.47 Response analysis of the spatial structure of Nanxiu Village and its surrounding communities. *Source* Own elaboration

The response of its public service facilities mainly originates from the following two aspects:

First, the location of the city with superior conditions Nanxiu Village and surrounding communities were originally located in the old town of Nanjing, with convenient transportation, dense population, and diversified functions, numerous colleges and universities, and there is a huge demand and market for various public service facilities. Moreover, it has been affected by the acceleration and new development of Nanjing's urban construction since the 1980s. Driven by the agglomeration effect of urban public service centers, such as Xinjiekou and Hunan Road (see Fig. 2.48), the commercial service industry in the area (including service facilities with international features) has developed rapidly, relying on the continuous improvement of road traffic conditions show a trend of belt development.

Second, a diverse and integrated humanistic environment International Student Settlements in Nanxiu Village is different from ordinary communities. Because of the existence of universities, such as Nanjing University and Nanjing Normal University, it has gradually formed a more diverse and active cultural atmosphere and an increasing number of venues for various activities; it is also because of the influx of international student groups that present a state of multicultural integration, which has spawned many types of commercial service facilities with international features in the area and which actually reflects a state of multi-national cultural blending in different places.

2.4.4 Community Response of Industrial Economy

2.4.4.1 Response from Distinctive Industries

With the increasing expansion of the international student groups, more and more industrial types with exotic features have been born in Nanxiu Village and surrounding communities, and the most distinctive exotic commercial street in

Table 2.31 The evolution of public service facilities in the international student settlements of Nanxiu Village

Phase	General layout of public service facilities	Layout of international service facilities	Phased response characteristics
Phase I: 1980–1999	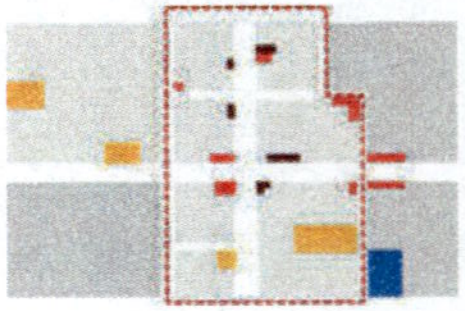	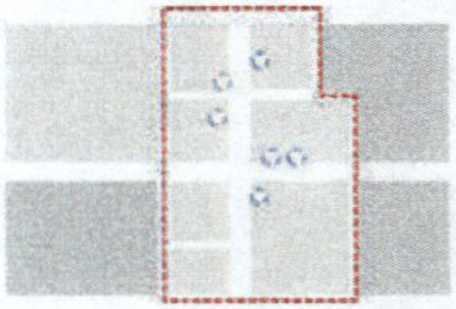	• In this stage, with the establishment of residential areas, a small number of supporting service facilities began to appear, schools and hospitals were equipped, and commercial facilities were mainly restaurants and shops, being scattered • There were very few international service facilities in this period
Phase II: 2000–2010	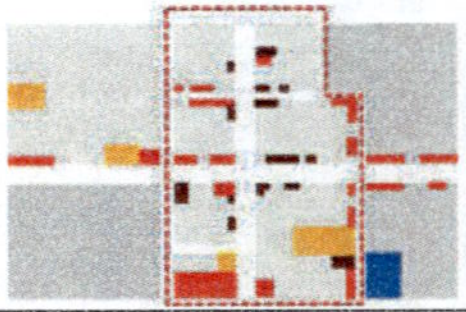	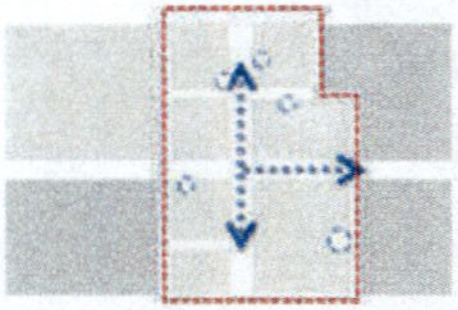	• In this stage, supporting service facilities began to increase, mainly along the street, but the layout was still scattered • During this period, service facilities with international characteristics began to appear and rapidly expanded, mainly along the street
Phase III: 2011 to now	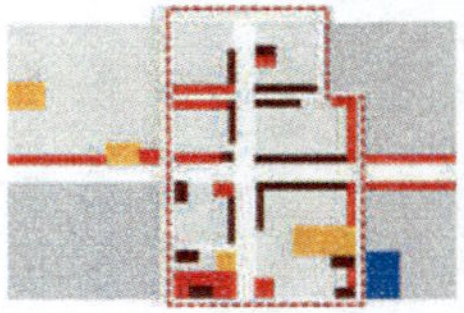	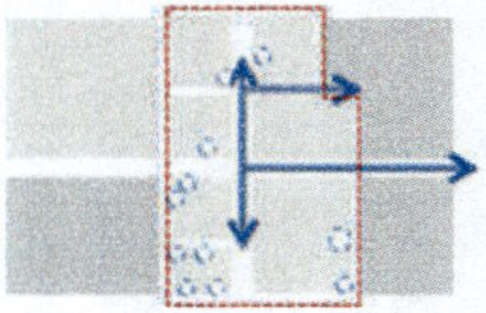	• In this stage, the supporting service facilities began to connect with each other, gradually forming the commercial service axis of the area, and the scale also expanded greatly • During this period, service facilities with international characteristics began to gather, forming a distinctive exotic food street
Legend	General facilities; Primary and secondary schools; Medical facilities; Foreign characteristic facilities; Site of service facilities for foreigners; Axis of featured businesses		

Source Sampling survey data of the International Student Settlements in Nanxiu Village by the research team (2017)

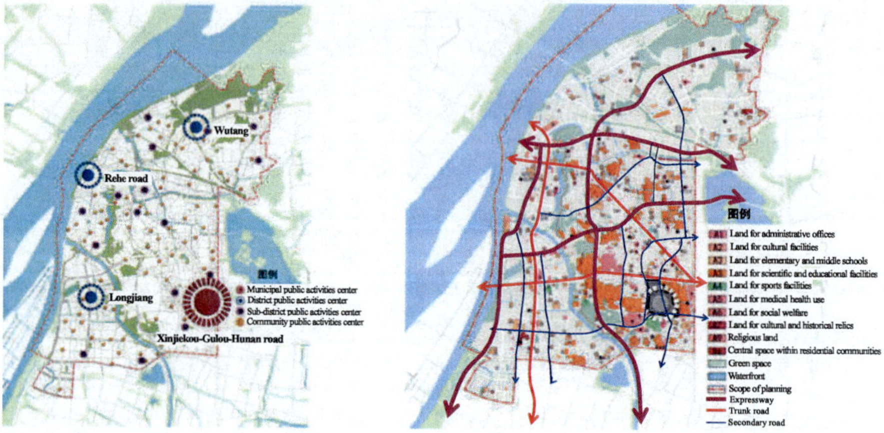

Fig. 2.48 Distribution map of public facilities center system and functions in Gulou District, Nanjing. *Source* Nanjing Gulou District Master Plan (2013–2030)

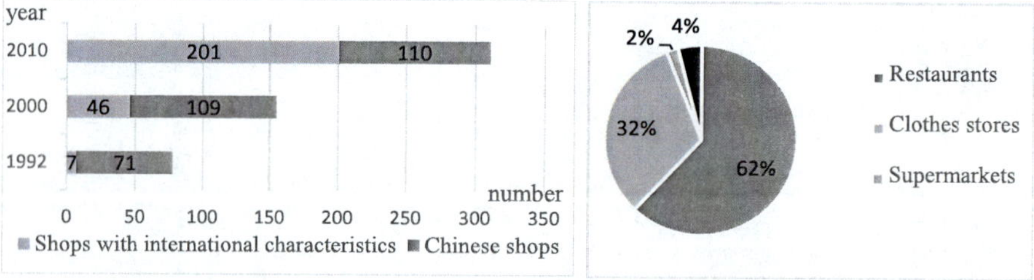

Fig. 2.49 Number of shops and business types in the international student settlements of Nanxiu Village. *Source* Sampling survey data of the research team on the International Student Settlements in Nanxiu Village (2017)

Nanjing has been formed (see Fig. 2.49). This represents the response characteristics "**in terms of the scale of the industry, the number and proportion of shops with exotic features have continued to rise; in terms of business type, it has shifted from a single catering type to a comprehensive type with catering and clothing stores as the mainstay as well as supermarkets and other entertainment facilities. Transition from European and American features to a diverse and inclusive international style with European, American, Japanese, Korean, and Southeast Asian features; in terms of space distribution, it presents a layout that extends from dots to lines and then to multi-axis**" (see Table 2.32).

The response of its distinctive industries may mainly derive from the following four aspects:

First, the strengthening of China-international education exchanges and cooperation The gathering of higher education schools and the country's increasingly opens policy to study in China have brought more and more foreigners to the region.

Table 2.32 Evolution of the industrial layout of the international student settlements in Nanxiu Village

Phase	Overall distribution of shops	Distinctive industry space structure	Phased response characteristics
Phase I: 1980–1999	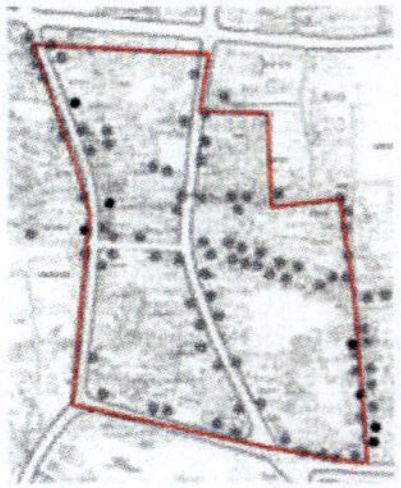	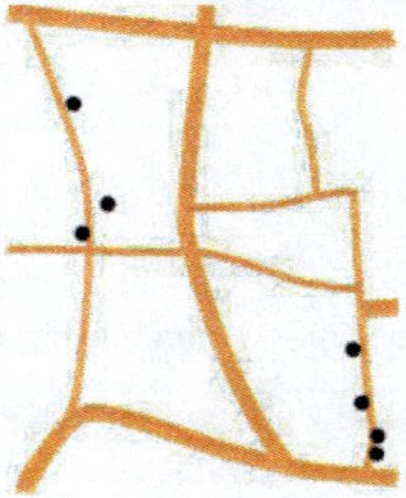	[Scale] The total number of shops is 78, of which only 7 are shops with exotic features [Types of Business] Generally, shops are mainly restaurants and small department stores, while shops with exotic features are all catering shops, and all of them are western restaurants [Distribution] The shops are mainly distributed along Ninghai Road, Shanghai Road, Hankou Road, Qingdao Road, and Guangzhou Road in a linear manner, while shops with exotic features are mainly distributed in dots on Ninghai Road and Qingdao Road
Phase II: 2000–2010	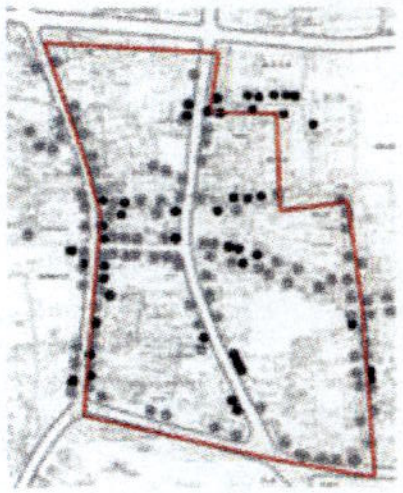	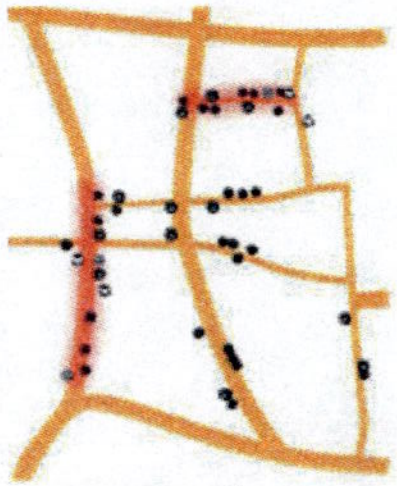	[Scale] The total number of shops has increased from 78 to 155, among which shops with exotic features have increased from 7 to 31, a nearly sevenfold increase [Business type] Catering is still the main business type, and clothing stores and supermarkets have begun to appear gradually, but exotic shops are still dominated by food and beverage stores, and a small number of clothing stores have appeared at the same time, and the trend of Japan and South Korea has also begun to prevail [Distribution] The number of shops on Ninghai Road has further increased, and exotic shops have also begun to gather on Jinyin Street and Ninghai Road in the north of Nanxiu Village, forming a partial linear distribution

(continued)

Table 2.32 (continued)

Phase	Overall distribution of shops	Distinctive industry space structure	Phased response characteristics
Phase III: 2011 to now		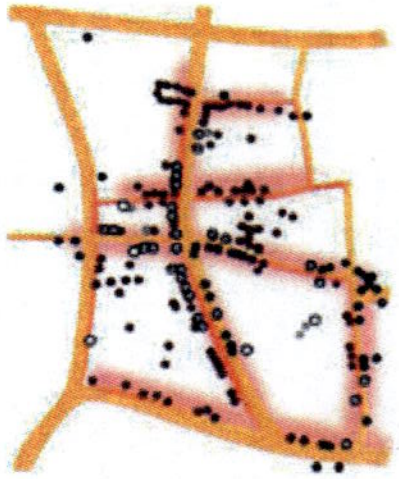	[Scale] The total number of shops has risen from 155 to 311, of which the total number of domestic shops is basically stable, while the number of shops with distinctive features in foreign countries continuing to maintain a rapid growth trend, which has increased by nearly five times [Business type] Catering and clothing stores have become the main business formats, with a small number of supermarkets and other entertainment facilities. Among them, exotic shops are mainly catering and clothing stores, and they have begun to present a blend of European, American, Japanese, Korean, and Southeast Asian styles [Distribution] The scale of shops along Ninghai Road has been significantly reduced, while Shanghai Road and Nanxiu Village surrounding exotic specialty shops have further amassed, and their proportion has increased significantly (29.6% to 64.6%). Generally, they have begun to show a multi-axis spatial layout
Legend	● Chinese shops ● Shops with international characteristics	● Restaurants ● Clothes stores ○ Supermarkets ● Other recreational facilities	

Source Sampling survey data of the International Student Settlements in Nanxiu Village by the research team (2017)

International students from developed countries such as Europe, America, Japan, and South Korea tend to rent out houses because they are not satisfied with the many constraints of the school's accommodation conditions; another part of the overseas students from underdeveloped countries chose to stay in China to start their own businesses, which has also brought about a chain migration of compatriots and work in China. The increasing number of foreigners and the gathering of shops that serve foreigners promote and influence each other, and the proportion of shops with exotic features in this area has been continuously increased and become the main body.

Second, the country's open policy to privately owned shops of foreigners The policy for foreigners to operate stores in China gradually opens, which has eliminated the differential treatment between foreign private owners and Chinese private owners. Consequently, foreign private owners have experienced an evolutionary process from "difficult to operate in China to operating a store with a Chinese partner to operating a store freely and equally." Among them, the scale of catering and clothing stores has expanded most rapidly. The single European and American style in the past has gradually transformed into the coexistence and tolerance of European, American, Japanese, Korean, and Southeast Asian styles (see Fig. 2.50).

Third, the renewal and transformation of old towns promoted by the government In the 1990s, Ninghai Road underwent an expansion. The narrow alleys between the walls of the original residential quarters were extended into urban streets with lots of shops. Because of the distribution of surrounding schools, shops gradually gather along Ninghai Road. In 2000, Ninghai Road had become the most lively and prosperous food shopping street in the area (including many international specialty restaurants). However, the demolition of Ninghai Road in 2000 forced many shops to

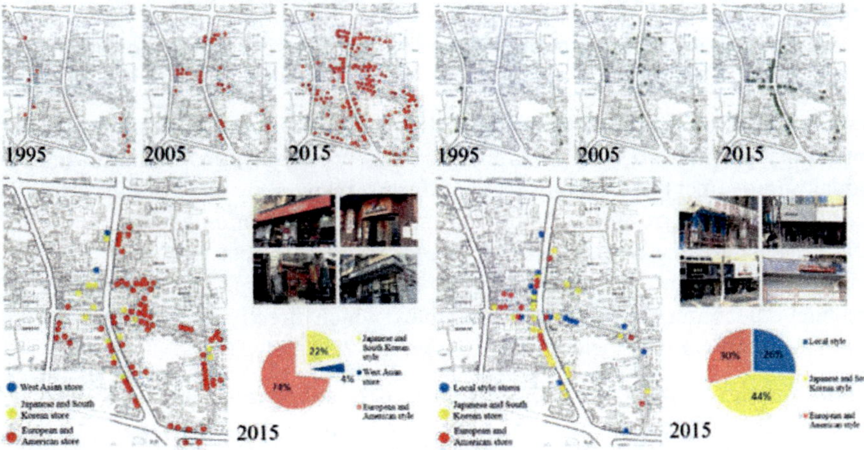

Fig. 2.50 Phased response schematic diagram of distinctive industries of international student settlements in Nanxiu Village (Left: restaurant; right: clothing store). *Source* Sampling survey data of the International Student Settlements in Nanxiu Village by the research team (2017)

move to the vicinity of Shanghai Road and Nanxiu Village. Objectively, it also caused the eastward movement and secondary agglomeration of foreign shops (mainly near Taogu New Village, Nanxiu Village, and Shanghai Road), which formed the current space layout of multi-axis extension.

Fourth, the changing consumption concepts of the local residents With the improvement of my country's economic level, community residents' consumption concepts have also changed, and they are willing to try and participate in newly sprouted things. In particular, the choices for food, shopping, festivals, and other activities have begun to be diversified and internationalized. To a certain degree, it also reversely promoted the development of distinctive industries in the region. Consequently, the gradually emerging and beginning to take shape of international business has promoted employment and tourism in the community and has become one of the most distinctive and recognizable international places in Nanjing.

2.4.4.2 The Response of Employment Structure

Nanxiu Village and its surrounding communities are based on the core location of Nanjing's scientific research and education. At the beginning of the community's formation, the employment structure was positioned to be dominated by education, scientific research (including student groups), and public management supplemented by other employment directions. Nowadays, a diversified and complicated employment structure with equal emphasis on education and scientific research, public management and business services has been formed. In other words, the employment structure of this area shows response characteristics "from a single industry dominance has gradually developed into a complex employment pattern with equal emphasis on multiple industries" (see Table 2.33).

The response of its employment structure mainly derives from the following two aspects:

First, the process of housing marketization brings about the deunitization of community population With the country's socio-economic transformation and the gradual improvement of housing system reforms, many units and courtyards in the area have separated from their units and moved toward marketization and socialization. Housing transactions have increased, property rights have been replaced frequently, and the number of renters has continued to rise. The composition of personnel is becoming more and more complex, which directly affects and mixes the employment structure of the community (see Fig. 2.51).

Second, the new employment opportunities brought by the increasing number of international students Since the twenty-first century, with the rapid growth of the number of foreigners coming to China, the country has continued to liberalize and loosen policies for the employment and business of international students and other foreigners in China. More and more foreigners settled down and operated shops in Nanxiu Village and its nearby communities offer business opportunities, vitality, and an increasing number of exotic shops for local communities. While forming a uniquely attractive and exotic commercial street, it also attracts more foreigners to settle here. Promoting a virtuous circle of the community's industrial economy further helped the increase of employment in various local commercial service industries.

Table 2.33 Evolution of the employment structure of the international student settlements in Nanxiu Village

Phase	Overall employment structure	Phased response characteristics
Phase I: 1980–1999	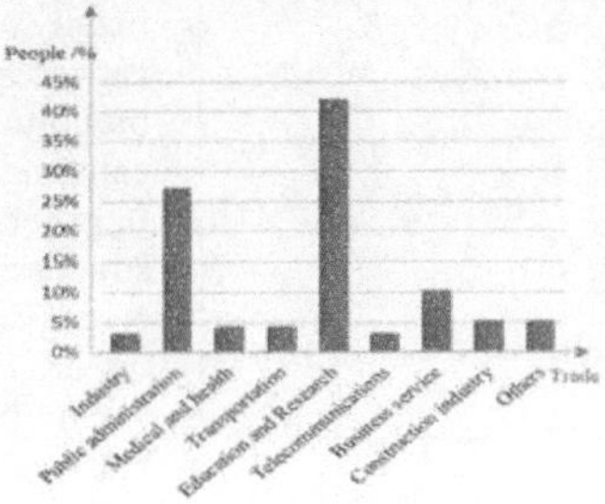	• Education and research at this stage dominate the occupational composition (43%) • As there are many surrounding enterprises and institutions, public management (27%) has therefore become one of the main employment directions • The employment ratio of other industries is below 10%, and the composition is relatively flat and decentralized • Overall, it shows an employment structure dominated by a single industry
Phase II: 2000–2010	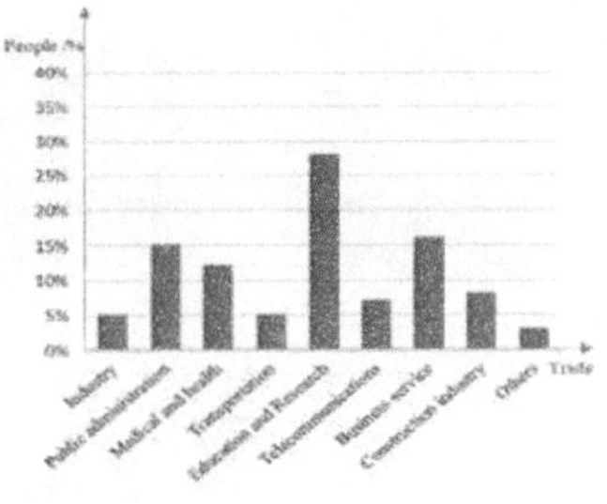	• At this stage, education and scientific research still dominate (28%), but the gap between leading industries and other employment structures is gradually narrowing • The business service industry (16%) began to rise rapidly, keeping pace with public management (15%), and the proportion of other industries also increased • Overall, a transitional structure from a single industry dominance to a multi-industry evolvement

(continued)

Table 2.33 (continued)

Phase	Overall employment structure	Phased response characteristics
Phase III: 2011 to now	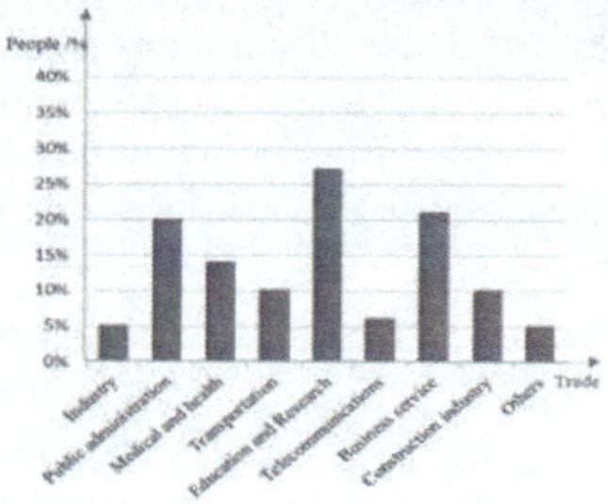	• The dominance of education and scientific research at this stage is no longer obvious (although the proportion of student groups including international students who rent here has risen sharply, the impact of the large-scale migration of the aborigines, mainly college faculty and staff, is even more significant). The proportion of other industries continues to rise • In general, an employment structure with equal emphasis on education and scientific research, public management, and business services and balanced development of various industries has gradually formed

Source Sampling survey data of the International Student Settlements in Nanxiu Village by the research team (2017)

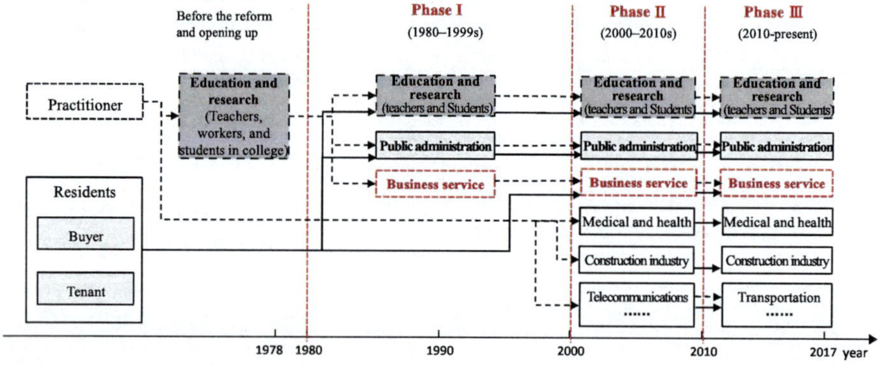

Fig. 2.51 The employment structure of the international student Settlements in Nanxiu Village in a phased response. *Source* Sampling survey data of the International Student Settlements in Nanxiu Village by the research team (2017)

2.4.5 Community Response to Identification and Interaction

2.4.5.1 Cultural Identity Response

This article mainly explores the response of local residents to the cultural identity of the foreign population, mainly international students, from both quantitative and qualitative aspects. In terms of quantitative analysis, starting from the survey data on the status quo of local people's identification with international students, we will conduct horizontal research on two levels of descriptive statistics and factor analysis; in terms of qualitative analysis, we will conduct a longitudinal analysis of the changes in local people's cultural identity through interviews with community residents and local media reports. The response characteristics of "**the local people's cultural identity with international students is generally low. Not only are there obvious differences in psychology, communication, and culture, there are also individual differences between different occupations.**" (See Table 2.34).

Descriptive statistics

Factor analysis

First, a correlation matrix analysis was performed on the above seven cultural identity indicators (attitudes toward the cluster of international students, relationship with international students, evaluation of international students, whether to communicate or not, whether to help each other or not, degree of cultural understanding, and cultural interest), with the findings indicating that all met the standards; afterwards, the exploratory factor analysis method was used to analyze the seven indicators by the principal component method, and the variance maximization method was used to orthogonally rotate the factor loads. The results are shown in the gravel diagram (see Fig. 2.52). It can be seen from the figure that the eigenvalues of the three main factors are greater than 1 (represented by F1, F2, and F3), and the commonality of all indicators is above 0.5 (when the commonality is greater than 0.4, it indicates that the main factor can be explained by the independent variable well) (see Table 2.35).

F1 main factor includes three indicators, which are "attitudes toward the cluster of international students," "relationship," and "evaluation of international students." They mainly reflect the psychological feelings of local people toward international students. This factor is named "psychological acceptance."

F2 main factor includes two indicators, which are "whether to communicate or not" and "whether to help each other or not." They mainly reflect the depth of communication between local people and international students. This factor is named "communication acceptance."

F2 main factor includes two indicators, which are "degree of cultural understanding" and "degree of cultural interest." They mainly reflect the degree of local people's absorption of foreign culture brought by international students. This factor is named "cultural acceptance."

Therefore, the cultural identity score takes the variance contribution rate of the three main factors as the weight and is converted into a value between 1 and 100

Table 2.34 Descriptive analysis of cultural identity indicators of residents in the international student settlements in Nanxiu Village

index	Degree of cultural identity	Feature description and summary
Attitudes toward international student clusters	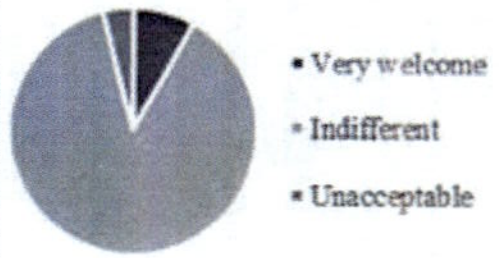	In terms of attitudes toward the cluster of international students, more than half of the interviewees felt indifferent to the fact that "international students live in clusters." Only six of the interviewees had more extreme views, of which four said they welcomed it and two said they couldn't accept it Overall, the respondents have a higher acceptance of the international students live in clusters here
Relationship	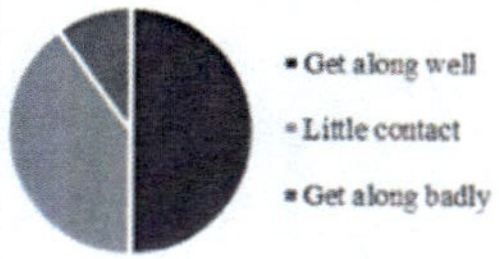	In terms of getting along with international students, 50% of the respondents "can get along well with international students," 10% said they "can't get along well with them," and 40% said they seldom contact with international students Overall, there is less disharmony with international students among the respondents, but it is mainly due to the lack of contact with each other
Evaluation of international students	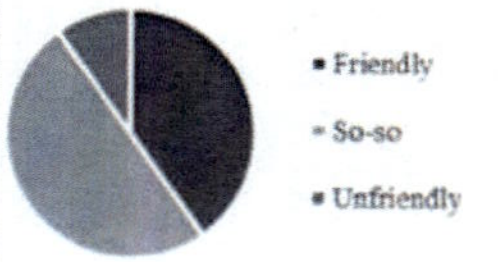	In terms of the evaluation of the attitude of foreign students, half of the respondents said that "the attitude of foreign students to them is average," 40% said that they are friendly, and only 10% thought that they are unfriendly Overall, the respondents' evaluation of the international student's group is relatively positive, while a few negative evaluations mainly think that the immigration of international students leads to the deterioration of public security and health in the surrounding communities

(continued)

Table 2.34 (continued)

index	Degree of cultural identity	Feature description and summary
Whether to communicate or not	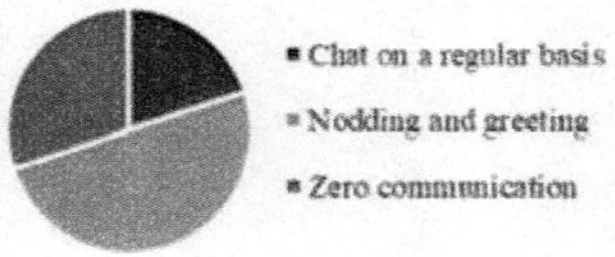	In terms of mutual communication, only 20% of the respondents said that they often chat with international students, while most of them only "nod to say hello," and 30% said that they completely lack communication Overall, there is little communication between the interviewees and international students, not to mention deep interaction
Help or not	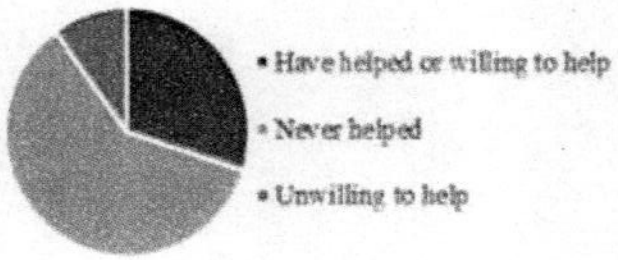	In terms of helping each other, 30% of the respondents said that they would help international students at ordinary times, but the majority (60%) had never provided any help, and another 10% said that they were not willing to help international students subjectively. On the contrary, some respondents mentioned that international students had a good attitude when seeking help and would return borrowed things in time Overall, there are great differences in the attitudes of the respondents toward providing help for international students, and most of them have reservations about this
Degree of cultural understanding	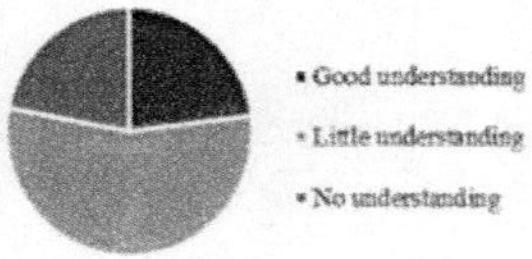	In terms of the degree of understanding of foreign culture, more than half of the respondents said they knew some of it, 22% said they knew it very well, but 22% said they didn't understand it at all Overall, the respondents have a certain understanding of foreign culture, but to a limited degree

(continued)

Table 2.34 (continued)

index	Degree of cultural identity	Feature description and summary
Degree of cultural interest	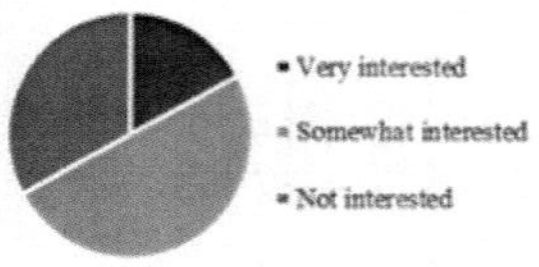	In terms of the degree of interest in foreign culture, most of the respondents expressed some interest, while 33% expressed no interest at all, and only 17% were very interested Overall, the degree of interest in foreign culture is not high, and there is a mismatch between the degree of interest and the degree of understanding

Source Sampling survey data of the International Student Settlements in Nanxiu Village by the research team (2017)

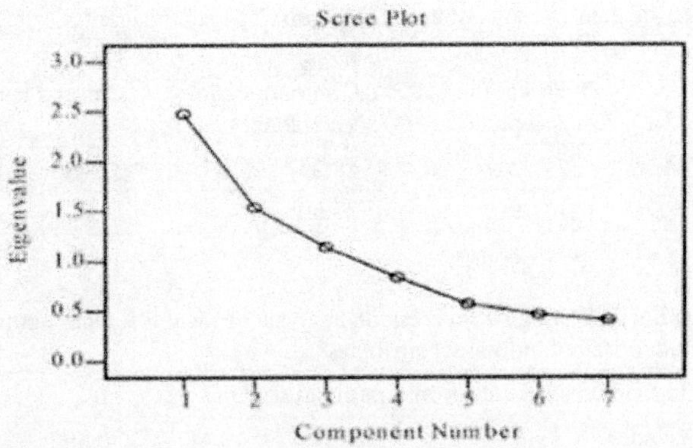

Fig. 2.52 Screenshot obtained by factor analysis method. *Source* Sample survey data of the International Student Settlements in Nanxiu Village by the research team (2017)

Table 2.35 Factor analysis results of cultural identity (Non-standardized Beta Value)

Independent variable	Main factor			
	F1 Psychological acceptance	F2 Communication acceptance	F3 Cultural acceptance	Commonality
Attitudes toward the cluster of international students	*0.698*	0.087	0.092	0.505
Relationship	*0.807*	0.068	0.069	0.661
Evaluation of international students	*0.702*	−0.059	−0.027	0.508
Whether to communicate or not	0.029	*0.805*	0.007	0.677
Whether to help each other or not	0.030	*0.629*	−0.150	0.544
Degree of cultural understanding	0.162	−0.014	*0.795*	0.682
Degree of cultural interest	−0.126	−0.111	*0.787*	0.655
Eigenvalues	1.908	1.864	1.767	
Variance devoting rates	17.342	16.950	16.059	
Accumulated variance contribution rates	17.342	34.292	50.351	

Table 2.36 Results of cultural identity of local resident

	F1 Psychological acceptance	F2 Communication acceptance	F3 Cultural acceptance	Cultural identity
Mean	54.72	48.25	37.45	46.81
S.D	13.99	23.60	18.31	18.63

Table 2.37 Logistic regression analysis of local residents' attitudes toward international atudents with different individual attributes

Individual attributes of international students	B	S.E	Sig
Gender	0.395	0.470	0.348
Jobs	0.296**	0.142	0.032
Age	−0.173	0.265	0.512
Education level	——	——	——
Income level	0.124	0.241	0.607
Constant	−2.019	1.478	0.170
Model Chi-square (df)	7.684 (5)		
−2Log Likelihood	117.879		

(Independent variable: 0 = accept 1 = exclude; significance: $^{*}p < 0.1$; $^{**}p < 0.05$; $^{***}p < 0.01$)

according to the standard score (see Table 2.37 for the basic distribution). It can be found from the table that the overall level of cultural identity of local people to international students is low (mean = 46.81, S.D = 18.63). Among them, **the speed of psychological and communication acceptance of local people is relatively fast, while the pace of cultural acceptance is relatively slow. Simultaneously, the individual difference of communication acceptance is the largest, and the individual difference of psychological acceptance is the smallest (see Table 2.36).**

Interview with community residents

Through interviews with local residents who have lived for a long time in this area, it can be found that local residents are experiencing the process of "unfamiliar, ignorant—acceptance, admission—contact and communication" for the increasing number of international students in the surrounding area. This is a process from rejection to gradual adaptation, but it is necessary to truly accept each other's culture and lifestyle still have a long way to go. In addition, the attitudes of community residents and store operators toward international students are also different. Community residents are mostly passive contacts, while store operators will take the initiative to interact with international students. It can be seen that there are professional differences in the attitudes of the locals.

Local media reports

Facing the emergence of international student groups and even settlements, the local media initially explored its freshness and novelty with a tentative attitude; in recent years, reports on the International Student Settlements in Nanxiu Village have mainly expressed acceptance and tolerance in their context. What is conveyed to people is that the degree of mutual understanding and communication between Chinese and international residents is increasing, and the community provides more and more comprehensive management and services for international students in recent years. In addition, there is also a small amount of news about the conflicts and disputes experienced by foreigners here.

It can be seen from the content and attitude of the media reports, Chinese and foreign residents in this area are in the process of adapting to each other and deepening their understanding, and the domestic public opinion generally accepts and supports international students.

The response of their cultural identity mainly derives from the following two aspects.

First, the inherent differences in thinking and habits of international students

The international students in Nanxiu Village come from all over the world. Before they moved to other cities, they have formed deep-rooted and firmly thinking mode and behavior habits after years of growth and learning. Therefore, the international students will show a kind of inertia and inertia of rejecting the changes of external forces, which will also reflect on the performance of local residents after being "invaded." However, this kind of "exclusion" is gradually entering the state of interactive adaptation, just as the local residents have shown. Therefore, the evaluation of international students is generally positive, but it is not deep enough.

Second, local residents of different occupations have different attitudes toward international students

To analyze the different attitudes of local residents with different individual attributes toward international students, the first-hand data can be processed and the weighted average is used to classify various indicators into two categories: "acceptance" and "exclusion" (the sample proportions are 52 and 48%, respectively). Then logistic regression analysis and SPSS were used (see Table 2.37), and the results found that, in addition to the "occupation" indicator ($p = 0.032$), other demographic and socioeconomic indicators are "accepting" or "rejecting" residents. There is no statistically significant influence on the attitude of international students. In other words, local people of different occupation types have very different views on the settlement of international students.

Further analysis also found that (see Fig. 2.53) shop owners, teachers, and students have a higher acceptance of international student clusters, reaching 88%, 66%, and 64% respectively. On the contrary, white-collar workers and workers have a higher degree of rejection of international students (86% and 65%, respectively). It is worth noting that different occupational groups have different opportunities and degrees of

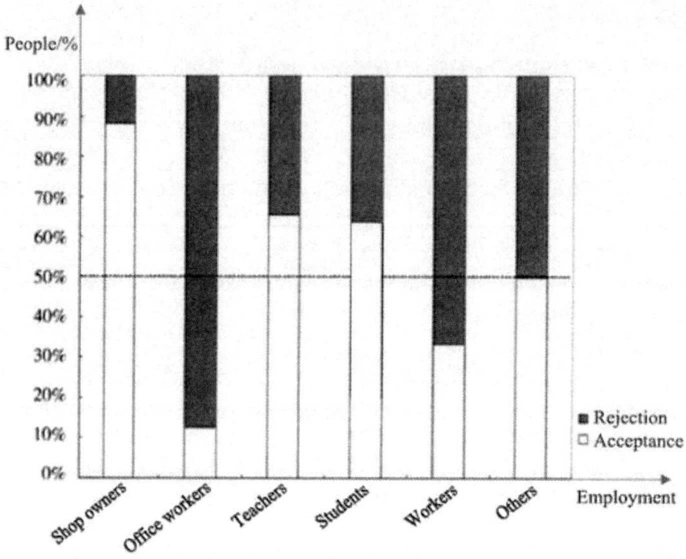

Fig. 2.53 Analysis of the attitudes of various occupational populations toward international students. *Source* Sample survey data of the International Student Settlements in Nanxiu Village by the research team (2017)

contact with international students. Among them, teachers, students, shop owners, and international students have more contact and understanding. This also shows that the acceptance degree of international students' living together is closely related to the level of contact (time and frequency). The higher the contact degree and the more understanding of the group, the higher the possibility of acceptance.

2.4.5.2 The Response of Social Interaction

Through the analysis of the personal communication, activity participation and organization between local people and international students, it is found that the social interaction of International Student Settlements in Nanxiu Village mainly presents the response characteristics where "**individual interaction began to increase but not deep enough, the age limit for participation in activities was large, the venue and content of activities gradually diversified, and the space distribution showed a model of 'multi-center plus commercial strip connection'**."

Personal communication

Most of the interviewees have a subjective willingness to interact with international students but objectively they have not had any experience of interacting; some interviewees have a strong willingness to communicate, and they also have at least 1–5 or even more than 5 international student friends; however, the interviewees, who are mainly elderly people, have no obvious subjective willingness to associate with foreign students, but compared with the past, the resistance has decreased year by year, and they no longer reject the interaction itself.

Activity participation

From the perspective of time stage, only a very small number of residents interacted and communicated with international students before the year of 2000. After that, the number of people participating in activities has gradually increased, but the participation is mostly in small parties, and there are also some festival celebrations and English corners. From the perspective of age structure, the communication channels between young people and international students are more diverse, and the forms of participation in activities are also more various. The communication methods are more subjective and active, while the interactions between the elderly and the international students are relatively passive, and they only have partial interactions by means of channels such as work and community activities.

Activity organization

In terms of venues and organizational structures, before 2000, the organizational units were limited to relevant foreign-related units, residents' committee, colleges, and universities, and the venues were mainly the Johns Hopkins University–Nanjing University Center for Chinese and American Studies and the Institute of Foreign Cultures of Nanjing Normal University; since 2000, in addition to the original event organizers, more Chinese and foreign residents have begun to spontaneously organize community activities, and the venues not only include the Nanxiu Village Foreign Service Station, the Huju Road Foreign Service Station, the Institute of Foreign Languages of Nanjing Normal University, the Johns Hopkins University–Nanjing University Center for Chinese and American Studies, the Literary Cafe, the Shanghai branch of the Women and Children's Center, and many other public club activities but also expanded to private clubs, bars, specialty restaurants, and other commercial and leisure venues.

As for the space distribution of venues, we can see from the space evolution of Chinese and foreign residents' assembly sites (see Fig. 2.54); in 1995, the international gathering places were limited to the exchange centers and research institutes of Nanjing University and Nanjing Normal University; in 2005, the location of community activities was still relatively sporadic, with foreigners' service stations set up by community neighborhood committees appearing along Shanghai Road and Qingdao Road, while there were few places for spontaneous activities; however, in the past 10 years, the space distribution of international community activities in Nanxiu Village and its surrounding communities has shown a cluster mode of "multi-center plus commercial strip connection." The public international activity sites form multi-center, while the commercial activity sites are mainly distributed along Shanghai Road and infiltrate into the community.

In terms of activity content, before 2000, although the scale of international students in Nanxiu Village and its nearby communities increased rapidly, in contrast,

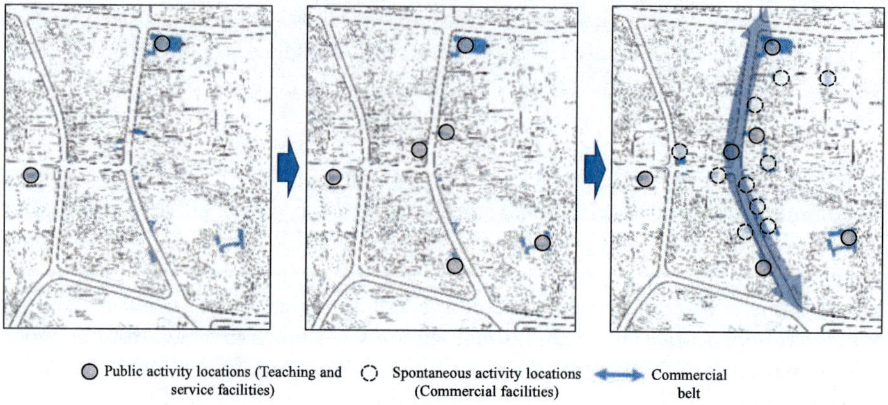

Fig. 2.54 The spatial evolution of gathering sites for Chinese and international residents (Left: 1995; middle: 2005; right: 2017). *Source* Sampling survey data of the International Student Settlements in Nanxiu Village by the research team (2017)

its daily activities were still organized and carried out in a relatively closed and independent group mode, and the content was mainly entertainment activities, such as festival parties, and there were almost no corresponding organizations to contact Chinese and foreign residents; since 2000, the content of international community activities held in Nanxiu Village and its surrounding communities has become richer, and entertainment activities have increased day by day which were mainly Chinese and foreign fraternity, daily party, and other spontaneous small activities, while the main communication activities were concentrated on the cultural level, such as English corner, language guidance, cultural club, and so on. Chinese residents were eager to learn foreign culture in these activities and hoped to communicate more closely with foreign international, reflecting the gradual recognition of international students and the gradual infiltration of foreign culture.

The response of his social interaction mainly derives from the following two influences (see Fig. 2.55).

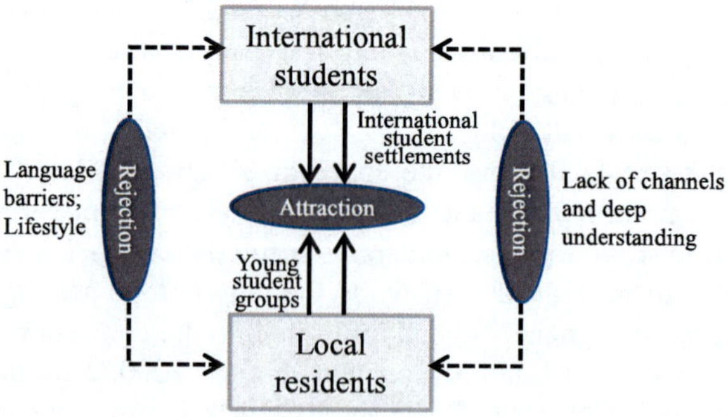

Fig. 2.55 The response mechanism of social interaction in the international student settlements of Nanxiu Village. *Source* Sample survey data of the International Student Settlements in Nanxiu Village by the research team (2017)

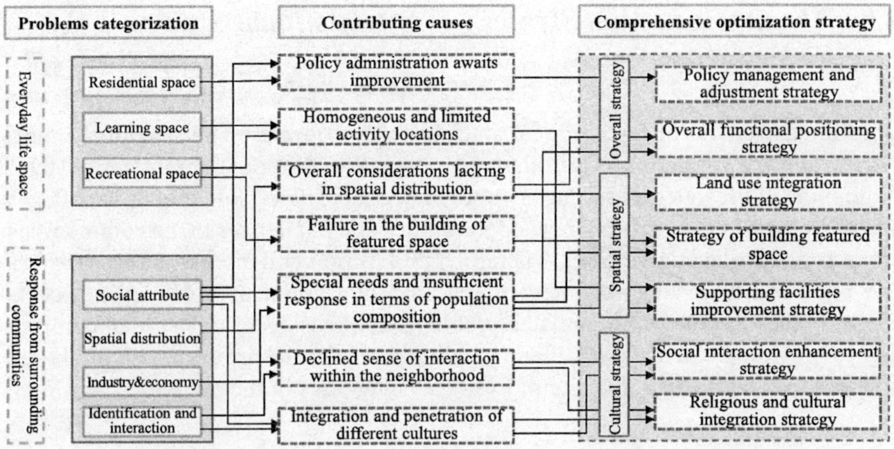

Fig. 2.56 Deduction of improvement strategies for the International Student Settlements in Nanxiu Village. *Source* Own elaboration

First, the increasing interest and tolerance of local residents in foreign cultures

Shanghai Road is an entertainment base where immigrants from Nanjing gather, while Nanxiu Village and its surrounding communities are communities where immigrants live. The degree of cultural penetration between the two must be higher than that of ordinary communities; simultaneously, the student groups brought by many universities in the area have more contact with foreign students due to study, work and other reasons, and are easy to understand and communicate with each other due to their similar age, so they have a higher degree of interest and tolerance in foreign culture.

Second, the lack of in-depth understanding of foreign culture

Local residents have gradually become interested in international students due to the increasing frequency of foreign cultural information (slogans, foreign stores, etc.), but they are unable to understand them in depth due to obstacles such as difficulty in language communication and large differences in living customs. Similarly, the access to culture has not changed due to the diversified cultural atmosphere brought about by international students, and it is still based on books and the internet; a small number of residents who understand foreign cultures also have intersections with international students due to work reasons or some similar living habits.

2.5 The Optimization Strategy of International Student Settlements

With the expansion of international students, a new group of transnational immigrants, and the formation of all kinds of international student communities, their differentiated lifestyles and values not only inject new blood into domestic colleges and universities but also bring new opportunities and challenges to the communities around colleges and universities. Therefore, this section will discuss the comprehensive optimization strategy of International Student Settlements from seven aspects based on the summary of the related problems and reasons of this special community to provide the diverse daily living environment for the group and highlight the internationalization and diversity of local characteristics of the region.

2.5.1 The Existing Problems in International Student Settlements

2.5.1.1 The Existing Problems in Daily Life Space

Through the research on the daily life space of International Student Settlements, from the aspects of living, learning, shopping, leisure space and daily life space–time trajectories, the existing problems and their reasons are summarized as follows (see Table 2.38).

2.5.1.2 The Existing Problems Responded by Surrounding Communities

Based on the research on the community response around International Student Settlements, this paper summarizes the existing problems and their causes from the aspects of social attribute, space layout, industrial economy, and community response of identity interaction (see Table 2.39).

2.5.1.3 Problem Summary and Strategy Deduction

As mentioned above, the International Student Settlements of Nanxiu Village is a coexistence of evolution and problems in daily life space and community response, and the causes of these social or space problems are often complex and intertwined with each other. Based on this, seven optimization strategies can be sorted out as follows (see Fig. 2.56).

Table 2.38 Summary of daily life space problems in the international student settlements of Nanxiu Village

Category	Main problem	Negative reasons	Category summary
Living space	The active choice of residence outside the school brings certain difficulties to the management of international students in colleges and universities	The contradiction between the differentiated living style of international students and the unified management of universities	Policy management
Learning space	The existing learning space cannot meet the flexible time needs and differentiated space needs of international students	The contradiction between the habit of autonomous learning and the limited learning space conditions of international students	Supporting facilities
Shopping space	——	——	——
Leisure space	The types of existing leisure venues are relatively single, especially those lacking the distinctive and personalized space required by international students with different cultural backgrounds	The contradiction between the diverse leisure needs of international students and a single leisure place	Featured space Supporting facilities
Space–time trajectories	——	——	——

Source Own elaboration

2.5.2 Optimizing Strategies for International Student Settlements

2.5.2.1 Policy Management Adjustment Strategy

With the increasing scale of studying broad in China, Nanxiu Village has become Nanjing's largest International Student Settlements. Improving the policy system related to international students is the fundamental guarantee for the educational development of international student and optimizing the management system and mode is an important way to accelerate the internationalization of my country's higher education. According to the above summary of the problems of the International Student Settlements in Nanxiu Village, improvement strategies can be proposed from two aspects: policy systems and management methods (see Fig. 2.57).

Table 2.39 Summary of response problems of surrounding communities in the international student settlements of Nanxiu Village

Category		Main problem	Negative reasons	Category summary
Social attribute	Nationality composition	The number of international students who actively choose to live around colleges and universities has increased	While bringing difficulties to community management, there have been difficulties in interaction between Chinese and international residents	Policy management Social interaction
	Living habit	Invasion and conflict of foreign culture in information age	It is difficult to meet a balance between spreading and promoting excellent culture and preventing purposeful cultural infiltration	Religious customs
Space layout	Space structure	The increasing tension of land use in old towns and the increasing difficulty of renewal	The functions are becoming more and more complex, and the space is developing continuously, but there is no large-scale spatial agglomeration core	Function position Land use integration
	Public service facilities	Lack of distinctive commercial service facilities for overall layout considerations	There are still some deficiencies in supporting scale, categories, and space layout	Land use integration Supporting facilities
Industrial economy	Distinctive industries	Lack of unified planning and guidance for the needs of special groups	Distinctive industry has gradually become a scale, but lack of space with international features	Function position Distinctive space
	Employment structure	——	——	——
Identity interaction	Cultural identity	There are interaction barriers in language, living habits, cultural background, etc	Cultural identity of the locals to international students needs to be improved	Religious customs Social interaction

(continued)

Table 2.39 (continued)

Category		Main problem	Negative reasons	Category summary
	Social interaction	Lack of activities and organizations to enhance interaction with each other	There are group differences in the local people's identification with international students	Social interaction

Source Own elaboration

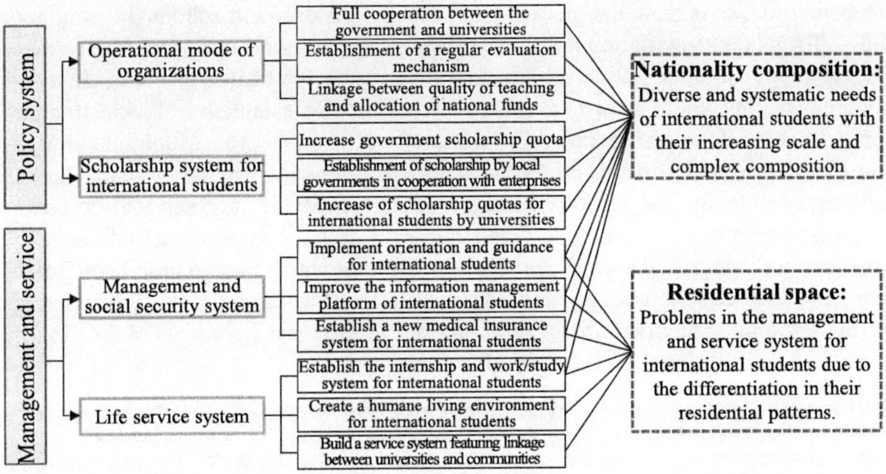

Fig. 2.57 Policy management and adjustment strategy for the international student settlements in Nanxiu Village. *Source* Own elaboration

Improve the international student policy system

Establishing and improving the operating model of the organization is a top priority. First, the government education authority and the enrollment colleges and universities cooperate with each other. The government education authority is responsible for the macro-monitoring of the teaching quality of international students and actively assists universities in establishing and improving the international evaluation and certification system for higher education academic degrees, while the enrollment colleges and universities are responsible for the implementation of various teaching monitoring tasks, adjusting corresponding teaching plans and training programs according to the characteristics of international students, and optimizing the education quality monitoring system on campus; second, establish a regular review mechanism, led by the government's education department, and specific responsibility by industry associations, organizing schools of similar level and size to form corresponding education quality evaluation committees. According to international students of different categories, levels, and disciplines, formulate classified and guided talent training quality and evaluation standards; third, the government education department should sign corresponding quality assurance agreements with universities that recruit international students, link the evaluation results with the relevant funds allocated by the state, and intensify rewards and punishments to fully mobilize the enthusiasm of universities in ensuring the quality of teaching.

Improving the scholarship system for international students is an important means and a sign to identify the level of internationalization of my country's higher education. First, to allow the world understand China more and increase the international competitiveness and attractiveness of our government scholarships, it is recommended to appropriately increase the number of government scholarship students. At the same time, to narrow the income gap between government scholarship students and self-financed international students, it is recommended to appropriately raise the per capita funding standard with reference to the fee standard of self-financed students; second, local governments are encouraged to establish a "Local International Student Scholarship Fund Committee" entrusted to the Municipal Education Commission or Provincial Department of Education for daily management and responsible for the various tasks of scholarships for international students. The funding sources can include "Outstanding international student scholarship," "research assistant scholarship," "professional scholarship," "social practice scholarship," and "course scholarship," which are used to feed back a part of the fees collected by international students to the development of international students' education in China to develop education through education.

Optimize the management and service system for international students

Improving the management services and social security systems for international students in universities is a strong guarantee. First, do the work of enrollment for international students, make full use of websites, bulletin boards, and the "Handbook for International Students in China" to introduce relevant regulations to international students in China in multiple languages. Anti-theft, fraud prevention, fire prevention, traffic safety, how to seek medical treatment, alarm and emergency assistance must be publicized, actively guiding students to regulate their own behavior, and laying a good foundation for their daily management in the future. Second, the establishment and improvement of the management information platform for international students, timely and effective information analysis, and processing are very important for the prevention and control of international students' emergencies; the construction of management information platform for international students need the communication and cooperation of management cadres, homeroom teacher, tutors, teachers, apartment management and service personnel, relatives, friends, and other personnel of international students. The management departments of international students should not only pay attention to timely and effective psychological intervention for international students in China but also pay attention to the construction of multi-level, multi-angle, multi-integrated management platform and adopt the mode of keeping the center of gravity downward to prevent and control the emergencies of foreign students; third, establish a new medical insurance mechanism for international students in China to enable more domestic and foreign insurance companies to participate in medical insurance services for international students in China, allowing international students to freely choose hospitals. In the medical insurance for international students in China the insured, the insurance company, and the hospital have their own different interests. Therefore, we must establish a new type of tripartite relationship that can achieve a balance of interests and fully protect the interests of policyholders without interfering with the normal operation of insurance companies.

Establishing a "people-oriented" international student life service system is an effective way. First, establish a tolerant part-work and part-study system for international students to allow international students to participate in the social life of our country to a certain extent and within a certain range. This is not only the need of one culture to tolerate and accept another culture but also the common need of international students and us in China. Simultaneously, there is an urgent need to introduce a special regulation for international students to work and study, so that universities, employers, international students, and foreign management departments have a law-based code of conduct; second, realize the humanization of international students' accommodation service. For international students who passively choose to live on campus, the dormitory is not only a place to rest and sleep but also a place for activities, learning, and entertainment. A set of dormitory management system for international students should be established to make it have a certain sense of belonging and identification with the dormitory. Third, strengthen cooperation with the community to create a harmonious environment. For international students who choose to live off campus, on the one hand, localities need to further regulate the operation of international student housing leasing agencies and clarify the rights and obligations between the lessor and the lessee. Colleges and universities should also actively assist international students to go through various procedures for off-campus accommodation and earnestly safeguard their legal rights and interests. On the other hand, university retention management departments need to maintain contact with the communities where international students live, visit communities regularly, link communities actively, help international students integrate into the community and solve practical difficulties, and create a service mechanism for international students as well as create a harmonious growth environment for international students through the service mechanism of "interaction inside the school and linkage outside the school."

2.5.2.2 Overall Function Positioning Strategy

City master planning analysis

According to the "Nanjing City Master Plan" (2011–2020), the International Student Settlements of Nanxiu Village is located in the center of the densely populated old town of Nanjing. It is also the core area that shows the historical characteristics of the old town, which is an important historical townscape.

Its comprehensive positioning is as follows.

In terms of land use function, educational and scientific research land and residential and commercial land coexist; in terms of industrial development, science and technology research and development should be carried out simultaneously with characteristic commerce; in terms of space style, university education buildings, modern large-medium and high-rise commercial buildings, and multi-storey residential buildings should be paid equal attention to. It can be seen that the International Student Settlements of Nanxiu Village is facing the challenge of transforming international education and housing.

Analysis of district level master plan

The Gulou District, where Nanxiu Village is located, is the core area of Nanjing's main city. It is rich in history, culture, and mountain water resources; developed in science, education, culture, and commercial services; and has concentrated medical and health resources. It is the seat of the Jiangsu Provincial Party Committee, the Provincial Government, and an urban area where Nanjing landscape city forest is integrated and represents Nanjing City's image. The "Master Plan for Gulou District in Nanjing" (2013–2030) compiled at the end of 2013 is centered on the direction of "striving for the best area and building a happy Gulou District" and builds the Gulou District into a core area of innovation and development in Nanjing and beyond, Binjiang Development Demonstration Zone, livable life model areas, and ancient capital distinctive areas.

Overall function positioning guidance

According to the guidance and positioning of the International Student Settlements in Nanxiu Village in the "Master Plan for Gulou District in Nanjing" (2013–2030), it can be seen that in the development of various systems in the settlement area, it is closely linked to the rich educational and scientific research resources and historical and cultural resources of the area; however, there is no specific explanation and emphasis on the internationalization and multi-ethnic cultural features of international students and other foreigners in the settlement area.

Therefore, based on the interpretation of the urban master plan and the Gulou District plan, combined with the field observations of the International Student Settlements in Nanxiu Village in this article, the existing positioning can be further supplemented and optimized: **mainly for education, scientific research, and residence, supplemented by administrative office and commercial facilities—a multi-ethnic fusion settlement area with internationalization and multi-ethnic distinctive culture as the highlight and traditional small-scale space features.**

2.5.2.3 Land Use Integration Strategy

From the city aspect, the International Student Settlements of Nanxiu Village is in the old city. The planned land is mainly for education and scientific research and residential land, as well as commercial land and administrative office land. This land layout emphasizes the complex function of the area and the comprehensiveness of

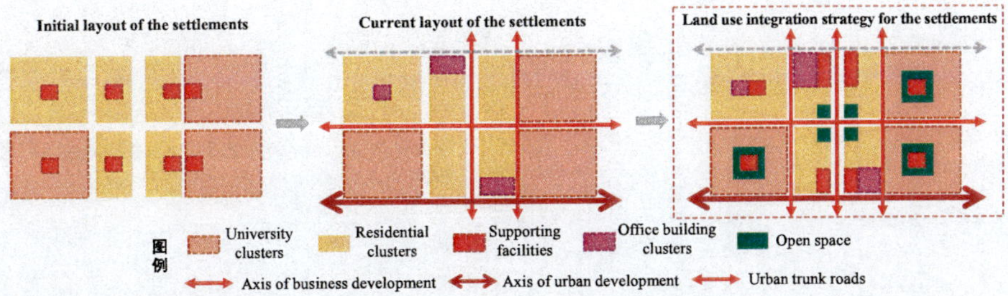

Fig. 2.58 Land use integration model of the international student settlements in Nanxiu Village. *Source* Own elaboration

future development. Therefore, the future land use should conform to the old city renewal stock development law, fully excavate and integrate various space resources, realize functional integration and replacement, and accelerate the transformation and utilization of idle space in colleges and universities, which is also the focus of the area renewal.

From the perspective of the community, based on the initial foundation, current structure and functional positioning of Nanxiu Village and its surrounding communities, the integration strategy for land use can be determined as follows. First, the corresponding service facilities are configured as a group and integrated into the special needs of international students. Second, adopt a modern commercial layout model, combined with the existing exotic feature businesses along the street in the settlement area to form a dynamic commercial axis to strengthen the relationship between the residential groups and the relationship with the city. Finally, set up an open space inside the settlement area and form a vibrant settlement core together with surrounding supporting facilities (see Fig. 2.58).

In short, the land use model of the International Student Settlements in Nanxiu Village should not only guide the land use function from the overall level but also coordinate the relationship between the settlement and the city. Emphasis is placed on the space layout model of high-schools and residential clusters in residential areas, which effectively utilizes the international commercial characteristics and encourages the compound development of functions to meet the differentiated needs of people of different nationalities, different occupations, and different cultures.

2.5.2.4 Shaping Strategy of Distinctive Space

Combined with the above overall functional positioning and land use integration strategy, this paper analyzes the typical space segments of the International Student Settlements in Nanxiu Village and provides strategic proposals for its function and space shaping (see Table 2.40).

Table 2.40 Shaping strategies for distinctive space fragments

	Space schematic	Typical section	Description
Node space	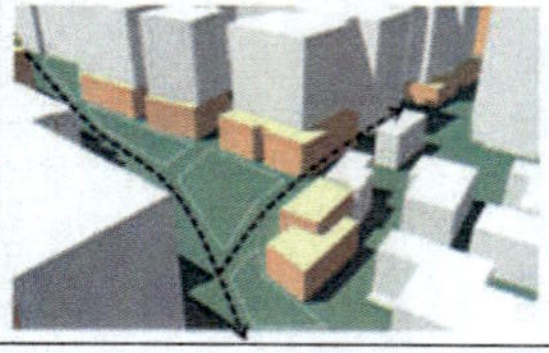	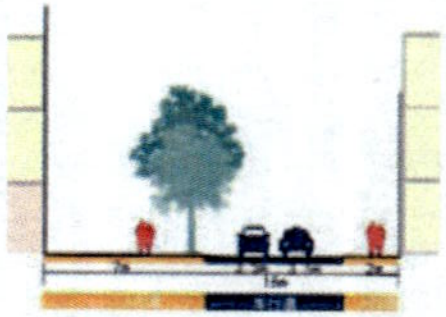	[Problem analysis] The international featured businesses on both sides of Nanxiu Village Street in the residential area are the most concentrated, but the street space separated by roads is fragmented and not effectively used
Improvement goal	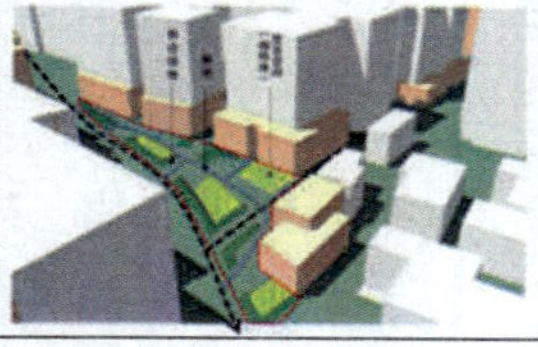	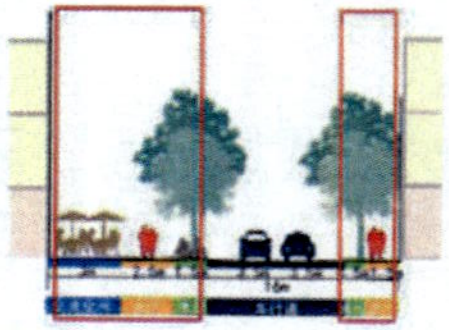	[Specific strategy] Increase the green space beside the street, improve the space environment while enclosing the front area of the building, and add communication spaces such as cafes to ensure the accessibility and attractiveness of street shops, which it can also ease it to a certain extent. The problem of shortage of open space in this area
Street space			[Problem Analysis] Many streets (Jinyin Street, Shanghai Road, Hankou Road, etc.) in the residential area have attracted a certain scale of distribution of exotic restaurants, but they are relatively scattered and have low space utilization

(continued)

Table 2.40 (continued)

	Space schematic	Typical section	Description
Improvement goal			[Specific strategy] Take Jinyin Street as an example, coordinate the distinctive industry formats on both sides of the street, and actively absorb more diversified and rich international featured commercial outlets to form a more continuous and concentrated commercial street; at the same time, the existing open space environment is optimized and transformed, and communication spaces are added to enhance the vitality of the street
Street space		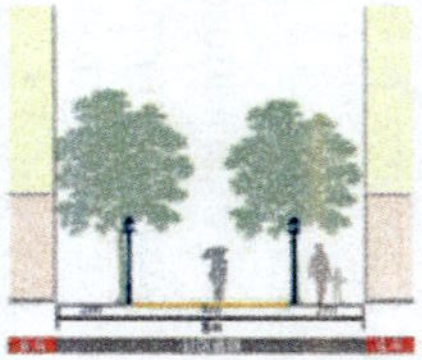	[Problem Analysis] The distinctive industries in the eastern part of the settlement area are relatively concentrated, while the western section of Hankou Road and Ninghai Road, which are connected to the nature of China, have few distinctive commercial distribution, especially the lack of large-scale distinctive industry cluster core areas
Improvement goal	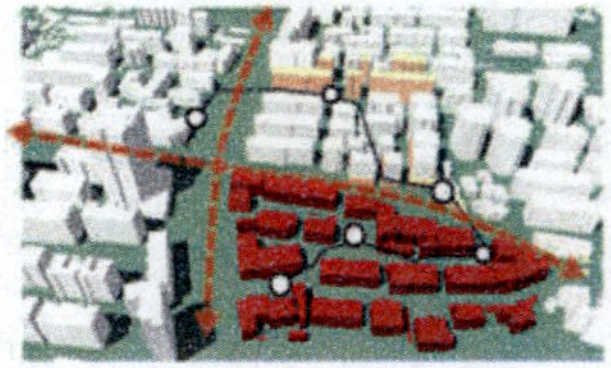		[Specific Strategy] Make use of the existing exotic commercial and cultural atmosphere in the settlement area, expand the distinctive commercial space along the road, and improve the overall layout of the distinctive industry in the settlement area; at the same time, build the core block of the international distinctive industry cluster and strive to improve the open space of the block. The vitality and quality of the pedestrian space system

Source Own elaboration

2.5.2.5 Supporting Facilities Improvement Strategy

In terms of supporting facilities, the International Student Settlements of Nanxiu Village is not only close to two city-level commercial centers—Xinjiekou and Hunan Road—but also has a wealth of community-level commercial service facilities and serves as an incubation base for technological innovation. The gathered educational resources are available from kindergartens to universities, and other facilities such as sports and medical treatment are relatively abundant. Under this positioning, the strategies for further improvement are as follows.

Establish supporting standards for public service facilities, and determine the category and scale ratio of international featured facilities

In the planning standard and policy formulation stage, the actual needs of residents should be put in the first place, and the living habits of international students should be taken into consideration to determine the type and scale of supporting facilities (see Table 2.41). While the government guarantees the allocation of basic public welfare facilities, it encourages private organizations to set up profitable institutions that not only serve specific international students in the community but also achieve recognition and acceptance among local residents, thereby forming a public welfare, quasi-public welfare, and profitability The coexistence of diversified facilities closely connects the prosperity of the international featured facilities in the settlements with the vitality of sustainable development.

Prioritize the layout of international featured facilities and improve the construction of general facilities

Reasonably constructing the configuration of international featured facilities can not only serve the international students in the settlement itself but also meet and radiate the needs of international students in the surrounding area and even the whole city, cultural identity seeking, and carry out the sustainable development of the diverse culture of the settlement. The main space strategies include the following (see Fig. 2.59).

First, optimize the characteristic service facilities. The international student community service station and activity center are arranged in the residential group in the residential area to provide international students with consultation and management of affairs, daily life services, and leisure activities space to meet their community-level service needs.

Second, shape the axis of distinctive industries. Relying on Hankou Road, Shanghai Road, Jinyin street, and other historical original commercial interfaces while optimizing the existing commercial formats, we should introduce more fresh and competitive international commercial formats, shape and enhance the international distinctive industry axis in the area, and further build an international distinctive service network.

Third, upgrade the characteristic commercial district. Relying on the commercial axis of Hankou Road, which has the highest degree of international characteristic commercial agglomeration, and the dot-shaped distinctive commercial in

Table 2.41 Recommendations for facilities in the international student settlements

Facility category	General facilities		International featured facilities	
	Facilities content	Recommended size (m^2)	Optimization of original facilities	New facilities
Higher education	Classrooms, study rooms, libraries, etc. in colleges and universities	——	Add bilingual logo guidance	24-h study rooms
Basic education	Kindergarten and elementary school	70 students/1000 people Service radius 500 m^2	Add international classes Introduce foreign teachers, etc	International kindergarten/school
Medical hygiene	Clinic, health station, hospital, etc	2500–3000	Set up a special window for foreign personnel; add bilingual logo guidance	——
Culture and sports	Cultural activity stations, libraries, gymnasiums, etc	4000–5000	Add foreign culture display window; setting up characteristic lectures; bilingual signs are added to guide all kinds of facilities	Foreigners' activity station
Commercial service	Various commercial facilities, meat, and vegetable markets, etc	2000–4000	Add foreign merchandise windows in supermarkets; add a window for foreign students in university canteens; bilingual signs are added to guide all kinds of facilities	A variety of exotic restaurants; exotic goods shops

(continued)

Table 2.41 (continued)

Facility category	General facilities		International featured facilities	
	Facilities content	Recommended size (m^2)	Optimization of original facilities	New facilities
Leisure and entertainment	Internet cafes, cinemas, parks, etc	4000–5000	Bilingual signs are added to guide all kinds of facilities	Exotic dance activity room
Financial posts and telecommunications	Bank, savings office, post office, etc	16,000–20,000	Bilingual signs are added to guide all kinds of facilities	—
Municipal utilities	Parking lots (garages), subway station, bus station, etc	1000–1500 m^2 bicycle parking lot; set up 20–30 motor vehicle parking lots	Bilingual signs are added to guide all kinds of facilities	—
Administrative management	Residents committee, sub-district office, police station, etc	3000–5000	Set up a service window for foreigners; equipped with special personnel to guide; bilingual signs are added to guide all kinds of facilities	Foreigners service station; foreigners service management system
other	Civil air defense facilities, street welfare facilities, etc	1500–2500	Bilingual signs are added to guide all kinds of facilities	—

Source Own elaboration with reference to relevant specifications (Based on the scale of a residential area of about 30,000 people)

Nanxiu Village, it extends to the inner part of the block, and finally upgrades to an international commercial district with functional agglomeration and distinctive features.

Fourth, improve the learning and leisure space. On the one hand, we should optimize the management mode of learning space (such as extending the opening hours, adding bilingual signs, updating teaching equipment, etc.) to fully consider the personalized needs of international students; on the other hand, we should add a variety of international leisure space with diverse space to meet the needs of international students' leisure activities.

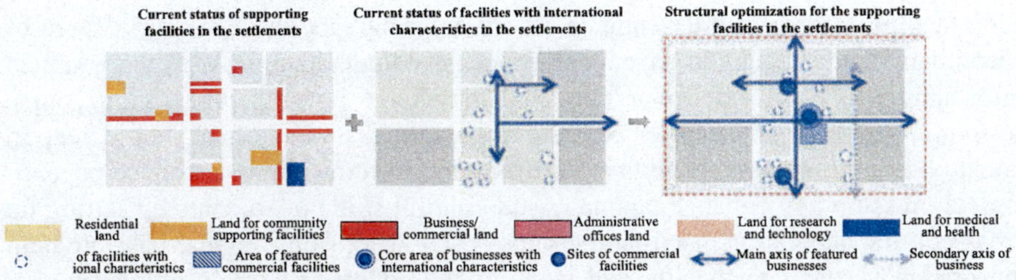

Fig. 2.59 Improvement of supporting facilities in international student settlements. *Source* Sampling survey data of the International Student Settlements in Nanxiu Village by the research team (2017)

Fifth, activate the core of distinctive business. Combined with the important spatial nodes such as urban road intersections, it is necessary to build a large-scale commercial core with international features and enhance the urban influence of the international distinctive facilities in the residential areas.

2.5.2.6 Social Interaction Promotion Strategy

Establish a psychological, language, and cultural support system for international students

The interaction and integration of international students in China is a cross-cultural adaptation process, which will be diversified with the time change and influenced and restricted by various variables from the individual and social level.

Generally, the major countries that accept international students will generally set up cross-cultural psychological consultation rooms and language training classes in the International Student Center to alleviate the psychological and socio-cultural adaptation problems of international students. For example, the International Student Center of the University of Toronto has set up cross-cultural counseling positions, which are held by professionals with psychological counseling or cross-cultural psychological counseling. Free language training classes are also provided for students who need language training; the University of Tokyo in Japan has set up a student consultation office, trouble consultation office, and an international student consultation room. There is a special fund to support the Japanese language teaching in the International Student Center, friendship, and cultural exchanges in the International Student Center to achieve the psychological and social and cultural adaptation of international students.

In contrast, China's colleges and universities are still lacking in systematic and effective solutions in this regard, so we need to establish cross-cultural psychological consultation and language and culture support system to improve the psychological adaptation and social and cultural adaptation of foreign students in China. Specifically, it includes providing pre-trip support for international students, such as providing paper or online introduction of Chinese social culture, university system, and teaching information to students before entering school; in the early stage of enrollment, students should have a comprehensive understanding of some practical information, such as residence permit, legal knowledge, housing, registration, course selection, daily life information, etc.; the international student office should keep in touch with international students at all times and introduce how to reduce psychological pressure and enhance language and cultural skills of students by setting up cross-cultural adaptation discussion groups or workshops on a regular basis; to help the graduates adapt to the life and work of the country of origin efficiently, we should carry out the cross-cultural adaptation education of "returning to the country of origin" among the graduates.

Provide an effective way for international students to interact with the locals

On the one hand, international students are encouraged to participate in the activities of Chinese students' clubs, which will help them better adapt to the study and life in different places and integrate into the local campus culture. According to the characteristics of international students, the types of clubs and organizations that are more suitable for participation include the following: first, the language and literature type, such as speech associations, cultural groups, poetry clubs, etc. This type of association is particularly attractive to international students studying Chinese language and culture majors; second, cultural and artistic types, such as calligraphy associations, photography associations, dance associations, martial arts associations, drama clubs, etc. The participation of cultural and artistic associations will give full play to the interests and hobbies of international students in China and enhance their understanding and love of Chinese culture; third, the type of sports competition, such as various ball clubs and chess associations.

On the other hand, encouraging international students to participate in local community activities is conducive to mutual understanding and interaction with local residents. First, provide more opportunities for Chinese and foreign exchanges and learning. For example, neighborhood committees and service stations can regularly organize some more formal large-scale lectures. The organization of activities should be more standardized so that residents can gain the popularization of basic foreign culture, and at the same time, it can also enhance the cultural understanding of foreign neighbors; second, it is recommended to set up more Chinese and foreign cultural activities so that Chinese and foreign friends who want to know and learn from each other have the opportunity to get to know each other; third, let international student groups participate in community management and promptly put forward their own demands and suggestions to the community.

2.5.2.7 Religious Custom Integration Strategy

(Religious) Cultural selective absorption strategy

Religion and culture have an inseparable endogenous mechanism. In a sense, human cultural history is a history of religion. Religion is not only a kind of belief but also a code of conduct that guides and influences all aspects of people's lives. Therefore, it is necessary to look at the foreign culture brought by the international student group with religion as a representative background from a rational point of view. While accepting the excellent foreign culture, it should prevent the infiltration of special purpose culture.

On the one hand, understand and exchange outstanding foreign cultures. First, it combines policy orientation with cultural identity to ensure the free and stable development of excellent culture. For example, improve the equipment of religious activity sites (such as Christians, mosques, etc.) around the settlements, consider the needs of international students in terms of language and staffing, and encourage the orderly development of legal religious activities; second, provide opportunities for international students to participate in community-level service facilities and use cultural propaganda and characteristic activities to further strengthen mutual understanding; finally, while the wave of modernization has brought about cultural globalization, we also need to face up to the promotion of cultural diversity, take this as an opportunity to fully absorb excellent foreign cultural resources, add cultural elements to the international characteristic industries in the settlements, and enrich the characteristics of the industries.

On the other hand, we should be wary of special purpose cultural penetration. Most international students come to China to study for the purpose of studying and cultural exchanges, but it is still necessary to prevent criminals from using the identity of international students as a cover to infiltrate religion through various channels, disrupting the normal teaching order and social security of the school. Measures include, for one thing, strengthening the management mechanism of international students in colleges and universities. First, publicity and education of my country's relevant laws and regulations should be put in place. On this basis, management personnel of various universities should correctly realize the harmfulness of foreign students' religious infiltration of university teachers and students and further improve related management systems and strengthen management of international students. For another, establish an awareness and prevention and control system for the risk of religious infiltration by international students. Strengthen the religious knowledge education for the management personnel of international students in China, increase the vigilance and sensitivity of religious infiltration behavior, and establish a cross-unit, cross-departmental, multi-faceted religious infiltration risk identification and prevention mechanism. Finally, strengthen the political and ideological education of college students. We should strengthen the education of Marxist theory for college students, guide their behavior and practice with materialism, enhance their legal awareness, and have a correct understanding of religion and religious freedom.

Moderate integration strategy of life customs

On the one hand, it provides the basic means of integration through mutual cultural exchanges. How to integrate the international students into the local environment, culture, climate, customs, laws and regulations, social system, etc. has become a practical problem that this group must face. Similarly, for local residents, the influx of international students also has a subtle impact on their life and culture. Therefore, to enhance the understanding of each other's living customs has become a prerequisite for further integration. Non-governmental organizations and voluntary agencies of local governments can work together with universities to carry out exchange activities with the participation of foreign students and local people, introduce Chinese cultural characteristics, traditional arts, folk crafts, ethics, laws and regulations, common sense of life, etc. while spreading the essence of Chinese culture and we should also understand the source culture and daily life needs of the majority of international students to promote the collision and integration of each other's culture and life.

On the other hand, international student social organizations provide the necessary support for integration. First, from the external guidance, we should change our thinking and accept a variety of cultures. The government departments should fully understand the needs of foreigners to establish their own social organizations, change the past ideas of control management of social organizations, properly guide and encourage the cultivation management of transnational social organizations, integrate the development of social organizations in the areas inhabited by transnational immigrants into the planning of basic system construction, and give some support in system construction and capital investment. Second, from the perspective of internal initiative, we should encourage foreigners to take the initiative to carry out modern democratic and legal transformation of transnational social organizations and ensure their effective participation in social organizations. Finally, taking the opportunity of modern urban renewal or the spatial transformation of the construction project of the settlement area, we should establish the social organization network of foreigners, timely guide the establishment of the social organization structure of the networked settlement area, and attract them to participate in the interaction of the modernization construction of the settlement area (see Table 2.42).

2.6 Summary of This Chapter

This chapter is a special discussion on the objective of "Integration-Response" in the field of "international migration—immigrating country." First, it defines the related concepts of International Student Settlements, daily living space, and community response and outlines the daily life and its mode of international students. Then, it take the International Student Settlements (including passive dormitories or apartments and surrounding communities where they live spontaneously) in Nanxiu Village, Nanjing City as the research object while starting from the two major issues of daily life space and community response to parallelly discuss the daily activities, life trajectories and impact effects of the typical group of international immigrants, "international students," on a community scale. Additionally, considering existing problems with regard to the International Student Settlements, it proposes a comprehensive strategy to optimize its external environment and reduce its integration barriers.

The main conclusions of this chapter are as follows.

Table 2.42 Suggestions for setting up social organizations in the international student settlements

Category	Content		Type	Function
General society organization	Residents committee		Political organization	Autonomous grass-roots organization
	Community Housekeeping Service Center		People's organization	Assist residents in family affairs management
School organization	International Students Office		Administrative machinery	International student management organization
	College Students' Union		Associations of students	Students' own mass organizations are the bridge and link between schools and students
Foreign personnel institutions and organizations	Culture Interaction	Institute of foreign languages	Scientific institution	Research institutions for international students that regularly organizes relevant activities to promote exchanges among China and foreign countries
		Cultural Exchange Center		
	Foreigners Service Station		People's organization	Organize various activities for foreigners, such as festivals, English corners, entertainment parties, etc.; organize communication activities with Chinese people, such as lectures on culture, etc
	Association of Foreigners		People's organization	Small-scale assembly activities are organized by foreigners of the same nationality or race

Source Own elaboration

(1) In terms of daily life space, the key point pertains to the classification discussion and motivation analysis of the rules of various daily activities in the settlements and the characteristics (mainly for residence, study, leisure, and shopping) of daily life space, which then superpose to generate the typical space–time trajectory of the daily life of the international student group as follows:

	Overall characteristics		Group differentiation		Space–time model
	Time characteristics	Space characteristics	Differentiation of social attributes	Differentiation of space attributes	
Residence space	The regularity of sleeping time is strong, while the time of housework is relatively scattered and takes up less time	There are two kinds of modes: the passive choice of apartments (dormitories) on campus and the active choice of communities off campus	[Nationality] The residence rates of international students in Europe, America, Southeast Asia, Africa, Japan, and South Korea are lower, higher, and equal to the average level, respectively; [Religion] The proportion of foreign students who believe in Christianity live on campus is the lowest, while the proportion of students who believe in Buddhism is the highest; the higher the frequency of religious activities, the higher the off-campus residence rate; [Level of integration] The proportion of international students with a high degree of integration who choose to live on campus is slightly higher than the average level, and the percentage of choosing to live off campus increases with the decrease in integration	[Set inside floor area] The area of the apartment on campus is smaller than off campus, single and double on campus, and one room for one person off campus [Facilities in the set] The off-campus community is relatively complete [Supporting facilities] Apartments on campus share school facilities, and there are only two types of facilities in the off-campus community: activity rooms and fitness equipment, and no learning space is available	"Single place, long-term event"

(continued)

(continued)

	Overall characteristics		Group differentiation		Space–time model
	Time characteristics	Space characteristics	Differentiation of social attributes	Differentiation of space attributes	
Study space	Working days are longer than rest days, and there are differences in the distribution of time periods, working days are concentrated and regular, and rest days are scattered and flat	There is a difference in travel distance between workdays and rest days, but the principle of proximity is generally followed, and the place of study is relatively fixed	[Degree type] The study time of graduate students is more than that of undergraduate students, and the regularity of undergraduate and nonacademic students is more obvious, and there is a big difference between working days and rest days; [Place of residence] The regularity of learning activities of students living on campus is stronger than that of students living off campus, and the proportion of students living on campus studying abroad is significantly higher than that of students living off campus	[Common features of attractive learning space] The number of people is relatively large, the learning atmosphere is relatively quiet, the layout type is mainly determinant and regular layout, and self-study is the main form	"Multi-location, average time distribution with limited distance"

(continued)

(continued)

	Overall characteristics		Group differentiation		Space–time model
	Time characteristics	Space characteristics	Differentiation of social attributes	Differentiation of space attributes	
Shopping space	There is a certain regularity, and there are obvious differences between working days and rest days	Follow the law of distance attenuation centered on the place of residence; there are differences between working days and rest days, and there are differences in the types of shopping within different distances	[Gender] Women have a higher proportion of shopping time, and the difference in shopping time between working days and rest days is greater; the space attenuation rate of men's shopping activities on working days and rest days is higher than that of women; the spatial structure of women's shopping activities is on working days. It is more different from rest days, and the types of shopping within different distances vary greatly [Level of consumption] The space distribution of shopping activities of international students with different consumption levels is roughly the same, while the decline rate of shopping activities of international students with high consumption level on working days and rest days is lower than that of international students with low consumption level, and the spatial structure of shopping activities is more different between working days and rest days	[Attractive shopping space commonality] Community-level business, within 1 km from the residential area, mainly in the middle and low grades, primarily selling daily necessities and shopping items	"The universality and decreasing outward of spatial distribution, and the outward increasing of time distribution."

(continued)

(continued)

	Overall characteristics		Group differentiation		Space–time model
	Time characteristics	Space characteristics	Differentiation of social attributes	Differentiation of space attributes	
Leisure space	There are significant differences in the time of various activities and obvious preferences. There are some differences in the time and types of activities between working days and rest days	Various types of leisure activities are more diverse in choice of venues, and there is a certain correlation between the locations and types of activities	[Gender] Female international students take up a higher proportion of leisure activities, and the difference between working days and rest days is greater, the location tends to be diversified during the day and close activities at night; while the proportion of men who have leisure activities within 1 km is slightly higher than that of women, the location presents diversified characteristics during the day and night [Level of integration] International students with a high degree of integration have a higher proportion of leisure activities and a richer variety, which tends to be outward-looking leisure activities; the low- and middle-integrated international students tend to single inward-looking leisure activities; the difference of space choice for leisure activities of middle-to-highly integrated international students on working days and rest days is even greater	[Attractive leisure space commonality] It is mainly indoor public space for communication activities, with a scale of about 10–30 people, close to the residence, and has a lively and free atmosphere	"The double-sided decrement of the location of the event and the randomness of the duration of the event"

(2) In terms of the time and space trajectory of daily life, there are obvious differences between international students on working days and rest days; different types of activities have different regularities in the use of time; after further spatiotemporal clustering, the daily life of international students can be divided into eight typical patterns, namely:
Working days: residence study, out study, out leisure;
Rest days: residence study, out study, shopping activity, residence leisure, out leisure.

(3) In terms of the evolutionary background and stage division—through the summary and analysis of the country's study abroad policy in China, the policy and scale of international students in Jiangsu Province, the scale of international students from Nanjing University and Nanjing Normal University, and the development history of Nanxiu Village and surrounding communities—the evolution of the International Student Settlements in Nanxiu Village can be divided into three stages, which represent the unformed stage of the International Student Settlements (Phase I: 1980–1999), the stage of development and growing (Phase II: 2000–2010), and the stage of maturity and perfection (Phase III: 2011 to present).

(4) In terms of community response and cause analysis, through the perspectives of residents, operators, and public service personnel in Nanxiu Village and surrounding communities, this paper selects four aspects that are closely related to international student settlements and can reflect the characteristics of "internationalization," including social attribute, space layout, industrial economy, identity and interaction, and the dynamic analysis of the response of its surrounding communities is as follows:

		Feature summary	Cause analysis
Social attribute	Nationality composition	The scale of international students is significantly increasing, the composition is diverse, and there are community differences	• The opening of the policy for international students in China • The influence of objective conditions of local community • The promotion of China's international influence
	Living custom	The composition of customs is becoming more abundant, and participation in customs is increasing, but religious activities are isolated from each other and weakly connected	• Cultural infiltration and integration in the context of globalization • The deep interaction barriers between community residents and international students

(continued)

(continued)

		Feature summary	Cause analysis
Space layout	Space structure	From a single function and a closed group to an open international community with points, lines, and multiple functions	• Community connection is becoming increasingly close • The development of commercial service industry • The highlight of international features • The dilemma of the renewal and development of the old town
	Public service facilities	From scattered distribution to belt-like extension, from localized commercial service industry outlets to a public service system with international features	• City location with superior conditions • A diverse and integrated humanistic environment
Industrial economy	Distinctive industry	In terms of scale, the number and proportion of exotic featured shops continue to increase in terms of business types, mainly restaurants and clothing stores, and features of Europe, America, Japan, and South Korea; in terms of space distribution, it shows from point to line to multi-axis extend	• The state has open policy to foreign private stores • The government promoted the renovation of the old city • The change of consumption concept of the locals
	Employment structure	From single industry dominance, it has gradually developed into a complex employment pattern with equal emphasis on multiple industries	• The advancement of housing marketization leads to the deunitization of community population composition • New employment opportunities brought by the increasing number of international students

(continued)

(continued)

		Feature summary	Cause analysis
Identity and Interaction	Cultural identity	The cultural identity of the locals toward international students is generally relatively low, but there are significant differences in psychology, interaction, and culture, and there are large differences between individuals in different occupations	• The inherent differences of thinking and habits of international students • The attitude of the locals in different occupations toward international students is different
	Social interaction	Individual interactions began to increase but were not deep enough, the age limit for participation in activities was large, the location and content of activities gradually diversified, and the space distribution showed a "multi-center plus commercial strip connection" model	• The increasing interest and tolerance of the locals in foreign cultures • Lack of access to in-depth understanding of foreign cultures

(5) In terms of policy management and adjustment strategies, on the one hand, it is necessary to improve the policy system of international students by improving the operating mode of the organization and the scholarship system. On the other hand, it is necessary to optimize the management and service system of international students to provide a strong guarantee for their study and life in China.

(6) In terms of overall function positioning strategy, it is recommended that the overall development positioning of the International Student Settlements in Nanxiu Village in the future is as follows: It is a multi-ethnic integrated settlement area that emphasizes on education, scientific research, and residence supplemented by administrative offices and commercial facilities, with internationalization and multi-ethnic distinctive culture as the highlight and traditional small-scale space features.

(7) In terms of land use integration strategies, it is necessary to guide land use functions at an overall level and coordinate the relationship between settlements and cities. Emphasizing the space layout model of high school and residential clusters in the settlement area and effectively exerting the international commercial features encourages the compound development of functions to meet the differentiated needs of people of different nationalities, occupations, and cultures.

(8) In terms of supporting facilities improvement strategies, on the one hand, it is necessary to establish supporting standards for public service facilities and determine the category and scale ratio of international featured facilities. On the other hand, it is necessary to give priority to the layout of international featured facilities and improve general facilities.

(9) In terms of social interaction promotion strategies, there is not only the need to improve the psychological and language and cultural support system for international students but also the need to provide international students with effective ways to interact with the locals.

(10) Regarding the integration strategy of religious customs, we must first look at the religious-based foreign culture brought by the international student group from a rational perspective. While accepting the excellent foreign culture, it prevents the cultural penetration with special purpose; then, by strengthening mutual cultural exchanges and guiding the establishment of social organizations for international students, the degree of integration between international students and the locals can be improved.

References

Agnes, H. 2010. Everyday life (Junqing Y, trans.). Chongqing: Chongqing Publishing House.

Bo L, Haiying W. A literature review of foreign researches. Journal of Gansu Administration Institute. 2010;2:18–27.

Chengbi L. New business operation type: Great change in urban consumption. Nanjing: Southeast University Press; 2004.

Chongqi W, Zhimou X, Weiqin W. Construction and examination of structural equation model of leisure participation, leisure satisfaction and subjective well being. J. Leisure Sports 2006;2006(5):153–165.

DESASD (Department of Economic and Social Affairs Statistics Division, UN). Recommendations on statistics of international migration. New York: United Nations; 1998.

Gang L. Cultural tradition and urban experience—Reflections on the Shanghai culture studies. J. Hangzhou Normal Univ. (social Science Edition). 2004;1:44–8.

Huntington SM. Clash of cultures. Beijing: Xinhua Publishing House; 2013.

Kim H. Ethnic enclave economy in urban China: The Korean immigrants in Yanbian. Ethnic Racial Stud. 2003;26(5):802–28.

Kosik K. Dialecties of the concrete. Boston: Reidel Publishing Company; 1979.

Longyu L. Study on the social interactions and cross-cultural adaptation of international students in China. Beijing: Beijing University; 2003.

Ma L, Xiang B. Native place, migration and the emergence of peasant enclaves in Beijing. China Q. 1998;155:546–81.

Portes A. The social origins of the cuban enclave economy of Miami. Sociol. Persp. 1987;30:340–72.

Rajan SI. International migration statistics and data sources India. ESCAP Regional Census Programme.

Smith P. Transnational urbanism. Malden: Blackwell; 2001.

Wu FL, Webber K. The rise of foreign gated communities in Beijing: Between economic globalization and institutions. Cities, 2004;21(3):203–213.

Xiaoming C, Ou Q, Wang Z. Investigation and analysis of the adaptability of international students in Beijing. Int. Chin. Lang. Educ. 2015;2015(1):19–32+198.

Xingzhong W. Research on the structure of urban social space in China. Beijing: China Science Publishing & Media; 2000.

Xuewei Z. Everyday life space. Shanghai: Tongji University; 2007.

Ying Z. Investigation and research on the current situation of foreign students' education in Universities of Jiangsu Province. Nanjing: Nanjing Normal University; 2016.

Yuan Z. Analysis of international students in China in terms of social interactions. Guangzhou: South China University of Technology; 2010.

Zhigang L, Xue D, Brown A et al. The African enclave of Guangzhou—A case study of Xiaobeilu. Acta Geographica Sinica 2008;62(2):207–218.

Zhigang L, Desheng X, Feng D, Ying Z. The local response of transnational social space under globalization in urban China: a case study of african enclave in Guangzhou. Acta Geographica Sinica, 2009;28(4):920–932

Chapter 3
Community Sample II: Spatial Resettlement and Social Integration: Research on Resettlement Space for Land-Lost Peasants

Abstract Following the focus on "international migrants" (international students) in the previous chapter, the author focuses on two groups of "internal migrants" that are closely related to the urbanization wave in China. This chapter first delves on "land-lost peasants," an emerging social group that has developed in conjunction with the rapid expansion of urban space and with the "land expropriation frenzy" in peri-urban areas. Different from migrant workers who went through urbanization actively under the "urban–rural" push and pull effect (see Community Sample V), this is a typical large group of people who went through urbanization passively, a group that was nurtured and generated due to land expropriation and relocation activities launched under the administrative order. In addition, within the social situation of land-loss and urbanization, the group passively clustered in various resettlement areas built by the government in a top-down manner. However, under the effects of a combination of factors including bias in urban development policies, lack of planning control standards, deficient land expropriation system, and the subject's own limited conditions, their resettlement areas face a series of problems: inappropriate site selection, limited housing conditions, community segregation, low compensation standards, single compensation method, difficulties in solving farmers' livelihood problems, etc. The citizenship of land-lost peasants has become one of the key points in the implementation of new urbanization. In the midst of the drastic social changes, whether they can transit from a familiar environment (the rural society) to an unfamiliar environment (the modern urban society) and whether they can actively complete the dual transformation of institutional and actual identities are of great public concern. Therefore, this chapter takes land-lost peasant resettlements as the research object, compares, and analyzes the social and spatial attributes of the resettlement space for this type of passive urbanized group and the reasons for the formation of this space. It goes on to explore the social integration patterns of land-lost peasants and their influencing mechanisms to further discuss the strategies and measures to improve the current space resettlement and social integration for land-lost peasants. The selection and examination of this community sample is essentially a specialized exploration of the theme of "integration-response" in the field of "land-lost peasants—land expropriation cities."

X. Wu and L. Wang, *Settlement Spaces: Urban Survival Prospects of China's Special Communities*, Springer Geography,
https://doi.org/10.1007/978-981-16-4892-2_3

Keywords Land-lost peasant resettlements · Spatial resettlement · Social integration · Nanjing · Strategy

3.1 Definition of Research Objects

3.1.1 Definition of Land-Lost Peasants

3.1.1.1 Land-Lost Peasants

Literature indicates that a clear understanding in terms of the concept of land-lost peasants has not yet been formed and unified. Since there is lack of corresponding concepts and connotations within foreign studies, the author mainly focuses on the directly relevant studies of domestic scholars. In addition, there is divergence in the understanding of land-lost peasants among the different studies, which mainly focuses on the causes of land-loss, peasant household registration, degree of land loss, spatial location, etc.

The Ministry of Human Resources and Social Security defines land-lost peasants in the "Guidance on Employment Training and Social Security Work for Peasants whose Land have been Expropriated" as peasants who, "on account of the government's unified expropriation of their collective land, have lost their land or most of their land, whereas their possession of land is insufficient to sustain their basic livelihood".

According to the causes of formation, there are two types of land-loss: active and passive (Yang et al. 2009). Active land-lost peasants are social workers who were originally rural registered with permanent residence and work in rural areas. They, however, actively gave up the right to cultivate land, left the countryside or the agriculture industry, and entered urban areas. It is considered that active land-lost peasants mainly comprise migrant workers staying in the city, students who failed the College Entrance Examinations, and certain administrators of rural communities. As for passive land-lost peasants, most scholars consider them as peasants who lost their land due to various reasons such as development and construction by governmental departments in the process of urbanization. According to the locations, Ding (2009a) believes that land-lost peasants can be divided into three types: those living in urban built-up areas, in urban fringe areas, and in deep rural areas. According to the degree of land loss, Jin and Lu (2007) suggests that farmers whose contracted land has been fully or partially expropriated by law can be defined as land-lost peasants, whereas Wang (2005) considers land-lost peasants as those who have lost all their land and occupy less than 0.3 mu of arable land per capita. Ni (2004) considers them as those who have lost all their land due to state expropriation. Depending on the degree of land-loss and the local policies regarding the transfer from agricultural to non-agricultural household, land-lost peasants can also be classified as rural (Deng 2004) and urban registered permanent residents (Su 2005; Lei 2005) according to their household registration status.

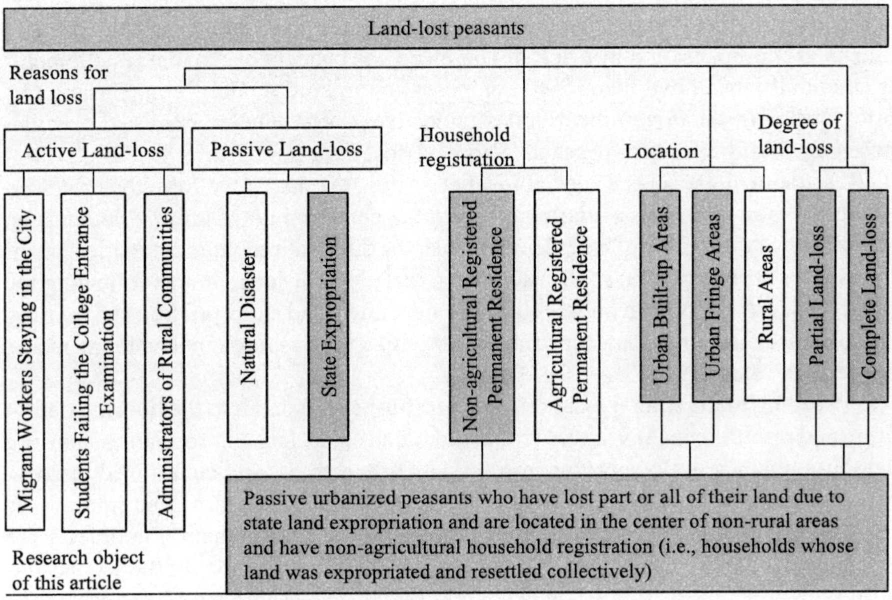

Fig. 3.1 Definition of land-lost peasants. *Source* Own elaboration

3.1.1.2 Land-Lost Peasants Studied in This Research

Based on the above definitions, the "land-lost peasants" studied in this chapter mainly include passive urbanized peasants who have lost part or all of their land due to state land expropriation in the urbanization process since the reform and opening-up policy was launched, are located in the center of non-rural areas, and have non-agricultural household registration (i.e., households whose land is expropriated and who are resettled collectively) (see Fig. 3.1).

3.1.2 Definition of Spatial Resettlement

3.1.2.1 Spatial Resettlement

At present, the compensation and resettlement for land-lost peasants mainly includes resettlement with one-time monetary compensation, resettlement with dividends, resettlement with affordable housing, and resettlement with compensation in kind. Of these, resettlement with monetary compensation refers to paying certain compensation to households whose land has been expropriated, such that they can use the money to buy other commercial houses or apply for resettlement houses at a preferential price (also known as "compensation by individual"). Conversely, compensation in kind refers to evaluating the approved building area of the expropriated land

and converting it into resettlement housing according to a certain percentage (also known as "compensation by brick"). Therefore, the concept of "spatial resettlement" is essentially the above-mentioned compensation in kind model: "compensation by brick" wherein the government builds collectively resettlement areas and provides necessary living space for peasants who have lost their land.

Resettlement areas, as a special form of living area that arose from the urbanization process in China, essentially belong to the construction of affordable housing. According to the causes of resettlement, it can be divided into three categories: resettlement for reconstruction in disaster areas, resettlement for migrants following the implementation of certain policies, resettlement after land expropriation and housing demolition caused by urban construction activities such as tourism development and property development (Zhang 2013).

According to the spatial location, it can be further divided into the housing demolition and resettlement of villages within old urban areas, land expropriation, housing demolition and resettlement to construct new urban areas, and centralized resettlement for the construction of new rural areas (multiple villages merged into one or the whole village being moved). For the first type of resettlement, the objects are residents living in run-down urban areas and urban villages, while for the second and third types, the objects of resettlement are rural residents (with land-lost peasants as typical representatives).

Spatial Resettlement Studied in This Research

On the basis of sorting out related concepts, the "spatial resettlement" studied in this chapter focuses on land-lost peasants settlements, namely, expropriated land and resettlement areas built for land-lost peasants as a result of urban construction activities during the urbanization process, which essentially belong to the type of resettlement areas built for disadvantaged communities by governmental organizations and constitute one of the important components of China's affordable housing policies (see Fig. 3.2). At present, there are three main types of resettlement housing for land-lost peasants in Nanjing: scattered resettlement areas in the center of the city, evenly distributed resettlement areas in industrial parks, and large mixed resettlement areas at the fringe of the city.

3.1.3 *Definition of Social Integration*

To date, domestic and foreign scholars have not reached a consensus on the definition of social integration due to differences in research perspectives, dimensional framework, and research priorities. However, there has been a consensus on the dynamic, multi-dimensional, and multi-level characteristics of social integration and as an acknowledgement that "social integration is at the same time an objective and

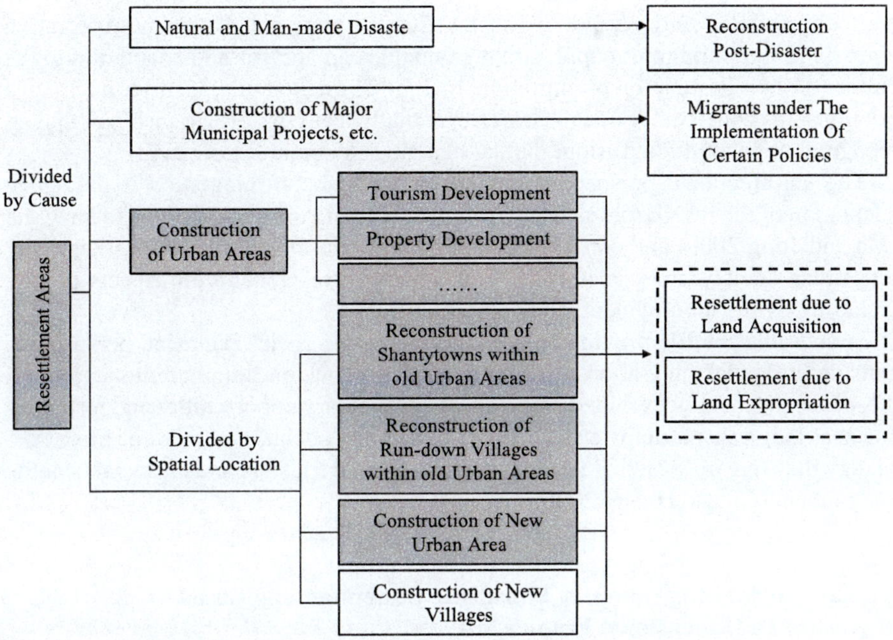

Fig. 3.2 Types of resettlement areas. *Source* Own elaboration

a means, a process and a result, and includes subjective psychological integration and objective institutional integration."

3.1.3.1 Social Integration: Multiple Understandings from Different Research Perspectives

Understanding of social integration can be divided into that perceived from a bidirectional perspective, from a unidirectional perspective, and from a social exclusion perspective. The specific understanding and relevant studies are as follows:

The bidirectional perspective emphasizes the joint role and mutual effect of the integrators and the community being integrated. The typical foreign study results are represented by Gordon (1964), who believe that social integration is the process of mutual penetration and interaction between migrants and local residents who share their respective cultural memories and adapt to each other. The migrants also interact with the local urban area and merge into a common cultural life with local residents (Park and Burgess 1924). Alba and Nee (2005) modified this to some extent, arguing that social integration is the reduction of ethnic differences and the reduction of cultural and social differences that result from ethnic differences . Coincidentally, domestic scholars Ren and Wu (2006) also believe that "social integration is the process of mutual cooperation and adaptation between different individuals, groups, and cultures (Ren and Wu 2006)." Zhou (2012) points out that social integration consists in the immigrant population's gradual acceptance of and adaptation to the culture of their destination of migration, making efforts to build positive, interactive relationships with local residents, which eventually leads to s a relationship of mutual recognition, penetration, fusion, mutual benefits, and complementation.

The unidirectional perspective emphasizes that social integration is the active adaptation of the integrator to the environment wherein the person wants to integrate (Ma and Tong 2008) and considers social integration as "the process of integrating into urban society and transforming into an urban resident in all aspects of life, including living, employment, and common values."

In addition, the EU provided a definition from the social exclusion perspective, considering social integration as the process of ensuring that communities at risk and those that are socially excluded have access to necessary opportunities and resources and that they can participate fully in the economic, social, and cultural life in the society they live in, such that they can enjoy a normal lifestyle and all social benefits due to them (Ga and Huang 2008).

3.1.3.2 Social Integration: A Pluralistic Understanding Based on a Dimensional Framework

Scholars tend to define social integration from a multidimensional perspective. For example, Ratna Omidvar argues that social integration should bring people who are isolated from each other closer together spatially, socially, and economically. In contrast, controversies and differences still exist in the research results on the theoretical framework and measurement dimensions of social integration in China. However, after reviewing certain international indicators, Lang et al. (2007) proposed that social integration should include material benefits, employment opportunities, equal access to education and health care, and social benefits. Yue and Li (2012) also argues that social integration refers to "(the integrators) reducing the differences with urban citizens in terms of culture, socio-economic status, and psychology." Zhu (2002) argues that the multiple dimensions of social integration are sequentially progressive, of which the economic dimension is the foundation of integrating into urban areas, the social dimension is the further requirement of urban life integration, and the psychological dimension is the integration on spiritual level. This indicates the degree of integrators' participation into urban life and that only when psychological adaptation is being accomplished can real social integration be achieved. In addition, some research results include institutional and spatial dimensions into the evaluation system of social integration and point that the institutional dimension provides the basic conditions and policy background for social integration, while the often-neglected spatial dimension provides important means and spatial carriers for social integration (Zhao and Li 2013).

3.1.3.3 Social Integration Studied in This Research

Based on the principle of maintaining the openness, inclusiveness, and neutrality of the concept, and for the social integration analysis to be operable, the "social integration" to be studied in this chapter refers to the unidirectional process of the land-lost peasant integrating into urban areas and transforming into urban residents (see Fig. 3.3, and the evaluation index system of this study is described in Sect. 3.4.1).

3.2 Overview of Land-Lost Peasants' Background

3.2.1 Background for the Emergence of Land-Lost Peasants

Since the reform and opening up, the population of land-lost peasants in China has exceeded 60 million due to urban expansion and non-agricultural construction activities and is expected to exceed 100 million by 2020. Nanjing, as an important central city in the Yangtze River Delta region, has been undergoing land expropriation and relocation since 1983. With the spillover of urban functions, promotion of new economic growth points, and continuous improvement of the urban transportation system, urban construction land has been expanding outward constantly. To meet the demand for urban development, a large amount of agricultural land has been expropriated, which resulted in a large number of peasants losing their land and entering urban areas.

Based on the urbanization level of 1 km^2 of land per 10 thousand people, for each percentage of advancement in the process of urbanization, 12.7 million peasants would be urbanized and 12.7×10^4 km^2 of agricultural land will be lost (Wang 2008). By analogy, in the future, along with the outward expansion of urban space and relocation of industrial land, it is expected that an additional 200,000 km^2 of

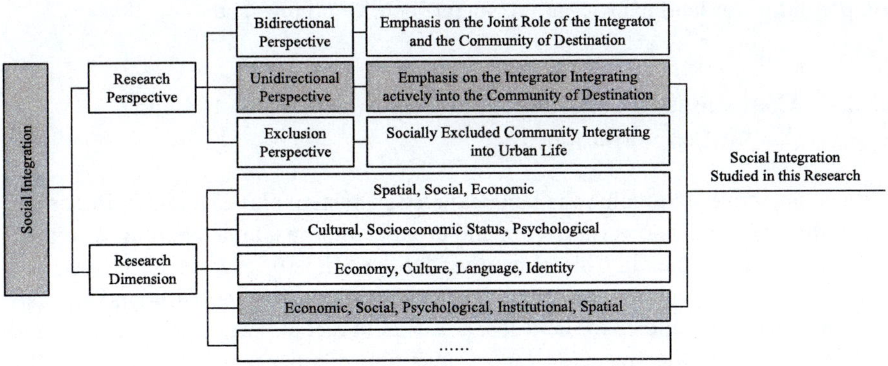

Fig. 3.3 Definition of social integration. *Source* Own elaboration

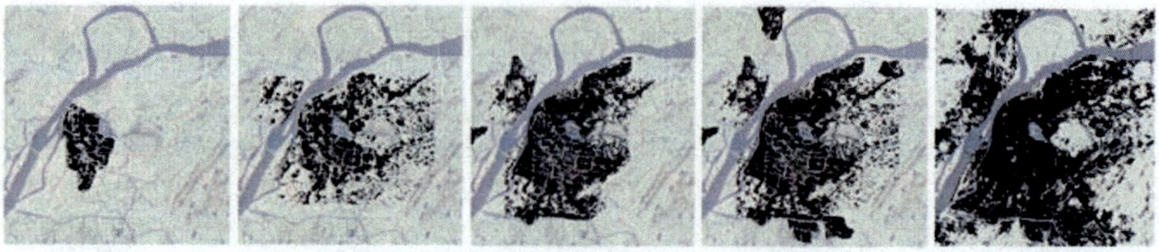

Fig. 3.4 Evolution of Nanjing's construction land expansion. *Source* Own elaboration based on relevant references

land will be acquired each year as a result of which 3.75–5 million people will lose their land. From 1978 to the end of the twentieth century, the built-up area of Nanjing city was mainly confined to the old urban area within the Ming City Wall. Nanjing entered a phase where the renewal of the old urban area and external expansion took place rapidly and simultaneously. From 1980 to 2010, the built-up area of Nanjing rapidly increased by about 400 km^2 (see Fig. 3.4), and the total number of land-lost peasants reached 512.8 thousand, involving more than 100 thousand families (http://news.163.com/13/0314/06/8PTJOB8F00014AED.html). With the influx of a large group of land-lost peasants, the spatial pattern of urban living areas will undoubtedly change significantly. However, the successful integration of land-lost peasants into the city depends to a large extent on relevant policies. Therefore, it is particularly important to sort the housing demolition and resettlement policies and compensation policies.

3.2.2 *Policies Related to Land-Lost Peasants*

Land-less peasants are one of the three groups difficult to work with considered in Nanjing's housing security policy, and policies related to them are focused on the compensation for land expropriation, social security policies, etc.

3.2.2.1 Compensation Policy for Land-Lost Peasants Due to Land Acquisition in Nanjing

From an objective standpoint, the land acquisition system played a highly important role in the process of industrialization and urbanization in China. Nanjing, as one of the first nine pilot cities in China identified by the Ministry of Land and Resources to reform the land acquisition system, has actively experimented with land acquisition compensation policies for land-lost peasants in recent years (see Table 3.1) and made three major adjustments to land acquisition compensation policies in 2004, 2010, and 2015 (see Table 3.2). In April 2004, Nanjing officially launched the new "Compensation and Resettlement Measures for Land Acquisition and Expropriation

Table 3.1 Evolution in the compensation system for land expropriation in Nanjing over the years

Stage	Year	Related Policies	Characteristics of the Stage
Stage I	Before 1996	–	The amount of expropriated land is small, conflicts were not prominent, and there were no formal policies and regulations on the compensation for land acquisition
Stage II	1996–2004	Notice No. 13 (1993) of the People's Government of Nanjing Municipality Notice No. 86 (2000) of the People's Government of Nanjing Municipality Notice No. 21 (2001) of the People's Government of Nanjing Municipality	Resettlement for land-lost peasants was completed by the acquiring organization or company
Stage III	2004–2010	Notice No. 93 (2004) of the People's Government of Nanjing Municipality Notice No. 61 (2004) of the People's Government of Nanjing Municipality Notice No. 100 (2004) of the People's Government of Nanjing Municipality	One-time monetary resettlement for peasants whose land has been expropriated
Stage IV	2010–2015	Notice No. 264 (2010) of the People's Government of Nanjing Municipality	Compensation for land acquisition is based on a comprehensive pricing model
Stage V	2015 up to now	Notice No. 15 (2015) of the People's Government of Nanjing Municipality	Land expropriation compensation policies were formulated separately for state-owned and collective land, and a relatively complete compensation system for land expropriation has been established

Source 1. Cao, Guilin., (2013). *Research on the Model of Compensation and Resettlement of Expropriated-land in Nanjing* [D]. Sichuan Agricultural University

2. Compensation Measures for Residential Houses built on acquired Collective Lands and Resettlement Measures for its Residents in Nanjing. [EB]. http://www.nanjing.gov.cn/xxgk/szf/201506/t20150617_3383966.html. (visited on May 29, 2015)

Note The area price compensation policy implements the principle of same price for same land. In other words, within a certain area designated by the government, compensation for land expropriation adopts a unified standard. Regardless of whether it is acquired for public welfare or commercial use, its compensation price will not change

Table 3.2 Comparison of compensation and resettlement policies for land expropriation in Nanjing over the years

Year	2004	2010	2015
Compensation and resettlement method	Monetary compensation, Self-demolition and Self-construction		Monetary compensation, centralized construction
Compensation standard	Comprehensive compensation standard	Area price compensation policy	
Compensation and resettlement fee for land acquisition	Land compensation fee, labor-force resettlement fee, crop compensation fee, attached buildings compensation fee, housing expropriation compensation fee	Area land acquisition compensation fee (combination of the resettlement fee and land compensation fee), comprehensive compensation fee for crops and attached buildings	
Housing compensation	Compensation for the original house, compensation for house purchase, compensation for location		
Size for resettlement housing	No explicit regulation		$\leq$ 30 m^2 per capita $\leq$ 220 m^2 per household

Source Nanjing Provincial People's Government. Compensation and Resettlement Measures for Land Acquisition and Expropriation in Nanjing. [EB]. (http://www.njgt.gov.cn/default.php mod = article&do = detail&tid = 220,334.2004–03–30)
Nanjing Provincial People's Government. Compensation and Resettlement Measures for Land Acquisition and Expropriation in Nanjing. [EB]. (http://www.nanjing.gov.cn/xxgk/szf/201012/t20101229_1182874.html.2010-12-27)
Nanjing Provincial People's Government. Compensation Measures for Residential Houses built on acquired Collective Lands and Resettlement Measures for its Residents in Nanjing [EB]. (http://www.nanjing.gov.cn/xxgk/szf/201506/t20150617_3383966.html. 2015–05–29)

in Nanjing" and the "Compensation and Resettlement Standards for Land Acquisition and Expropriation in Nanjing." In 2010, Nanjing completed outlining a comprehensive pricing strategy for the acquisition of land areas, developed the land acquisition and housing expropriation working system, implemented a new approach wherein the management of land acquisition and land expropriation are separated, which led to innovation in the housing expropriation model for collective land areas. In 2015, Nanjing also issued the "Implementation Rules in terms of the Expropriation Measures for Residential Houses built on State-owned Land and the Resettlement Measures for its Residents in Nanjing," and the "Compensation Measures for Residential Houses built on acquired Collective Lands and Resettlement Measures for its Residents in Nanjing." The compensation and resettlement for collective land expropriation has been defined in a more targeted and reasonable manner. The evolution of the compensation system for land expropriation in Nanjing can be divided into five stages as follows:

Notably, in 2010, Nanjing adopted a new comprehensive area price compensation policy for land acquisition. This method no longer uses the multiples of land output value to measure compensation fees but takes into account various factors such as land use, land locations, local economic, and social development level, and the relation between land supply and demand while setting the specific price for a certain area. It also fully considers the value of the land as a means of production and social security for peasants. Thus, under this policy, the land of Nanjing was divided into three levels: first-degree expropriated land, seconddegree expropriated land, and third¯degree expropriated land, and the corresponding area prices were 148.4 thousand yuan/mu, 111.1 thousand yuan/mu, and 80 thousand yuan/mu, respectively (2016)[21]. This policy not only innovated the compensation method but also realized a gradient improvement in the compensation amount.

3.2.2.2 Social Security Policy for Land-Lost Peasants in Nanjing

The social security system for land-lost peasants in Nanjing is also in the process of being adjusted continuously. Due to the economic conditions and relevant policies, land-lost peasants have been basically excluded from the dualistic social security system since the founding of the country. Since the reform and opening up of the country, more attention has been brought to land-lost peasants along with the rapid development of the social economy. After several explorations, the urban and rural social security systems have basically completed the reform process from "resettlement through training and enterprise recruitment" to ""monetary resettlement" and then to "basic living security resettlement" (2010). The security and resettlement method for land-lost peasants has also become increasingly complete and humane. It is expected that theoretically (or policy-wise), its social security system will be fully integrated with the system in urban areas. The current social security system applies different standards for different age groups (see Table 3.3).

Different from previous policies that only solved immediate problems without tracing back to remaining problems from the past, the policy implemented in Nanjing since 2010 aimed to fully solve the social security problems of nearly 500,000 peasants from 1983 to present. A series of bridging policies ensured that people whose land had been expropriated at all stages would be able to find specific channels to enter into the insurance system. Once the new and old policies were implemented, the land-lost peasants would have obtained full coverage of social security, and the minimum living allowance and welfare benefits would have increased substantially (see Table 3.4).

In general, the specific group of land-lost peasants received greater focus in terms of compensation and social security, but there many problems remain. First, the basic living security policies are yet to be improved, with medical, employment injury and maternity insurance being inadequate in particular. Second, the development needs are not met. For land-lost peasants, the strengthening of their vocational skill training will be one of the key measures to avoid being marginalized in the fiercely competitive work environment in urban areas and ensure their long-term security.

Table 3.3 Social security policy for land-lost peasants in Nanjing by age group

Age of land-lost peasants	Social security item	Yes/No
1st age group	One-time monetary compensation	√
	Minimum living security	√
2nd and 3rd age groups	Pension insurance	•
	Health insurance	O
	Unemployment insurance	•
	Employment Injury Insurance	O
	Maternity insurance	O
	Minimum living security	√
	Vocational skills training	√
	Education allowance	O
4th age group	Pension insurance	•
	Health insurance	O
	Minimum living security	√

Source Nanjing Provincial People's Government. Social Security Measures for Land-lost Residents in Nanjing [EB]. (http://rsj.njgl.gov.cn/art/2015/7/15/art_13301_2270299.html, 2010–12–16)
Note 1.1st age group: under 16 y/o; 2nd age group: females between 16 y/o and 45 y/o, males between 16 y/o and 50 y/o; 3rd age group: females between 45 y/o and 55 y/o, males between 50 y/o and 60 y/o; 4th age group: females older than 55 y/o, males older than 60 y/o
2. √—included • existence of similar policies O—not included

Third, the dynamic adjustment mechanism is not complete. Although the amount of social security for land-lost peasants in Nanjing has been increased, it will still be insufficient compared with the actual living expenses in the area.

3.2.3 Overview of Spatial Resettlement for Land-Lost Peasants

Nanjing, one of the key cities in the empirical studies on poverty space, has been promoting the construction of subsidized houses on a large scale since 2002. The subsidized housings in Nanjing can be specifically divided into four categories: economically affordable housing, resettlement housing for land-lost individuals, low-rent housing and medium- and low-priced commercial housing. In terms of the target population of Nanjing's affordable housing, there are two major groups: families whose housing was expropriated and who were relocated in the process of urban development and construction, and middle- and low-income families with housing difficulties in the city.

Table 3.4 Comparison of social security policies of land-lost peasants in Nanjing over the years

Year	2004	2010	2015
Group	People with basic living security	Urban population with social insurance	Urban population with social insurance
Minors (<16 y/o)	6 k/person	10 k/person	The minimum standard of resettlement allowance stipulated by the provincial government
Working age group (16 y/o ≤ Female ≤ 55 y/o) (16 y/o ≤ Male ≤ 60 y/o)	I. When you have not reached the pension age, the second age group enjoys a 2-year living allowance of 320 yuan/month, and the third age group enjoys a 10-year living allowance of 200 yuan/month II. When reaching the pension age, each individual can enjoy a basic living security payment of 440, 500, 557, 616, 675 yuan per person per month according to the individual's contribution	III. When you have not reached the pension age, unemployment insurance of approximately 634 yuan/month for 24 months; 130 yuan/month of medical insurance allowance; 400 yuan/quarterly of commodity price allowance; 1,000 yuan/year of Spring Festival condolence payment IV. When reaching the pension age, 745 yuan/person/month	V. When you have not reached the pension age, the second age group enjoys a 2-year living allowance of 300 yuan/month, and the third age group enjoys a 10-year living allowance of 280 yuan/month VI. When reaching the pension age, each individual can enjoy a basic living security payment of 460, 515, 566, 620, 673 yuan per person per month according to the individual's contribution
Retirement age group (Female ≥ 55 y/o) (Male ≥ 60 y/o)	440, 470 yuan/person/month	Around 700–710 yuan/person/month	Around 790–900 yuan/person/month

Source Nanjing Provincial People's Government. Basic Living Security Measures for Land-lost Peasants in Nanjing (for trial implementation) [EB] (http://www.nj.gov.cn/xxgk/qzf/jyq/jyqrmzfxlbsc/200803/t20080328_2088147.html, 2004–04–09)
Nanjing Provincial People's Government. Social Security Measures for Land-lost Individuals in Nanjing [EB] (http://rsj.njgl.gov.cn/art/2015/7/15/art_13301_2270299.html, 2010–12–16)
Nanjing Municipal Human Resources and Social Security Bureau. Notice on the Adjustment of Social Security Benefits for Land-lost Individuals in 2016 [EB]. (http://www.nanjing.gov.cn/xxgk/bm/rsj/201505/t20150504_3293108.html, 2015–04–27)

3.2.3.1 Housing Types in the Resettlement Area for Land-Lost Peasants in Nanjing

As mentioned above, there are many types of compensation and resettlement for land-lost peasants, such as one-time monetary compensation, resettlement with dividends,

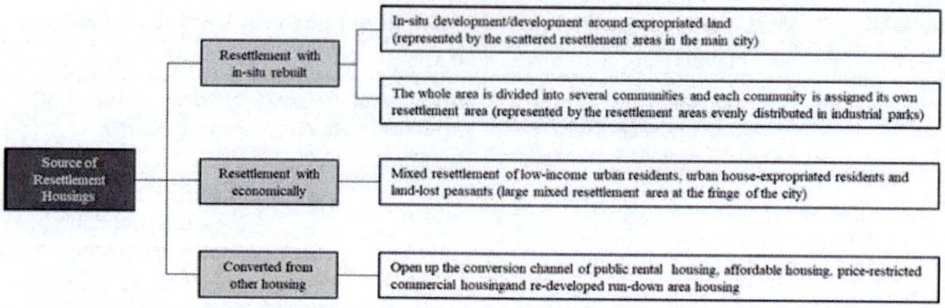

Fig. 3.5 Resettlement Housing Types in Nanjing. *Source* Wang et al. (2009). Ministry of Land and Resources. Notice on Optimizing the Supply Structure of Housing and Land in 2015 to Promote the Stable and Healthy Development of the Real Estate Market [EB]. (http://www.mlr.gov.cn/zwgk/zytz/201503/t20150327_1346321.htm, 2015–03–27)

resettlement with affordable housing and resettlement with compensation in kind. The "resettlement area for land-lost peasants" focused on in this study is essentially a centralized manifestation and result of the above-mentioned "resettlement with compensation in-kind" model and has been differentiated into various types during the policy exploration and implementation process by the government.

In the case of Nanjing, there are three main types of resettlement housings provided for land-lost peasants up to date. In terms of land acquisition mechanisms, resettlement housing can be divided into two types: resettlement with in-situ rebuilt housing and centralized resettlement and economically affordable housing. In addition, major cities are exploring actively new supply modes of resettlement housing by converting other housing sources that meet certain conditions. This is gradually becoming a third type of option for resettlement housing supply (see Fig. 3.5).

The first type of housing (resettlement with in-situ rebuilt housing) can be subdivided into two types. 1 In the earlier process of suburbanization, while commercial housings were developed around the expropriated land, part of the land was set aside for the building of resettlement housing. In other words, developers were appointed by the government to build housings at a small profit (i.e., the scattered rehousing areas in the main city). 2 The second model has been more commonly adopted in industrial park projects in the process of industrial suburbanization since the twenty-first century, wherein the whole area is divided into several communities and each community is assigned its own resettlement area. The land is administratively allocated, and the government designates or adopts the bidding form to determine the developer for house-building at a small profit (i.e., the resettlement areas evenly distributed in industrial parks).

The second type of housing (integrated into the economically affordable housing category) is a model that has been more commonly adopted in the twenty-first century along with the industrial suburbanization process. The government's construction plan for economically affordable housing included resettlement housing for low-income urban residents, urban house-expropriated residents and land-lost peasants and the land for construction is allocated by the local government gratis (i.e., large

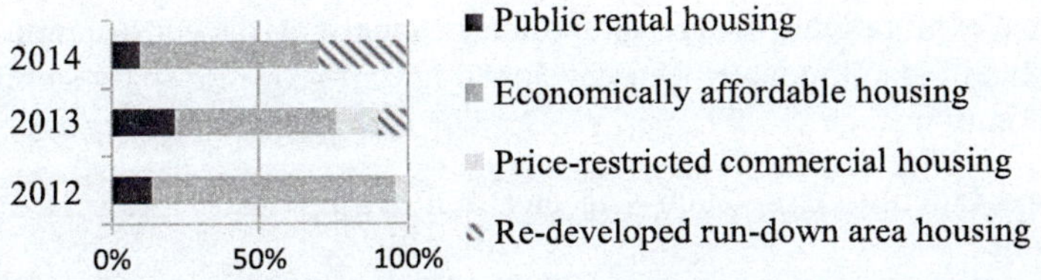

Fig. 3.6 Percentage of different housings in the subsidized housings that went into construction in 2012–2014. *Source* Nanjing Department of Subsidized housing: http://www.njszjw.gov.cn/ywtd/zfbz/bzfzt/

mixed resettlement area at the fringe of the city). According to statistics, economically affordable housings accounted for 82, 54, and 60% of the new subsidized housing constructions in Nanjing from 2012 to 2014 (see Fig. 3.6), while resettlement housing for land expropriated residents accounted for 86% of the total economically affordable housings. Among the resettlement housings, 70% were for residents living on collective land, 20% were for residents on urban land, and the remaining 10% were for urban low-income residents with housing difficulties (Guo et al. 2011a).

In addition, the third type of housing (converted from other small housing sources that meet certain conditions) is also a source of resettlement housing. Considering the long construction period of resettlement housing, opening up the conversion channels from commercial housing, economically affordable housing and resettlement housing and further integrating public rental housing, economically affordable housing, price-restricted commercial housing, and re-developed run-down area housing into resettlement housing plans can not only alleviate the short supply problem of resettlement housing but also revitalize stock properties and promote healthy development of the real estate market. However, given the large uncertainty and liquidity of such resettlement housing sources, it is not the focus of this chapter for the time being.

In summary, in addition to in-situ re-built housing, economically affordable housing, as an important form of subsidized housing, can also solve the resettlement problems for house-expropriated residents (including land-lost peasants). Therefore, the first two types of resettlement housing mentioned above will be the main objects of sampling research and comparative analysis in this study.

3.2.3.2 Construction History of the Resettlement Area for Land-Lost Peasants in Nanjing

The commercialization of housing has stimulated the rapid growth of the local economy and maintained housing prices at a high level. The People's Government of Nanjing Municipality is committed to solving the problem of spatial resettlement of land-lost peasants and has, therefore, implemented a series of institutional reforms and practical explorations. As a result, policies concerning the construction of resettlement areas in Nanjing have been developed and improved step by step from the initial exploration stage.

Reform and Opening Up—2000—accelerated urbanization and scattered resettlement

In the 1990s, Nanjing entered the phase of accelerated urbanization, resulting in land-loss for a large number of peasants. Nanjing's resettlement housing projects can be traced back to the construction of "comfort housing" between 1995 and 1998, which mainly focused on the housing problems of specific groups, i.e., land-expropriated households (including land-lost peasants). In 1998, with the deepening of the housing system reform and booming of the real estate industry, Nanjing formally launched affordable housing construction projects to provide basic living conditions for land-expropriated residents, but the resettlement housing during that period was mostly rebuilt in situ in the main urban area and their spatial distribution was more scattered.

2000–2010—Renovation project of the old city initiated; vigorous development of centralized resettlements

Since 2000, along with the accelerated urbanization, Nanjing's old city renovation project was officially launched, but only 24 families whose houses were expropriated were resettled in the same year. The rapid development period reached its peak in 2006, when 3,716 families (including land-lost peasants) whose houses were expropriated were resettled in the same year. From 2008 to 2010, Nanjing had another peak of land expropriation, with a total expropriation area of 3.8 million m^2 in three years and 3 million m^2 of dilapidated houses in the city demolished.

During this period, Nanjing planned to build 54 subsidized housing communities with a total land area of 1280 ha to solve the housing difficulties of low-income people in the city, i.e., provide centralized resettlement housing for residents whose houses were located on state-owned land and collective land.

2010 to present—Decline in land expropriation and large scale of mixed resettlement

With the number of land expropriation decreasing on a yearly basis, only 313 households whose land was expropriated were resettled in 2010 and 2011, marking the gradual decline of urban renewal and renovation of dilapidated houses in the inner city of Nanjing. By April 2016, Nanjing had established the "New Five-Year Plan" for the transformation of run-down urban areas and is expected to renovate about 15 million m^2 of land, involving 197 urban villages and nearly 100 thousand residents (including land-lost peasants) (http://www.tuliu.com/read-24869.html). To resolve the contradiction between supply and demand of housing, Nanjing prepared the "Twelve Five-Year Plans for Housing Security in Nanjing." From 2010 to 2012, four major housing projects with a total area of about 9.7 million m^2 and 82,800 housing units (Hang and Qian 2013) were built in the urban fringe area: Maigaoqiao Innovation Park, Huagang, Xishanqiao Daishan West, and Jiangning Shangfang. They showed an obvious and rapid growth in the quantity and scale of resettlement housing, which were developed in the mixed resettlement mode in urban fringe areas.

3.2.3.3 General Distribution of the Resettlement Areas for Land-Lost Peasants in Nanjing

As mentioned above, the resettlement space for land-lost peasants in Nanjing mainly includes two types of subsidized housing: housing rebuilt in-situ and economically affordable housing. Therefore, there is a high coupling correlation between the resettlement areas of land-lost peasants and subsidized housing in terms of location and spatial distribution (see Fig. 3.7). To discuss the general distribution of resettlement areas for land-lost peasants, overall observation of subsidized housing is also necessary. Furthermore, the spatial distribution of resettlement areas for land-lost peasants is closely related to their quality of life. Improper location selection often leads to real problems such as employment difficulties, long commuting distances, difficulties in sharing facilities, and low-level social integration of residents. Therefore, it is necessary to first study the general distribution characteristics of resettlement areas (or subsidized housing) for land-lost peasants in Nanjing. The findings are as follows:

Unevenly distributed space between administrative regions
Through wide mapping and preliminary research of the resettlement areas of Nanjing, it was found that the construction of resettlement areas for land-lost peasants between different administrative districts had not evenly progressed in the city due to the

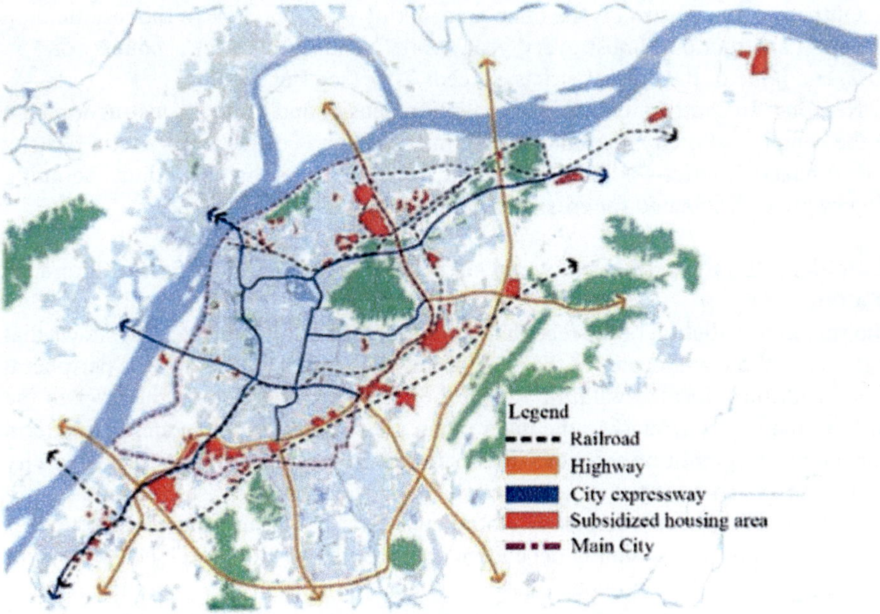

Fig. 3.7 Spatial Distribution of Subsidized Housing in Nanjing in 2015. *Source* Chen, Shuangyang. *Study on the Spatial Distribution of Large-Scale Subsidized Resettlement Housings in Eight Districts Located South of Yangtze River of Nanjing*[D]. Nanjing: Southeast University, 2012:31

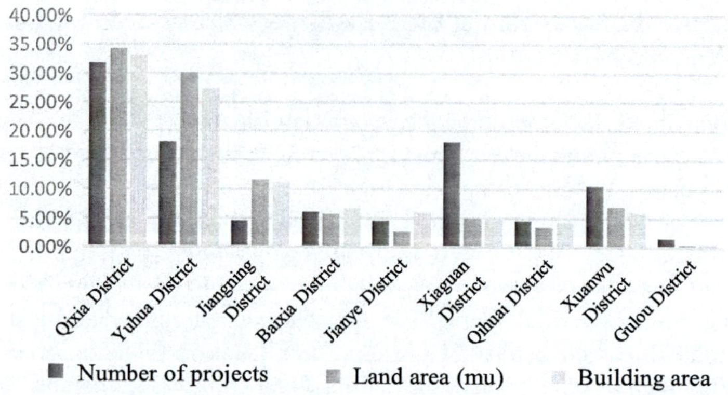

Fig. 3.8 The distribution ratio of resettlement housing and economically affordable housing in each district of Nanjing. *Source* Zhang et al. A Study on the Spatial Structure Evolution of Subsidized Housing in Nanjing Based on Historical Data [J]. Modern Urban Research. 2013(3):108

different stages of urbanization and operation modes of each administrative district. From the viewpoint of the entire city, Qixia District, Yuhuatai District, and Jiangning District have always been the main resettlement areas for land-lost peasants in Nanjing and the number of constructed resettlement houses in these three districts accounts for more than two-thirds of the total volume (Guo et al. 2011a).

Of them, Qixia District is the main area, taking up about 50% of the resettlement housing construction almost every year, while Yuhuatai District accounts for 15–30%, and none of the other districts exceeds 20% (see Fig. 3.8).

Reason—In contrast to the high population density and limited construction land in the center of old city, the land supply in fringe areas of the city—Qixia District and Yuhuatai District—is relatively abundant and carries low cost, thus becoming the key areas designated for cross-district resettlement in the city.

Built along the bypass highway and surrounding the main city in the shape of a fan

The spatial distribution of the resettlement areas in Nanjing (see Fig. 3.9) shows that majority of them are located at the edge of the main urban area and in the peripheral areas, with 60% located within 3 km of the city bypass and along the river in the north of the city, surrounding the main urban area of Nanjing in an annular pattern. The intensive demolition activities in a dispersed dot pattern on the streets of the city are in sharp contrast to the large-scale housing development constituting clusters in the peripheral and suburban areas of the city. This is because the construction of resettlement housing is mainly dependent on government-allocated land, which will inevitably reduce land revenue.Reason—Under the current compensation mechanism for land expropriation, the commercial housing developed on expropriated land often carries a high selling price, considerable increase in the living area, and enhancement in the environment compared with the old housing of residents. For

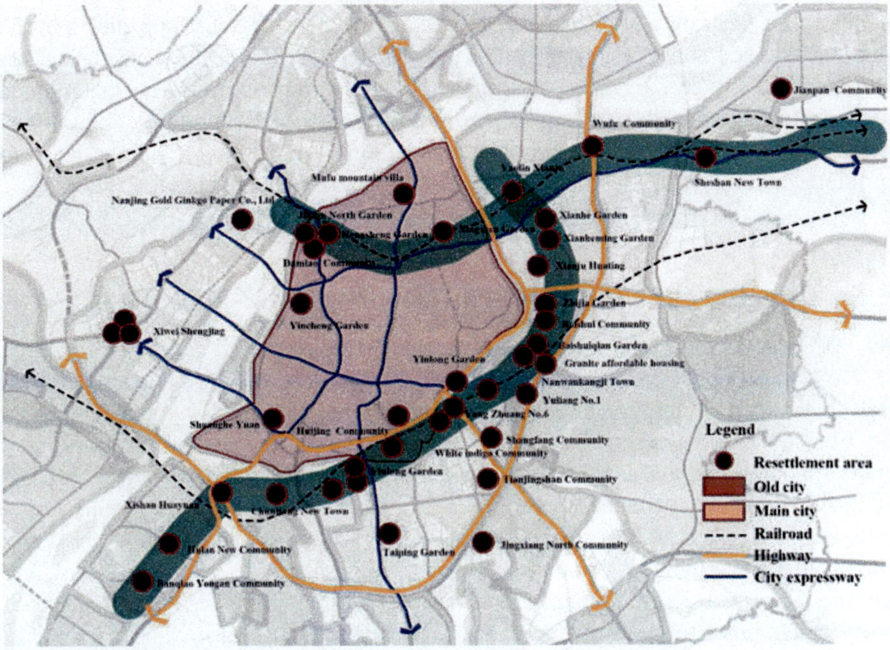

Fig. 3.9 Spatial distribution of major resettlement areas for land-lost peasants in Nanjing (2016). *Source* Own elaboration

land-expropriated households urging to move back, these housings are unfortunately unaffordable, and they often have to move to centralized resettlement areas.

Located in a dispersed dot pattern within the main city and constituting clusters in peripheral areas
The scale and distribution of resettlement areas inside and outside the main city of Nanjing indicate that the resettlement housing projects located in the main city are generally small in dimension and distributed in a scattered manner (e.g., Hensheng Jiayuan in Gulou District covers an area of only 5.5 ha). In contrast, the resettlement housing projects constructed around the main city and in peripheral areas are mostly distributed in a centralized and continuous manner, while large resettlement areas often contain several communities with an obvious cluster-like feature as a whole. In recent years, the trend of building centralized housing on a large scale has become more obvious. For example, Shangfang District, Huagang Happy City, and Daishan Star City occupy more than 1.4 million m^2, while the four major housing projects planned in 2010 have a construction area of more than 1.5 million m^2, the largest of which, Daishan Star City, even reached an area size of 3.4 million m^2 (Guo 2010).

Reason—The scale of resettlement housing varies mainly because of the difference in the construction area and land price variance between the inside and outside of the main city. As the periphery area has a more abundant construction space and is cheaper than the main city, the construction cost is lower. At the same time, to make up for the shortage of urban facilities of resettlement housing projects in the past, centralized construction promotes the sharing of resources, which maximizes economic effectiveness. For this reason, resettlement housings of the peripheral area are mostly constructed on a large scale in clusters. Notably, while the centralized construction of resettlement housing brings benefits from its scale, it also leads to problems and risks such as poverty concentration and social differentiation.

3.3 Study on the Spatial Resettlement of Land-Lost Peasants

Based on the above analysis of the spatial distribution and general layout characteristics of resettlement areas in Nanjing, by choosing several representative resettlement areas for land-lost peasants as the research samples, this chapter focuses on and analyzes the social attributes (including age structure, education level, family structure, occupational structure, composition of income, and consumption pattern) and spatial attributes (including spatial layout, housing use, employment and travel, and service facilities) of resettlement areas one-by-one to explore the current status of resettlement areas for land-lost peasants.

3.3.1 Research Scope and Data

3.3.1.1 Selection of Research Objects

The overall distribution analysis above shows that, due to the expansion of urban construction and land price variance, the resettlement areas that do not aim for economic benefits show the following characteristics: unevenly located between districts, mostly located in peripheral areas, and distributed in a scattered way inside and centralized way outside the main city. Barring a limited number of small-scale resettlement areas constructed in the main city in the early days, most of the resettlement areas are situated at urban fringe areas in the south, east, and north of the city along the bypass road and are shaped like a fan around the main city.

At present, the resettlement areas of land-lost peasants in Nanjing are basically divided into three categories: scattered resettlement areas in the main city, evenly distributed resettlement areas in industrial parks, and large mixed resettlement areas at the fringe of the city. To comprehensively study the spatial resettlement realities of land-lost peasants, this chapter selects one to two samples from each category for in-depth research. The selection of samples is based on the following basic principles:

① The selected samples need to reflect the general distribution characteristics of the resettlement areas for land-lost peasants in Nanjing, covering not only the regions in the north, east, and south of the main city but also the three levels of the city in terms of urban spatial structure: main city, sub-city, and new city.

② The selected samples of resettlement areas for land-lost peasants should be representative in terms of type and characteristic. First, it should cover a variety of locations such as the main city, the fringe area of the main city, and industrial parks. Second, it should cover areas of different dimensions and scales. Third, it should cover areas of different spatial agglomeration characteristics such as scattered and centralized distribution. Fourth, it should cover different construction stages such as early-stage, development stage, and maturation stage.

③ The selected sample of resettlement areas for land-lost peasants should be mainly distributed in the main city or the surrounding neighborhoods to facilitate in-depth research.

Considering the above principles of sample selection of resettlement areas for land-lost peasants and spatial distribution of the resettlement areas in Nanjing, the following four resettlement area samples were selected: Mufu Mountain Villa (one of the scattered resettlement areas in the main city: early-stage—north of the city—main city—scattered distribution), Yaolin Xianju (evenly distributed resettlement area in the industrial park: development stage—north of the city—sub-city—centralized distribution), Tianjingshan Community (large-scale resettlement area at the fringe area of the city: development stage—east of the city—sub-city—centralized distribution), and Xishan Huayuan (large mixed resettlement area at the fringe area of the city: maturation stage—south of the city—new city—centralized distribution) (see Fig. 3.10).

3.3.1.2 Research Data Collection

This chapter studies the spatial resettlement and social integration of land-lost peasants. Data concerning spatial resettlement are mainly obtained through field observation, questionnaire statistics, and thematic interviews. The content design of the questionnaires mainly includes two aspects: (a) basic personal information, which focuses on understanding the comprehensive information of the residents living in resettlement areas, including gender, age, education level, income level, family structure, occupational structure, consumption structure, etc., and (b) economic, social, spatial, psychological, and institutional information that corresponds to the social integration evaluation index system aimed at understanding the degree of integration of the respondents in each of the above dimensions and exploring the degree of correlation between the variables and their influence mechanisms. The primary information and basic data about the resettlement areas for land-lost peasants thus obtained can be used as the basic data for analysis and research in this paper.

In terms of specific questionnaire statistics, this paper adopts the sampling method of classification and segmentation, taking the four resettlement areas selected as the research objects, from which 860 respondents were randomly selected. The sampling

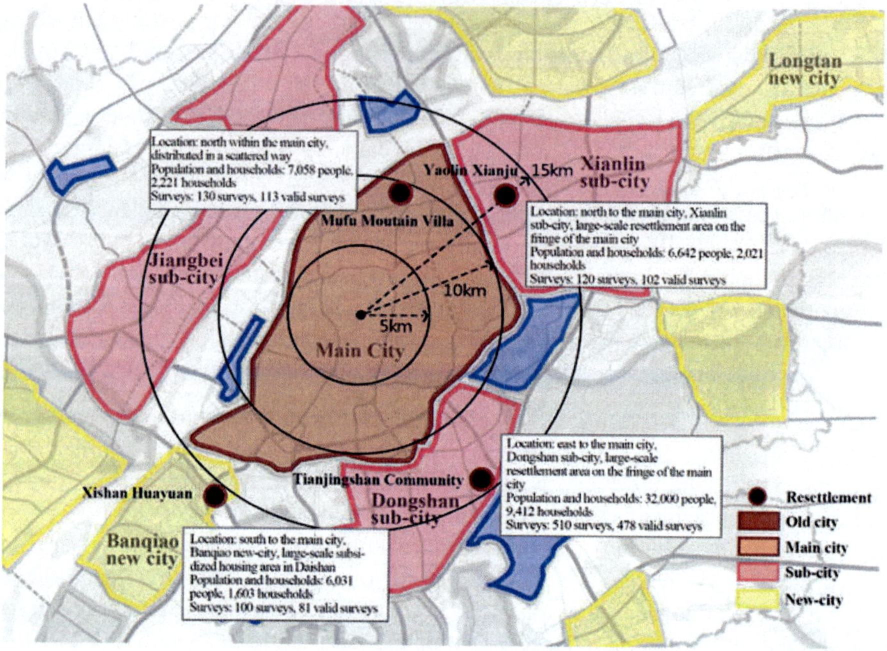

Fig. 3.10 Four resettlement areas for land-lost peasants selected as samples. *Source* Own elaboration based on relevant information and actual research

ratio was mainly determined based on the total number of households in each resettlement area. Due to the large dimension of resettlement areas, the built-up space of each resettlement area was divided into different zones, and an equal number of questionnaires were distributed in each zone. Considering the complexity of the types of resettlement areas (including mixed resettlement areas) and the composition of the population (which includes tenants and resettled households whose expropriated land was state-owned, etc.), the survey was conducted in the form of random sampling and carried out through face-to-face and one-to-one questionnaire filling method. The questionnaires need to be screened and targeted only at the heads of households or their family members who resettled due to the expropriation of their housing built on collective land. After sorting and reviewing the questionnaires, it was finally determined that 774 valid questionnaires were returned, with an effective rate of 90.1% (see Table 3.5). The interviews were completed simultaneously with the questionnaires, and random interviews were conducted during the questionnaire filling process. A total of 26 qualitative interviews were completed. In addition, other data sources used included professional literature, news reports, etc.

Table 3.5 Population and household size and number of questionnaires sampled in each resettlement area for land-lost peasants

Name of the resettlement area	Population and households		Number of questionnaires		Sample rate
	Total population size (people)	Total household size (households)	Number of questionnaires distributed	Number of questionnaires returned	
Mufu Mountain Villa	7058	2221	130	113	5.07%
Yaolin Xianju	6642	2012	120	102	5.06%
Xishan Huayuan	6031	1603	100	81	5.08%
Tianjingshan Community	32,000	9412	510	478	5.08%

Source Own elaboration based on relevant information and actual research

3.3.1.3 Overview of Study Sample

Based on the characteristics and distribution of the four resettlement areas selected as samples, the detailed explanation of their location conditions and specific characteristics is as follows (see Table 3.6).

3.3.2 Analysis of Social Attributes of Resettlement Areas for Land-Lost Peasants

The composition of people in resettlement areas for land-lost peasants is complicated due to the different construction time and scales. Generally speaking, they can be divided into two categories according to the source of housing: home buyers (land-expropriated households on collective land and state-owned land and economically affordable house buyers) and tenants (low-cost housing tenants, newly employed people, migrant workers, and jobless urban citizens). To ensure the relevance and depth of the research, this paper only analyzes the social attributes of the house buyers whose previously expropriated house was built on collective land (i.e., land-lost peasants).

Based on the typical samples from the various resettlement areas for land-lost peasants selected in this study, we analyzed the social characteristics of these areas in Nanjing by exploring the different social attributes such as age structure, education level, family structure, occupational structure, income composition, and consumption pattern. Furthermore, through the cross-sectional comparison among various types of resettlement areas, we examined the general patterns and differentiated characteristics of resettlement areas in terms of social attributes in Nanjing in a general and segmented manner.

Table 3.6 Overview of study sample in the resettlement area for land-lost peasants

Name	Mufu Mountain Villa	Yaolin Xianju	Tianjinshan Community	Xishan Huayuan
Type	Scattered resettlement area in the main city	Evenly distributed resettlement area in the industrial park	Large mixed resettlement area at the fringe area of the city	
Relation with the main city	North of the inner main city	North of the main city	East of the main city	South of the main city
Relation with urban space	Main city	Xianlin sub-city	Dongshan sub-city	Banqiao new city
Completion time	2000	2004	2012	2008
Distance to city center	9.6 km	12.5 km	13.4 km	16.6 km
Scale	Land area: 16 ha, Total building area: 190,000 m^2	Land area: 34.5 ha, Total building area: 470,000 m^2	Land area: 238 h, Total building area: 2,860,000 m^2	Land area: 37.3 ha, Total building area: 543,000 m^2
Current location		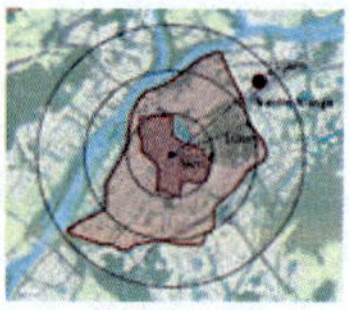	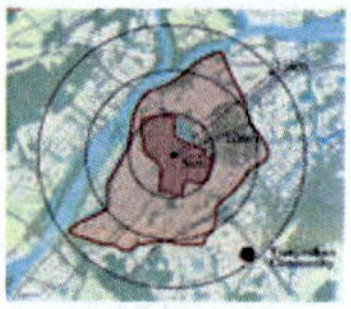	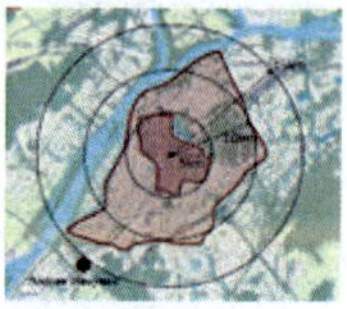
Overview of current situation	Located on Heyan Road of Qixia District at the foot of Mufu Mountain, it is a resettlement area built to be in line with the construction of Nanjing Economic and Technological Development Zone. The current average price of second-hand houses is 12,848 yuan/m^2	Located in Yaohuamen area of Xianlin University City in Qixia District, it is the first batch of centrally built resettlement areas constructed in line with the construction of Xianlin University City. The current average price of second-hand houses is 10,045 yuan/m^2	Located in Jiangning District, on the west of the intersection of Changshen and Ning-Hang Expressway, on the south of Longmian Avenue Station of Metro Line 1. The current average price of second-hand houses is 6141 yuan/m^2	Located on the south of Qinhuai New River in Yuhuatai District, on the east of the intersection of Hulong Expressway and Ningwu Highway, it was used to resettle land-lost peasants living in Xishanqiao Town. The current average price of second-hand houses is 8140 yuan/m^2

(continued)

Table 3.6 (continued)

Name	Mufu Mountain Villa	Yaolin Xianju	Tianjinshan Community	Xishan Huayuan
Population composition	The composition of residents is relatively balanced, with land-lost peasants accounting for the highest percentage and relatively more tenants	The residents are mainly land-lost peasants, including a small number of tenants	A centralized resettlement area wih a mixed composition of residents, mainly land-lost peasants. Tenants also account for a large proportion	Mixed resettlement with residents living in low-rent housing, economically affordable housing, and other subsidized housing. The composition of residents is relatively complex, including a large number of land-lost peasants (i.e., land-expropriated households living on collective land), land-expropriated households who used to live on stated-owned and tenants
Layout	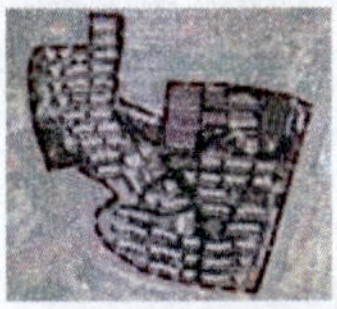	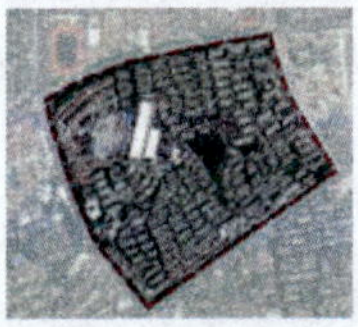	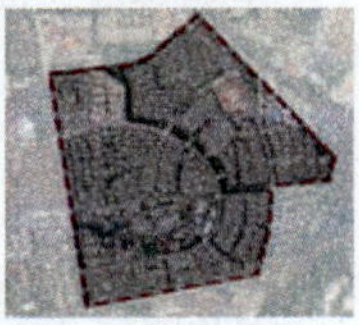	

Source Nanjing Housing Price Website http://nj.fangjia.com

3.3.2.1 Age Structure

As a specific group emerging in the urbanization process, land-lost peasants present an "inverted V-shaped, generally older age structure." The overall age span of the group is large, and the segmented clustering feature is obvious. Land-lost peasants mainly comprise middle-aged and young people, and the aging tendency is also significant. At present, their age is mainly concentrated between 40 and 70 years old, with 66.7% of them being over 40 years (see Fig. 3.11). The reason for the older age structure is mainly due to different age groups having different preferences when choosing their residence. Many newly employed people prefer to rent in the main urban area that provides a more convenient lifestyle, while unemployed or jobless

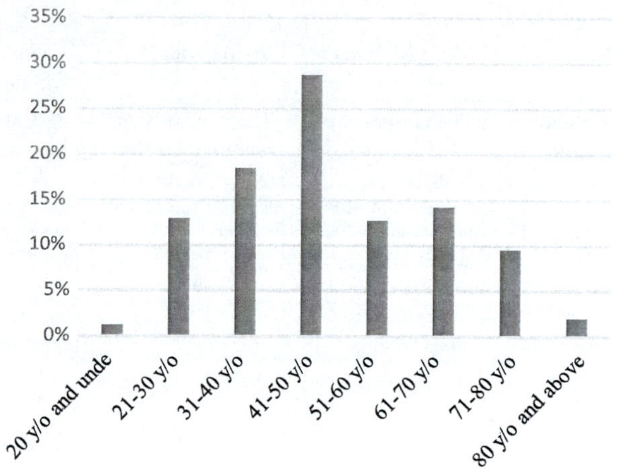

Fig. 3.11 Overall age structure of residents living in resettlement areas for land-lost peasants. *Source* Sample data collected by the research team on resettlement areas for land-lost peasants in Nanjing (2016)

middle-aged and elderly people have a higher acceptance and tolerance to remotely located resettlement areas because of good environmental conditions.

A further horizontal comparison across different resettlement areas reveals the following (see Fig. 3.12): there are certain differences in the age structure of the various resettlement areas. The residents of Mufu Mountain Villa are mainly young and middle-aged, with more than half (55.5%) of them aged between 30 and 50, while the residents of Yaolin Xianjiu and Xishan Huayuan present a more obvious aging trend, with 38.7%[28] and 36.5% of them aged 60 years old and older, respectively. As 42.8% of the residents of Tianjingshan Community are aged under 40 years, the overall age structure is young. The slight difference in age structure across different resettlement areas is mainly due to residents of all age groups having a similar degree of acceptance toward resettlement areas that have more convenient transportation

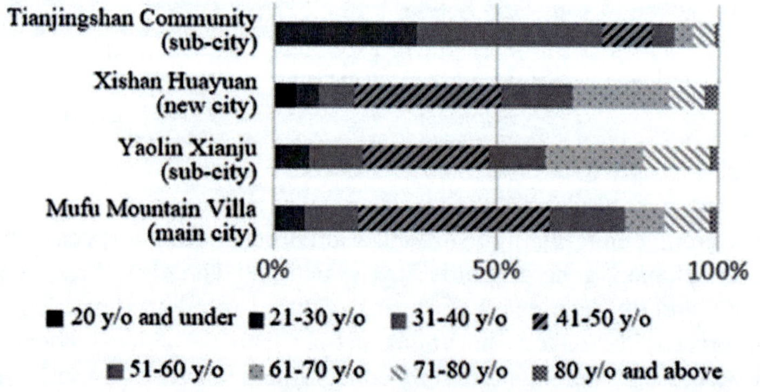

Fig. 3.12 Age structure of each resettlement area for land-lost peasants. *Source* Sample data collected by the research team on resettlement areas for land-lost peasants in Nanjing (2016)

conditions and employment opportunities (e.g., Mufu Mountain Villa and Tianjingshan Community), while remotely located resettlement areas with fewer employment opportunities (e.g. Yaolin Xianjiu and Xishan Huayuan) are less attractive to residents of working age due to high commuting costs.

3.3.2.2 Education Level

Land-lost peasants have in general a "low education level and mainly hold a junior high school diploma or has received less education." The overall education level of land-lost peasants living in the four resettlement area samples is not high. Additionally, the education level of the working-age population is also low in general, with 73.9% of the land-lost peasants having only junior high school education or below, 16% have never attended school, 23% have only received elementary school education, and very few have pursued post-secondary or higher education (see Fig. 3.13). The low education level of land-lost peasants is mainly due to the general attributes of peasants as social subjects. Although the education system and facilities in rural areas are becoming increasingly standardized, the problems of low quality of education, limited teacher resources, and poor information in rural areas are still not promising compared with those in urban areas.

A further horizontal comparison across different resettlement areas reveals the following (see Fig. 3.14): the difference in education level between the various resettlement areas is not obvious. In every area, majority of the residents have of junior high school education. In the resettlement areas of Mufu Mountain Villa, Yaolin Xianju, and Xishan Huayuan, residents mainly have a low or middle educational level, with 72% of them having junior high school education or below. In comparison, residents of Tianjingshan community have the lowest education level, showing a pyramid structure where the lower the education level, the greater is the proportion

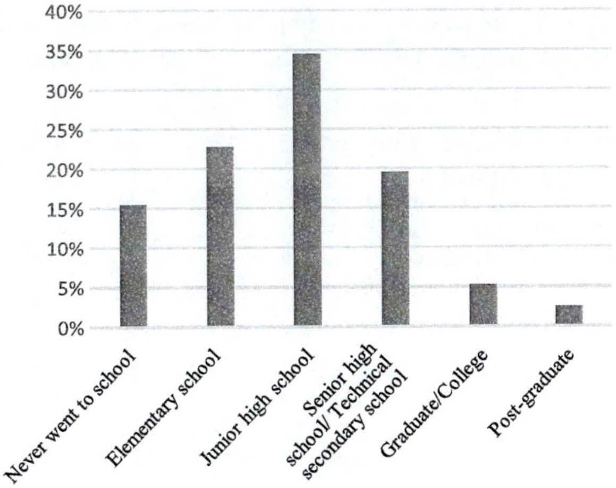

Fig. 3.13 Overall education level of land-lost peasants. *Source* Sample data collected by the research team on the resettlement areas for land-lost peasants in Nanjing (2016)

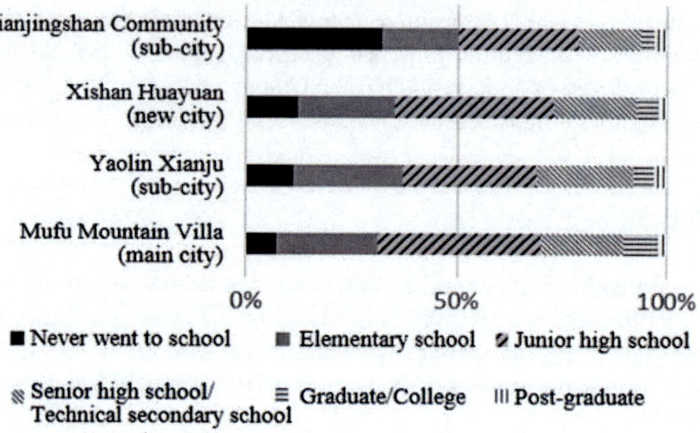

Fig. 3.14 Education level of each resettlement area for land-lost peasants. *Source* Sample data collected by the research team on the resettlement areas for land-lost peasants in Nanjing (2016)

of the number of people. Residents with an education level of junior high and below account for 79.6%.

3.3.2.3 Family Structure

As a marginal group between peasants and urban residents, the family structure of land-lost peasants has the characteristics of an "urban–rural transitional family structure where the majority of families consist of at least two generations." A high proportion of families have three to five members from two or more generations, accounting for 61% of the total households (see Figs. 3.15 and 3.19). These transitional characteristics are due to the one-child policy implemented nationwide in the 1970s. Within the context of this policy, the fertility rate of most households in the study generally declined after a long period of stability and thus showed more obvious transitional characteristics in terms of family structure.

Fig. 3.15 Overall family structure of land-lost peasants. *Source* Sample data collected by the research team on the resettlement areas for land-lost peasants in Nanjing (2016)

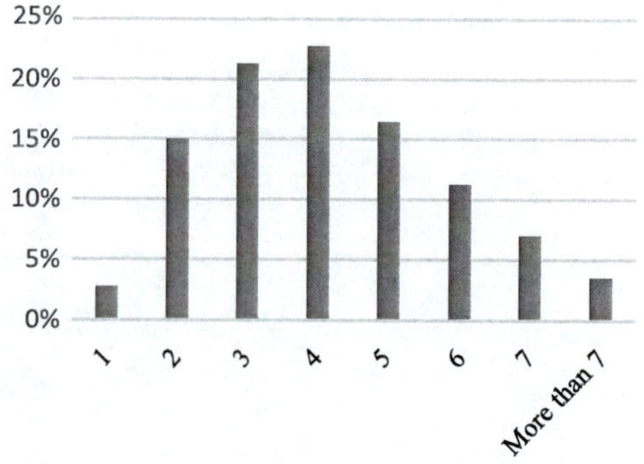

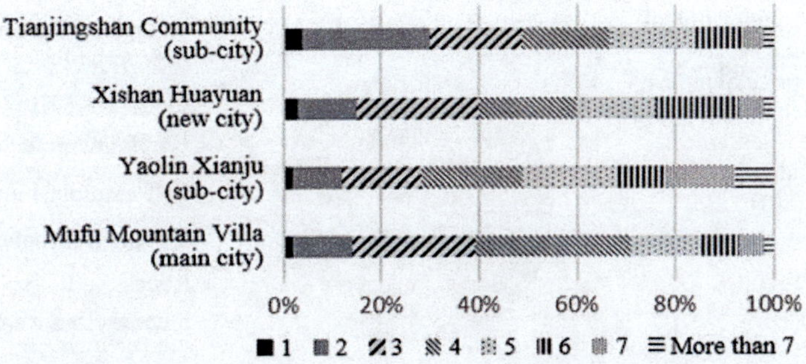

Fig. 3.16 Family structure of each resettlement area for land-lost peasants. *Source* Sample data collected by the research team on the resettlement areas for land-lost peasants in Nanjing (2016)

A further horizontal comparison across different resettlement areas reveals the following (see Fig. 3.16): there are some differences in family structure across various types of resettlement areas. The maximum number of family members investigated in the survey is 10. Therefore, the family types can be divided into three: small families of one to three members (two-member families are usually elderly couples, and three-member families are mostly nuclear families), medium-sized families of four to six members (four-member families mainly comprise husband, wife, their children or daughters-in-law, while families with five or more members are mostly three-generation families or relatives co-living in a house), and large families with seven or more members. Of the four resettlement areas, residents of Mufu Mountain Villa and Yaolin Xianjiu are mostly medium-sized families with four to six members, while Xishan Huayuan consists mostly of small families of one to three members, accounting for 34.4% of the total households. Moreover, majority of the households in Tianjingshan community are small nuclear families of three people.

3.3.2.4 Income Composition

In terms of income composition, the characteristic of "high proportion at both ends and low in the middle, with transfer income and income from property as the main sources" is presented in all types of resettled peasants (see Fig. 3.17). Transfer income (which mainly refers to the compensation for resettlement and social security received by land-lost peasants in the process of land expropriation) and income from property (which mainly refers to the one-time or long-term stable income obtained through transaction practices in the market such as selling or renting houses) account for 31% and 35% of their total income, respectively, and have become the most important source of income for land-lost peasants. The overall monthly household income of the group presents an obvious "V-shaped structure." Low-income households that earn 500–1500 yuan per month account for 27% of the total households, while high-income households that earn 5000–10,000 yuan represent 29% of the total (see Fig. 3.18). Within this V-shaped structure, the low-income is due to the presence of a

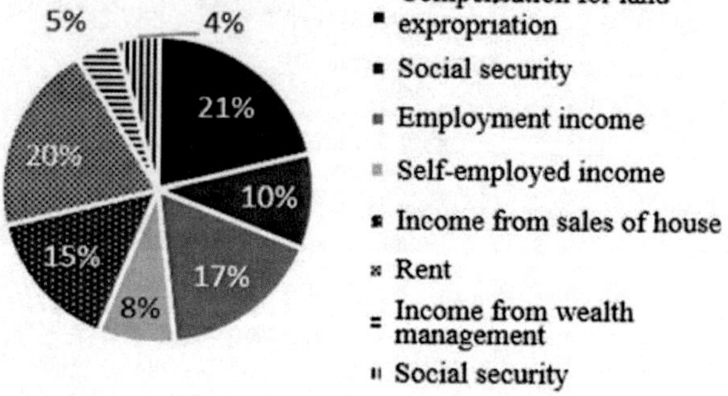

Fig. 3.17 Composition of Income of Land-lost Peasants. *Source* Sample data collected by the research team on the resettlement areas for land-lost peasants in Nanjing (2016)

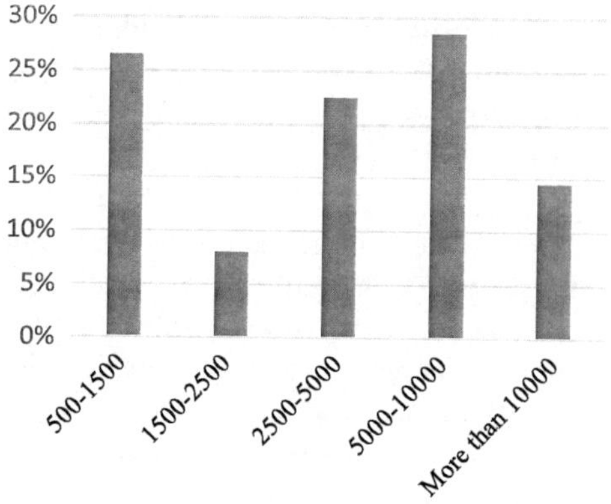

Fig. 3.18 Overall monthly income composition of land-lost peasants. *Source* Sample data collected by the research team on the resettlement areas for land-lost peasants in Nanjing (2016)

large proportion of jobless people in resettlement areas (e.g., Tianjingshan community), and the income source of such households is mainly dependent on and limited by the above-mentioned transfer income and property income.

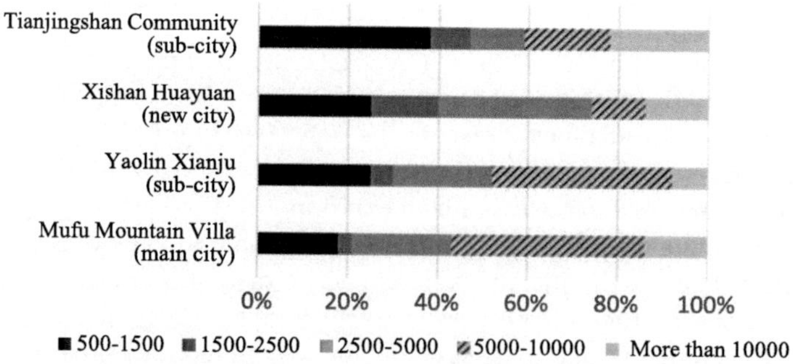

Fig. 3.19 Monthly income composition of each resettlement area for land-lost peasants. *Source* Sample data collected by the research team on the resettlement areas for land-lost peasants in Nanjing (2016)

A further horizontal comparison across the different resettlement areas reveals the following (see Fig. 3.19): the average monthly household income of residents living in Tianjingshan Community, Mufu Mountain Villa, and Yaolin Xianju has surpassed 6000 yuan and stands at 6758 yuan, 6523 yuan, and 6200 yuan, respectively. Of them, majority of the residents of Mufu Mountain Villa and Yaolin Xianju have an average income of 5000–10,000 yuan. In Tianjingshan Community, low-income households (whose monthly income is below 1500 yuan) account for more than 38% of the total households, while high-income households (whose monthly income is above 10,000 yuan) account for more than 20% of the total. In the meantime, the unemployment rate of these residents is the highest, while more households rely on housing rent to maintain a higher income. Therefore, the income composition presents a "high proportion at both ends and low in the middle" characteristics. Furthermore, the average monthly household income of residents living in Xishan Huayuan is only about 4200 yuan, the unemployment rate of this community is higher, and the proportion of low-income households (500–2500 yuan) is the largest (about 24.1%). The high unemployment rate and low-income level of employed persons are the main cruces of the low average income of residents of this entire resettlement area.

3.3.2.5 Occupational Structure

The occupational composition of all types of land-lost peasants in resettlement areas presents an "underemployment" characteristic. To be specific, residents have a "low employment rate and are usually employed in jobs with low technical content and stability" (see Fig. 3.20). Underemployment is different from unemployment. In 1995, the statistical definition of underemployment was clarified in the country: underemployment refers to people who work less than 20 h a week, for impersonal reasons, and are willing to work more (Zhang and Wang 2006). Land-lost peasants

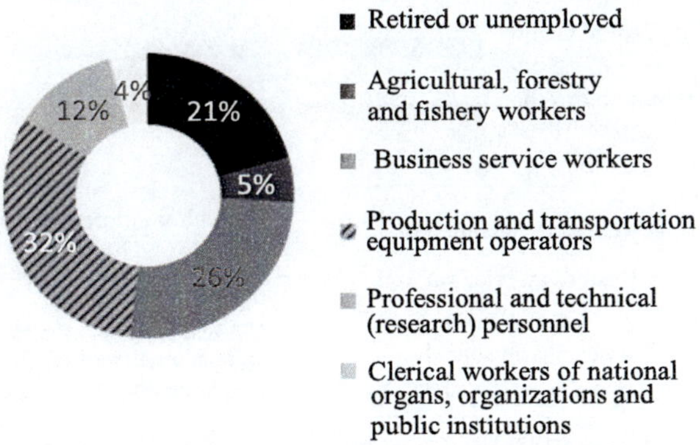

Fig. 3.20 Overall occupational composition of land-lost peasants. *Source* Sample data collected by the research team on the resettlement areas for land-lost peasants in Nanjing (2016)

who are jobless and idle at home meet these three conditions. The employment rate of land-lost peasants in the four sample areas is generally low, with a large proportion of retired or jobless people (21.7%), and even those who are employed mostly work in labor-intensive commercial and domestic service fields such as catering, retail, sanitation, and cleaning services due to limited self-quality and low level of labor skills. They mainly include temporary workers (71%) and ordinary workers who have poor job stability. This can be traced to the fact that most residents living in resettlement areas already have a certain transfer and property income gained in the process of land expropriation and resettlement. As a result, they do not have high expectations and demands for engaging in new jobs in the city and generally lack a sense of competition. Thus, there are fewer self-employed people. On the whole, the overall employment situation is not optimistic.

A further horizontal comparison across different resettlement areas reveals the following (see Fig. 3.21): there is no significant difference in the occupational composition of residents in each resettlement area, most of whom are operators of production and transportation equipment and commercial service workers that have a low occupational threshold. In Mufu Mountain Villa and Yaolin Xianju, 62% and 65% of the residents fall in these two categories, respectively. In Xishan Huayuan, residents are mostly operators of production and transportation equipment and retired or unemployed persons, each accounting for 32% and 19% of the total population. In case of Tianjingshan community, the percentage of retired or unemployed people is 24%. In terms of occupational composition, in addition to commercial service workers that represent the highest percentage (31%) and operators of production and transportation equipment (23%), 15% of the residents work in professional and technical (research) jobs, which is the highest among all four resettlement areas. Thus, while this settlement area has the highest unemployment rate, its residents also have the

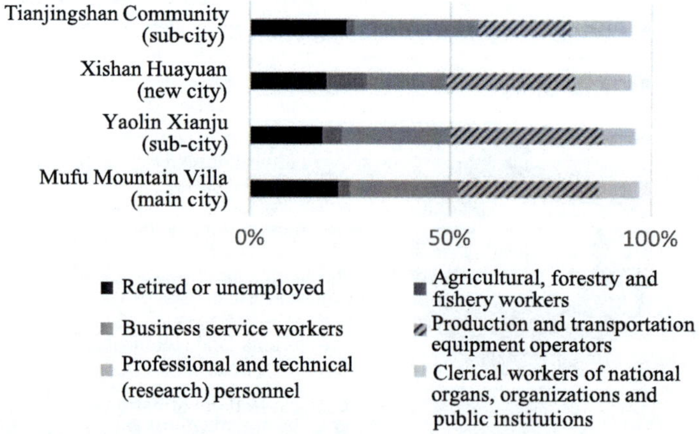

Fig. 3.21 Occupational composition of each resettlement area for land-lost peasants. *Source* Sample data collected by the research team on the resettlement areas for land-lost peasants in Nanjing (2016)

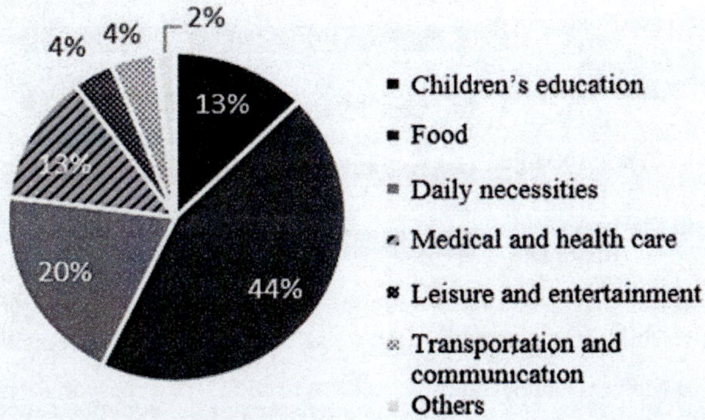

Fig. 3.22 Overall consumption pattern of land-lost peasants. *Source* Sample data collected by the research team on the resettlement areas for land-lost peasants in Nanjing (2016)

most diversified job types as well as the highest percentage of people working in jobs that have high technical requirements and relatively high occupational threshold.

3.3.2.6 Consumption Pattern

The consumption pattern of land-lost peasants in all resettlement areas shows a "consumption pattern dominated by food and daily necessities" (see Fig. 3.22). The overall consumption pattern of each resettlement area was measured by Engel's coefficient[1] (the proportion of total food expenditure to total consumption expenditure), and it was found that although an Engel's coefficient of 44% basically reached the well-off level, their consumption in education, transportation, entertainment, and cultural activities was still limited (only 21%). The consumption pattern of land-lost peasants was generally homogeneous, and their consumption level and quality were not high.

A further horizontal comparison across different resettlement areas reveals the following (see Fig. 3.23): there is a greater commonality in the consumption pattern of the four sample resettlement areas, i.e., consumption expenditure is dominated by food and daily necessities (all accounting for more than 50%). Of them, the residents of Xishan Huayuan spend up to 52% on food alone (having the largest Engel's coefficient), indicating that their lives are just at the subsistence level. While the cost of children's education is the second most important expense for residents after daily expenses, 71% of the residents feel that they are unable to afford the cost of their children's education. The expenditure on leisure and entertainment is almost zero and the spending on their own cultural education is minimal, indicating that the motivation and initiative for land-lost peasants to receive new knowledge and improve their own cultural quality is not high.

[1] Engel's coefficient of 59% or more is poverty, 50–59% is subsistence, 40–50% is well-off, 30–40% is rich, and below 30% is the richest.

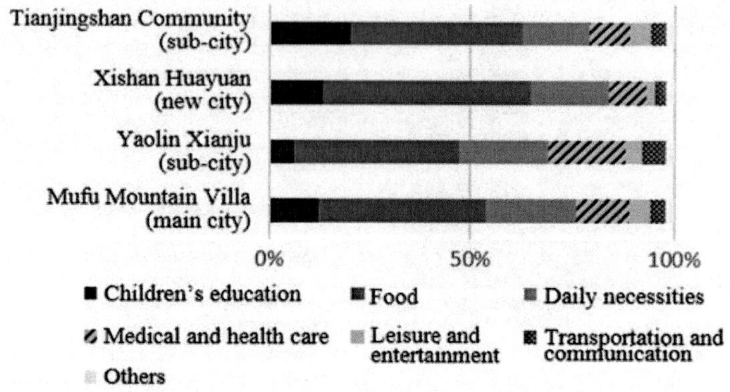

Fig. 3.23 Consumption pattern of each resettlement area for land-lost peasants. *Source* Sample data collected by the research team on the resettlement areas for land-lost peasants in Nanjing (2016)

3.3.2.7 Summary

This section presents the social attributes of land-lost peasants in terms of age structure, education level, family structure, income composition, occupational structure, and consumption pattern, and analyzes some of their causes, on the basis of which their current characteristics are summarized as follows (see Table 3.7).

3.3.3 Analysis of Spatial Attributes of Resettlement Areas for Land-Lost Peasants

This section takes the different types of selected resettlement areas for land-lost peasants as samples and analyzes the current spatial characteristics of these areas in Nanjing in terms of spatial layout, housing use, transportation to work, and service facilities. Through a horizontal comparison of different types of samples, the general patterns and differentiated characteristics of the attributes of resettlement areas for land-lost peasants are analyzed in a general and segmented manner.

3.3.3.1 Spatial Layout

Characteristics of Spatial Layout

In terms of the surrounding environment, there is no significant difference between the resettlement areas for land-lost peasants and the general commercial-residential areas, both of which show the layout characteristics of "long-distance to work and mixed land use." The basic facilities of these areas are relatively complete, and the residents are satisfied with the surrounding environment of the community.

Table 3.7 Summary of social attributes of residents of resettlement areas for land-lost peasants

Common characteristics		Differentiated characteristics			
		Mufu Mountain Villa	Yaolin Xianju	Tianjingshan Community	Xishan Huayuan
Age structure	"Inverted V-shaped older age structure"	The majority are young and middle-aged	The aging trend is obvious	The overall age structure is young	The aging trend is obvious
Education level	"Low education level, with mainly a junior high school diploma or less education"	The majority have medium to low level of education	The majority have medium to low level of education	Pyramid structure with the lowest education level	The majority have medium to low level of education
Family structure	"Urban–rural transitional family structure where the majority of families consist of at least two generations"	The majority are medium-sized families with four to six members	The majority are medium-sized families with four to six members	The majority are nuclear families with three members	The majority are small-sized families with one to three members
Income composition	"High proportion at both ends and low in the middle, with transfer income and income from property as the main sources"	The income of the majority of households range between 5000 and 10,000 yuan, with a relatively large proportion of middle- and high-income households		The polarization is more serious, with a relatively large proportion of low- and high-income households	High unemployment rates and the majority are low-income employed persons
Occupational structure	"Low employment rate, low technical content, and low stability"	The majority are operators of production and transportation equipment and commercial service workers		The majority are operators of production and transportation equipment, with a high proportion of retired or unemployed persons	The majority are operators of production and transportation equipment and retired or unemployed persons

(continued)

Table 3.7 (continued)

Common characteristics		Differentiated characteristics			
		Mufu Mountain Villa	Yaolin Xianju	Tianjingshan Community	Xishan Huayuan
Age structure	"Inverted V-shaped older age structure"	The majority are young and middle-aged	The aging trend is obvious	The overall age structure is young	The aging trend is obvious
Consumption pattern	"A homogenous consumption pattern dominated by food and daily necessities"	With food and daily necessities as the mainstay, Engel's coefficient lies between 40 and 50%, basically reaching the well-off level			The largest expenditure is on food, with Engel's coefficient of 52%, is at the subsistence level

Source Own elaboration

An elaborated comparison on the surroundings of the various resettlement areas reveals the following (see Table 3.8 for the summary): there are significant differences across the four resettlement areas. Mufu Mountain Villa, a resettlement area built in the main city and located in Qixia District, is close to the commercial center and mid-level residential areas. The community is 2.6 km away from the nearest subway station, has relatively convenient transportation and is equipped with complete surrounding facilities. It also has a high degree of shared public equipment with the surrounding communities. Yaolin Xianju, a resettlement area built within an industrial park, is located in Qixia District. This area is adjacent to Xianlin University City and is part of the Nanjing Economic and Technological Development Zone. The surrounding factories and enterprises are concentrated but there is no subway. Although the bus stops are reasonably set up, the accessibility is poor and public facilities are shared with the surrounding communities to a low degree. Tianjingshan Community, a resettlement area built in the urban fringe region, is located in Jiangning District. The community is closely located to subway Line 1 (Longmian Avenue Station) to the south and has convenient public transportation and complete surrounding facilities. It also has a high degree of shared public equipment with the surrounding communities (mostly mid-level communities). Xishan Huayuan, a resettlement area built in the urban fringe region, is located within the subsidized housing area of Daishan in Yuhuatai District. Among the four resettlement areas, this community is located the farthest from the city center (16.6 km) and has extremely inconvenient public transportation services. However, it does have a high degree of shared public facilities with the surrounding communities and better environmental quality.

In terms of layout, the resettlement areas for land-lost peasants have shown the characteristics of "different in sizes and scale, but mostly built by rows and located in periphery region." Although the physical and spatial forms of resettlement areas

Table 3.8 Analysis of surrounding environment of each resettlement area for land-lost peasants

Type	Scattered resettlement area in the main city	Evenly distributed resettlement area in the industrial park	Large mixed resettlement area at the fringe area of the city	
Resettlement area	Mufu Mountain Villa	Yaolin Xianju	Tianjingshan Community	Xishan Huayuan
Administrative location	Qixia District	Qixia District	Jiangning District	Yuhuatai District
Distance from city center	9.6 km	12.5 km	13.4 km	16.6 km
Schematic diagram of transportation location	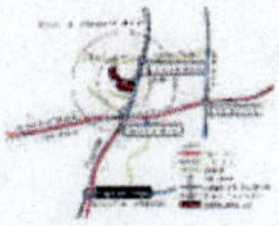		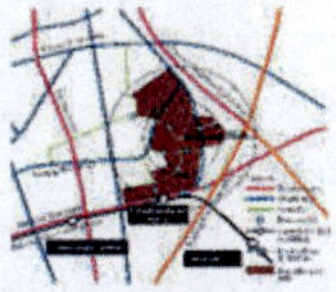	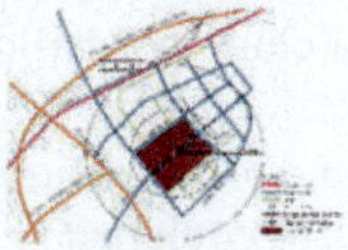
Proximity to subway	Line 1 (completed) Line 7 (planned)	Line 7 (planned)	Line 1 (completed) Line 5 (planned)	Line 7 (planned)
Distance to the nearest subway station	≥ 2.6 km	≥ 8.2 km	≤ 1 km	≥ 5.4 km
Number of bus stops	500 m: 3 1000 m: 12	500 m: 5 1000 m: 20	500 m: 5 1000 m: 16	500 m: 7个 1000 m: 12个
Transportation features	Relatively high transportation accessibility	Relatively low transportation accessibility	Relatively high transportation accessibility	Relatively low transportation accessibility
Surrounding sites	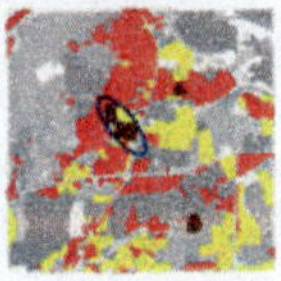	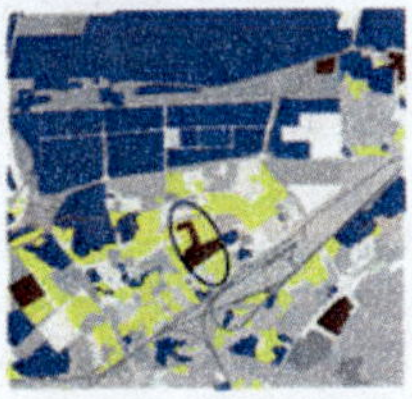		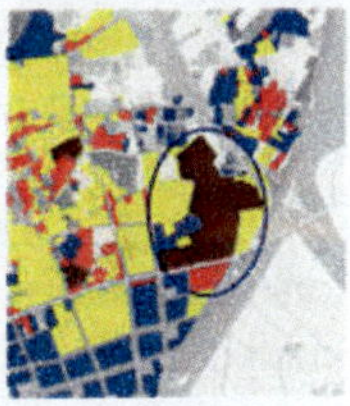
Surrounding site features	Close to commercial centers and mid-level residential areas	Close to the Economic and Technological Development Zone, Xianlin University City, Longtan Deepwater Port, and other emerging industrial parks	The south side is concentrated with factories and enterprises	It is located in the subsidized housing area of Daishan and mixed with public rental housing, low-cost housing, and economically affordable housing

Source Sample data collected by the research team on the resettlement areas for land-lost peasants in Nanjing (2016)

have improved in recent years with the construction of a large number of this type of community, due to the objective conditions, the design of transportation location, spatial forms, community creation, and specific functionality of resettlement areas lack systematic consideration.

An elaborated comparison of the various types of resettlement areas reveals the following (see Table 3.9 for the summary): Mufu Mountain Villa, a resettlement area built in the main city, is the smallest (14.3 ha) independent resettlement area, which adopts a building grouping form that combines staggered linear grouping and a cluster of buildings around a central space. On the outside, there are commercial residential buildings constructed along the street, forming an enclosed neighborhood, and on the inside, there is a public square and many central green spaces. This area is also equipped with recreational areas and facilities. The homogeneously distributed resettlement area located in the industrial park, Yaolin Xianjiu, was built at the same time as Mufu Mountain Villa and has a moderate scale (24.9 ha). There are commercial facilities along the street, and the three sub-communities of this area are grouped by rows, which results in a relatively high building density and low quality of public space. Tianjingshan Community and Xishan Huayuan are large mixed resettlement areas built at the fringe of the city. Of them, Tianjingshan Community contains eight sub-communities, which represent a typical large-scale resettlement area (296 ha). Overall, the buildings in this community are grouped by rows and show a certain homogeneity. Its planning and design of public greenery, activity space, and parking lot are also complete. Xishan Huayuan, which is part of the large subsidized housing area in Daishan, is surrounded mostly by public rental housing, economically affordable housing, low-cost housing, and other types of subsidized housing. In addition, due to its short construction time, the road network layout for "mixed small neighborhood" was adopted, and the buildings are grouped in rows.

Analysis of Spatial Layout

Due to the rapid development of the city, the differential land rent, and land-lost peasants' passiveness in selecting their resettlement areas, these regions are usually far from the transportation facilities, and there is usually a mixed-use of land of its surroundings. In the meantime, the "differences and commonalities in the layout" of the various types of resettlement areas are mainly cross-influenced by when the settlements were built and how land-lost peasants were settled.

Regarding the surroundings of the resettlement areas for land-lost peasants, the differences across the various types are due to the location of the area on the surface. However, in essence, they are subject to graded distribution of land values. The spatial resettlement for land-lost peasants is a government-led investment and construction activity, and the passive choice of housing is an indisputable fact that most peasants face after losing their land. This activity underwent two major stages: In the early urban expansion period, the collective land that was close to the cities was classified as urban construction land, from which a number of small-scale resettlement areas emerged in a scattered way at the fringe of the city (which are now part of the main city). The location and transportation advantages of these areas are relatively obvious, and the use of their surrounding land was usually explored in a diverse

Table 3.9 Analysis on layout of each resettlement area for land-lost peasants

Type	Scattered resettlement area in the main city	Evenly distributed resettlement area in the industrial park	Large mixed resettlement area at the fringe area of the city	
Resettlement area	Mufu Mountain Villa	Yaolin Xianju	Tianjingshan Community	Xishan Huyuan
Layout				
Land area	16 ha	34.5 ha	238 ha	37.3 ha
Building area	190,000 m^2	470,000 m^2	2,860,000 m^2	543,000 m^2
Composition of the community	–	Qingshanyuan, Cuilinyuan, Bishuiyuan	Xinrongyuan, Bihuayuan, Huajinyuan, Ruitingyuan, Yongfuyuan, Jingxiuyuan, Shengleyuan, Chunxiuyuan	Jiaheyuan, Xiangheyuan, Wanfuyuan, Wanxingyuan
Building grouping form	The combination of staggered grouping, point grouping, and grouped along the street	Mainly grouped by rows		
Traffic	Three entrances and exits, the roads are open but not unobstructed	One to two entrances and exits at each community; the roads are open but not unobstructed		Road network distribution mode for mixed small neighborhood
	Motor vehicle parking is divided into ground-level scattered parking and overhead parking on the building's ground floor			
Public space	The public green space is mainly planted with turf and shrubs. A small number of pavilions, seats and other leisure and recreational facilities are installed			

Source Sample data collected by the research team on the resettlement areas for land-lost peasants in Nanjing (2016)

manner. In addition, these areas usually have a high degree of community resource sharing. At a later stage, on account of the continuous expansion of urban areas, development of new cities, construction of industrial parks and development zones, and the government's policy guidance for subsidized housing, a large number of concentrated constructions of subsidized settlements emerged in the urban fringe areas. Although accessibility to transportation in the areas is relatively low as they are far from the city center (only part of them can be reached by subway), the overall environmental conditions are still good.

As far as the layout of the resettlement area for land-lost peasants is concerned, the building layout, external environment, and public service facilities of the areas follow the conventional design pattern of modern communities. The differences across the various types are due to two factors: the difference in resettlement mode leads to different sizes of communities. The Mufu Mountain Villa was built to accommodate house-expropriated residents living in the Xingang Development Zone, while the Yaolin Xianju was built because of the emergence of the Nanjing Economic and Technological Development Zone and the Xianlin University City. Additionally, in consideration of the constraints on construction land, original peasants mostly resettled in a decentralized manner. As a result, the scale of a single area was relatively moderate. In contrast, Tianjingshan Community and Xishanhuayuan are resettlement areas built along with urban expansion and new city development. The Tianjingshan Community resettled 30,000 peasants from Shangfang (2 administrative villages), Chunhua (4 administrative villages), and Fangshan (12 administrative villages) at one time, and Xishan Huayuan resettled more than 5,000 peasants living in Shanjia Village, Heng Village, Meishan Village, and Qiu Village. The communities are larger in scale. Conversely, the different construction time also led to differences in the layout of each community. The planning and design of Yaolin Xianju constructed earlier is problematic. The planning of public green space, building density, distance between houses, and parking facilities fall behind the current standards. In comparison, Xishan Huayuan and Tianjingshan Community that were built in recent years have basically met or even partially exceeded the construction specifications of residential communities and the current standards of Nanjing.

3.3.3.2 Use of Housing

Characteristics of the use of housing

In terms of house types, most land-lost peasants applied for subsidized housing based on the compensation standard of land expropriation. Therefore, majority of the housing includes small to medium-sized units. Two bedrooms with one living room and three bedrooms with one living room units, which are mainstream housing types in urban areas, account for 80.3% of the total, while one-bedroom with one living room units, two-bedroom with one living room and one dining room units, and three-bedroom with one living room and one dining room units account for only 19.7% (see Fig. 3.24). Most of the multi-story buildings have two units per

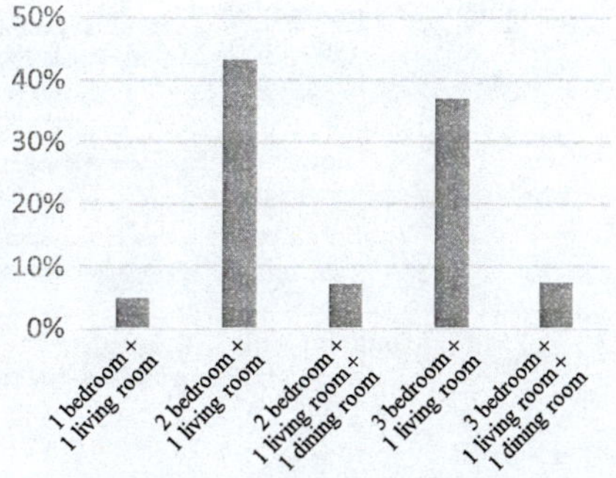

Fig. 3.24 Overall unit types in the resettlement area for land-lost peasants. *Source* Sample data collected by the research team on the resettlement areas for land-lost peasants in Nanjing (2016)

floor, medium-rise buildings usually have two or three units per floor, and high-rise buildings usually have four units per floor. At the same time, land-lost peasants generally pay more attention to the ventilation, orientation, and configuration of the house than to the interior decoration of the house to meet the basic needs of living.

A comparison across the various types of resettlement areas reveals (see Fig. 3.25) the following: the units within each type of resettlement area possess different unit type characteristics due to the differences in construction time, standard, and size (see Table 3.10). The buildings in Xishan Huayuan and Tianjingshan communities, which were constructed in recent years and are located at the fringe of the city, basically conform to the standards. They both contain skip floor design on the top floors as well as basements. In comparison, Mufu Mountain Villa and Yanlin Xianju, which were built earlier, present certain problems in terms of design: the living room has limited daylight, the entrance faces the toilet, the hallway is narrow, some rooms have poor lighting and bad ventilation, or the bathroom of certain units are not equipped with windows. In addition, as the buildings are close to each other, the privacy of residents cannot be guaranteed due to windows being poorly designed to be closing facing another unit. Another problem of these communities is that there are many spontaneous reconstructions and expansions carried out by ground floor residents, resulting in an overall disordered environment of the community.

The overall unit size of housings in the resettlement area for land-lost peasants is small. Additionally, their extended family structure also leads to the problem of "insufficient living area per capita" in general. According to the Floor Area per Person proposed by the United Nations (see Table 3.11), the minimum living area for a family of three to five people should be no less than 59.5 m^2, i.e., 11.9–19.8 m^2 of living area per capita. However, research indicates that 28.7% of the land-lost peasants have a per capita living area of less than 15 m^2, including 5.5% of those with less than 12 m^2, while only 6.7% have a per capita area of more than 50 m^2, and these people are mostly land-lost peasants living alone (see Fig. 3.26).

In the context of overall shortage of living space, the situation of each resettlement area for land-lost peasants is as follows: Tianjingshan community has the best condition in terms of living space. There are three types of units (65, 90, and 120 m^2), and a moderate floor area per capita is also ensured for nuclear families, of which 51%

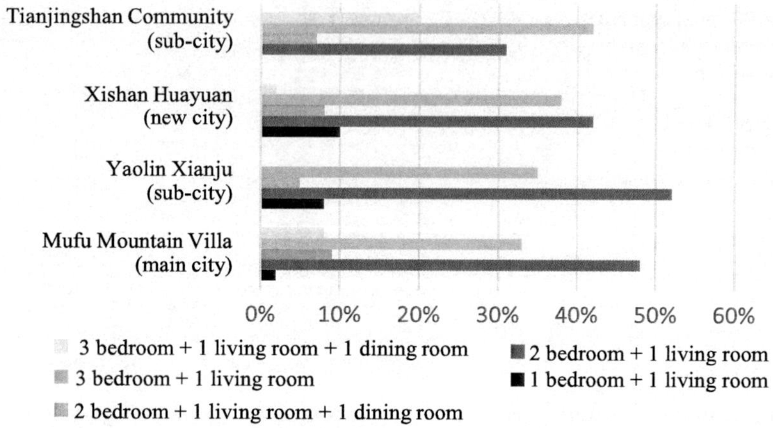

Fig. 3.25 Unit types of each resettlement area for land-lost peasants. *Source* Sample data collected by the research team on the resettlement areas for land-lost peasants in Nanjing (2016)

Table 3.10 Typical unit types of each resettlement area for land-lost peasants

Type	Scattered resettlement area in the main city	Evenly distributed resettlement area in the industrial park	Large mixed resettlement area at the fringe area of the city	
	Mufu Mountain Villa	Yaolin Xianju	Xishan Huayuan	Tianjingshan Community
Type 1	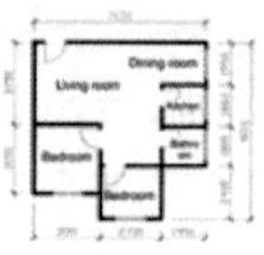			
Type 2				
Type 3	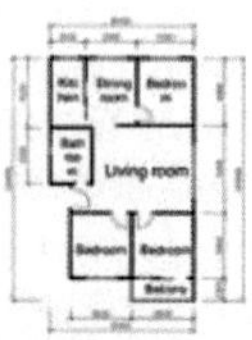		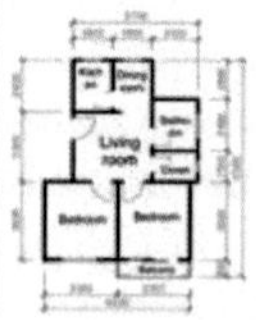	

Source Sample data collected by the research team on the resettlement areas for land-lost peasants in Nanjing (2016)

Table 3.11 Floor area per person for families of three to five prople proposed by the United Nations

Rooms	Floor area(m^2)
Living + Dining space	18.6
Kitchen	7.0
Major bedroom	13.9
2nd bedroom	12.0
3rd bedroom	8.0
Total area	59.5
Floor area per capita	11.9–19.8

Source Xu, Qing. *Planning and Architectural Design of Resettlement Area in the Context of Urban–Rural Integration* [D]. Post-graduate Thesis. Hangzhou: Zhejiang University of Technology, 2009:24

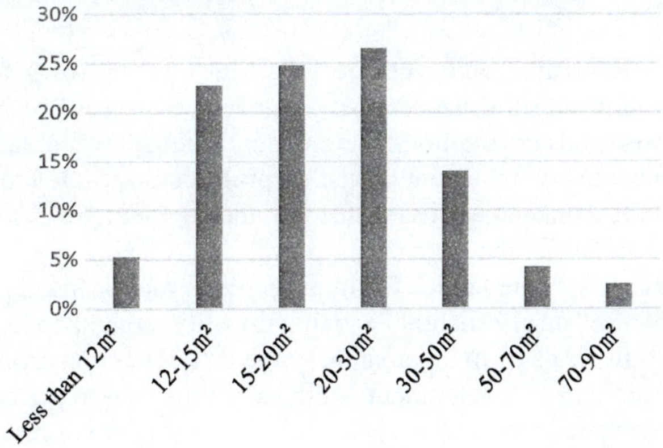

Fig. 3.26 Overall living area per capita in the resettlement area for land-lost peasants. *Source* Sample data collected by the research team on the resettlement areas for land-lost peasants in Nanjing (2016)

have a per capita living area of 30–40 m^2; the second best is Xishanyuan (60–90 m^2 units), which is located at the fringe of the city and where half of the households of this community have a per capita living area of 20–30 m^2; communities that face the worst situation are Yaolin Xianju and Mufu Mountain Villa (52–80 m^2 units), where 43.2% of the households have a per capita living area of less than 20 m^2. Additionally, residents have limited choice in unit types and per capita living area. The only advantage of the areas is the convenient transportation location and relatively complete service facilities (see Fig. 3.27).

Analysis of the use of housing

The "predominance of small and medium-sized units and insufficient living area per capita" characteristics of the resettlement areas for land-lost peasants are mainly due to the consideration of a combination of factors such as living costs, service facilities, compensation policies, family structure, and whether the residents choose to make a profit from these housings.

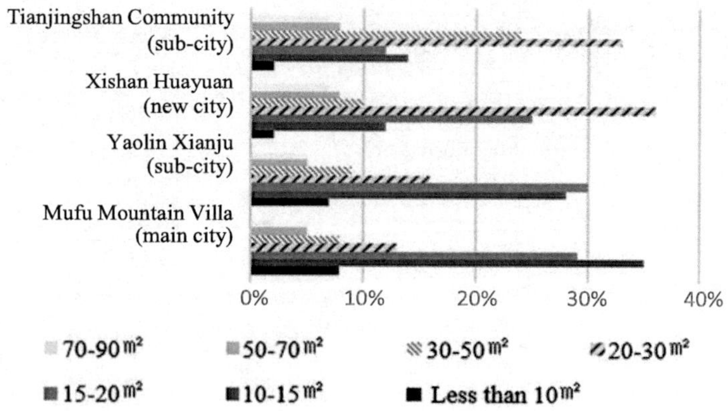

Fig. 3.27 Living area per capita of each resettlement area for land-lost peasants *Source* Sample data collected by the research team on the resettlement areas for land-lost peasants in Nanjing (2016)

Regarding unit types, resettlement housing for land-lost peasants is one of the non-profit subsidized housings funded by the government. Considering the development and construction costs and purchasing power of land-lost peasants, resettlement housing, which is not driven by profit, is positioned to be small and medium-sized units to meet basic living needs at the beginning of community planning and design. In addition, the purchasing power and housing needs of land-lost peasants determined that the service facilities and construction standards of this type of housing are focused on affordability and functionality rather than luxury and excessive standards.

In terms of living area, first, the different compensation policies for land expropriation and resettlement at different times led to a great difference in the total area purchased by land-lost peasants. The first batch of resettlement areas, Mufu Mountain Villa and Yaolin Xianju, have a small compensation area. The compensation standard of 30 square meters per capita can only ensure one to two resettlement units per household, while residents living in Tianjingshan Community and Xishan Huayuan, which were built in recent years and benefit from a new compensation policy, are able to purchase two to three resettlement units per household. Some can even purchase seven to eight units. Moreover, the complexity of the family structure also brings about a shortage of living area per capita. When arranged with units that are generally suitable for small to medium-sized households, a medium-sized household of four to six people usually has to live with a smaller living area per capita. In addition, as land-lost peasants have autonomy over the usage of their resettlement units, many families chose to leave more rooms for rent at the cost of lowering their living standards and private spaces to gain more profit. Consequently, for certain households, seven to eight people have to be packed into a small room, whereas for certain other households, the children live on the upper floors and the elderly in the basement.

3.3.3.3 Work Commuting

Characteristics of Commuting

In terms of commuting, majority of the land-lost peasants choose to work outside the community. It is found that the residents "have a commuting time of 40 min, choose buses as the major means of transportation, and mostly tend to be economical on commuting costs." In this study, the commuting distance is defined as the straight-line distance between the resettlement area and workplace. Based on the average speed of public transportation (the dominant choice of land-lost peasants), the research and measuring results indicate that the one-way commuting time of 30–50 min accounts for as high as 67.4% (see Fig. 3.28), which indicates that the worksites of most land-lost peasants is more than 5 km away from their resettlement areas (i.e., the place of residence). In terms of commuting mode, buses constitute the main choice of majority of the land-lost peasants whose worksite is far away from the resettlement area, while those who work nearby mainly walk or use bicycles or electric bicycles (see Fig. 3.29). As for the commuting cost, land-lost peasants do not spend more than 5 yuan for one-way trips, and only 13.4% spend more than 3 yuan, indicating that land-lost peasants usually carefully plan their budget in commuting and there is no significant difference among residents (see Fig. 3.30).

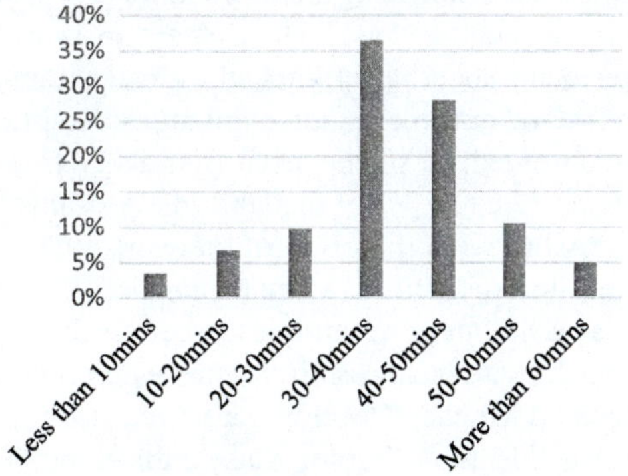

Fig. 3.28 Overall commuting time of land-lost peasants living in resettlement areas. *Source* Sample data collected by the research team on the resettlement areas for land-lost peasants in Nanjing (2016)

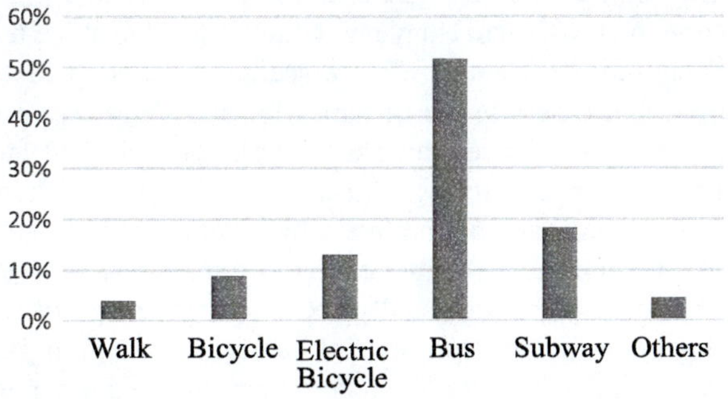

Fig. 3.29 Overall commuting mode of land-lost peasants living in resettlement areas. *Source* Sample data collected by the research team on the resettlement areas for land-lost peasants in Nanjing (2016)

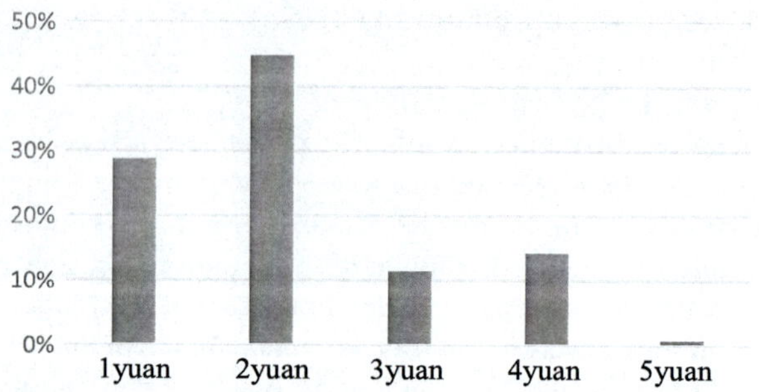

Fig. 3.30 Overall commuting cost of land-lost peasants living in resettlement areas. *Source*: Sample data collected by the research team on the resettlement areas for land-lost peasants in Nanjing (2016)

A comparison of the various types of resettlement areas reveals the following: the commuting time, mode and cost are highly correlated with the spatial distribution and transportation accessibility of resettlement areas (see Table 3.12). Residents living in Mufu Mountain Villa have more flexibility in terms of commuting time and mode as the area is located in the main city, and their travel cost is not high. As Yaolin Xianju is located in an industrial park, residents spend less time and cost in commuting. However, there are fewer choices related to the means of transportation. Long-distance travel, in particular, would have to depend on buses. Residents of Tianjingshan community, a resettlement area located at the fringe of the city, spend less time commuting and have more diverse options due to proximity to the subway. On the contrary, as Xishan Huayuan is located the farthest from the central city, residents spend a significantly greater time and cost commuting than the residents of the other three resettlement areas. Additionally, a combined transportation mode is observed. Residents need to transfer from one mode to another to arrive at their workplace (11.4% of the land-lost peasants take "both bus and subway").

A further comparison of the worksite of land-lost peasant reveals the following: land-lost peasants of Mufu Mountain Villa, which is located in the main city, work in places that are concentrated to a certain degree. The surrounding Yanziji, Xiaozhuang, and Xinyao business circles are where the majority of them work, and the closely located Xingang Development Zone also provides certain employment opportunities for residents. In Yaolin Xianju, the resettlement area located in an industrial park, land-lost peasants are mainly employed in industrial and research institutions such as Xingang Development Zone, Longtan Deep Water Port, and Xingang Development Zone. As for Xishan Huayuan, which is located at the fringe of the city, the worksites of land-lost peasants are more scattered, some at the closely located urban fringe areas and others at the distantly located main city. However, relatively speaking, worksites are concentrated at two places: scattered industrial parks to the west of the resettlement area and business circles of the main city. Land-lost peasants of the Tianjingshan community, which is also located at the fringe of the city, have a broader and more flexible choice of worksites. With the community as the center, their workplaces are usually set in a larger range of area and along two directions in general: nearby commercial centers or Jiangning Economic and Technological Development Zone or Xinjiekou, Fuzimiao and other urban commercial centers that

Table 3.12 Commuting time, mode, and cost of each type of resettlement area for land-lost peasants

	Comparison by resettlement area
Commuting time	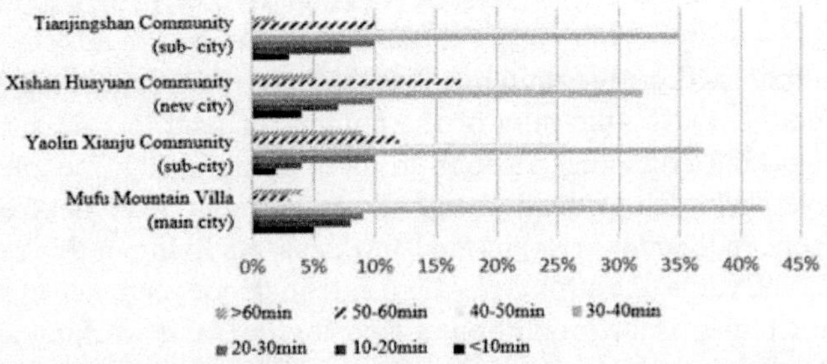
Commuting mode	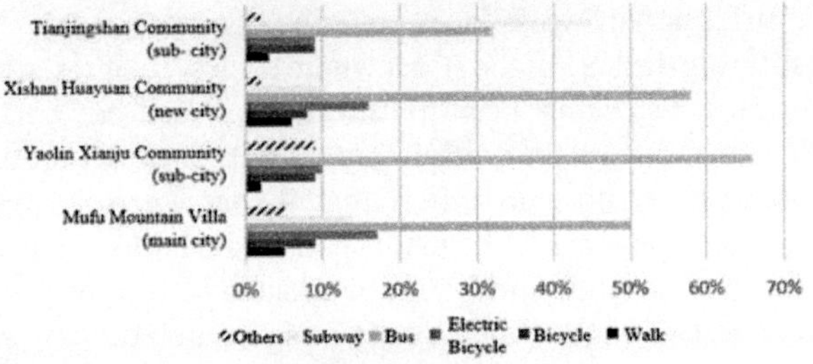
Commuting cost	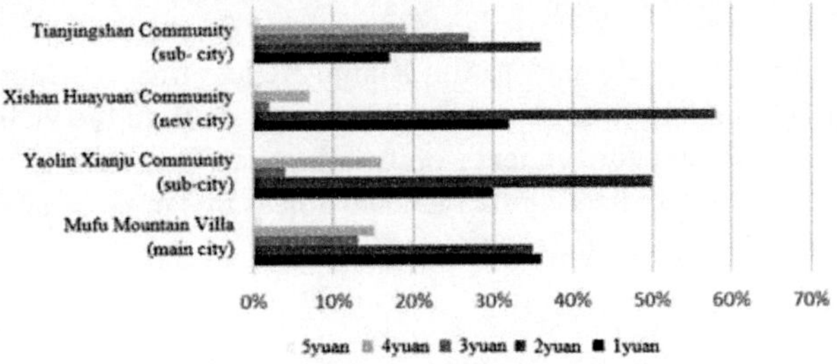

Source Sample data collected by the research team on the resettlement areas for land-lost peasants in Nanjing (2016)

can be reached conveniently through public transport such as subways. In general, "urban centers at all levels and areas adjacent to the resettlement areas" are still the principal choices for land-lost peasants in terms of worksite.

Self-sufficiency Measurement

Self-sufficiency is a quantitative analysis of land-lost peasants living and working in resettlement areas and those living resettlement areas but working at a different place to reflect the equilibrium of the work-home spatial structure of land-lost peasants. Considering the residential dominance of resettlement areas, this section uses streets as the statistical unit and focuses on the "residential independence index" to measure and analyze the self-sufficiency of each resettlement area. The "Residential Independence Index" (Rsi) of land-lost peasants in street unit i is measured as follows:

$$Rs_i = Ni/Avg(Ni)(i = 1, n)$$

In other words, the ratio of the residential independence level of a certain resettlement area (Ni—the ratio of the number of land-lost peasants living in the area and working at another location to the total number of people living in area (i) to the average value of this level of all resettlement areas is taken as the "residential independence index" (Rs) of land-lost peasants living in this resettlement area. The higher the value, the stronger is the residential independence of that resettlement area and the higher is the proportion of land-lost peasants working at another location. In addition, the work-home separation model is more inclined toward separation and a scattered distribution of workplaces could be observed.

By substituting the data of the street where each type of resettlement area is located into the formula and using 1 as the threshold, it can be seen that, based on self-sufficiency measurement, the work-home spatial separation of the resettlement areas for land-lost peasants presents differentiated characteristics: residents living in the scattered resettlement areas in the main city and the industrial park resettlement areas mostly work in the vicinity, while the worksites of residents living in resettlement areas located at the fringe of the city are more scattered.

By comparing the self-sufficiency of various types of resettlement areas for land-lost peasants (see Fig. 3.31), it can be found that the residential independence indexes of Mufu Mountain Villa and Yaolin Xianju are less than 1, which indicates that land-lost peasants in these resettlement areas mostly work in the vicinity. Conversely, the residential independence indexes of Tianjingshan Community and Xishan Huayuan are greater than 1, indicating that land-lost peasants in these areas are more inclined to choose worksites that are dispersed.

Commuting Frequency Measurement

Based on the sampling data of the three types of resettlement areas for land-lost peasants in Nanjing, the work-home commuting frequency in each direction between the locations of the resettlement areas for land-lost peasants and the streets of their worksites is calculated to discover the distribution characteristics of the streets that connect worksites and residence areas. Suppose that the number of residents living in X street and working at Y street is Ai (i.e., commuting from the same starting point and in the same direction), and the total number of residents commuting on the streets of this resettlement area is Bi, the commuting frequency Ci is calculated as follows:

$$\mathbf{Ci = Ai/Bi.}$$

The commute-related streets were classified into three categories: high commuting frequency ($Ci > 10\%$), average commuting frequency ($5\% < Ci \leq 10\%$), and low commuting frequency ($Ci \leq 5\%$). By substituting the current work-home commuting

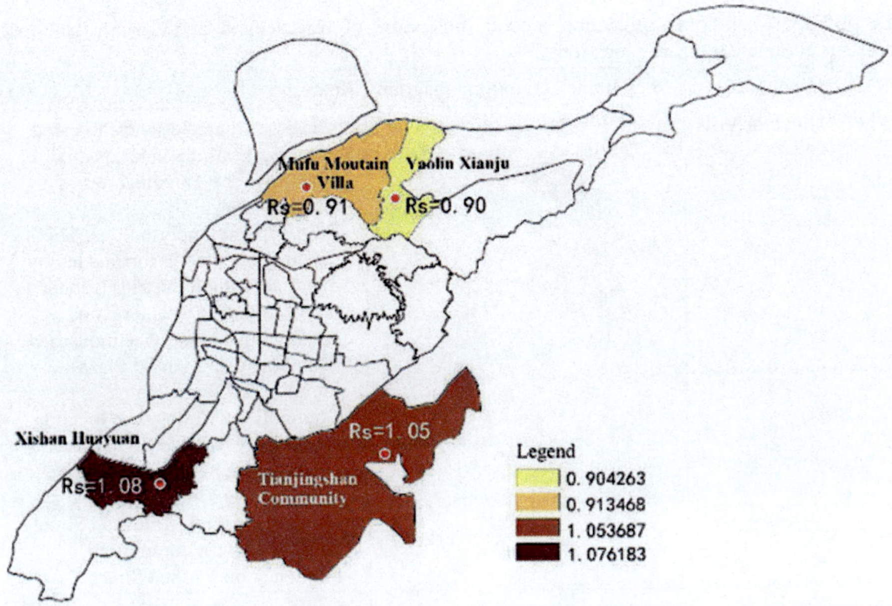

Fig. 3.31 Spatial distribution of residential independence index for each type of resettlement area. *Source* Sample data collected by the research team on the resettlement areas for land-lost peasants in Nanjing (2016)

data of the different types of resettlement areas into the above formula, the following results were obtained.

The commuting frequency of all types of resettlement areas presents a differentiated characteristic where residents "either commute toward the center (urban centers at all levels), to nearby locations, or to scattered locations." That said, due to the different work-home spatial separation models, the work-home commuting frequency between the streets of the resettlement area and the worksite is also different. Among them, residents living in the resettlement areas where the work-home separation model features working nearby from home tend to "commute to closely located neighborhoods," and residents living in a work-home separation model where the work and residence space is at a distance have a more diverse choice, "some commute to distant locations, some commute toward the center, and some commute to nearby locations."

Based on the above-mentioned work-home spatial separation model of land-lost peasants, a graphical analysis and comparison of the "residence-worksite" correlation among each type of resettlement area for land-lost peasants were conducted (see Tables 3.13 and 3.14). This includes illustrations and overview of the street location characteristics, residential opportunities, employment opportunities, and work-home commuting of the resettlement areas, thus presenting the distribution characteristics of the streets that are connected to each resettlement area because of commuting. The following was found.

Table 3.13 Work-home separation model: illustration of resettlement areas where land-lost peasants commute to nearby locations

Resettlement area	Work-home commuting diagram	Analysis
Mufu Mountain Villa	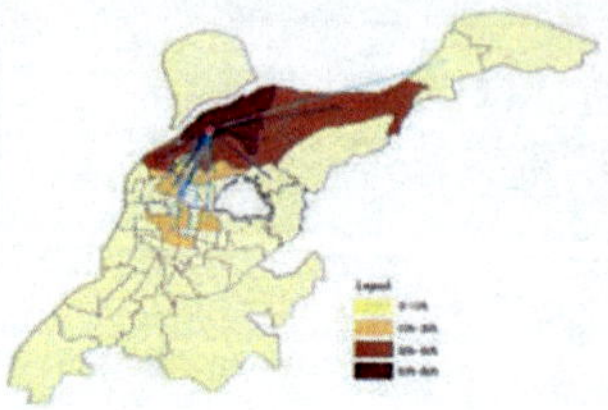	• Location Characteristics: Located to the north of the inner city, it belongs to the Yanziji Street of Qixia District • Employment Opportunities: The resettlement area is surrounded by manufacturing industries (rubber products industry) and related commercial and service industries (Rainbow Commercial Plaza) • Summary of Work-Home Commuting: Work-home location is relatively balanced. The work locations are mainly concentrated in the nearby Yaohua Street, Maigaoqiao, and Mofushan Street, with the highest commuting frequency on Yaohua Street
Yaolin Xianju	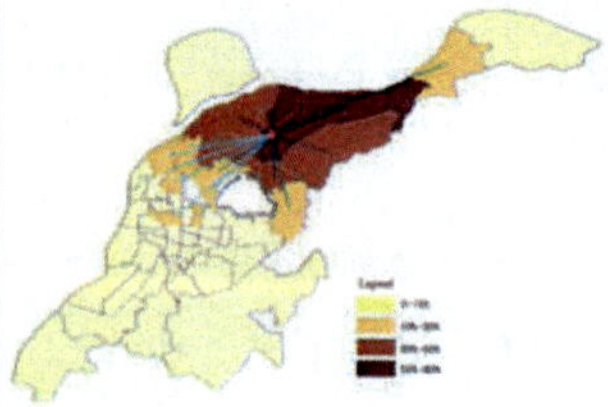	• Location Characteristics: Located in the northern suburban industrial park, it belongs to the Yaohua Street of Qixia District • Employment Opportunities: To the north, it is adjacent to many enterprises such as Xingang Development Zone and Longtan Deep Water Port, and to the south, it is adjacent to Xianlin University City and many other scientific research and educational institutions • Summary of Work-Home Commuting: Work-home location is relatively balanced, and streets connected to the area are closely located. The work locations of the residents are mainly concentrated in the streets adjacent to the industrial park, such as Yanziji, Maigaoqiao and Xianlin streets, of which Yaohua street, where the Xingang Development Zone is located, has the highest commuting frequency

Table 3.14 Work-home separation model: illustration of resettlement areas where land-lost peasants commute to scattered locations

Resettlement area	Work-home commuting diagram	Analysis
Tianjingshan Community	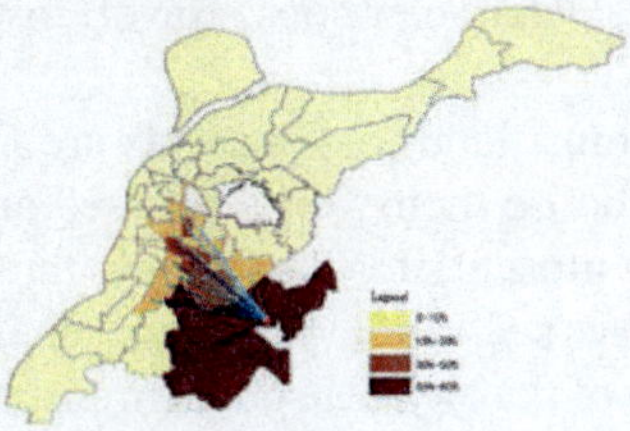	• Location Characteristics: Located in the southeastern fringe of the city, it belongs to Chunhua Street, Jiangning District • Employment Opportunities: The surrounding area is dominated by food service and retail business (Yiwu small commodity city), with many colleges and universities such as China Pharmaceutical University to the south and Jiangning Economic and Technological Development Zone to the west • Summary of Work-Home Commuting: Work-home separation is obvious, and streets related to commuting are scattered. The residents' work locations include Xinjiekou and Fuzimiao streets, which are rather distant from the community, as well as Honghua and Ningnan streets, which are located very close. Dongshan street has the highest commuting frequency
Xishan Huayuan	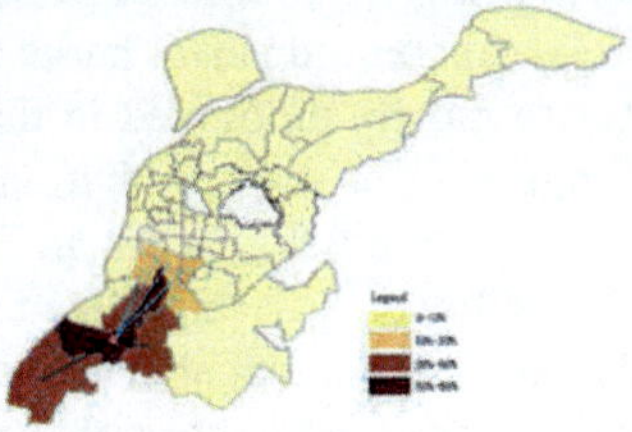	• Location Characteristics: Located at the fringe area of the city, it belongs to Xishanqiao Street of Yuhuatai District and is one of the four major subsidized housings in Nanjing • Employment Opportunities: Small businesses are scattered in the west, and the surrounding area is dominated by small commercial services, with limited employment opportunities • Summary of Work-Home Commuting: Work-home balance is low in terms of location, and the related streets are scattered. The work locations include the nearby Banqiao Street and Saihongqiao Street as well as the main city streets such as Zhonghua Gate and Fuzimiao, which are relatively remote. The highest commuting frequency is on Saihongqiao Street

* *Source* ①Sample data collected by the research team on the resettlement areas for land-lost peasants in Nanjing (2016); ②Administrative maps of Nanjing's main city, Qixia District, Yuhuatai District, and some streets in Jiangning District

Analysis of Work Commuting

In terms of commuting, land-lost peasants of the resettlement areas tend to "take buses, travel to dispersed workplaces, and work either in city centers or in the vicinity." This conclusion is mainly derived from a comprehensive analysis of employment opportunities, occupational characteristics, travel costs, commuting conditions, and other factors.

First, in terms of commuting costs, most land-lost peasants are limited by their skills and financial ability and tend to choose their worksite based on their occupational characteristics and travel costs. At present, in Nanjing, low-threshold employment and labor-intensive employment areas such as development zones and industrial parks are mainly located in the periphery of the city. The public transportation system in the main city is relatively complete, and bus and subway fares are within the affordable range. Therefore, the commuting time has become the primary consideration for land-lost peasants. They tend to choose distant work locations that can be arrived using one-stop" transportation or nearby work locations.

As for the self-sufficiency test, the dominant function of resettlement areas for land-lost peasants is residential, and majority of the residents choose to work outside the resettlement area. Therefore, the work-home separation phenomenon in large mixed resettlement areas at the fringe of the city is particularly apparent. Since the resettlement areas that are dispersed in the main city are surrounded by large and small business circles, there are plenty of employment opportunities and transportation is rather convenient. The employment options for land-lost peasants of this area are thus more flexible and extensive. Based on the commuting cost, the work-home separation model of this community features primarily working at a nearby location. As the evenly distributed resettlement areas in the industrial parks have a spatial advantage in their employment, peasants are mostly employed in the vicinity. For large resettlement areas located at the fringe area of the city that have convenient transportation conditions, where residents can choose to work in the vicinity or at a distant location by taking the expedite subway, a decentralized work location characteristic can be observed. As for the large resettlement area located at the fringe of the city that has poor transportation conditions, it is not possible for the residents to work in the vicinity due to the scarcity of employment opportunities. They can only travel to the main city for more job opportunities at the cost of time and energy spent on the long-distance and multi-transportation means, thus showing an obvious work-home spatial separation pattern.

In the end, as far as commuting frequency is concerned, since most commercial services (such as catering, retail, sanitation, and cleaning), wherein majority of the land-lost peasants of working age are employed, are dispersed on the streets of the main city, the worksites of residents living in the resettlement areas of the main city are mostly "proximate to their residence and diverse." Similarly, the worksites of residents living in the resettlement areas of the industrial parks are also proximate. The connected streets generally have higher commuting frequency and are mostly concentrated and closely situated. On the contrary, residents of large mixed resettlement area at the fringe of the city, which is disadvantaged in terms of spatial location, have limited employment options due to many factors such as location, traffic condition, and employment opportunities. Consequently, the commuting frequencies to nearby and distant streets hardly differ. Of them, Xishan Huayuan, which is located at the fringe of the city, is the farthest from the main city and has fewer employment opportunities in the neighborhood. Therefore, as a large number of land-lost peasants have no choice but to work in the main city, the commuting frequency between this resettlement area and the main city is relatively high.

3.3.3.4 Service Facilities

Characteristics of Service Facilities (at municipal level)

Public service facilities at the city level serve a wide range of people and have a vast variety of facilities. According to the definition of land for urban public service facilities provided in the Code for Urban Public Facilities Planning, the land for urban public service facilities could be divided into seven categories: land used for administrative offices, commercial and financial services, cultural and recreational facilities, sports arena, medical and health care facilities, education and research design institutions, and social welfare services. Taking into account the social attributes of land-lost peasants and their specific needs for urban public service facilities, this study focuses on four types that have a greater impact on land-lost peasants: commercial services, education, medical and health care facility, and cultural and recreational facilities. This study also covers service facilities at two levels, municipal and regional, and determines the public service status of resettlement areas mainly through the location of existing facilities.

The supply of public service facilities at the municipal level is "category-specific and non-homogeneous" across different resettlement areas for land-lost peasants, and the accessibility of these facilities is generally low.

A comparison across the different types of resettlement areas reveals the following (see Fig. 3.32): except for commercial facilities, which are distributed equally across the different resettlement areas, education institutions, medical and healthcare services, as well as cultural and recreational facilities exhibit an obvious spatial deviation from the different resettlement areas, and the number of these facilities also decrease gradually from the old city to the peripheral areas. For land-lost peasants living in the resettlement areas distributed dispersedly in the main city (Mufu Mountain Villa), the accessibility and selection of various service facilities are well ensured, while for residents living in resettlement areas located at the industrial park and the fringe of the city (large mixed resettlement areas such as Xishan Huayuan and Tianjingshan Community), it is difficult to have equal access to urban commercial, educational, medical, and cultural services, all of which show a serious imbalance in quality and quantity.

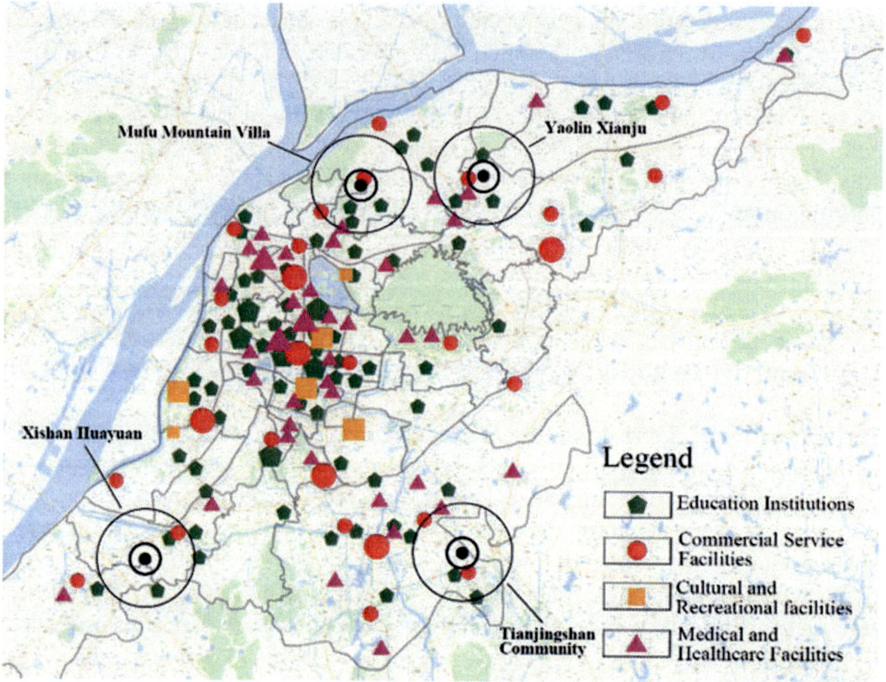

Fig. 3.32 Map of urban public service facilities around the resettlement area of land-lost peasants. *Source* ① Sample data collected by the research team on the resettlement areas for land-lost peasants in Nanjing (2016); ② Master plan of Nanjing (2011–2030)

Characteristics of Service Facilities (at community level)

The public service facilities involved at the community level are equally complex and diverse. With practical function as the classification standard, the "Code of Urban Residential Area Planning and Design" (GB50180—93) divided the public service facilities in residential areas into eight categories: facilities for education, health care, culture and sports, commercial services, finance and post and telecommunications, municipal utilities, and public administration and others. Considering the unique attributes and needs of land-lost peasants, this study focuses on the public service facilities that have a greater impact on the living space quality of resettlement areas for land-lost peasants. According to the understanding of Jin et al. (2007) and Yuan and Ma (2012) on the classification of public service facilities in indemnificatory residential community, which was based on the theory of public goods, the abovementioned eight categories of facilities are screened and reclassified into six categories accordingly: facilities for education, medical and healthcare, culture and recreation, commercial services, post and telecommunications and municipal administration, and community services (see Table 3.15).

Table 3.15 Classification of public service facilities in resettlement areas for land-lost peasants

Type	Facility	Location		
		District of the resettlement area	Community of the resettlement area	Housing group
Education	Kindergarten		√	√
	Primary school		√	
	Secondary school	√		
Medical and Healthcare	Community medical service center	√		
	Clinic		√	
	Health station		√	
Culture and recreation	Cultural activity center	√		
	Cultural activity stations			√
	Fitness facilities for residents		√	√
	Teahouse, chess room, etc		√	
Commercial services	Open fair	√		
	All kinds of financial institutions, bookstores, saving banks		√	
	Grain and oil store, subsidiary food store, restaurants, etc		√	
Post and telecommunications and municipal administration	Post office, business office	√		
	Vehicle parking lot (Garage)		√	
	Bus stops	√		
Community services	Community service center	√		
	Homes for the elderly (nursing home)	√		

(continued)

Table 3.15 (continued)

Type	Facility	Location		
		District of the resettlement area	Community of the resettlement area	Housing group
	Property management	√	√	

Source Yuan, Qifeng, Ma, Xiaoya. Supply of Public Facilities in Indemnificatory Residential Community: A Case Study of Guangzhou. *Urban Planning*, 2012, 36(2):24–30

In terms of the categories and scales of public service facilities at the community level, resettlement areas for land-lost peasants possess common characteristics wherein "education, commercial services, cultural and recreational and community service facilities are of different quantities and qualities, post and telecommunications and municipal administration facilities are well equipped, while medical and healthcare facilities are insufficient" (see Table 3.16). As for the spatial layout of various facilities, the characteristics are "public welfare facilities are distributed in points and blocks, while commercial facilities are built along the road" (see Fig. 3.33).

A comparison across the different types of resettlement areas reveals the following: there is no significant difference in the category and scale of public service facilities: (1) Education facilities—There is no kindergarten in the resettlement area of the industrial park (Yaolin Xianju), and the one in the resettlement areas located at the urban fringe region (Tianjingshan Community) is also deficient in quantity and scale. However, the education facilities in the two other resettlement areas are well equipped. (2). Medical and healthcare facilities—There is deficiency in quantity and scale in all three types of resettlement areas. (3) Cultural and recreational facilities—Higher-level cultural centers and stations are largely vacant, while lower-level recreational facilities are numerous (especially chess and cards rooms). (4) Commercial services—The facilities can meet the basic demands of the resettlement areas. However, as the Tianjingshan community is extremely large and the distribution of commercial facilities is highly concentrated, the accessibility for the purchase of certain wholesale goods is poor. (6) Community services—the basic community service and property management of all resettlement areas can meet the basic need. However, there is lack of considerations for nursing houses in general (among the samples, only Xishan Huayuan, which is located at the fringe of the city, is equipped with home care centers).

Analysis of Service Facilities

The "category-specific and non-homogeneous" characteristics of service facilities in resettlement areas for land-lost peasants are mainly due to differences in the construction of service facilities at different levels and their different modes of supply.

Table 3.16 Facilities configuration of each resettlement area for land-lost peasants

Type		Scattered resettlement area in the main city		Evenly distributed resettlement area in the industrial park		Large mixed resettlement area at the fringe area of the city			
Resettlement Area		Mufu Mountain Villa		Yaolin Xianju		Tianjingshan Communicty		Xishan Huayuan	
		Number	Note	Number	Note	Number	Note	Number	Note
Education	Kindergarten	**1**	Private	–	–	**2**	One Pubic And One Private	**1**	Private
	Primary school	**1**	Provincial	**1**	Public, Provincial	**1**	Municipal	**1**	Public
	Secondary school	–	–	**1**	Public High School	**1**	Public Secondary School	**1**	Public Secondary School
Medical and Healthcare	Community Medical Service Center	–	–	–	–	–	–	–	–
	Clinic	**1**	Chinese Medicine Clinic	–	–	**5**	–	**1**	–
	Health Stations	**1**	Yanziji Hospital And Health Service Station	**1**	Yaolin Xianju Community Health Service Station	–	–	–	–
Culture and Recreation	Cultural Activity Center	–	–	–	–	–	–	–	–

(continued)

Table 3.16 (continued)

Type		Scattered resettlement area in the main city		Evenly distributed resettlement area in the industrial park		Large mixed resettlement area at the fringe area of the city			
Resettlement Area		Mufu Mountain Villa		Yaolin Xianju		Tianjingshan Communicty		Xishan Huayuan	
		Number	Note	Number	Note	Number	Note	Number	Note
	Cultural Activity Stations	–	–	–	–	–	–	**1**	
	Fitness Facilities for Residents	**Multiple**	–	**Multiple**	–	**Multiple**	–	**Multiple**	–
	Teahouse, Chess Room, Etc	**Multiple**	–	**Multiple**	–	**Multiple**	–	**Multiple**	–
Commercial Services	Open Fair	**1**	Mufu Mountain Villa Farmer's Market	**1**	Tiandun Farmer's Market	**1**	Wholesale Farmer's Market	**1**	Xishan Huayuan Farmer's Market
	All Kinds of Financial Institutions, Bookstores, Saving Banks	**Multiple**	–	**Multiple**	–	**Multiple**	–	**Multiple**	–

(continued)

Table 3.16 (continued)

Type		Scattered resettlement area in the main city		Evenly distributed resettlement area in the industrial park		Large mixed resettlement area at the fringe area of the city			
Resettlement Area		Mufu Mountain Villa		Yaolin Xianju		Tianjingshan Communicty		Xishan Huayuan	
		Number	Note	Number	Note	Number	Note	Number	Note
	Grain and Oil Store, Subsidiary Food Store, Restaurants, etc	**Multiple**	–	**Multiple**	–	**Multiple**	–	**Multiple**	–
Post and Telecommunications and Municipal Administration	Post Office, Business Office	**Multiple**	–	**Multiple**	–	**Multiple**	–	**Multiple**	–
	Vehicle Parking Lot (Garage)	**2**	–	**2**	–	**1**	–	**5**	–
	Bus Stops	–	–	–	–	–	–	**1**	Xishan Huayuan Bus Stop

(continued)

Table 3.16 (continued)

Type		Scattered resettlement area in the main city		Evenly distributed resettlement area in the industrial park		Large mixed resettlement area at the fringe area of the city			
Resettlement Area		Mufu Mountain Villa		Yaolin Xianju		Tianjingshan Communicty		Xishan Huayuan	
		Number	Note	Number	Note	Number	Note	Number	Note
Community Services	Community Service Center	**1**	Resident Committees, Labor Protection Centers And Community Service Stations	**1**	Community Service Centers and Guard Rooms	**1**	Community Service Centers and Guard Rooms	**1**	Community Labor and Social Security Service Stations
	Homes for The Elderly (Nursing Home)	–	–	–	–	–	–	**1**	Runle Home Care Center
	Property Management	–	–	–	–	**1**	Chunghua Property Management	**1**	Lianxing Property Management

Source Sample data collected by the research team on the resettlement areas for land-lost peasants in Nanjing (2016)

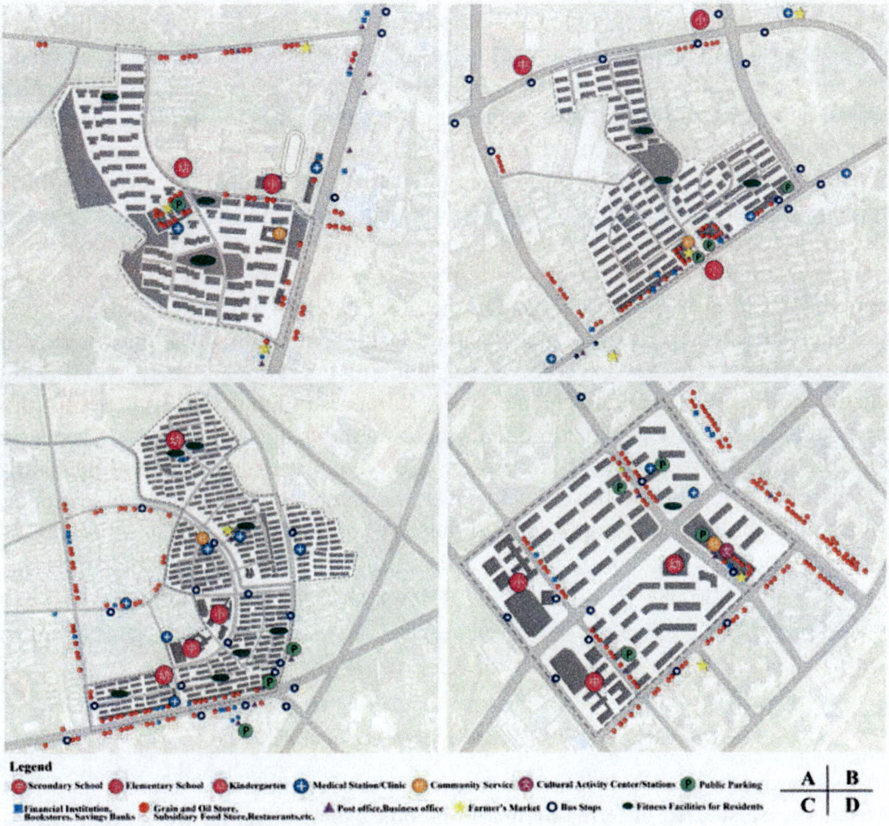

Fig. 3.33 Distribution of facilities in each resettlement area (A. Mufu Mountain Villa; B. Yaolin Xianju; C. Tianjingshan Community; D. Xishan Huayuan). *Source* Sample data collected by the research team on the resettlement areas for land-lost peasants in Nanjing (2016)

At the municipal level, the investment, construction, and management of public service facilities are mainly undertaken by the government and related public administration agencies. Thus, the choice of their location is determined by the local government. Due to the limited budget and efficiency considerations, the "equalization of basic public service facilities" is often constrained, resulting in the over-concentration of public service facilities in the old city of Nanjing. As the resettlement areas built at various stages gradually move from the center to the peripheral areas, the "bottom-up" competition for locations based on market economy and the "top-down" location arrangement of the resettlement areas by the municipal government led to the spatial deviation and misalignment between the locations of urban public service facilities and resettlement areas.

At the community level, the supply mode of various public service facilities is different. Resettlement areas are residential areas built by the government for special groups. Therefore, barring commercial service facilities whose construction is market-driven, the supply mode of other kinds of service facilities mainly follow the principal of "whoever developed the resettlement area needs to build the facilities and be in charge of the transfer," i.e., the municipal government is responsible for the supply. As a result, the resettlement areas built under different construction backgrounds possess different characteristics in the construction of facilities. In case of Mufu Mountain Villa and Yaolin Xianju, which were built earlier, there is a delay in the construction of large-scale stand-alone facilities or insufficient land reserved for them (education, medical and healthcare, parking lots, etc.). Specific examples include insufficient planning of courts for sports and little land reserved for the construction of elementary schools. On the other hand, in Xishan Huayuan, a large mixed resettlement area completed at a mature stage, the immediate construction of service facilities and the land reserved for planning are more adequate, showcasing an advanced facility planning characteristic. Compared with the other three resettlement areas, the Tianjingshan community built on the fringe of the city is a recently constructed large resettlement area. Thus, the type and scale of supporting facilities were often planned uniformly. However, the actual construction of educational, commercial, cultural, and medical facilities does not meet the predetermined standards. In this way, the facility construction results differ across resettlement areas due to the different facility supply patterns and construction backgrounds.

3.3.4 Summary

In this section, the spatial attributes of various types of resettlement areas for land-lost peasants were studied in terms of spatial layout, use of housing, work commuting, and service facilities. The reasons are also analyzed accordingly, on the basis of which their characteristics are summarized as follows (see Table 3.17).

3.4 Study on Social Integration of Land-Lost Peasants

Based on the above analysis of the social and spatial attributes of resettlement areas for land-lost peasants in Nanjing, this section constructs a social integration evaluation index system through the analytic hierarchy process to explore and compare the integration characteristics of resettlement areas in terms of economic, social, psychological, institutional, and spatial dimensions (i.e., to explain "how it is") following a "community overview—individual community" analytical pattern. Next, using structural equation modeling, this study analyzes the integration mechanism from multiple dimensions, including individual characteristics, economic factors, social factors, psychological factors, institutional factors, and spatial factors (i.e., to explain "why it is") following an analytical pattern based on "overall—individual dimension."

Table 3.17 Summary of spatial attributes of resettlement areas for land-lost peasants

Spatial attributes		Common characteristics	Differentiated characteristics	Reasons
Spatial layout	Surrounding	"Long travel distance and mixed-use of land"	The resettlement areas dispersed in the main city and the ones located at the fringe area of the city have a relatively advantageous location and convenient transportation. There is also a high degree of community resource sharing	The planning and location of resettlement areas for land-lost peasants are essentially constrained by the graded distribution of land values. In addition, the construction time, government policy, and passive choice of housing by land-lost peasants are the decisive factors
	Layout	"Different in scale and size, mostly grouped by rows and constructed by street"	The recently built resettlement areas are larger and more reasonably planned and designed than the resettlement areas built earlier	Due to the differences in objective factors such as construction time and resettlement methods, resettlement areas of different types vary in terms of scale and other economic and technical indicators
Use of housing	Unit types	"Mainly small and medium-sized units"	The recently built resettlement areas are more complete than the earlier resettlement areas in terms of housing facilities	Resettlement areas are designed to be a kind of economic housing in the first place and land-lost peasants are mainly forced to resettle. Given their limited housing purchasing power, the planning and design of resettlement areas is conducted in conformity with the principle of being able to meet basic living needs at the beginning
	Living area	"Insufficient living space per capita"	Residents of earlier resettlement areas for land-expropriated peasants are living with a particularly inadequate living space	Due to the relatively low compensation standard for land expropriation and resettlement, the medium-sized family structure and the relatively free disposition of resettlement housing for land-lost peasants, there is insufficient housing area per capita

(continued)

Table 3.17 (continued)

Spatial attributes		Common characteristics	Differentiated characteristics	Reasons
Work commuting	Commuting cost	"Average commuting time is 40 min, buses constitute the major means of transportation, and the cost is relatively flat."	Spatial distribution and accessibility have an impact on the time, mode, and cost of their work commuting	Travel costs and travel time are the primary considerations in choosing a place of employment for land-lost peasants because of their limited skills and economic capacity and the relatively well-developed public transportation system in the main city of Nanjing
	Self-Sufficiency Measurement	–	The work-home spatial separation model in the resettlement areas for land-lost peasants shows a differentiated characteristic among different areas where "residents living in the scattered resettlement areas in the main city and the industrial park resettlement areas mostly work in the vicinity, while the worksites of residents living in resettlement areas located at the fringe of the city are more scattered."	Because the employment options in the main city are wide and diverse, and there are also plenty of employment opportunities in the industrial park, land-lost peasants living in these two types of areas mainly work in the vicinity. The urban peripheral resettlement areas are still characterized by scattered work locations of the residents due to the limited employment opportunities and transportation conditions

(continued)

Table 3.17 (continued)

Spatial attributes		Common characteristics	Differentiated characteristics	Reasons
	Commuting frequency	"Differentiated commuting mode: commute toward the center (urban centers at all levels), to nearby locations, and to scattered locations"	Residents living in the resettlement areas where the work-home separation model is choosing a nearby worksite tend to "commute to closely located neighborhoods," and residents living in a work-home separation model where the work and residence space are far apart have a more diverse choice	Land-lost peasants living in the main city have a diverse employment selection and mostly work in nearby locations, while residents living in the industrial park resettlement area typically choose to work in the vicinity. Therefore, the streets featuring high commuting frequency within the two resettlement areas are concentrated and close to each other. While for resettlement areas located at the urban fringe region, due to the influence of various factors such as location, transportation convenience and employment opportunities, closely and distantly located streets are similar in terms of commuting frequency
Service facilities	Urban public service facilities	"Category-specific and non-homogeneous"	–	The "bottom-up" competition for locations based on market economy, and the "top-down" location arrangement of the resettlement areas by the municipal government led to the spatial deviation between the resettlement areas which were gradually built from urban center to peripheral areas, and urban public service facilities which are over-concentrated in the old city

(continued)

Table 3.17 (continued)

Spatial attributes		Common characteristics	Differentiated characteristics	Reasons
	Residential service facilities	"In terms of scale and category, education, commercial services, cultural and recreational and community service facilities are of different quantities, post and telecommunications and municipal administration facilities are well equipped, while medical and healthcare facilities are insufficient."	The recently constructed resettlement area is more complete than the earlier built resettlement area in terms of service facilities support	Due to the significant differences in the supply mode of various public service facilities, the construction background of various resettlement areas for land-lost peasants and their construction scale, there are differences across various areas in terms of the categories and scale of the facilities

3.4.1 Research Strategy and Methodology

3.4.1.1 Research Methodology

The quantitative analysis method of the analytic hierarchy process (AHP) is adopted in this paper to measure the social integration characteristics of resettlement areas for land-lost peasants. The AHP method is one of the most frequently used factor-based comprehensive assessment methods in urban research[2] and has obvious advantages in solving complex multi-level and multi-dimensional problems in urban studies. It is thus a persuasive method and widely applied in social integration studies. Specifically, based on the primary data collected through questionnaires and interviews, this study constructs a hierarchical structure of three levels—goal, criteria, and alternative—to decompose the complex issue of social integration into several ordered levels. By comparing the relative materiality of various factors at each level, we carry out statistical and weighted calculations of all indicators and analyze the integration characteristics from multiple dimensions and two levels, including "communities overview and individual community."

Structural equation modeling (SEM) is chosen as the main quantitative analysis method in the study to analyze the social integration mechanism of resettlement areas for land-lost peasants. This modeling is essentially a confirmative model analysis and a multivariate statistical analysis technique for building, estimating, and testing causal models. As SEM integrates factor analysis, path analysis, and multiple linear regression analysis (Wu 2009), it holds significant advantages in latent variable analysis and path analysis. It is thus known as one of the three major advances in applied statistics in recent years (Qiu 2009). In recent years, this method has been applied to impact factor, effect level, performance, satisfaction, and competitiveness (Xin et al. 2014), from which a certain number of research results have been obtained. Structural equation models contain observed variables that can be directly measured and latent variables that cannot be directly observed. The model that analyzes the relationship between observed and latent variables is the measurement model, and the one that analyzes the relationship between latent variables is the structural model (Yi 2008). Since most impact factors related to social integration studied in this paper are not directly measurable variables, the introduction of structural equation modeling is an effective way to analyze the social integration mechanism, i.e., using AMOS 22.0 software to build and interpret the model for analyzing the social integration mechanism following the steps of "theoretical hypothesis—model construction—parameter estimation—model correction."

---

[2] Xu, Shubai. *Principles of Analytic Hierarchical Process: A Practical Decision-Making Method* M]. Tianjin: Tianjin University Publishing House, 1995.

3.4.1.2 Overall Strategy

Since the analytic hierarchy process has been used more commonly, this paper only elaborates the relatively complex structural equation modeling. The modeling process of structural equation modeling can be roughly divided into two stages, model development and model estimation and evaluation, with seven steps (see Fig. 3.35). The first stage contains theoretical hypothesis, model construction, and model identification, wherein the researcher proposes a basic default model according to the relevant theory. The second stage is the model estimation and evaluation stage, wherein the basic model is validated with sample data, and if the default model is not found to fit sample data well, the default model must be modified and tested again. This "trial and error" process is repeated until finally a model that has satisfactory goodness-of-fit and can reasonably explain "what factors influenced the different degrees of integration in resettlement areas for land-lost peasants" is obtained (see Fig. 3.34).

3.4.2 Analysis of Social Integration Characteristics of Resettlement Areas for Land-Lost Peasants

First, this study uses the analytic hierarchy process to analyze the characteristics of social integration. Taking the selected resettlement areas for land-lost peasants as samples, a multi-dimensional evaluation index system (5 dimensions, 10 criteria, and 26 specific alternatives) containing economic, social, psychological, institutional, and spatial factors was established based on relevant research on social integration. Next, this study standardizes the indicators and calculates their weights to obtain the weighted sum of the questionnaire statistics at each criteria level and sub-goal level (dimensions), namely, the evaluation results. Finally, the overall integration characteristics of each resettlement area and differentiated characteristics among them were analyzed from two perspectives: "community overview and individual community." Notably, this section uses the objective weighting method (weighted standard deviation decision-making method) rather than the subjective weighting method (such as the expert survey method) to calculate the weights of the five dimensions. In addition, the weights of each indicator at the alternatives, criteria, and sub-goal levels were measured objectively to obtain the weighted overall evaluation value and evaluation values of each dimension in terms of integration of each resettlement area. The specific operation steps are as follows.

3.4.2.1 Establishment of Index System

Evaluation system

As mentioned above, domestic scholars and foreign scholars have not yet reached a consensus on the assessment index system of social integration but have conducted a large number of empirical studies on different frameworks of social integration, from

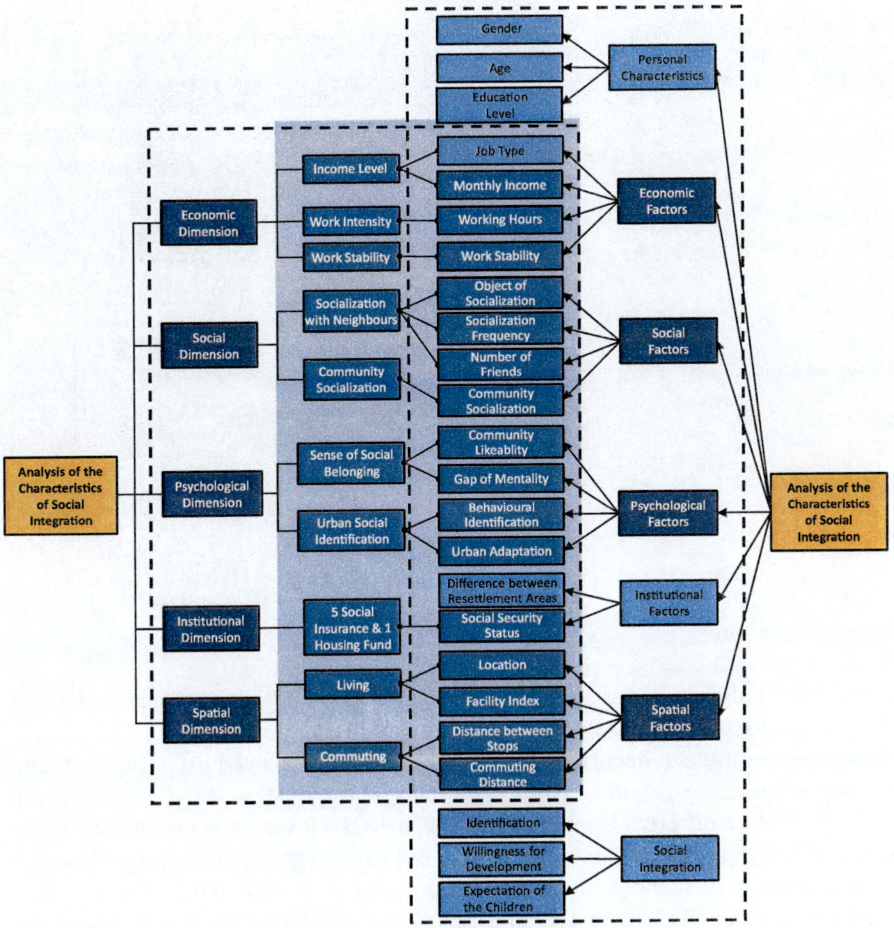

Fig. 3.34 General technical illustration. *Source* Own elaboration

which they have obtained abundant academic results. Objectively speaking, although there are some differences in their standpoints, scholars were consistent in taking four dimensions into account: economic, social, institutional, and psychological dimensions.

Nevertheless, the studies (especially in sociology) have in general disregarded "spatial integration" as a measurement dimension of social integration. However, for land-lost peasants, a special group that underwent major transformation in their living space, spatial integration is not only an important aspect of social integration but also an important factor that can hinder or promote social integration. Many domestic studies on subsidized housing also found spatial segregation in such resettlement areas in cities (Lu and Pang 2013), and certain advanced concepts as well as planning and design strategies such as mixed housing and mixed communities have been proposed (Long 2013; Li 2015). The social integration dilemma caused by spatial

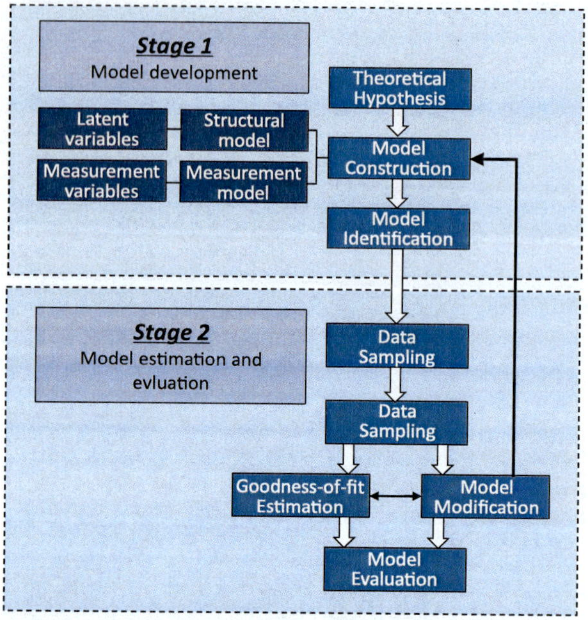

Fig. 3.35 Analysis process of structural equation modeling. *Source* Wu (2006)

segregation is also an important topic in foreign social integration research (City of Chula Vista 2010). Spatial segregation, as a physical manifestation of social stratification in urban space, reflects the degree of alienation between land-lost peasants and ordinary urban citizens and constitutes an indispensable perspective for measuring social integration different from the other dimensions. However, at present, the study of spatial integration (or spatial segregation) of land-lost peasants is rarely present in the discussion of their social integration. At the same time, spatial integration indicators are rarely included in the overall index system of social integration. Therefore, this study supplements the analysis of spatial dimensions on top of the other four dimensions (economic, social, institutional, and psychological), which is five in total, to comprehensively study the overall social integration degree of land-lost peasants. The specific establishment of the index system is as follows (see Fig. 3.36).

The sub-goal level includes economic, social, psychological, institutional, and spatial dimensions (i.e., the main dimensions of social integration evaluation). The criteria level is a subdivision of the above dimensions into 13 more specific indicators. Economic dimension includes income and expense level, work intensity, and work stability. Social dimension includes family migration, object of socialization, and socialization frequency. The psychological dimension includes a sense of community belonging and urban and social identification. The institutional dimension includes the five categories of social insurance and house provident fund. The spatial dimension includes living and commuting, and the alternative level is a subdivision of the above-mentioned criteria level. The alternatives are directly measurable indicators

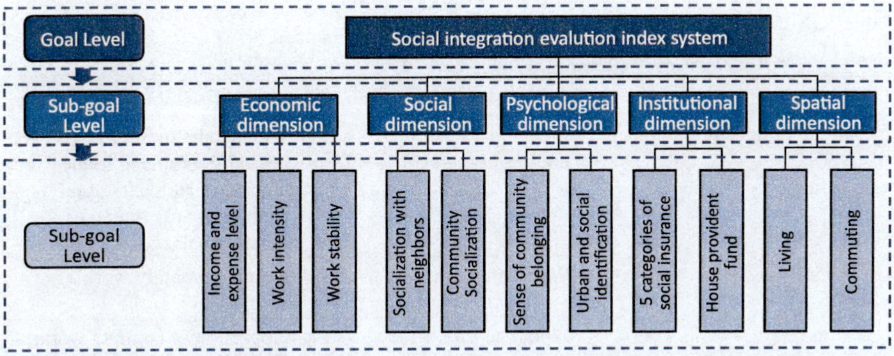

Fig. 3.36 Social integration evaluation index system. *Source* Own elaboration

and data that could be collected and usually correspond directly to the question set in the questionnaire design.

Selection of Criteria and Alternatives

As mentioned above, social integration involves five dimensions, each of which covers several criteria, and each criterion contains several alternatives that can be measured directly. The economic, social, institutional, and spatial dimensions are all objective indicators reflecting the objective integration status of land-lost peasants in the city, while the psychological dimension is a subjective indicator reflecting the psychological integration status related to psychological satisfaction, sense of community belonging, and urban social identification. For each dimension, five to six alternatives are selected and (see Table 3.18) explained in detail as follows:

Criteria of economic dimension Three criteria, including income and consumption level, work intensity, and work stability, are used to reflect the gap between land-lost peasants and ordinary urban citizens in terms of economy, work, and others. The smaller the gap, the higher is the integration degree. Income and consumption level is represented by relative indicators such as the ratio of income of land-lost peasants to that of other residents in the area and monthly balance. Work intensity is represented by monthly working hours calculated as daily working hours multiplied by the number of days in a month. Work intensity is represented by the proportion of contract workers and average length of each job.

Criteria of social dimension Two criteria, socialization with neighbors and community socialization, were used to measure the degree of social integration. The more land-lost peasants socialize with ordinary urban citizens, the more ordinary urban citizen friends they have and the more frequently they visit surrounding communities. Additionally, the more they trust residents of surrounding communities, the higher is the degree of social integration.

Criteria of the psychological dimension Two criteria, sense of community belonging and urban and social identification, were used to measure social integration from the psychological dimension. The more the land-lost peasants like a

Table 3.18 Social integration evaluation index system

Social Integration Evaluation Index System		
Goal level	Criteria level	Alternative level
Economic dimension	Income and consumption level	(1) The income ratio of land-lost peasants compared to the income of residents of the community (2) Average monthly balance
	Work intensity	(3) Average monthly working hours
	Work stability	(4) Proportion of contract workers (5) Average time frame per job
Social dimension	Socialization with neighbors	(1) Object of socialization (2) Whether socialization with urban residents is frequent (3) Number of citizen friends that they socialize with regularly
	Community socialization	(4) Socialization frequency with surrounding communities (5) Whether land-lost peasants trust residents of surrounding communities
Psychological dimension	Sense of community belonging	(1) Whether land-lost peasants like the community (2) Gap they perceive to have with urban residents
	Urban social identification	(3) Whether the land-lost peasants acknowledge the behaviors of urban citizens (4) Whether they adapt to the urban lifestyle and working mode (5) Whether they get along with urban residents
Institutional dimension	Social insurance	(1) Whether they benefit from pension insurance (2) Whether they benefit from medical insurance (3) Whether they benefit from unemployment insurance (4) Whether they benefit from employment injury insurance (5) Whether they benefit from maternity insurance (6) Whether they benefit from housing provident fund system
Spatial dimension	Living	(1) Type of population in the surrounding areas (2) Living area per capita (3) Community facilities index

(continued)

Table 3.18 (continued)

Social Integration Evaluation Index System		
Goal level	Criteria level	Alternative level
	Commuting	(4) Distance to the nearest bus or subway station (5) Distance to the worksite

Source Own elaboration

community, the smaller is the gap they perceive with urban citizens, the more they acknowledge urban citizens' behavior, and the better they adapt to the urban lifestyle and working mode. Additionally, the better they get along with urban residents, the higher is the degree of social integration from the psychological perspective.

Criteria of the institutional dimension This dimension includes two criteria: social insurance and house provident fund. The first means whether they benefit from the city's social security, including the five categories of pension insurance, medical insurance, unemployment insurance, employment injury insurance, and maternity insurance. The second means whether they benefit from the city's housing provident fund system.

Criteria of the spatial dimension This dimension includes two criteria: living and commuting. The living perspective is reflected by the type of population in the surrounding area, the living area per capita, and the community facility index. Of them, the type of population in the surrounding area refers to whether the resettlement area is encircled by ordinary residential communities, while the larger the living area per capita and the more the complete the community facilities, the higher is the degree of social integration. The distance to the bus or subway station and the distance to the workplace are the two alternatives of commuting.

3.4.2.2 Questionnaire Indicator Value Assignment

Since the questionnaire for individual land-lost peasants involves many dimensions, the answers set up are complicated and varied in nature, meaning that the questions of different categories cannot be compared with each other. To accurately reflect the social integration status of land-lost peasants in various resettlement areas, different methods were used to assign values to questions according to categories.

Fill-in-the-blank-with-numbers questions: For this type of question, land-lost peasants can directly answer in specific numbers, such as age, income, and average monthly hours worked. For questions that have a quantitative dimension, the mean of the sample from each resettlement area is used to assign the value.

Multiple choice questions: Land-lost peasants can select from the alternative options. Some of the options are specific numbers such as social security, rent, salary, income from self-employment work, financial income, and others, while others are judgments on the level and degree. For example, for the question of whether they are adapted to the urban lifestyle, the five options are: very well adapted, adapted, average, not adapted, not adapted at all. For this type of question, the proportion of the most selected option is usually used for value assignment (indicator value assignment is shown in Table 3.19).

3.4.2.3 Standardized Processing and Weighting

The measurement units and orders of magnitude for each social integration assessment indicator are different. To avoid weight exaggeration due to the difference in the value range of the indicators, this paper adopts the extreme value subtraction method to nondimensionalize each indicator.

In the comprehensive evaluation of multiple indicators, the determination of weights is a key part of the entire evaluation process. There are several research results on the determination of the weights and many relevant methods. Based on the sources of original data when calculating the weights, the methods of weight determination can be broadly divided into two categories: subjective weighting methods (such as DARE and Delphi) and objective weighting methods (such as principal component analysis, deviation maximization method, and mean variance weight decision method). The subjective methods are mainly derived from experts' subjective judgments based on their experiences. Although they can be more mature, they are less objective. Therefore, this study selects the mean-variance weight decision method proposed by Wang (1999), which is often used in the comprehensive comparison of different regions and cities, to weight each indicator via the objective assignment method.

Standardized processing

The multiple options involved in social integration include leading and lagging indicators, which are gauged in different orders of magnitude. Therefore, this study uses the method of extreme value subtraction to turn forward the lagging indicator X3 before calculation. Next, all the leading indicators are subtracted from the mean value and then divided by the standard deviation for nondimensionalization to obtain the final standardized index.

$$Z_j = \frac{(X_j - X_i)}{\sigma}$$

$$\sigma = \sqrt{\frac{\sum_{i=1}^{n}\left(X_j - \overline{X}^2\right)}{n}}$$

The formula is where, $i = 1, 2, \ldots, n$, and $n = 26$; $j = 1, 2, \ldots, m$, $m = 4$. X_i is the mean value of the four resettlement areas in terms of the indicator, Z_{ij} is the indicator

Table 3.19 Indicator value assignment for social integration of land-lost peasants in each resettlement area by dimension

Dimension	Indicator	Indicator code	Mufu Mountain Villa	Yaolin Xianju	Tianjingshan Community	Xishan Huayuan
Economic dimension	The income ratio of land-lost peasants compared with the income of residents of the community (%)	X1	62.3	51.3	47.3	35.2
	Average monthly balance (yuan)	X2	1230	1135	1026	958
	Average monthly working hours (h)	X3	164.3	168.4	112.3	156.6
	Proportion of contract workers (%)	X4	27.9	45.4	25.1	20.3
	Average time frame per job (m)	X5	24.6	28.6	22.3	18.2
Social dimension	Objects of socialization are mainly urban citizens (%)	X6	13.2	12.6	5.1	12.3
	Socialize frequently with urban citizens (%)	X7	18.6	15.3	16.4	14.2
	Number of urban citizens friends frequently visited (person)	X8	5.6	5.2	4.7	5.1
	Frequency of visiting surrounding communities (%)	X9	24.3	19.3	21.1	17.5
	Trust residents of surrounding communities (%)	X10	80.4	90.3	86.7	79.3
Psychological dimension	Like this community (%)	X11	60.3	32.3	66.8	52.3
	Do not perceive gap with urban citizens (%)	X12	42.3	12.4	34.4	22.7

(continued)

Table 3.19 (continued)

Dimension	Indicator	Indicator code	Mufu Mountain Villa	Yaolin Xianju	Tianjingshan Community	Xishan Huayuan
	Acknowledge the behavior of urban citizens (%)	X13	84.2	34.4	68.2	62.6
	Adapt to urban lifestyle and work mode (%)	X14	75.2	40.3	82.2	63.2
	Get along with urban residents (%)	X15	68.1	22.4	36.5	30.9
Institutional dimension	Benefit from pension insurance (%)	X16	53.4	38.6	51.9	43.5
	Benefit from medical insurance (%)	X17	80.3	62.3	66.7	68.4
	Benefit from employment injury insurance (%)	X18	26.2	25.2	26.4	40.6
	Benefit from unemployment insurance (%)	X19	41.2	35.4	42.4	27.5
	Benefit from maternity insurance (%)	X20	33.2	31.2	36.7	25.6
	Benefit from housing provident fund system (%)	X21	48.3	39.4	42.8	36.9
Spatial dimension	Residents living in surrounding communities are mostly urban citizens. (%)	X22	38.2	29.3	25.2	36.2
	Living space per capita is more than $16m^2$. (%)	X23	57.1	63.2	84.2	82.3
	Complete facilities in the community (%)	X24	86.3	85.2	84.6	90.3
	Proportion of people arriving at subway or bus station within 15 min by walk (%)	X25	79.6	78.5	70.3	82.3

(continued)

Table 3.19 (continued)

Dimension	Indicator	Indicator code	Mufu Mountain Villa	Yaolin Xianju	Tianjingshan Community	Xishan Huayuan
	Proportion of people arriving at work within 40 min (%)	X26	85.3	77.6	59.6	45.2

Source Sample data collected by the research team on the resettlement areas for land-lost peasants in Nanjing (2016)

obtained after nondimensionalizing the 26 indicators of each resettlement area, and X_{ij} is the forwarded indicator before the nondimensionalization is carried out.

Weighting

The basic principle of the mean-variance decision method is as follows: the magnitude of the relative weight coefficient of each indicator depends on the relative dispersion of the attribute values of each alternative under this indicator. The greater the dispersion, the more is the relative weight coefficient of the indicator and vice versa. The most important and commonly used indicator of the degree of dispersion of a random variable is the mean-variance deviation of that random variable.

Therefore, the basic idea of the mean-variance decision method is as follows: take each evaluation indicator as a random variable and nondimensionalize the attribute values of each alternative A_j under indicator B_i as the value of this random variable; calculate the mean-variance of these random variables, and unify the mean variance. The result is, therefore, the weight coefficient of the indicator. The calculation steps of this method are:

1. Calculate the sample mean $E(B_i)$ of the variable Z_{ij}.

$$E(B_i) = \frac{1}{m}\sum_{i=1}^{m} Z_j \tag{3.1}$$

2. Calculate the mean–variance δ (B_i), which is the mean variance of the 26 indicators Z_{ij} of the four resettlement areas.

$$\delta(B_i) = \sqrt{\frac{1}{m}\sum_{j=1}^{m} (Z_j - E(B_i))^2} \tag{3.2}$$

3. Calculate the weight coefficients $W(B_i)$ of Z_i, which are the different weights of the 26 indicators Z_{ij}.

$$W(B_i) = \frac{\delta(B_i)}{\sum_{i=1}^{n} \delta(B_i)} \tag{3.3}$$

4. Perform multi-indicator weighting and ranking.

$$F_j = \sum Z_j W(B) \times \Phi\% \tag{3.4}$$

In the formula, F_j denotes the values of the four resettlement areas at the criteria level, Z_{ij} denotes the standardized value of each indicator, and $W(B_i)$ is the weight of each indicator. The weights of the indicators at each level of the social integration evaluation system are as follows (see Table 3.20).

According to the abovementioned method, taking the resettlement areas as the alternatives set A1 = Mufu Mountain Villa, Yaolin Xianju..., and the criteria set as B1 = revenue ratio, average monthly balance..., this study conducts nondimensionalization on the indicators of the five dimensions, including economic, social, psychological, institutional, and spatial and calculates their weights to obtain the evaluation value and ranking of each resettlement area in terms of the five dimensions. Of them, the evaluation values and ranking within the economic dimension can be found below as an example (see Table 3.21).

After calculating the evaluation value and ranking of each dimension, the indicator decision was reconstructed and the resettlement areas were taken as the alternatives set A2 = Mufu Mountain Villa, Yaolin Xianju..., while the criteria set is taken as B2 = economic dimension, social dimension..., to calculate the overall social integration evaluation value of land-lost peasants in each resettlement area. The evaluation values of land-lost peasants in each resettlement area for each dimension and overall social integration evaluation are shown in Table 3.22.

3.4.2.4 Analysis of Social Integration

Overall Integration Analysis

As evident from the above analysis, the overall social integration degree of land-lost peasants in each resettlement area differs significantly: there is a 54.87-point difference between Mufu Mountain Villa, which has the highest overall social integration degree, and Yaolin Xianju, which has the lowest integration degree. A further comparison of each community shows that the integration degree of Mufu Mountain Villa in all dimensions is the highest except the economic dimension (where it is ranked second) and the institutional dimension, while Yaolin Xianju has the lowest evaluation value in all dimensions except the economic and social dimensions (see Fig. 3.37).

Integration Analysis in Economic Dimension

The integration degree in economic dimension is the basis for land-lost peasants to establish themselves in the city. The integration degree of the four sample resettlement areas in this dimension varies greatly, with a difference of 14 points between Tianjingshan Community, which has the highest value, and Xishan Huayuan, which has the lowest value, while the values of Mufu Mountain Villa and Yaolin Xianju are

Table 3.20 Social integration evaluation index system and index weights

Goal level	Sub-goal level		Criteria level		Alternative level		Weight to sub-goal level	Weight to goal level
	Name	Weight	Name	Weight	Name	Weight		
Social integration	Economic dimension	0.18	Income and consumption level	0.45	Monthly income	0.62	0.2790	0.0502
					Monthly balance	0.38	0.1710	0.0308
			Work intensity	0.22	Average working hours	1	0.2200	0.0396
			Work intensity	0.33	Proportion of contract workers	0.59	0.1947	0.0350
					Average time frame per job	0.41	0.1353	0.0244
	Social dimension	0.14	Socialization with neighbors	0.62	Objects of socialization	0.35	0.2170	0.0304
					Socialization frequency	0.36	0.2232	0.0312
					Number of urban citizens friends	0.29	0.1798	0.0252
			Community socialization	0.38	Community socialization	0.55	0.2090	0.0293
					Degree of community trustworthiness	0.45	0.1710	0.0239
	Psychological dimension	0.33	Sense of community belonging	0.58	Degree of community likeness	0.52	0.3016	0.0995
					Mental Gap	0.48	0.2784	0.0919
			Urban social identification	0.42	Behavioral identification	0.33	0.1386	0.0457
					Urban adaptation	0.36	0.1512	0.0499
					Degree of harmony	0.31	0.1302	0.0430
	Institutional dimension	0.11	5 Categories of social insurance	0.83	Pension insurance	0.26	0.2158	0.0237
					Medical insurance	0.25	0.2075	0.0228

(continued)

Table 3.20 (continued)

Goal level	Sub-goal level		Criteria level		Alternative level		Weight to sub-goal level	Weight to goal level
	Name	Weight	Name	Weight	Name	Weight		
					Employment injury insurance	0.18	0.1494	0.0164
					Unemployment insurance	0.14	0.1162	0.0128
					Maternity insurance	0.17	0.1411	0.0155
			House provident fund	0.17	Housing provident fund	1	0.1700	0.0187
	Spatial dimension	0.24	Living	0.41	Population living in surrounding communities	0.21	0.0861	0.0207
					Living area	0.42	0.1722	0.0413
					Facility Index	0.37	0.1517	0.0364
			Commuting	0.59	Distance to stations	0.48	0.2832	0.0680
					Distance to commute	0.52	0.3068	0.0736

Source Sample data collected by the research team on the resettlement areas for land-lost peasants in Nanjing (2016)

Table 3.21 Social integration evaluation value and ranking of land-lost peasants in each resettlement area within economic dimension

Resettlement area	Income ratio (%)	Average monthly balance (yuan)	Average monthly working hours (h)	Proportion of contract workers (%)	Average time frame per job (m)	Evaluation value	Ranking
Mufu Mountain Villa	62.3	1230	164.3	27.9	29.6	5.35	2
Yaolin Xianju	51.3	1135	170.4	45.4	33.2	4.26	3
Tianjingshan Community	47.3	1026	112.3	25.1	26.3	10.07	1
Xishan Huayuan	35.2	958	156.6	20.3	22.4	3.62	4

Source Sample data collected by the research team on the resettlement areas for land-lost peasants in Nanjing (2016)

in the middle and have little difference, which are 8.35 and 7.26, respectively (see Fig. 3.38).

In terms of income and consumption level, the average monthly income ratio of land-lost peasants in Mufu Mountain Villa compared with that of urban residents in this area is 62.3%. With an average monthly balance of 1230 yuan, they have the smallest gap with other residents in this regard. In comparison, the average monthly income ratio of land-lost peasants in Xishan Huayuan compared with that of urban residents in the area is only 32.5%, which is not even half of the latter.

In terms of work intensity, there is not much difference in the average monthly working hours across the three resettlement areas of Mufu Mountain Villa, Yaolin Xianju, and Xishan Huayuan (which are all around 160 h). It is also basically comparable with the 40 working hours per week for enterprise employees as stipulated in the "Regulations of the State Council on Working Hours of Employees" in 2016. However, the average monthly working hours of land-lost peasants in Tianjingshan is only 112.3 h, indicating that their work intensity is apparently lower than residents of the three other resettlement areas.

The job stability of land-lost peasants in all resettlement areas is not ideal. The average length of each of their jobs is only 23 months, which shows that the job change frequency is generally high.

Integration Analysis in Social Dimension

Positive communication and socialization with other residents of the city is an important criterion of whether residents in resettlement areas can achieve social integration. All four resettlement areas have a low level of integration in the social dimension and do not differ significantly from each other. The area with the highest evaluation value,

Table 3.22 Evaluation value and ranking of social integration of land-lost peasants in each resettlement area

Resettlement area	Economic dimension	Ranking	Social dimension	Ranking	Psychological dimension	Ranking	Institutional dimension	Ranking	Spatial dimension	Ranking	Overall social integration	Total Ranking
Mufu Mountain Villa	8.35	2	5.83	1	10.23	1	1.94	3	10.86	1	39.4	1
Yaolin Xianju	7.26	3	4.74	2	−20.52	4	1.32	4	−8.27	4	−15.47	4
Tianjingshan Community	11.07	1	−2.78	4	8.96	2	4.13	2	6.02	3	27.4	2
Xishan Huayuan	−3.62	4	3.23	3	−3.58	3	9.28	1	9.16	2	14.47	3

Source Sample data collected by the research team on the resettlement areas for land-lost peasants in Nanjing (2016)

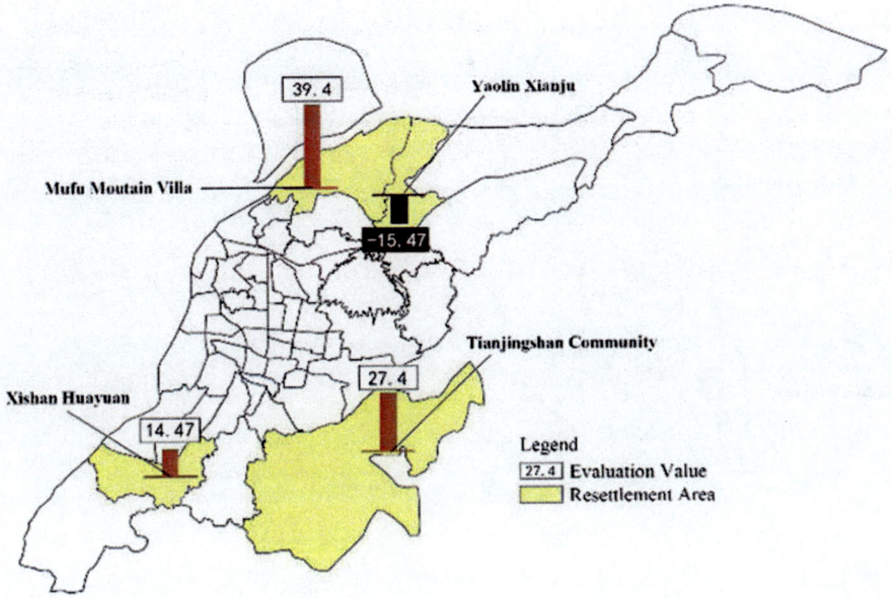

Fig. 3.37 Overall integration degree of land-lost peasants in each resettlement area. *Source* Sample data collected by the research team on the resettlement areas for land-lost peasants in Nanjing (2016)

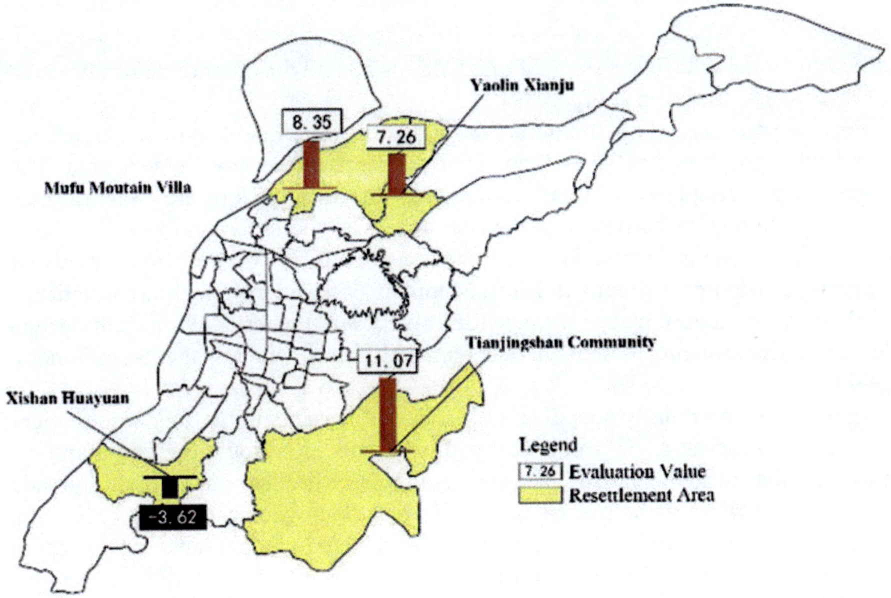

Fig. 3.38 Integration degree of land-lost peasants in each resettlement area in economic dimension. *Source* Sample data collected by the research team on the resettlement areas for land-lost peasants in Nanjing (2016)

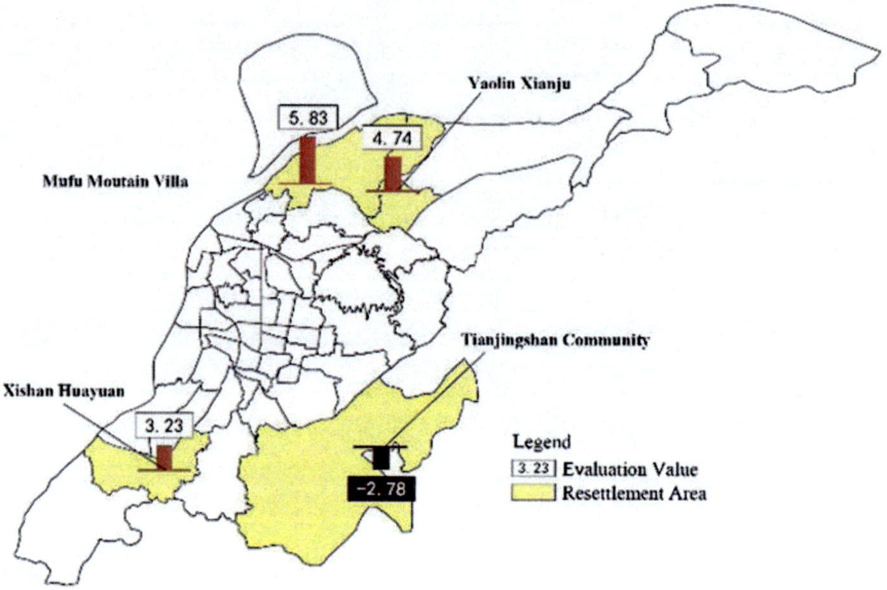

Fig. 3.39 Integration degree of land-lost peasants in each resettlement area in social dimension. *Source* Sample data collected by the research team on the resettlement areas for land-lost peasants in Nanjing (2016)

Mufu Mountain Villa, has a score of mere 5.83, while all the other three resettlement areas score less than 5 (see Fig. 3.39).

From the perspective of community socialization, most land-lost peasants interact with land-expropriated villagers and other residents in the resettlement area. The proportion and frequency of their interaction with urban citizens are generally low due to psychological barriers and the remoteness of the area, with most land-lost peasants only having four to five urban citizen friends. In comparison, a relatively higher proportion of residents in Mufu Mountain Villa interact with urban citizens (13.2% of the residents interviewed), while only 5.1% of the residents of Tianjingshan Community communicate with citizens on a daily basis, which is the lowest among the four.

In terms of community socialization, land-lost peasants in the four resettlement areas also interact at a low frequency with residents in surrounding communities, indicating that residents' participation in community life and contact with outside areas is also low. Nevertheless, when interviewed about whether they trust residents of surrounding communities, more than 80% of the land-lost peasants answered yes on average, indicating high level of trust.

Integration Analysis in Psychological Dimension

The integration degree in the psychological dimension is a subjective measure of land-lost peasants' sense of community belonging and their identification with urban

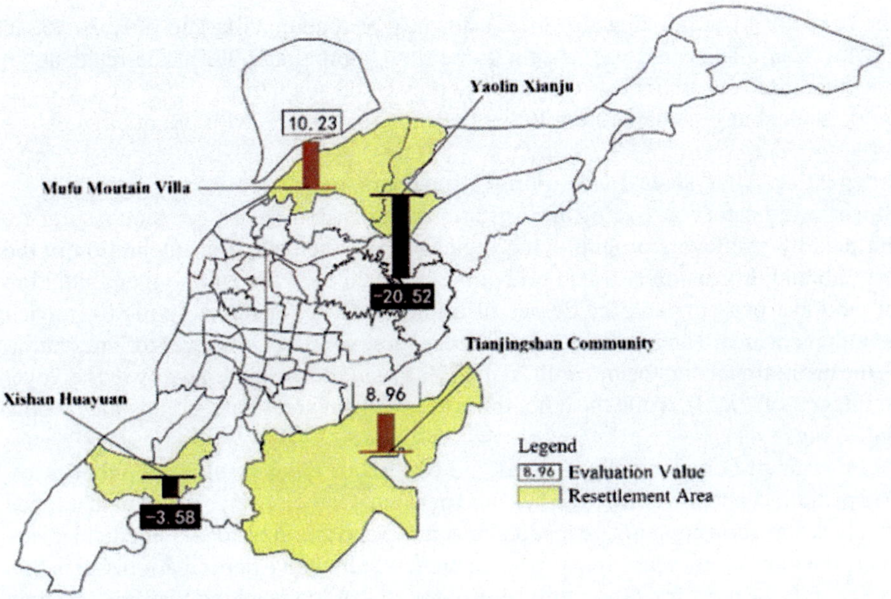

Fig. 3.40 Integration degree of land-lost peasants in each resettlement area in psychological dimension. *Source* Sample data collected by the research team on the resettlement areas for land-lost peasants in Nanjing (2016)

society. The four resettlement areas show the greatest differences in the integration degree in the psychological dimension, where there is a 30-point difference between Mufu Mountain Villa, the area with the highest evaluation value, and Yaolin Xianju, the area with the lowest evaluation value (notably, Yaolin Xianju has the lowest value for all sub-indicators of this dimension). Tianjingshan and Xishan Huayuan, which are ranked in the middle, have an evaluation value of 8.96 and − 3.58, respectively (see Fig. 3.40).

In terms of the sense of belonging, more than half the people in all three resettlement areas, except Yaolin Xianju, tend to like their own community (the number even reaches up to 66.8% in Tianjingshan Community), whereas in terms of the gap they perceive to have with other urban citizens, most land-lost peasants show a disorder in self-identification and evaluate their current living condition by the living conditions of the surrounding citizens. In particular, land-lost peasants whose housing was expropriated and were resettled at an early stage due to the not-yet-perfect compensation policies tend to feel a strong sense of deprivation and imbalance. Even Mufu Mountain Villa, the area boasting a relatively high value in this regard, has only 42.3% of its residents feeling no different from other urban citizens.

In terms of urban social identification, significant differences are seen in the values for all evaluation indicators across resettlement areas. The values for the three indicators of Mufu Mountain Villa are higher than those of the other three resettlement areas, while those of Yaolin Xianju are the lowest (e.g., in terms of acknowledging the behavior of urban citizens, 84.2% in Mufu Mountain Villa and only 34.4% in Yaolin Xianju answered yes). At the same time, more than 60% of the residents in the other three resettlement areas said that they could adapt to the urban lifestyle and work mode, but less than 40% felt that they could get along with urban citizens.

Integration Analysis in Institutional Dimension

Since the indicators measuring the institutional dimension, such as pension insurance, are paid by employers or employing organizations, the degree of integration in the institutional dimension can also reflect the differences in the employment situation of land-lost peasants and the degree of implementation of related policies in each resettlement area. The four resettlement areas show similarly low levels of integration in the institutional dimension, with Xishan Huayuan showing a relatively higher level of integration (9.28), while the other three resettlement areas have rating value below 5 (see Fig. 3.41).

In terms of being socially insured, the two major types of insurance that resettled peasants benefit from are medical and pension. Of them, since medical insurance involves new rural cooperative medical insurance and urban residents' medical insurance, the four resettlement areas have the highest cumulative participation rate in this indicator (all exceeding 60%, with Mufu Mountain Villa reaching 80.3%), which is undoubtedly a great benefit for land-lost peasants when they need medical care. At

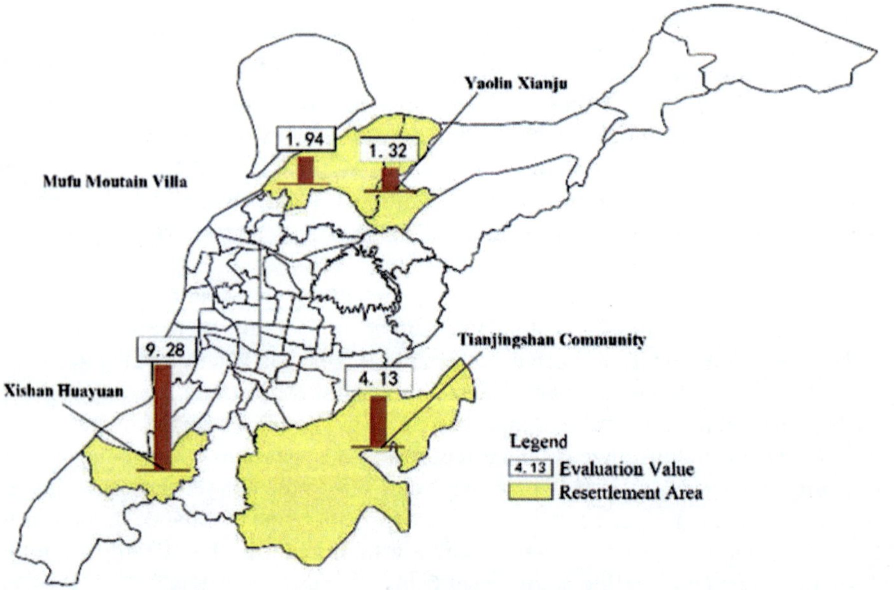

Fig. 3.41 Integration degree of land-lost peasants in each resettlement area in institutional dimension. *Source* Sample data collected by the research team on the resettlement areas for land-lost peasants in Nanjing (2016)

the same time, the participation rate of land-lost peasants in pension insurance is also approximately close to or even more than 50%.

In contrast, the participation rate of land-lost peasants in other types of insurance (such as employment injury insurance, unemployment insurance, maternity insurance and housing provident fund) is generally low, with less than 50% of the residents insured in these categories among all resettlement areas.

Integration Analysis in Spatial Dimension

The degree of integration in the spatial dimension not only reflects the residential isolation between land-lost peasants and surrounding urban citizens but also reflects whether the public transportation facilities are complete and what the work-home commuting relationship is like in all resettlement areas. The degree of integration of the four sample resettlement areas in the spatial dimension varies greatly, of which the integration degree of Mufu Mountain Villa and Xishan Huayuan is higher, with an evaluation value of 10.86 and 9.16, respectively, followed by Tianjingshan Community, while Yaolin Xianju has the lowest integration degree of 8.27 (see Fig. 3.42).

From the housing perspective, the phenomenon of the large-scale concentration of land-lost peasants is obvious, and the residential isolation from urban residents is also apparent. Some citizens have settled in the ordinary commercial housing near the resettlement areas. Mufu Mountain Villa, which has the highest proportion, is

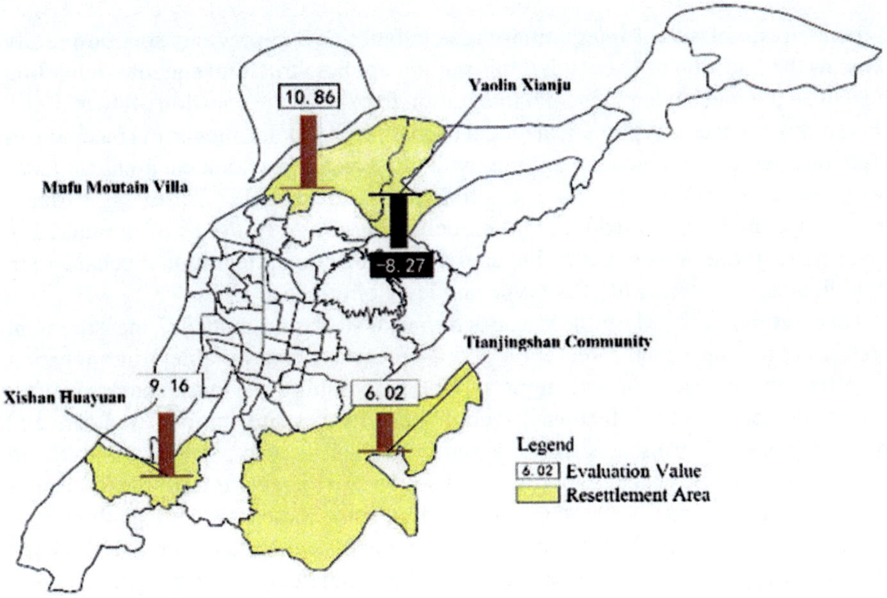

Fig. 3.42 Integration degree of land-lost peasants in each resettlement area in spatial dimension. *Source* Sample data collected by the research team on the resettlement areas for land-lost peasants in Nanjing (2016)

only at 38.2%. In terms of per capita living area, there are still a large number of land-lost peasants in each resettlement area whose living area per capita is below the minimum proposed by the United Nations (i.e., 16 m^2), especially in Mufu Mountain Villa, where only 57.1% of the land-lost peasants have a per capita living area of 16 m^2 or more. However, in terms of the types of community facilities, the residents are reasonably satisfied (the proportion is above 84%). Although the quality of the various facilities vary, residents think that the ones in resettlement areas can meet their daily needs.

Finally, in terms of commuting, residents in all resettlement areas mainly take public transportation, and more than 70% of the land-lost peasants can reach the nearest subway station or bus stop within 15 min on foot. However, the balanced distribution of transportation stations is not equal to the fact that the transportation routes are reasonable, and it is still inconvenient to travel in some resettlement areas (e.g., in large mixed resettlement areas at the fringe of the city). Therefore, the work-home commuting mode between residents of different areas is diversified. In terms of whether the residents can arrive at worksites in 40 min, Mufu Mountain Villa has the highest proportion (85.3%) and Xishan Huayuan has the lowest (only 45.2%).

3.4.3 Analysis of the Social Integration Mechanism of Resettlement Areas for Land-Lost Peasants

If the analysis of social integration characteristics in the previous section mainly explains the question of "how it is," this section applies structural equation modeling to analyze the mechanism of social integration to explain the question of "why it is." This study proposes a hypothetical model to analyze the social integration mechanism based on a review of relevant literature, with the selected resettlement areas for land-lost peasants as the samples. After multiple rounds of model construction, variable processing, model modification, and model fitting to verify the fit of the model, a final revised model was generated to analyze the different integration mechanism in each dimension in terms of effect type and level of effect.

This method is based on the construction of a structural model (7 measurement models containing social integration and 24 measurement variables) in analyzing the effect mechanisms of six latent variables, including personal characteristics, economic factors, social factors, psychological factors, institutional factors, and spatial factors, in terms of social integration (which is also a latent variable). In this section, this study makes appropriate deletion and merges some of the indicators in the five dimensions discussed above (deleting some indicators with small effects, adding two single indicators of "difference between resettlement areas" and "occupational categories" and two main factors of "personal characteristics" and "social integration") to obtain 24 directly measurable indicators in the end. Different from the analytic hierarchy process adopted in the previous section (calculating the evaluation value of each level of indicators by first determining their weight), the structural equation model essentially explores the effect type of each latent variable on social integration, the effect type of each measurement indicator on the corresponding latent variable, and the magnitude of the impact level. The specific steps are as follows.

3.4.3.1 Theoretical Hypothesis

Empirical research follows the logic of "hypothesis proposal—hypothesis testing," and the formulation of hypotheses must be supported by solid theoretical foundations and literature. Some studies show that social integration (citizenization) refers to the transformation of peasants into citizens in terms of ways of production, thinking, lifestyle, identification, etc. (Mao 2009). The adaptation of land-lost peasants to the city is a social process of continuously adjusting their behavior with "urban citizens" as the reference group (Jiang 2003). It is actually a process of re-socialization (Tian 1995). In addition, the adaptation in the economic, social, and psychological dimensions is progressive (Zhu 2002).

Based on this, the author proposed the following hypothesis:

Hypothesis 1 The economic dimension provides the material basis, the social dimension is an important factor in urban integration, and psychological dimension indicates the highest stage of integration; economic, social, and psychological dimensions have a direct and significant impact on overall integration.

Hypothesis 1.1 The higher the degree of integration in the economic dimension, the higher is the overall integration.

Hypothesis 1.2 The higher the degree of integration in the social dimension, the higher is the overall integration.

Hypothesis 1.3 The higher the degree of integration in the psychological dimension, the higher is the overall integration.

Hypothesis 1.4 Economic and social factors influence social integration by affecting psychological factors.

Related studies show that when focusing on the special group of land-lost peasants, the institutional dimension (social security participation) is also an important factor influencing their overall integration, and integration in the institutional dimension provides the basic conditions and contextual support for the integration of land-lost peasants in the city; therefore, hypothesis 2 was inferred:

Hypothesis 2 The institutional dimension has a direct impact on the overall integration and its level of effect is relatively small compared with economic, social, and psychological factors.

The previous study also showed that integration in the spatial dimension, including living and commuting, is the first step in the integration of land-lost peasants in the city and an important reason for the varying levels of social integration; thus, hypothesis 3 was inferred:

Hypothesis 3 The spatial dimension is the spatial vehicle and specific means of integration for land-lost peasants, and the spatial dimension has a direct and significant effect on overall integration.

In addition, it is shown that personal characteristics (Ye 2010) (e.g., gender, age, education level) also have an intrinsic effect on the degree of social integration, which leads us to hypothesis 4:

Hypothesis 4 The differences in personal characteristics of land-lost peasants also have a direct impact on the overall integration of the group with a relatively greater level of effect.

3.4.3.2 Model Construction

Based on the above hypotheses, this study first constructs the structural equations of social integration to determine which of the economic, social, psychological, institutional, and spatial factors (directly or indirectly) affect social integration and the magnitude of their effect levels.

To test hypotheses 1.1–1.3, the author set up the paths between economic, social, and psychological factors and social integration. According to hypothesis 1.4, the author set up the paths between economic, social, and psychological factors; for testing hypotheses 1.1 and 1.3, the paths between institutional factors, spatial factors, and social integration were set up with reference to the multidimensional evaluation index system of social integration constructed in the previous section. For hypothesis 1.4, this study added personal characteristics (including gender, age, and education level) and set up the paths between them and social integration. The hypotheses are shown as paths in the model in Fig. 3.43.

The five measurement models, including those for economic, social, psychological, institutional, and spatial factors, were constructed with reference to the corresponding dimensions and specific indicators of the social integration evaluation index

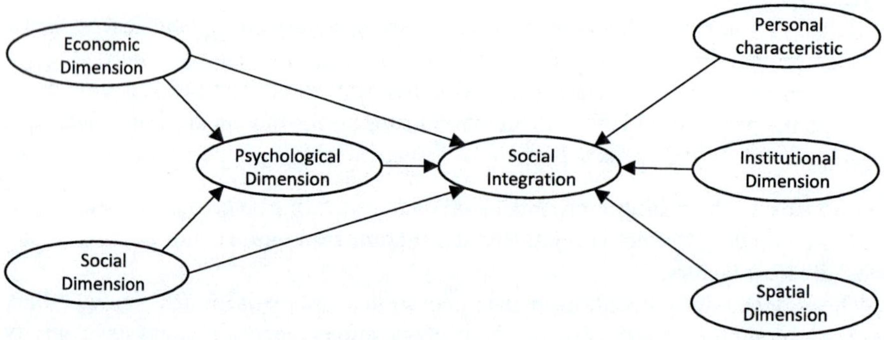

Fig. 3.43 Hypotheses shown as paths in the model of social integration of land-lost peasants. *Source* Own elaboration

system discussed in the previous section. In measuring social integration, this study draws on Huntington's (2005) definition and classification of social identification and takes into account the research results of Zhang and Lei (2009) and Lu and Wei (2011). It ultimately uses three indirect variables: "identity," "intention to develop," and "expectation toward children" for the measurement. Specifically, they correspond to answers to the following three questions: "Do you consider yourself a rural or an urban citizen?" "Are you willing to progress in the city?" and "Are you willing to let your children progress in the city?" The answers reflect the self-identification of the sampled land-lost peasants, the identification of the sampled lost-peasants with cities, and the willingness of land-lost peasants to truly embed and integrate into the city. In addition, since the samples come from different types of representative resettlement areas, it is necessary to introduce the variable of "difference between resettlement areas" in the measurement model of institutional factors to test the impact of different types of resettlement areas on the social integration of land-lost peasants. The measurement variables corresponding to each latent variable were incorporated into the model to form seven measurement models (see Fig. 3.44).

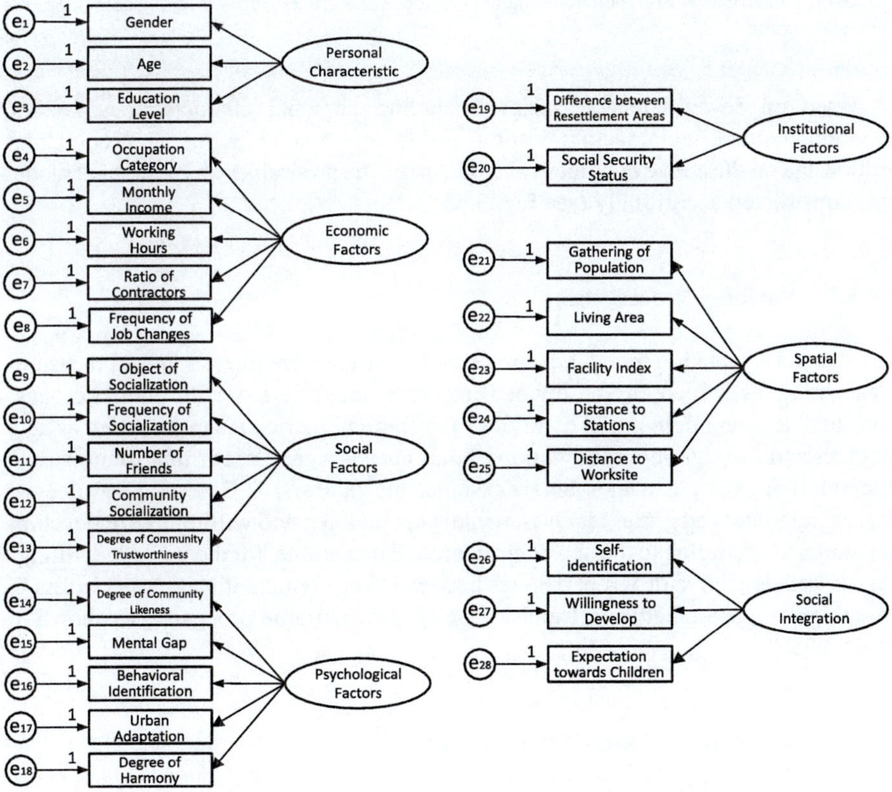

Fig. 3.44 Construction of measurement model for social integration of land-lost peasants. *Source* Own elaboration

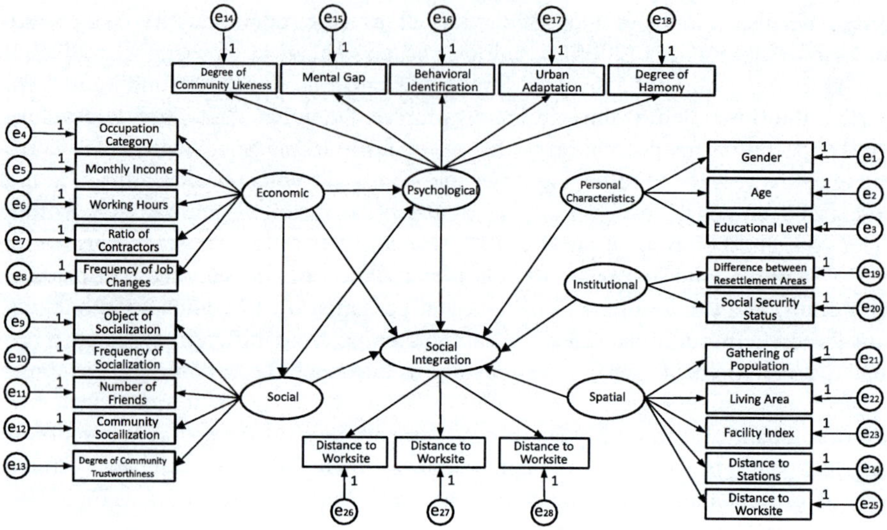

Fig. 3.45 Construction of hypothetical model for social integration of land-lost peasants. *Source* Own elaboration

Based on seven latent variables, including personal characteristics factors, economic factors, social factors, institutional factors, spatial factors, and social integration and on the above hypothetical model path, the preliminary hypothetical model was constructed accordingly (see Fig. 3.45).

3.4.3.3 Variable Processing

The structural equation model here involves 7 measurement models and 28 measurement variables, which can be divided into two categories: specific numeric questions and degree judgments. The values of specific numeric categories (such as age, income, and average monthly working hours) are assigned according to numerical intervals. For example, n intervals are assigned the values "1, 2, 3 ... n." The values of degree judgment categories (such as mental gap, identity, and willingness to develop) are assigned according to the perceived degree. For example, for the variable 'willingness to develop," five different degrees between "very reluctant" and "very willing" are assigned a score between 1 and 5. The specific variable processing is shown in Table 3.23.

3.4.3.4 Estimation of Model Parameters

SEM uses the maximum likelihood estimation model rather than the usual least-squares method for parameter estimation, while the AMOS 22.0 software provides

Table 3.23 Description of variables in the measurement model of social integration of land-lost peasants

Model	Variables	Measurement variables	Description of variable's value assignment
Measurement model 1	Personal characteristics	Gender	Female = 1; Male = 2
		Age	Under 20 y/o = 1; 21–30 y/o = 2; 31–40 y/o = 3; 41–50 y/o = 4; 51–60 y/o = 5; 61–70 y/o = 6; Over 70 y/o = 7
		Education level	Below primary school = 1; junior high school = 2; High school or technical secondary school = 3; College = 4; University and above = 5
Measurement model 2	Economic factors	Occupational category	Retired or unemployed = 1; Agricultural, forestry and fishery workers = 2; Production and transportation equipment operators = 3; Commercial service workers = 4; professional and technical (research) personnel = 5; Management or clerks of institutions and organizations = 6
		Household monthly income	Below 1,500 yuan = 1; 1,501–2,500 yuan = 2; 2,501–5,000 yuan = 3; 5,001–10,000 yuan = 4; above 10,000 yuan = 5
		Average monthly working hours	Below 80 h = 1; 81–120 h = 2; 121–160 h = 3; 161–200 h = 4; 200 h or more = 5
		Ratio of contract workers	Retired or unemployed = 0; temporary workers = 1; contract workers = 2; office workers = 3

(continued)

Table 3.23 (continued)

Model	Variables	Measurement variables	Description of variable's value assignment
		Time frame of each job	Below 10 months = 1; Below 11 months = 2; 21–30 months = 3; More than 31 months = 4
Measurement model 3	Social factors	Object of socialization	Residents of the original village who were resettled with them = 1; Other residents in the community = 2; Citizens outside the community = 3; Complex socialization = 4
		Frequency of socialization	Basically no communication = 1; Do not know them, only greetings = 2; Little communication = 3; Frequent = 4; Very frequent = 5
		Number of urban citizen friends	None = 0; 1–3 = 1; 4–6 = 2; More than 6 = 3
		Frequency of Socialization with Surrounding Communities	Basically no communication = 1; Do not know them, only greetings = 2; Little communication = 3; Frequent = 4; Very frequent = 5
		Degree of trust in surrounding communities	Highly distrustful = 1; Distrustful = 2; Fairly trustful = 3; Trustful = 4; Highly trustful = 5
Measurement model 4	Psychological factors	Degree of community likeness	Highly dislike = 1; Dislike = 2; Fairly Like = 3; Like = 4; Highly fond of = 5
		Mental gap	No gap = 1; Very small gap = 2; A little gap = 3; Relatively large gap = 4; Large gap = 5

(continued)

Table 3.23 (continued)

Model	Variables	Measurement variables	Description of variable's value assignment
		Behavioral identification	Strongly disapprove = 1; Disapprove = 2; Fairly approve = 3; Approve = 4; Strongly approve = 5
		Urban adaptation	Highly unadapted = 1; Unadapted = 2; Averagely adapted = 3; Adapted = 4; Highly adapted = 5
		Degree of harmony in the relationship with urban citizens in the relationship with urban citizens	Highly inharmonious = 1; Inharmonious = 2; Fairly Harmonious = 3; Harmonious = 4; Highly harmonious = 5
Measurement model 5	Institutional factors	Difference between resettlement areas	In an ascending order of the construction time of each resettlement area, Mufu Mountain Villa = 1; Yao Lin Xianju = 2; Xishan Huayuan = 3; Tianjingshan Community = 4
		Social security status	Each insurance category is 1 score, and the value of the social security indicator is the total of the six categories (5 categories of social insurance and 1 housing fund), with a range of 0–6
Measurement model 6	Spatial factors	Type of population living in surrounding communities	Mostly land-lost peasants = 1; Mostly citizens = 2
		Living area	Below 15 m^2 = 1; 16–30 m^2 = 2; 31–50 m^2 = 3; 51 m^2or more = 4
		Facility index	Calculate the total score for each sample of 19 housing facilities, with a score of 10 or less = 1; 11–15 = 2; 16–19 = 3

(continued)

Table 3.23 (continued)

Model	Variables	Measurement variables	Description of variable's value assignment
		Distance to nearest bus or subway station	Within 15 min by foot = 1; 16–20 min by foot = 2; 21–30 min by foot = 3; More than 30 min by foot = 4
		Distance to worksite	Within 15 min by car = 1; 16–30 min by car = 2; 31–40 min by car = 3; More than 40 min by car = 4
Measurement model 7	Social integration	Self-identification	Hard to say = 0; Peasants = 1; Between peasants and urban residents = 2; Urban residents = 3
		Willingness to develop	Highly unwilling to = 1; Unwilling to = 2; Fairly willing to = 3; Willing to = 4; Highly willing to = 5
		Expectation toward children	Highly unwilling to = 1; Unwilling to = 2; Fairly willing to = 3; Willing to = 4; Highly willing to = 5

Source Own elaboration

two ways to test the parameter estimates: output visual parameter estimations via path diagrams and use a table containing the overall model fit statistics.

The standardized coefficients put out in the first-round test of the initial hypothetical model are shown in Table 3.24. It was found that in the hypothetical structural model, the loading coefficients of the standardized factors, including the path coefficient of social integration pointing to personal characteristics, frequency of job changes pointing to economic factors, degree of trust in surrounding communities pointing to social factors, degree of harmony in the relationship with urban citizens in the relationship with urban the citizens pointing to psychological factors, and the gathering of population pointing to spatial factors did not reach significance requirements (*p*-values did not reach ***). Therefore, the hypothetical model needs to be revised. It should be noted that, given that the saturated model in AMOS is the least restricted model to fit, and the independence model is one of the most restricted models that can be fit, this paper mainly refers to the default model for the parameter estimation of the hypothetical model.

Table 3.24 Parameter estimation for hypothetical model of social integration of land-lost peasants

Model	Name of Path			Estimation of the hypothetical Model Parameters				
				Unstandardized parameters	S.E	C.R	P	Standardized parameters
Structural model	Social integration	←	Personal Characteristics	0.167	0.059	2.831	0.236	0.006
		←	Economic	0.135	0.033	4.091	***	0.132
		←	Social	0.11	0.042	2.619	***	0.102
		←	Institutional	0.136	0.051	2.667	***	0.124
		←	Spatial	0.172	0.031	5.548	***	0.159
	Psychological	←	Economic	0.177	0.096	1.844	***	0.166
		←	Social	0.159	0.052	3.058	***	0.153
Measurement model 1	Age	←	Personal Characteristics	1.000	–	–	–	0.821
	Gender	←		0.034	0.042	0.810	***	0.029
	Education level	←		0.666	0.041	16.244	***	0.654
Measurement model 2	Occupational category	←	Economic	1.000	–	–	–	0.614
	Monthly income	←		0.759	0.025	30.360	***	0.798
	Working hours	←		0.229	0.076	3.013	***	0.224
	Ratio of contract workers	←		0.487	0.164	6.571	***	0.462
	Frequency of job changes	←		0.013	0.106	0.123	0.259	0.012
Measurement model 3	Object of socialization	←	Social	1.000	–	–	–	0.714
	Frequency of socialization	←		0.485	0.064	7.578	***	0.362

(continued)

Table 3.24 (continued)

Model	Name of Path			Estimation of the hypothetical Model Parameters				
				Unstandardized parameters	S.E	C.R	P	Standardized parameters
	Number of friends	←		0.262	0.058	4.517	***	0.251
	Socialization with communities	←		0.524	0.053	9.887	***	0.555
	Degree of trust in surrounding communities	←		0.018	0.152	0.118	0.212	0.017
Measurement model 4	Degree of community likeness	←	Psychological	1.000	–	–	–	0.326
	Mental gap	←		0.695	0.128	5.430	***	0.702
	Behavioral Identification	←		0.274	0.042	6.524	***	0.269
	Urban Adaptation	←		0.672	0.122	5.508	***	0.685
	Degree of harmony in the relationship with urban citizens	←		0.021	0.147	0.143	0.321	0.020
Measurement model 5	Difference between resettlement areas	←	Institutional	1.000	–	–	–	0.885
	Social security status	←		0.214	0.103	2.078	***	0.209
Measurement model 6	Type of population living in surrounding communities	←	Spatial	0.016	0.067	0.239	0.159	0.015

(continued)

Table 3.24 (continued)

Model	Name of Path			Estimation of the hypothetical Model Parameters				
				Unstandardized parameters	S.E	C.R	P	Standardized parameters
	Living area	←		1.000	–	–	–	0.354
	Facility index	←		0.644	0.069	9.333	***	0.633
	Distance to nearest bus or Subway Station	←		0.211	0.126	1.675	***	0.203
	Distance to worksite	←		0.581	0.136	4.272	***	0.523
Measurement model 7	Self-identification	←	Social Integration	1.000	–	–	–	0.812
	Willingness to develop	←		0.359	0.365	0.984	***	0.374
	Expectation toward children	←		0.817	0.024	14.086	***	0.803

Note S.E. indicates standard error, C.R. is the critical ratio (unstandardized parameter/standard error), and *** in the *p*-value indicates significance at the 0.001 (two-tailed) level

3.4.3.5 Model Modification and Fitting

Fit index and evaluation

The overall evaluation indices commonly used for estimating and fitting the model output are the chi-square value and degrees of freedom (df). However, notably, many studies consider the conventional chi-square test to be highly influenced by sample size and model complexity. Additionally, using "whether chi-square value is significant or not" as the criterion may lead to the rejection of all models. Therefore, this study uses the chi-square/df ratio (NC) (Qiu and Lin 2009).

The fitting test of the model output generally includes two types: absolute fit test and relative fit test. Combining the existing research results on the selection of fit index and the determination of index value, this study focuses on the review of Wen et al. (2004) of the seven fit indices for testing structural equation models and selected four fit indices: absolute fit indices (including adjusted goodness of fit index—AGFI and root mean square error of approximation—RMSEA) and relative fit indices (including Tucker-Lewis index TLI/NNFI and comparative fit index CFI) for model testing (see gray bands in Table 3.25).

Model Modification and Fitting

When the fit of the hypothetical model is unsatisfactory, the structural equation model can be adjusted according to the parameter significance results of the initial model using two specific approaches: (a) Instead of adjusting the paths of structural equations, obtain the partial correlation between residuals of each measurement variable. The idea is valid only when the structural equation paths meet certain basic theoretical conditions. This option is considered the first choice for simple model correction as it entails the least correction to the original model and has a greater effect on improving model fit. (b) Redefine the model and directly adjust the model paths. There are more

Table 3.25 Related fit indices and their evaluation criteria

Indices		Evaluation criteria
Absolute fit indices	NC (Chi-square/df)	1 < NC < 3
	GFI	> 0.9
	RMR	< 0.05, the smaller the better
	SRMR	< 0.05, the smaller the better
	RMSEA	< 0.05, the smaller the better
	AGFI	> 0.9, the closer to 1 the better
Relative fit indices	NFI	> 0.9, the closer to 1 the better
	TLI	> 0.9, the closer to 1 the better
	CFI	> 0.9, the closer to 1 the better

Note The table lists the optimal criteria for the fit indices. Take RMSEA for example. A value less than 0.05 indicates a good model fit and 0.05–0.08 indicates a fair model fit. Thus, it can be analyzed according to the specific situation in the actual study

Table 3.26 Calculation results of the selected fit indices

Fit index	Chi-square/df	AGFI	RMSEA	TLI	CFI
Initial model	1.256	0.752	0.061	0.842	0.802
Revised model in the second round	2.152	0.852	0.043	0.748	0.824
Fit criteria	1 < NC < 3	> 0.9	< 0.05	> 0.9	> 0.9

Source Own elaboration

adjustments to make in this approach as it adds or deletes certain factors and their corresponding paths mainly based on theoretical experience and field observations.

First, the study considered the first approach for model modification, namely, residual correlation. Based on the above parameter estimation results, it is known that the path coefficients and loading coefficients of the standardized factors of two models (structural model and measurement model) in the hypothetical model do not reach the significance level. Thus, a simple correction to the model, namely, residual correlation was attempted, from which revised models were generated. However, results showed that the model fit still did not improve much (see Table 3.26).

Second, the study considered the second approach, which is modifying the model paths. Based on the output of the visual path diagram, the steps are as follows: (a) examine the structural equation paths (single path) and consider eliminating non-significant paths and latent variables that are highly self-correlated while reconsidering the correlation (mutual effect of variables A and B) and recurrence relationship between latent variables (variable A influences variable C by influencing variable B) and (b) examine the measurement model paths (single paths) and consider eliminating non-significant paths or highly self-correlated measurement variables.

Based on the output results of the first and second rounds of the hypothetical model mentioned above as well as a comprehensive assessment of the impact of the index parameters under different main path conditions, it is suggested that the model be redefined. As can be seen, the process is a cyclic validation process through continuous correction and fitting, and a structural model of satisfactory fit may be obtained only after multiple rounds of correction. The specific steps are as follows:

First, adjust direct influence factors into indirect influence factors. As personal characteristics influence social integration by influencing economic, social, and psychological factors, modifications were made on the basis of the first two rounds. The main path between personal characteristics and social integration was deleted, and the main paths between personal characteristics and economic, social, and psychological factors were added.

Second, adjust one-way influence factors into mutual influence factors. The hypothetical model overemphasizes the linear progressive relationship between economic, social, and psychological factors (unidirectional influence) and may result in the disregard of the interactive relationship between these three (mutual influence). Therefore, the two unidirectional influence paths between economic, social, and psychological factors were adjusted to three mutual influence paths between the three.

Table 3.27 Parameter estimation of the modified model of social integration of land-lost peasants

Model	Name of path			Parameter estimation of the modified model				
				Unstandardized parameters	S.E	C.R	P	Standardized parameters
Structural model	Social integration	←	Economic	0.133	0.033	4.030	***	0.142
		←	Social	0.105	0.042	2.500	***	0.091
		←	Psychological	0.132	0.025	7.680	***	0.212
		←	Institutional	0.152	0.051	3.627	***	0.189
		←	Spatial	0.192	0.031	10.065	***	0.174
	Economic	←	Personal characteristics	0.341	0.054	6.315	***	0.332
	Social	←	Personal characteristics	0.356	0.086	4.140	***	0.242
	Psychological	←	Personal characteristics	0.245	0.047	5.213	***	0.232
	Economic	←	Social	0.178	0.075	2.373	***	0.165
	Economic	←	Psychological	0.269	0.052	5.173	***	0.251
	Social	←	Psychological	0.135	0.062	2.177	***	0.129
Measurement model 1	Age	←	Personal characteristics	1.000	—	—	—	0.912
	Gender	←		0.029	0.041	0.707	***	0.021
	Education level	←		0.702	0.041	17.122	***	0.687
Measurement model 2	Occupational category	←	Economic	1.000	—	—	—	0.629
	Monthly income	←		0.852	0.024	35.500	***	0.802
	Working hours	←		0.254	0.076	3.342	***	0.236
	Work stability	←		0.143	0.083	1.723	***	0.159

(continued)

Table 3.27 (continued)

Model	Name of path			Parameter estimation of the modified model				
				Unstandardized parameters	S.E	C.R	P	Standardized parameters
Measurement model 3	Object of socialization	←	Social	1.000	—	—	—	0.724
	Frequency of socialization	←		0.382	0.351	5.969	***	0.363
	Number of friends	←		0.259	0.241	4.466	***	0.255
	Community socialization	←		0.521	0.052	10.019	***	0.563
Measurement model 4	Degree of community likeness	←	Psychological	1.000	—	—	—	0.324
	Mental gap	←		0.699	0.129	5.419	***	0.712
	Behavioral identification	←		0.271	0.045	6.022	***	0.265
	Urban adaptation	←		0.683	0.123	5.553	***	0.690
Measurement model 5	Difference between resettlement areas	←	Institutional	1.000	—	—	—	0.845
	Social security status	←		0.284	0.103	2.757	***	0.216
Measurement model 6	Living area	←	Spatial	1.000	—	—	—	0.362
	Facility index	←		0.717	0.068	10.544	***	0.621
	Distance to stations	←		0.226	0.112	8.268	***	0.212

(continued)

Table 3.27 (continued)

Model	Name of path			Parameter estimation of the modified model				
				Unstandardized parameters	S.E	C.R	P	Standardized parameters
	Distance to worksite	←		0.569	0.129	4.411	***	0.529
Measurement model 7	Self-identification	←	Social integration	1.000	—	—	—	0.793
	Willingness to develop	←		0.345	0.362	0.953	***	0.381
	Expectation toward children	←		0.825	0.058	14.224	***	0.812

Source Own elaboration

Note S.E. indicates standard error, C.R. is the critical ratio (unstandardized parameter/standard error), and *** in the *p*-value indicates significance at the 0.001 (two-tailed) level

Third, eliminate certain measurement variables. Eliminate variables with insignificant paths or that are highly self-correlated (meaning a higher repetition rate in terms of analysis and interpretation) from the measurement model, such as "monthly balance," "frequency of job change," "degree of trust in surrounding communities," "type of population living in surrounding communities," and "degree of harmony in the relationship with urban citizens." In the end, 24 variables such as "monthly income," "object of socialization," "community socialization," "mental gap," "urban adaptation," "living area, and "distance to worksite" remained.

The final modified structural model is shown in Fig. 3.46. In terms of the model fit indices, the chi-square/df ratio is 2.83, indicating that the model can be identified and passes the overall assessment. In addition, AGFI = 0.9013, RMSEA = 0.0612, TLI = 0.9123, and CFI = 0.9038, indicating that the overall model fit is good and the results are acceptable. Meanwhile, according to the analysis table of this model, the parameter estimates of the structural model and the seven measurement models meet the significance requirements (see Table 3.27).

3.4.3.6 Model Output and Model Interpretation

After the hypothesis, measurement, testing, correction, and fitting of the model, the overall fit of the modified model is noted to be good, and the path coefficients in the structural model and loading coefficients of standardized factors in the measurement model are meaningful. The factors with greater levels of effect (11 measurement indicators) were screened out to obtain the simplified model (see Fig. 3.47).

The path diagram and path coefficient values were further used to simplify the expression of the impact type and effect level of each latent variable on social integration. The impact (or effect) can be categorized as direct and indirect impact as well as one-way impact and mutual impact. In the simplified path of the social integration model, the solid lines represent direct impact, the dashed lines represent indirect impact, the arrow indicates the direction of the impact (one-way arrow indicates one-way impact, and two-way arrow indicates mutual impact), and the thickness of the line and path coefficient values indicate the actual level of effect (see Fig. 3.48).

Model output

First, economic, social, psychological, institutional, and spatial factors influence the degree of social integration directly; personal characteristics influence the degree of social integration indirectly; and economic, social, and psychological factors influence each other.

From the model output, it can be seen that the path coefficients of five paths, namely, economic factors pointing to social integration, social factors pointing to social integration, psychological factors pointing to social integration, institutional factors pointing to social integration, and spatial factors pointing to social integration, reached the significance level. Hypotheses 1.1–1.3 and 2 and 3 are strongly supported by the data. However, it is shown that economic, social, and psychological factors affect each other (correlated) rather than unidirectionally (progressive),

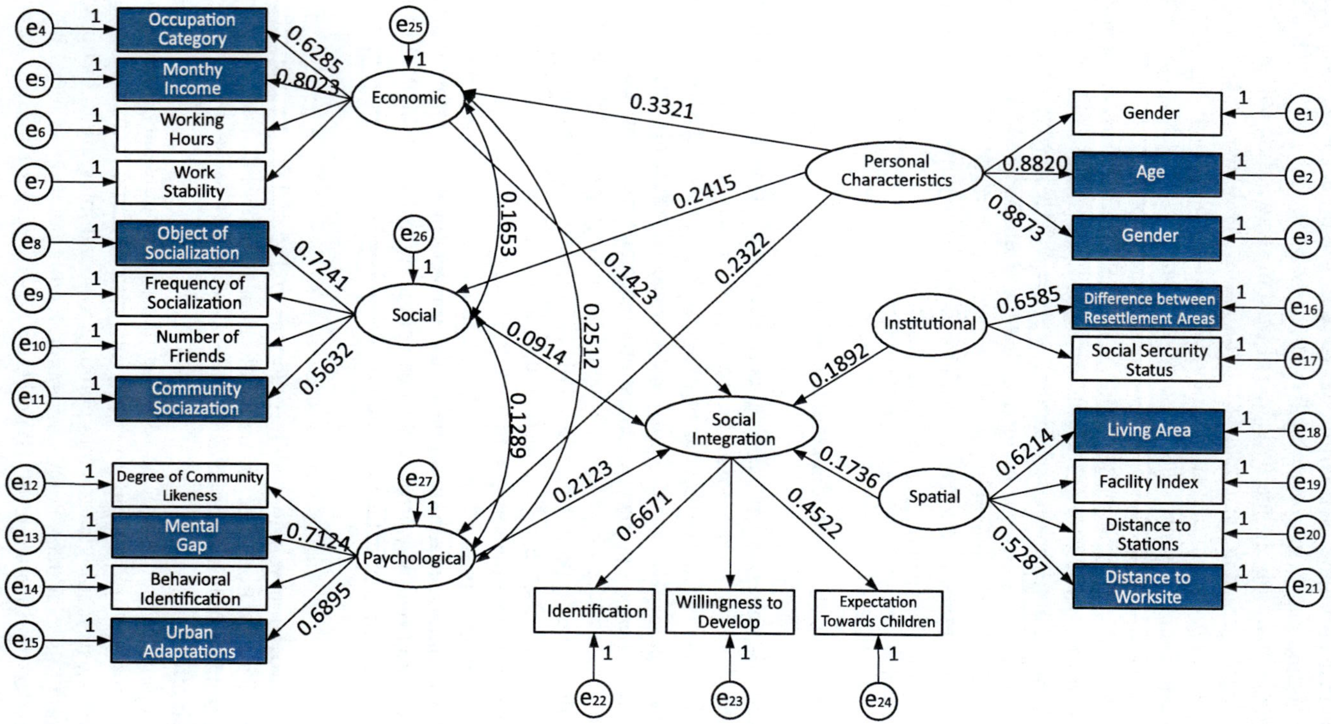

Fig. 3.46 Results output of the modified model of social integration of land-lost peasants (initial version). *Source* Own elaboration

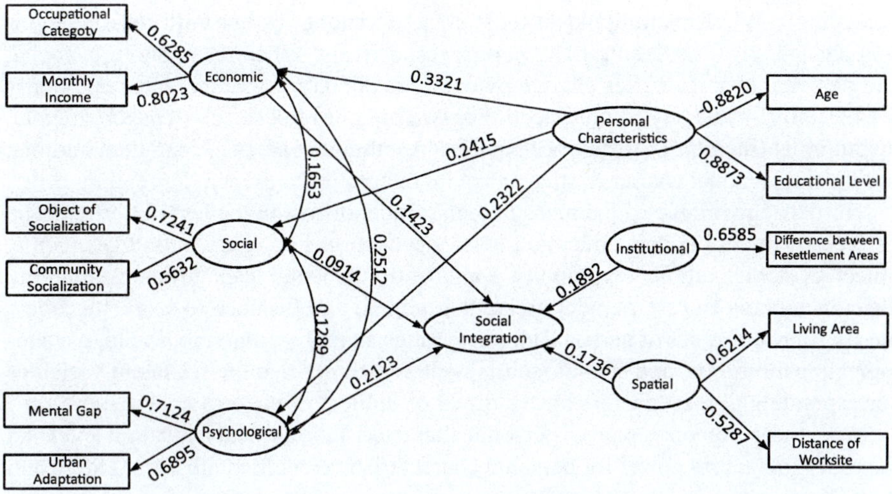

Fig. 3.47 Results output of the modified model of social integration of land-lost peasants (simplified version). *Source* Own elaboration

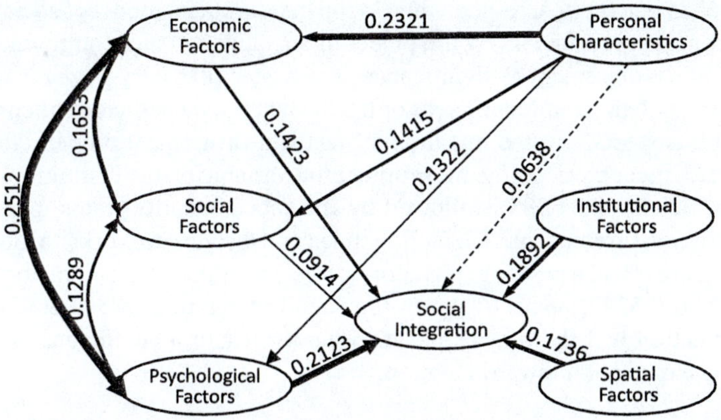

Fig. 3.48 The effect paths of social integration of land-lost peasants. *Source* Own elaboration

making hypothesis 1.4 invalid. In addition, the personal characteristics factors are found to have indirect effects on social integration by affecting economic, social, and psychological factors, which means that hypothesis 4 is also invalid.

Second, psychological factors have the greatest level of effect (with most related factors), followed by institutional and spatial factors, while economic, social, and personal characteristics have less influence.

Differences in the value of the path coefficients indicate differences in the degree of correlation between variables and their level of effect. For example, when the psychological factor changes by one unit, social integration changes by 0.2123 units accordingly. When institutional and spatial factors change by one unit, social integration changes by 0.1892 and 0.1736 units, respectively. When the economic, social, and personal characteristics change by one unit, social integration changes by only 0.1423, 0.0914, and 0.0361 units accordingly. Thus, the effect levels of psychological, institutional (including "difference between resettlement areas"), spatial, economic, social, and personal characteristics on social integration present a descending order.

Third, each variable of the measurement model differs in the level of effect. The 11 factors, including age, education level, occupational category, monthly income, object of socialization, community socialization, mental gap, urban adaptation, difference between resettlement areas, living area, and distance to worksite reflect the six latent variables of personal characteristics, namely, economic, social, psychological, institutional, and spatial factors well, indicating that the six latent variables corresponding to these factors have a direct or indirect effect on social integration.

In the measurement model of personal characteristics, age and education level had the best explanatory power for personal characteristics (coefficients were 0.8820 and 0.8873, respectively). In the measurement model of economic factors, the reflection of economic factors by occupational category and monthly income met the requirement of significance. Two variables had the best explanatory power, occupational category, and monthly income, whose standardized factor loading coefficients reached 0.6285 and 0.8023, respectively. In the measurement model of social factors, the reflection of social factors by the object of socialization and community socialization met the requirement of significance. The best explanatory power is shown by the object of socialization (coefficient of 0.7241), and the coefficient of community socialization is 0.5632. In the institutional factor measurement model, "difference among resettlement areas" is the most influential variable on the institutional factors (with a coefficient of 0.6585), followed by the social security status index. In the spatial factor measurement model, facility index and distance to worksite index of the resettlement areas had a higher explanatory power, with standardized factor loadings of 0.6214 and 0.5287, respectively. In the social integration measurement model, self-identification had the best explanatory power, (with a coefficient of 0.6671), followed by expectation toward children.

Model interpretation

In terms of the factors with greater direct influence, psychological integration is the highest stage and level of social integration, and individual characteristics, economic factors, and social factors influence social integration by exerting influence on psychological factors. Therefore, psychological factors are the most associated factors that have the greatest level of effect, and the highest explanatory power of social integration. Institutional factors are ranked second in this regard.. The reason for the larger coefficient of institutional factors among all paths of social integration lies in the introduction of the variable of "difference between resettlement areas." In other words, the difference in compensation policy due to the different time of land expropriation is an important factor affecting the degree of social integration of land-lost peasants. In addition, the spatial factors (living and commuting) constitute the most basic survival guarantee for land-lost peasants and directly affect whether land-lost peasants can live and work in peace and contentment. It is, therefore, an important influence factor of social integration.

In terms of factors with indirect influence, personal characteristics, including age and education level, are indirectly associated with social integration. Studies have shown that the younger and more educated the land-lost peasants, the easier it is for them to integrate into the city in terms of self-identification due to following. First, compared with their parents' generation, these people have lived in the rural society for a shorter period of time and have less identification with local customs, lifestyles, and social interactions and do not have an excessive attachment to traditional xiangtu concepts from the social perspective. Second, younger land-lost peasants are stronger in adapting to a new environment and accepting new knowledge, and have higher possibilities in participating in job training and obtaining employment opportunities. Hence, their degree of integration in the economic dimension is relatively high. Third, due to growing up in the information age, younger land-lost peasants tend to have an open mind, stronger resistance to stress, and higher aspirations for the future, all of which enable them to adapt more quickly to the transformation from peasants to urban citizens in ways of production and lifestyle despite the impact of governmental land expropriation. Thus, they are more inclined to identify themselves as urbanites psychologically.

In terms of mutual (two-way) influence factors, economic integration can affect social and psychological integration while being limited by the degree of integration in other dimensions. For example, if land-lost peasants adapt to urban behaviors more quickly and actively build social networks to gain more social capital, they are likely to accelerate the speed and extent of their economic integration. Another example is that, among the economic factors, the degree of social integration is likely to be higher for land-lost peasants whose jobs are of higher vocational and technical thresholds (professional and technical staff or management and common clerks at state organs, organizations, and institutions) and those with higher monthly income levels. This is because the degree of economic integration directly determines the material basis of whether they can integrate into the city, and the relevant path coefficients linked to it are larger. Within the social factors, land-lost peasants who interact more often with urban citizens and have more community interactions have a higher level of social integration. However, the mutual effect level of the two factors is not significant, which may be related to the fact that the heterogeneous urban social networks of land-lost peasants are not established. In other words, the original geographical and kinship relationships of land-lost peasants are dissevered, but the new social relationships and secondary interactions have not yet been established through effective means, resulting in lack of social support system for their integration into urban social life.

3.5 Strategies to Improve Spatial Resettlement and Social Integration of Land-Lost Peasants

The previous analysis shows that there are major problems in the spatial resettlement and social integration within the resettlement areas for land-lost peasants, which is a newly emerging gathering place for disadvantaged groups. Therefore, in the following sections, we discuss the existing problems of the resettlement areas for land-lost peasants in terms of spatial resettlement and social integration as well as their improvement strategies, in accordance with the findings of the previous sections and referring to the specialty of land-lost peasants as research samples.

3.5.1 Strategies to Improve Resettlement Areas for Land-Lost Peasants

3.5.1.1 Existing Problems of Resettlement Areas

In light of the research on resettlement areas for land-lost peasants in Nanjing in the previous section, the main problems faced by different types of resettlement areas for land-lost peasants and their causes were sorted out and summarized to provide a guiding basis for the following strategy discussions (see Table 3.28).

This study further explored the spatial resettlement problems in the four aspects mentioned above and their influence factors, from which six most important causes of these problems and three core improvement strategies to address these problems are proposed (see Fig. 3.49).

3.5.1.2 Improvement Strategies for Spatial Resettlement

According to the previous analysis, the resettlement areas for land-lost peasants in Nanjing are mostly concentrated in sub-cities and new cities at the fringe of the city. Therefore, the following discussion on the improvement strategies for the spatial resettlement for land-lost peasants in Nanjing will focus on new cities and sub-cities where the resettlement areas for this group are concentrated and elaborate from the macro-, mid-, and microscope aspects.

Macro-scope: Integration into the overall urban planning system

As a type of subsidized housing, the spatial layout of resettlement areas for land-lost peasants has to be considered within the context of the development and construction of subsidized housing in a broader sense. To ensure that the spatial distribution of this type of housing is fairer and more reasonable, and to solve the problems such as difficulty in employment due to the choice of location and high cost of living due to the incomplete service facilities, the planning of resettlement areas should be

Table 3.28 Brief summary of problems in spatial resettlement for land-lost peasants in Nanjing

Location of resettlement area				Main city	Industrial park	Fringe areaof the City	
Name of the resettlement area				Mufu Mountain Villa	Yaolin Xianju	Tianjingshan Community	Xishan Huayuan
Common problems				Differentiated problems			
Spatial resettlement	Spatial layout	Surrounding environment	–	–	Marginalized resettlement area long-distance transportation		
		Layout	The scale is too large and has a single form of building layout	As a resettlement area built in the early stages, its planning and design are not reasonable		As a recently constructed resettlement area, it is even larger in scale	
	Housing use	Unit types	Single unit type and design, with insufficient ventilation and lighting	The support facilities of this resettlement area, which was built in the early stages, are not complete		–	–
		Living area	Insufficient living area per capita	The living area of this resettlement area, which was built earlier, is particularly inadequate		–	–
	Work and commuting	Commuting costs	Commute by bus to save commuting cost	–	–	Composite transportation mode with relatively high commuting costs	
		Commuting mode	–	–	–	Obvious scattered worksites (limited employment opportunities and transportation conditions)	
	Services facilities	Urban public service facilities	Spatial mismatch between public service facilities and resettlement areas	–	–	–	–

(continued)

Table 3.28 (continued)

Location of resettlement area				Main city	Industrial park	Fringe areaof the City	
Name of the resettlement area				Mufu Mountain Villa	Yaolin Xianju	Tianjingshan Community	Xishan Huayuan
Common problems				Differentiated problems			
		Community public service facilities	Inadequate facility types and small in scale			–	–

Source Own elaboration

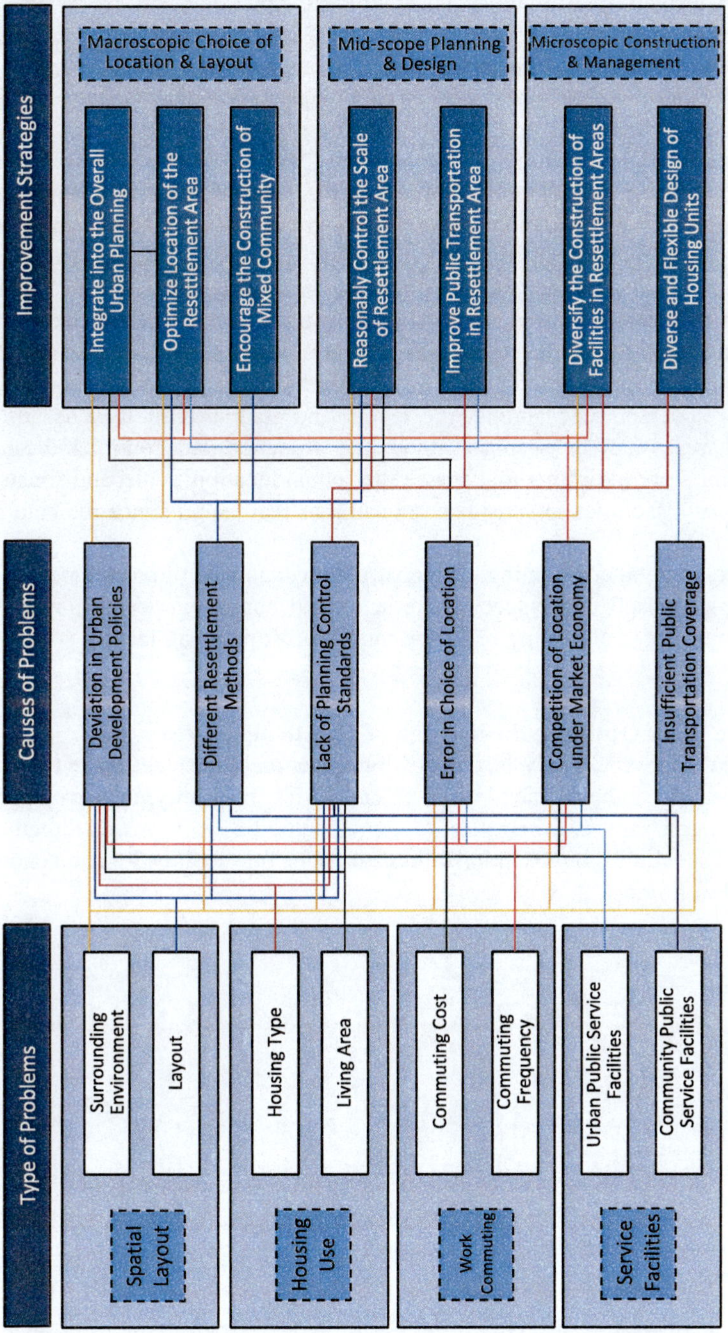

Fig. 3.49 Illustration of improvement strategies in spatial resettlement for land-lost peasants. *Source* Own elaboration

incorporated compulsorily into the system of urban planning. In addition, it is urgent to formulate reasonable and effective plans for the construction of subsidized housing (including resettlement areas for land-lost peasants) and coordinate the special plans of relevant departments. The guidance and control can be conducted at the following levels.

At the level of overall planning, the organization pattern of subsidized housing shall be regulated in the form of special planning. Drawing on the concept of "residential community" in the "Planning Guidelines for Public Facilities in Newly Built Areas in Nanjing," the "residential community—grassroots community" organization pattern can be adopted for the subsidized housing in Nanjing (see Fig. 3.50). Using part of the residential land use zone for subsidized housing (rather than the greenbelt between old and new urban areas) can ensure residents benefit from the supporting facilities of the residential communities and alleviate the financial pressure and reduce the operating costs for constructing subsidized housing.

At the level of zoning planning, the layout and standards of land use in the resettlement areas shall be proposed under the guidance and control indicators, including mandatory indicators that ensure the spatial location, scale, and construction of non-profit facilities in subsidized housing as well as guidance indicators at the employment level.

At the level of detailed planning, the construction standards of resettlement areas shall be also proposed under specific guidance and control indicators, including mandatory indicators concerning unit type ratios and supporting facility standards, and guidance on the level of mixed housing.

Macro-scope level: Optimize the location of resettlement areas

According to the previous analysis, the location selection of the resettlement area for land-lost peasants can be divided into three categories: main city, industrial parks, and urban fringes, each of which has its own specificities in terms of directional strategies and specific layouts. Optimization strategies for the three location choices are proposed as follows:

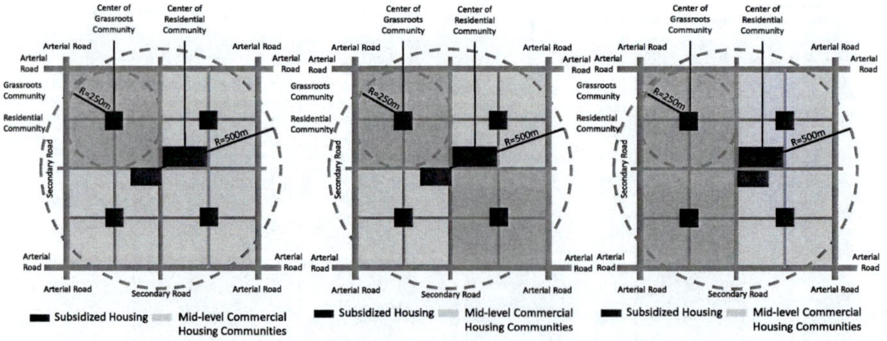

Fig. 3.50 Organization pattern of residential community—grassroots Community. Guo et al. 2011a

The first deals with location selection in the main city. Theoretically, for the main urban area of Nanjing, which is highly built up and where the land value is high, obstruction of the construction of micro-profit resettlement housings (for urban village residents) by means of urban renewal, such as renovating urban rundown areas, is immeasurable. However, from the perspective of social equity, we should avoid making the main urban area exclusive for the rich and resettle the special group of land-lost peasants to the fringe area of the city where urban resources are most scarce. Therefore, within the main city, a certain proportion of the land shall be allocated to build resettlement areas for land-lost peasants (mainly due to land-expropriation), and the four location selection strategies are as follows: 1 Drawing on the experience of Tianjin, where at least one piece of land in the expropriated area must be utilized for constructing resettlement housing during urban renewal on a yearly basis. 2 Draw on the experience of the United Kingdom where part of the mid-level commercial housings is designated as resettlement housing. 3 Promote mixed land use and leverage the upper space of large farmer markets, supermarkets, and bus terminals to build high-floor resettlement housing. 4 Renew partially old urban areas inhabited by low-income groups and build resettlement areas in the area.

The second deals with location selection in the industrial park. Drawing on the experience of Hong Kong (Jiang and Zheng 2011), the resettlement area for land-lost peasants can be built in combination with the layout of industrial land (by a very small portion). Consider mixing effectively residential areas with non-polluting and tertiary industries and setting aside part of the land in the industrial park for constructing resettlement areas for land-lost peasants. On one hand, the balance of employment and living can be ensured. There will be sufficient employment opportunities in the vicinity, implying that the housing and employment problems of land-lost peasants can be solved on a small scale while promoting the industrial development of the new city at the same time. On the other hand, housing rental means additional income for land-lost peasants resettled in the industrial park. They can rent out a large number of vacant units to local employees and migrant workers of the enterprises in the industrial park. This approach not only solves the problem of spatial resettlement for land-lost peasants but also resolves accommodation issues of employees and migrant workers of neighboring enterprises.

The third deals with location selection in urban fringe areas (new city and sub-city). In terms of macro location layout, we can learn from Singapore's (Zheng and Zhang 2010) model of developing public housing[3] in conjunction with the construction of new towns. The new towns are not only well equipped with facilities but also have 10–20% of the land reserved for industrial facilities to resolve the employment problem of residents. The resettlement areas for land-lost peasants could thus be concentrated and marginalization can be avoided to further drive the development

[3] The whole country of Singapore (whose area is about 1/3 of Shenzhen) is divided into 5 regions: East, Northeast, North, West and Central. There are 55 planning areas (Development Guide Plan, DGP) in these 5 regions, 33 of which are designed with 23 new towns (new town), and each new town has a total of about 40,000–100,000 households. The grading of the public housing is generally at three levels, new town, neighborhood, and neighborhood group, with different land and building scales corresponding to the residential scale of the service area.

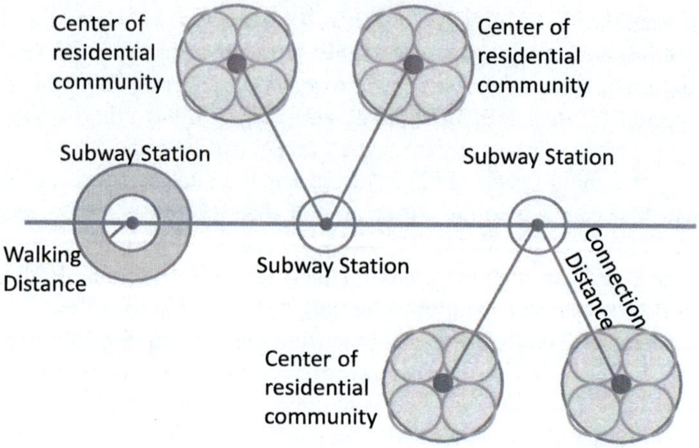

Fig. 3.51 Schematic representation of the location selection of subsidized housing along the subway line. *Source* Yang et al. (2009)

of new towns. In terms of micro location layout, routes of fast, high-capacity public transportation (e.g., subway) connecting the urban fringe area and the main city can serve as the main axis, and short-distance minibuses (e.g., BRT) or light rails shall be utilized to connect the center of the residential communities to the subway stations to resolve the last mile transportation issue of residents. Moreover, the TOD development model can be adopted for residential area planning to build high-intensity and high-density resettlement housings located within a 10 to 15-min walking distance from the metro stations (see Fig. 3.51).

Macro-scope level: Encourage the construction of mixed communities
Based on previous research, the resettlement areas for land-lost peasants have the characteristics of "unevenly located between districts, mostly located at peripheral areas, and being distributed in a scattered way inside and centralized way outside the main city," which will lead to residential differentiation, spatial segregation, and high-level of homogeneity (especially in large mixed resettlement areas at the fringe of cities). For this reason, many scholars have proposed the model of "mixed housing layout on a large scale and clustered housing layout on a small scale," implying that the resettlement areas are constructed at an appropriate scale in the same greater zone with other commercial housing while maintaining a certain distance. According to the different spatial scales of mixed housing, their layout pattern can be divided into two categories:

The first one is mixing residential areas according to gradient classification. The housing area is mixed with residents of two types or grades. One is to mix the housing of middle-income and low-income residents and the other is to mix the housing of middle-income and high-income residents. Additionally, residential integration is achieved through the mixed use of public space and public service facilities (see Fig. 3.52). According to this approach, it is recommended that the resettlement

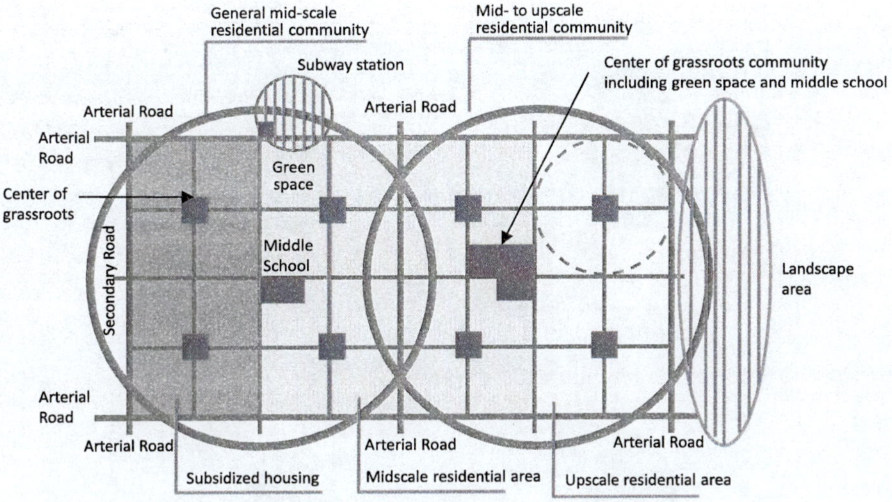

Fig. 3.52 Map of mixed housing model of different residential areas. *Source* Yang et al. (2009)

housing for land-lost peasants are built in adjacent to the general midscale commercial housings for middle-income residents. In addition, since land-lost peasants are highly concerned about the living convenience and employment conditions, the site selection should focus more on the accessibility of public transportation and less on the luxury of landscape resources. There should also be supporting service facilities and public space, the spatial scope of which should not be too large or small. A service radius equivalent to that of secondary schools is recommended (see Fig. 3.53).

The second is scattered mixed housing within residential areas based on a certain proportion. This pattern allows residents of different social stratifications to live together within a residential area, and there are two approaches of mixed housing according to the strata of the residents: (1) Build resettlement housings in the commercial housing area—Drawing on the incentive zoning and inclusive planning policies of the United States, where resettlement housings are built compulsively within the new commercial housing areas of reasonable sizes according to a certain ratio or area as a result of bonus policies such as tax exemption and floor area ratio incentive, several cities in China have started exploration along this line (e.g., Guangzhou). (2) Build subsidized housings led by public rental housing. The types of subsidized housing may vary but they belong to the mixed housing model. However, since residents who benefit from subsidized housing have smaller class gaps between each other, these types of areas are still close to a type of homogeneous community. The positive aspect about public rental housing and housing with limited price is that they are meant for the citizens of in-between social classes, which include intellectuals,

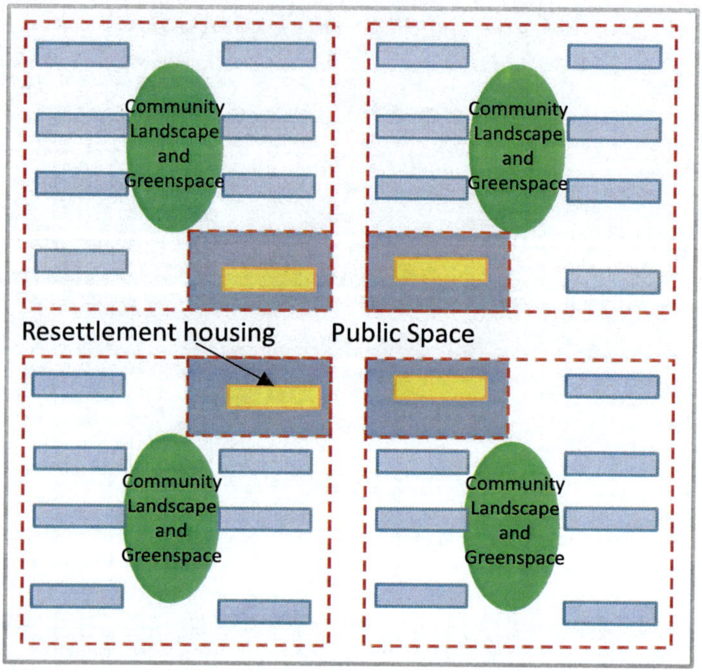

Fig. 3.53 Map of mixed housing model within residential areas. *Source* Own elaboration

university graduates, and technical talents. Thus, the introduction of public rental housing can play the role of a "blender" in the originally homogeneous community.

Mid-scope level: Reasonably control the scale of resettlement area
Residential areas that are too large or small are not conducive to the development of resettlement areas for land-lost peasants. If the scale of the resettlement area at the fringe of the city is too large, pressure on transportation and public facilities increases and social isolation is noted due to high homogeneity. Conversely, if the resettlement area is too small, public facilities cannot be shared with surrounding communities. It is also not recommended to provide a large proportion of public facilities for it as it may easily lead to uneconomical construction and even unsustainable operation.

Regarding the reasonable scale of resettlement areas for land-lost peasants, domestic and foreign scholars have conducted certain research. Alexander, a foreign scholar, takes the neighborhood cognitive range as the standard to determine the scale of the community, and believes that the diameter of the range should not exceed 300 yards (an area of about five hectares) (Wang 2007). Zhou Jian, a domestic scholar, after taking full account of the current situations of the natural environment and infrastructure, believes that the scale of a single residential area should be controlled at around 10 ha (Zhou and Zhang 1999). The scholar further proposes that the scale of residential areas in China not exceed a spatial area of 150 m in diameter or 4 ha (Zhou et al. 1999) in area. According to the concept and scale of "residential community" in the “Planning Guidelines for Public Facilities in Newly Built Areas in Nanjing,” Guo et al. (2011b) suggests that subsidized housing can adopt the organization model of “residential community—grassroots community,” the scale of which is suggested to be about 1500 households (about 5000 people), roughly equivalent to the scale of a grassroots community with a land use area of 10–20 ha.

Mid-scope level: improve public transportation in resettlement areas
The smoothness of the road system and completeness of the public transportation system directly affect the level of convenience in terms of work commuting for residents in resettlement areas for land-lost peasants. To facilitate travel to worksite for land-lost peasants, two aspects can be considered:

At the level of urban road system, first, additional urban expressways or main roads connecting the new city and the main city should be built to ensure fast and convenient traffic connections of the resettlement area with the main city. Second, the internal road network system of the new city should be improved. It is especially necessary to ensure that the road traffic between the districts and the public center and large public service facilities of the new city is smooth. In addition, according to the “Planning and Design Guidelines for Areas along Urban Rail Lines,” the distance between the light rail stations located at the periphery of the city and nearby neighborhoods should be controlled within 200 m, and the density of the road network branches should reach at least 6–8 km/km^2.

Regarding the public transportation system, forming a system with light rail transportation as the skeleton and conventional means of transportation as the principal parts is recommended. The establishment of the system includes three aspects in particular: (1) establish rapid transit lines (such as BRT and light rail) linking the main urban area with the new city; (2) optimize the layout of transportation stations to ensure that residents can arrive at their stops within five minutes, increase the number of bus lines between large residential areas and the main city, public centers of the new city and the large public service facilities, and arrange public service facilities in line with the light rail stations and bus transit hubs to form public service centers; (3) The layout of subsidized housing areas can be planned in combination with public transportation stations to form a public transport community. At the same time, construct bicycle facilities and slow walking trails (see Fig. 3.54), so that the residents of the new city can have a more diverse choice of transportation modes (public transportation, by bicycle and by foot), thus a more relaxed and comfortable urban living environment.

Micro-scope: Diversify the Construction of Facilities in Resettlement Areas
At present, the functionalist planning concept of estimating the scale of public service facilities of residential communities based on the traditional methods of measuring service radius and referring to the indicators of “floor area and land area of each public service facility per 1000 residents” is somewhat outdated for the “emerging group” of land-lost peasants. Although the number of services and facilities thus allocated can meet the overall demand, they are relatively homogeneous in terms of the types of facilities (lack of non-profit public facilities), which directly leads to the

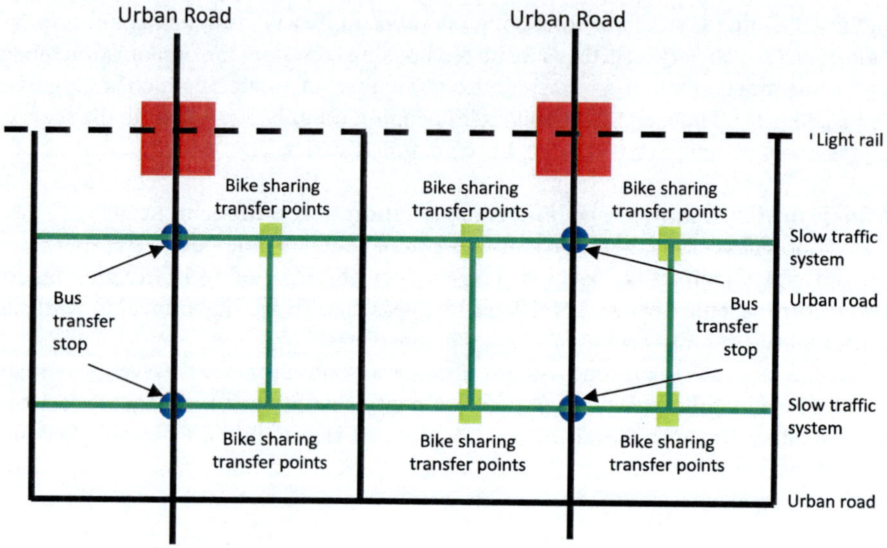

Fig. 3.54 Bus, bicycle and pedestrian transfer system. *Source* Hua and Wang (2012)

contradiction between the single-supply pattern of public services and the diversified needs of land-lost peasants. Therefore, to alleviate this contradiction, improvement strategies in the pattern and standard of facilities supply need to be considered:

The first strategy is to adopt a diversified supply approach based on the characteristics and attributes of the facilities (see Fig. 3.55). The facilities are first classified into essential facilities (educational, medical, commercial, transportation, etc.) and enhancement facilities (cultural, sports, etc.) according to the type of residential facilities. Another breakdown method is to classify them into profit-making (commercial, for-profit cultural, sports, etc.) and non-profit (educational, medical, non-profit social, etc.) based on whether they are profitable. The public interest in essential facilities must be strictly controlled and supervised, whereas that in enhancement-type facilities require government investment or policy support such as subsidies and concessions. For the profitability of essential facilities, the government should carry out macro regulations and guidance to avoid insufficiency in the scale of facility supplies due to market mechanism. The for-profit enhancement facilities would also need government regulations and guidance to avoid inadequate scale of facilities caused by reduced investment in the market.

The second strategy is to optimize the configuration standards of residential facilities according to the diversified needs of residents. The current standards need to be appropriately added/reduced in number or optimized based on the actual needs of land-lost peasants, and socio-economic attributes such as population density, age composition, education level, and economic status should be used as the corrective parameters of supporting standards to build differentiated configuration standards

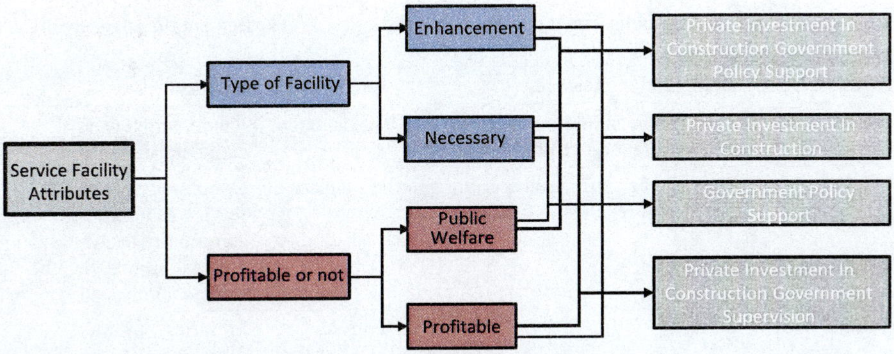

Fig. 3.55 Supply Mode of public service facilities. *Source* Own elaboration

according to the needs of land-lost peasants. At the same time, the current configuration of facilities also needs to be considered to propose targeted optimization measures (e.g., in some resettlement areas, residents have to "break down certain walls to open stores" due to insufficient commercial facilities). For example, peasants are more concerned about medical and health care, commercial services, cultural and recreational, and postal and municipal facilities, and the degree of demand for each of them varies (see Table 3.29).

Micro-scope: Diverse and Flexible Design of Housing Units

The housing type and unit design in resettlement areas are often influenced by the government's policy and product supplies from developers, and there are fewer unit type design options to choose from (for example, there are only three unit sizes to choose from in Tianjingshan Community). However, the specific social characteristics and living habits of land-lost peasants as a group determine the economy, efficiency, and adaptability in their living needs. Thus, the design of housing units for them should strive to achieve the following:

Reasonable functions and suitable area At present, most unit designs within the resettlement areas for land-lost peasants are replicated from the early urban unit designs, where the functional division of floor plans is insufficient and there is lack of necessary functional spaces. Therefore, as long as certain reasonable functions are ensured, storage space can be increased appropriately. For example, the proportion of storage space in Japanese houses is very large and meticulously classified, generally reaching about 1/10 of the residential use area. At the same time, when designing the units, it is necessary to reasonably determine the bay width and depth of the units to ensure sufficient lighting and ventilation of each unit.

Adapt to diversified usage patterns As a class undergoing transitions, land-lost peasants are largely different from traditional peasants and urban citizens in terms of ways of production and lifestyle. The changes in family structure and residents' needs as well as the unique economic form of land-lost peasants (i.e., rent economy)

Table 3.29 Table of the public service facilities optimization in resettlement areas for land-lost peasants

Facility type	Optimization measures		Facility
Education	They can be configured according to the standards of the current residential facilities		Kindergarten and elementary school
Medical and healthcare	They can be configured according to the standard of the current residential facilities		Community health service centers, community health service stations, health centers, clinics, etc
Cultural and recreational	Cultural	Reduce the current configuration indicators for cultural facilities such as the library and integrate them into composite cultural facilities	Libraries, cultural centers, etc
	Recreational	Improve the current sports facility configuration standards and expand the types of recreational facility configurations in a targeted manner	Sports fields, ping pong tables, chess rooms, gyms, etc.
Commercial service	Lifestyle	Improve the current configuration standards and enrich the variety of lifestyle service facilities, with a special need to increase the area and service functions of supermarkets	Farmer's market, 24-h supermarket, barber store, laundry, bathing center, photo studio, electrical maintenance department, etc.
	Restaurants	On the basis of the current supporting standards, open new low-cost restaurants that meet the characteristics of land-lost peasants	24-h restaurants, take-out restaurants, casual restaurants, pastry stores, beverage stores, grocery stores, etc
	Financial	On the basis of the current supporting standards, expand self-service financial facilities	Banks, savings offices, credit unions, etc.
Postal and municipal	Lower post office facility configuration standards and set up bus stops rationally		Supporting facilities such as post office, bus stop, and shades

(continued)

Table 3.29 (continued)

Facility type	Optimization measures	Facility
Community service	They can be configured according to the standard of the current residential facilities	Community service centers, nursing homes, property management centers, etc

Source Own elaboration

should be taken into account in unit type design. For example, a large unit can be divided into two small units that can be used independently, namely, the smaller unit of the two can be rented out to earn profit. By transforming the uneconomic living space to a means of production, two birds can be killed by one stone as this approach can avoid space waste and realize economic benefits at the same time. In short, the design of residential units should not only meet the current demand for housing but also maintain the flexibility of housing to adapt to diverse usage patterns (see Table 3.30).

3.5.2 *Strategies for Improving Social Integration of Land-Lost Peasants*

3.5.2.1 Existing Problems of Social Integration

In light of the research on social integration for land-lost peasants in Nanjing in the previous section, the main problems faced by different types of resettlement areas for land-lost peasants and their causes were sorted out and summarized to provide a guiding basis for the following strategy discussions (see Table 3.31).

The problems in social integration in the above five aspects and their influencing factors were further explored and summarized, and seven main causes were concluded. This study thus proposed four core improvement strategies to address the causes of the problems (see Fig. 3.56).

3.5.2.2 Improvement Strategies for Social Integration

According to the results of the previous analysis, different types of resettlement areas and land-lost peasants with different personal characteristics have different degrees of integration in each dimension. In response, the government can formulate countermeasures according to the current integration characteristics of different types of resettlement areas. Resettlement areas scattered in the main city should focus on integration in economic and institutional dimensions. Evenly distributed resettlement area in the industrial park should focus improving integration in economic, spatial, institutional, and psychological dimensions, while large mixed resettlement area at the fringe area of the city should focus more on integration in spatial, economic, and social dimensions.

Social integration is influenced by a combination of factors, including personal characteristics, economic, social, psychological, institutional, and spatial factors.

Table 3.30 Diversified types and characteristics of floor plans in resettlement areas for land-lost peasants

Type	Application scope	Case	Characteristics	Floor plan
"For rent" unit design	Extra space is available for rent	Hangzhou Sandun Lanyun Tiancheng residential community	Comprised two small units One is a large three-bedroom, two-bathroom suite in which the residents live The other one is a small one-bedroom, one-bathroom suite with a separate kitchen and bathroom, which can be rented out to the public	
"For multi-generation" unit design	Young couples and elderly people living together forming "multi-generation" families	Feoso Yayuan Unit	The elderly and their children live in the same housing unit. There are common zones for the entire family and independent zones for each small family, in addition to specific bathrooms and simple kitchen facilities in the elderly's living space. The two households have independent entrances and exits	

(continued)

Table 3.30 (continued)

Type	Application scope	Case	Characteristics	Floor plan
"For profit" unit design	Residents' own home used for business activities	"1 LDK" apartment in Japan	It is a new type of collective housing similar in terms of layout to the traditional houses that have a store in the front and a house in the back or a store on the bottom and an apartment on the top. In other words, a separate entry is set up in one of the rooms of the apartment as a direct entrance to either a study, gym, or even a store to meet the different needs of residents	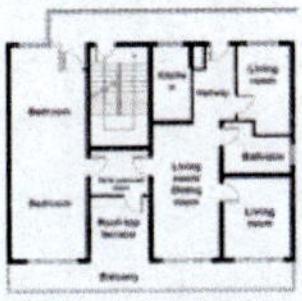

Source Ding (2009b)

Since spatial factors have been discussed in the spatial resettlement strategies of the previous section, this section does not deal with their improvement strategies. The mutual influence mechanisms between factors (dimensions) and social integration (goals) are complex (direct/indirect influences of factors on goals of different degrees and mutual influences between factors of different degrees). Therefore, comprehensive guidance measures need to be formulated in line with the diversity and complexity of the social integration process, and the joint efforts of local governments, communities, society, and land-lost peasants are required. Therefore, based on the principle of implementation and operability, this study discusses the promotion of progressive and multidimensional integration of land-lost peasants by combining the special needs and demands of this group with discussions on policy guidance, community work, social acceptance, and subject integration (see Fig. 3.57).

Policy guidance: Reform the existing land expropriation system

Different problems within the existing land expropriation system such as low compensation standards, inadequate resettlement of peasants, and irregular land expropriation procedures often violate greatly and deeply the land rights and interests of land-lost peasants whose housings were built on collective land. Therefore, how to protect and guarantee the land rights and interests of land-lost peasants by

Table 3.31 Brief summary of problems in social integration for land-lost peasants in Nanjing

Location of Resettlement Area				Main city	Industrial park	Fringe area of the city	
Name of resettlement area				Mufu Mountain Villa	Yaolin Xianju	Tianjingshan Community	Xishan Huayuan
Common problems				Differentiated problems			
Social integration	Economic dimension	Income and consumption level	Low income level and single consumption pattern	–	–	–	Comparatively speaking, income level is the lowest
		Work intensity	Work intensity is generally high	–	–	–	–
		Work stability	Poor socialization with community	–	–	–	–
	Social dimension	Socialization with neighbors	Lack of positive attitude toward neighborhood socialization	–	–	Socialization is relatively poor	Comparatively speaking, socialization is the poorest
		Socialization with community	Poor socialization with community	–	–	Socialization is relatively poor	Comparatively speaking, socialization is the poorest
	Psychological dimension	Sense of community belonging	Low sense of community belonging and self-identification disorder	The identification disorder was most severe in the resettlement area built earlier		–	–
		Urban social identification	Low social identification	–	–	–	–

(continued)

Table 3.31 (continued)

Location of Resettlement Area				Main city	Industrial park	Fringe area of the city	
Name of resettlement area				Mufu Mountain Villa	Yaolin Xianju	Tianjingshan Community	Xishan Huayuan
Common problems				Differentiated problems			
	Institutional dimension	Social security status	Low participation and incomplete type of social security coverage (medical and pension insurance)	–	–	–	–
		House provident fund	Participation is generally low	–	–	–	–
	Spatial dimension	Living	–	–	–	Residential segregation, remotely located, poorly equipped facilities	
		Commuting	–	–	The layout of transportation facilities is not complete		

Source Own elaboration

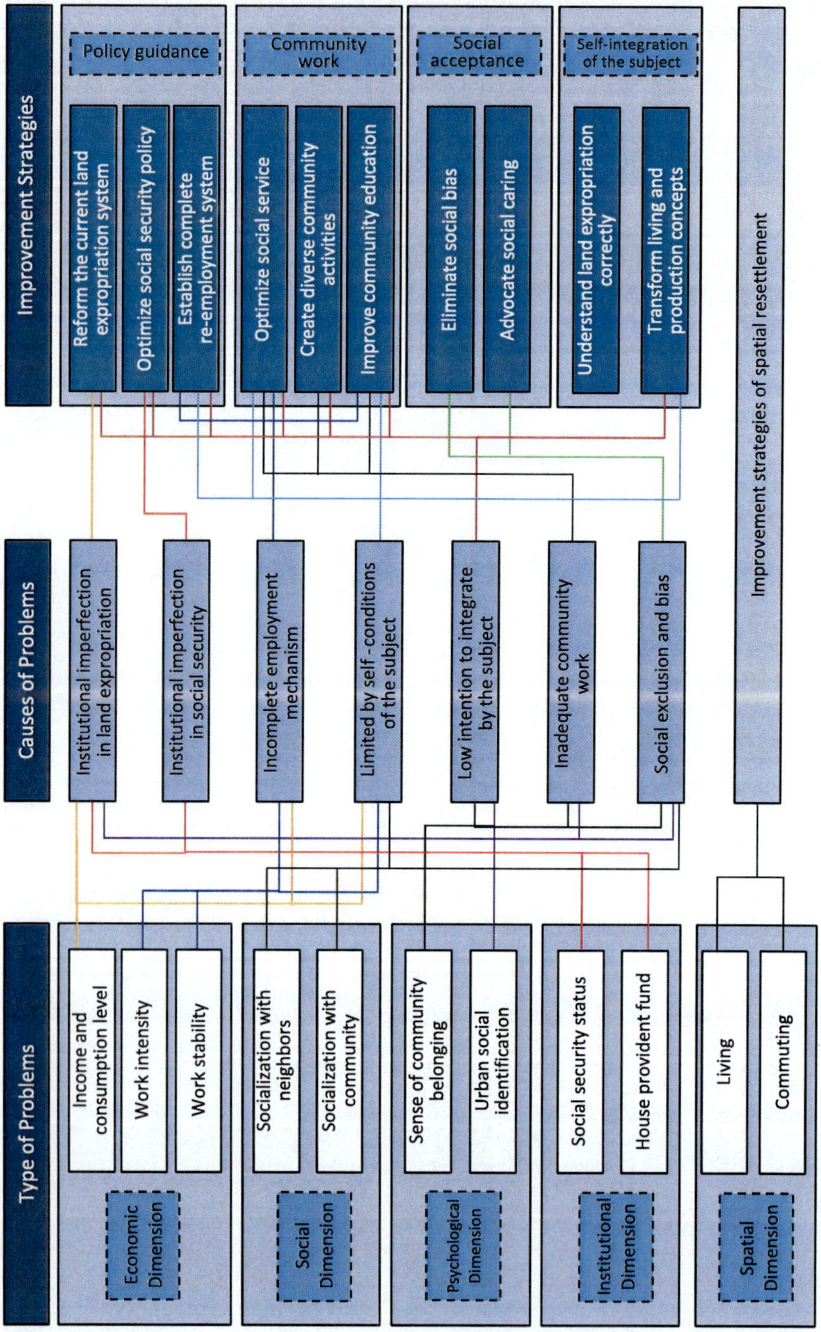

Fig. 3.56 Illustration of Strategies to Improve the Social Integration of Land-lost peasants. *Source* Own elaboration

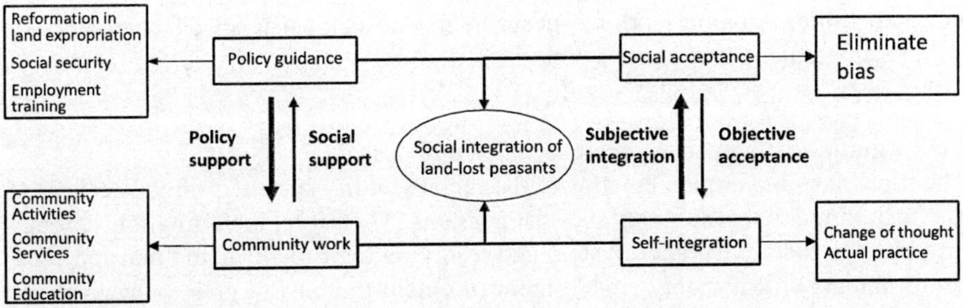

Fig. 3.57 Schematic diagram of strategies to improve the social integration of land-lost peasants. *Source* Own elaboration

reforming the land expropriation system and establishing a high compensation and high security expropriation system are of key significance to promote the institutional and economic integration of land-lost peasants.

Establish reasonable compensation standards Although the current zone pricing compensation and resettlement model in Nanjing improved the security of resettled land-lost peasants to a certain extent, the zone pricing standard still lags far behind the level of social and economic development. Therefore, it is necessary to establish a scientific and reasonable asset evaluation system and supplementary standards and fully consider the possibility of land-lost peasants sharing the value-added income of land. The fundamental interests of land-lost peasants can be guaranteed only by improving compensation standards by referring to market prices.

Explore differentiated compensation methods First, draw on international experiences. For example, in France, Canada, and some other countries, in addition to monetary compensation, governments also provide compensatory compensation in lieu of land, while Japan adopts a combination of monetary compensation and compensation in kind (land development, residential land development, transfer agent, engineering compensation). Second, draw on local practices. One approach is to implement the land stock cooperative system, where companies can develop the expropriated land for operational purposes (such as building roads, railroads, and airports) and compensate peasants with shares based on the area of their land, meaning that they could earn dividends as per the company's profit. The other approach is to implement a long-term compensation system, which adopts a continuous compensation policy instead of a one-time compensation policy: companies are suggested to consider taking out part of the operating income after land expropriation as dividends, distributing the amount to land-lost peasants directly or transferring it to their social security accounts.

Establish a regulatory system Drawing on international experience, a special independent agency should be established to mediate and negotiate disputes between local governments, land acquisition companies, collective-owned companies, and land-lost peasants in the process of land expropriation to help resolve conflicts of interest in a fair and just manner. Such conflicts also include the large differences in compensation among land-lost peasants due to inconsistency of compensation policies implemented in various periods.

Policy guidance: Improve social security policies

The soundness and implementation of the social security system is directly related to the institutional integration of land-lost peasants. Therefore, including this community into the social insurance system and resolving their medical and pension security problems will not only enable them to obtain the basic rights to survival and development but also promote stable development of the society. The social security system for land-lost peasants in Nanjing has gone through a development process from monetary compensation system to a basic livelihood security system and then to a social security system for quasi-urban residents (Zhou et al. 2014). At present, the social security system for land-lost peasants in Nanjing should focus on pension and medical insurance as well as the minimum living allowance. Additionally, work-injury, maternity, and unemployment insurance should also be gradually incorporated into the system. The specific measures include the following (see Fig. 3.58).

(1) Establish a social security fund for land-lost peasants: This method is a common practice in the social security systems of various countries. In terms of fund sources, the compensation and resettlement fees for land expropriation and the incremental income gained from it can be used as part of the social security fund. Additionally, government financial allocation, operating income of the fund, and donations from social charity organization are also possible sources. Regarding fund operation, drawing on international experience, social security fund operators (banks or non-banking financial institutions) are set up separately from management institutions, with the former responsible for activities such as fund raising, investment, operation, etc., and the latter responsible for supervising the former and regulating the social security market.

(2) Complete the social security items for land-lost peasants: At this stage, the basic security items are the basic social insurance programs (pension, medical

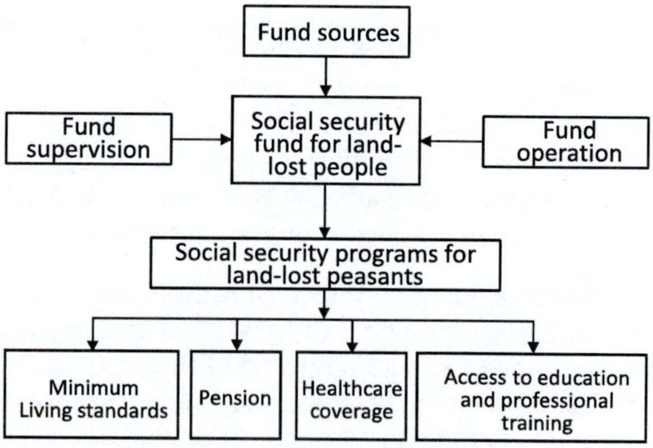

Fig. 3.58 General framework of social security system for land-lost peasants. *Source* Bao and Wu (2002)

care, unemployment, work injury, and maternity) and minimum living security system. Additionally, programs such as education and professional training should be improved to guarantee the right to development of land-lost peasants in addition to the basic right to survival, and specific differentiated protection standards can be proposed for peasants of different ages and genders.

Policy guidance: Establish a sound re-employment mechanism
Economic factors are the most important and basic factors affecting the social integration of land-lost peasants, and the level of income and work stability will largely affect economic and psychological integration and social integration of this community. Only by effectively improving the income level of land-lost peasants' families can their integration be truly promoted, and such a stable economic source must come from the full employment of this group of people. Since the employment situation of land-lost peasants before land expropriation varies greatly, a categorized guidance strategy is needed (see Fig. 3.59).

Develop employment channels and provide employment opportunities This strategy is applicable to land-lost peasants of working age who are only engaged in agricultural work before land expropriation. The strategy can be carried out through the vigorous development of two industries: manufacturing and service. In terms of manufacturing industry, select certain enterprises around the resettlement areas based on the self-quality and educational level of land-lost peasants and enable them to sign employment agreements with these residents. This will allow land-lost peasants to be transformed into industrial workers and they can, therefore, work in the

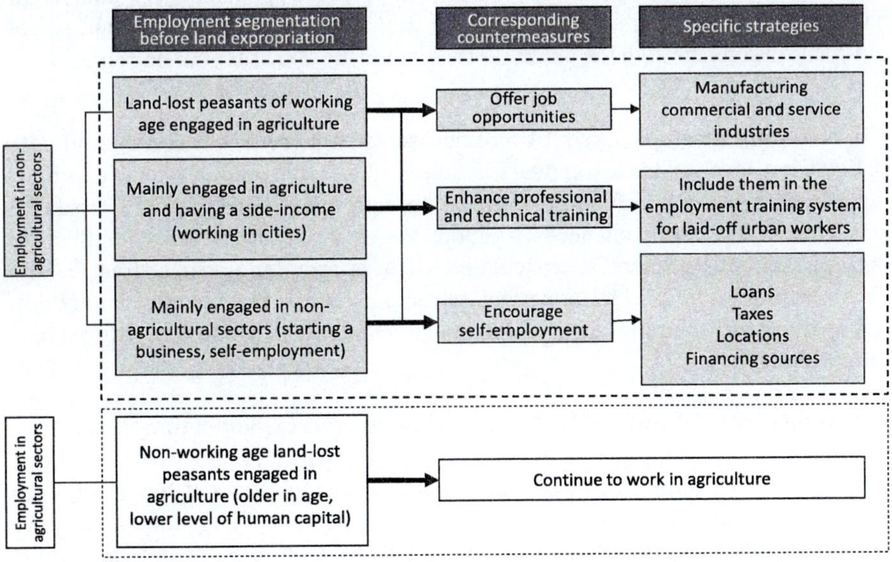

Fig. 3.59 Schematic diagram of re-employment mechanism for land-lost peasants. *Source* Own elaboration

vicinity after resettlement. In terms of service industry, it is one of the most important channels in providing employment opportunities for land-lost peasants. Experts estimate that China's service industry will absorb at least 450 million people by 2020 (Yan 2013). As long as they can be fully employed in the city, land-lost peasants will have the necessary economic basis to integrate into the city.

Strengthen professional training and increase employment opportunities This strategy applies to working-age land-lost peasants who were not only engaged in agricultural work before land expropriation but also have a side income from family business (working in the city). Government departments should establish a comprehensive employment training system for land-lost peasants or integrate them into the re-employment training system for laid-off urban workers. In addition, assign the planning of different training programs for land-lost peasants of different ages and education levels to district technical schools, vocational training centers, and adult education centers to improve the ability of land-lost peasants to change jobs and be re-employed.

Encourage land-lost peasants to start their own businesses and choose their own careers This strategy applies to working-age peasants who are engaged in non-agricultural sectors (who start their own business or are self-employed, etc.) before land expropriation. For land-lost peasants who start their own businesses, government departments should formulate active credit policies to provide them preferential policies in terms of application for individual management, loans, taxes, worksites, etc. Furthermore, it is recommended that the government introduce incentives and support measures for entrepreneurship funds for land-lost peasants, implement fully entrepreneurship support policies above the district level, spontaneously launch new channels for entrepreneurship financing, and stimulate entrepreneurial enthusiasm of land-lost peasants through the publicity and demonstration effects of entrepreneurial figures.

In addition, the employment of older peasants engaged in agriculture and with relatively low levels of human capital is hit the hardest in the process of land expropriation. Since it is highly difficult to transfer these groups to work in non-agricultural categories, the government needs to adopt targeted support policies for them to continue their engagement in agriculture, such as arranging special transfer and subcontracting of the land for this group or recommending them to relevant agricultural enterprises for employment, which can alleviate their employment difficulties to a certain extent.

Community work: Improve community services and exploit community potential

Community is the primary place for land-lost peasants to carry out social activities and plays an important role in guiding this group toward social integration. Community service refers to providing welfare and non-profit services for residents of the community through social organizations to meet the increasing material and cultural needs of residents. The specific approaches are as follows:

Improve the cultivation of community service workers Establish community social work service stations centered on community resident committees, bring in professional community work service experts, and set up professional service platforms for different groups according to their various needs. For example, establish a professional service team comprising full-time and part-time social workers for elderly people living at home, people with disabilities, and teenagers.

Establish community service institutions and organizations Communities should create day-care centers based on their resources for empty-nest elderly people and abandoned children and build fitness centers to address the needs of community residents in studying and exercising. At the same time, through effective consolidation of resources in the community, communities are encouraged to set up "benevolence supermarkets" and "40 & 50 y/o Party members' workshops," to create ways for peasants keen on community welfare to help each other and provide substantial services for peasants who have lost their land.

Create new funding sources for community service On one hand, attract social capital and rely on social organizations such as companies and charity organizations for financial support through investment in social assistance for land-lost peasants. On the other hand, charge user fees for some social services to generate income. Provide social services in response to the different needs of residents (e.g., day-care centers for the elderly and after-class tutoring centers for children) and charge user fees according to the actual household income levels of residents to ensure the financial functioning of service centers.

Community work: Enrich community activities and promote resident participation

Community cultural activities also significantly influence the social integration process of land-lost peasants. Promoting interactions and psychological adaptation of the group through a series of socializing community support and cultural activities can help enhance community cohesion and their sense of belonging.

The first approach is to carry out community cultural activities. The community can set up various interest groups and associations as well as cultural activities to enhance the psychological adaptation of land-lost peasants. For example, it is suggested that they hold "urban cognition contest," "urban culture competition," and other contests and organize regular activities for senior citizens' association or training and performances for dance associations. Such cultural and recreational activities will not only enrich the leisure life of land-lost peasants but also allow them to form a sense of community identification and belonging when participating in collective activities. The cultural ambiance of the community can also be enhanced in the process.

The second approach is to encourage land-lost peasants to participate in the self-governance of the community. Utilize actively the management system of the community in terms of organizational structure and guide land-lost peasants to participate in community activities and self-governance in terms of public affairs. On one hand, functional interdependence among residents can be formed through division of work within the community. On the other hand, through participation in community's self-governance, the potential of land-lost peasants can be fully utilized and their ability to participate in community management can be nurtured.

Community work: Strengthen community education and cultivate sense of ownership

Community education is a kind of non-formal education aimed at educating all community residents and meeting the learning needs of all community members. For land-lost peasants past their educational age, community education in urban communities—the new social organizations that are home to them—is an important means of promoting their social integration. The specific points are as follows:

Focus on vocational skills training The government shall invest funds and use educational resources such as colleges and vocational schools in the community to mobilize the enthusiasm of land-lost peasants in acquiring academic education or qualification certificates. The government can also provide vocational education, unemployment training, and citizenship education for peasants in a systematic manner to enhance their vocational quality.

Enhance cultural literacy education Land-lost peasants who have just entered the city do not have enough knowledge of urban socialization; thus, they often find it difficult to consciously adjust their own psychological status and make correct behavioral choices. Therefore, community departments can help land-lost peasants build their awareness of social subjects including legal consciousness, time consciousness, and contractual spirit to adapt to the urban society via various forms of community education. Some examples include radio programs, bulletin boards, community newspapers, in-person behavioral education, free lectures in community schools, psychological interviews, and selection of civilized household. At the same time, the community can also carry out services such as "advocating new civilization trends for residents to share a better life," which may include formulating a "Residents' Covenant" in areas where land-lost peasants live to regulate such bad habits and uncivilized behaviors as destroying greenery to plant vegetables and piling up personal belongings randomly in the community.

Social acceptance

The social integration of land-lost peasants is not only about the willingness of land-lost peasants to integrate into the city on their own initiative but also about the objective acceptance of this group by other citizens in the urban society. Only by eliminating inherent social prejudices and promoting social care can we truly achieve bilateral social integration.

Eliminate social bias The closure between urban and rural areas caused by the dual structure in China is the primary cause of the gap between peasants and urban citizens. The discrimination and misunderstanding of citizens toward peasants largely originate from the negative reports of mass media on this group. Therefore, mass media plays a decisive role in guiding public opinion in this regard. Media such as radio, TV, and newspapers are called upon to establish the concept of fairness and use the same criteria in evaluating citizens and peasants. At the same time, they should increase the coverage of peasants' contributions to the city to guide citizens to objectively evaluate peasants and eliminate social prejudices, thus gradually accepting land-lost peasants from a psychological viewpoint.

Advocate social care Land-lost peasants have made great sacrifices for the development of urbanization in China, and their production and living conditions in cities deserve the care, support, and assistance of all the sectors of society. The government should fully consider and address the needs of land-lost peasants when designing the institutional system, and enterprises should give priority to solving their employment problems when recruiting talent. At the same time, urban citizens should be more tolerant and open-minded in accepting and interacting with land-lost peasants.

Self-integration

The social integration studied in this paper mainly refers to the active integration of land-lost peasants into the city. The government's land expropriation has a huge impact on the traditional ideology of land-lost peasants as a group, who should face the reality, actively change their ideology, and truly enhance the awareness of active integration from the psychological viewpoint.

Land-lost peasants should correctly understand the government's act of land expropriation Only by correctly understanding the urbanization process can they be expected to actively accept the drastic changes in lifestyle brought about by land expropriation. Only by changing their mindset and consciousness can they be expected to transform their integration from passive to proactive, thus becoming invincible in the wave of urbanization.

Land-lost peasants should change their traditional concepts in terms of lifestyle and production In terms of lifestyle habits, they should change their casual, free-spirited, self-centered way of life and actively adapt to the urban lifestyle. In terms of production, they should also break the "passive, dependent, and demanding" mentality and avoid waiting passively for employment opportunities, using up all their savings but not working, and being choosy about their jobs and actively seek job opportunities. At the same time, they should abandon the ideology they have as traditional smallholder peasants, when they are content with small achievements and get self-complacent easily, and instead develop their own entrepreneurial awareness, market awareness, and risk awareness.

3.6 Summary

This chapter is a special discussion on the theme of "integration-response" in the field of "land-lost peasants—land-expropriation cities." First, it defines the concepts of land-lost peasants, spatial resettlement, and social integration; explores the background of land-lost peasants and related policies; and outlines the spatial resettlement and social integration realities of this group. Next, through the selection and research of three typical cases of resettlement areas for land-lost peasants in Nanjing—a scattered resettlement area in the main city (Mufu Mountain Villa), an evenly distributed resettlement area in the industrial park (Yaolin Xianju), and two large mixed resettlement areas at the fringe of the city (Tianjingshan Community and Xishan Huayuan), the current characteristics of resettlement areas for land-lost peasants and their causes are analyzed in terms of social and spatial attributes. Next, the characteristics and mechanisms of social integration of land-lost peasants in Nanjing are quantitatively analyzed and systematically studied by using analytic hierarchy process and structural equation modeling. Finally, the study systematically and purposefully proposes comprehensive improvement strategies for the existing problems in spatial resettlement and social integration.

The main conclusions of this chapter are as follows:

(1) In terms of object definition, "land-lost peasants" refer to passively urbanized peasants who have lost part or all of their land due to state expropriation and are located in the center of non-rural areas (i.e., households whose land was expropriated and were resettled collectively). Resettlement areas for land-lost peasants can be divided into three categories according to their location and distribution characteristics: scattered resettlement areas in the center of the city, evenly distributed resettlement areas in industrial parks, and large mixed resettlement areas at the fringe of the city. To ensure the operability of the measurement, the concept of "social integration" is defined as the unidirectional process of land-lost peasants integrating into urban areas and transforming into urban residents, which includes integration in five dimensions: economic, social, psychological, institutional, and spatial dimensions.

(2) In terms of social attributes, land-lost peasants in resettlement areas present an "inverted V-shaped older age structure." They have in general a "low education level, mainly holding a junior high school diploma or having received less education," and their family structure has the characteristics of an "urban–rural transitional family structure where majority of the families comprise at least two generations." The income composition of the group has a "high proportion at both ends and low in the middle, with transfer income and income from property as the main sources" characteristic, and the residents have a "low employment rate, engaged in jobs of low technical content and low stability" occupational structure, whereas their consumption pattern shows a characteristic of "domination by food and daily necessities."

(3) In terms of spatial attributes, the overall surrounding environment of resettlement areas for land-lost peasants have "a long travel distance and mixed

use of land" characteristics, and buildings in resettlement areas are "different in scale and size and mostly grouped by rows and constructed by street" in terms of layout. The housings units of resettlement areas are "mainly small and medium-sized units" and have an "insufficient living space per capita" in general for residents. As for the work commuting cost for land-lost peasants, their "average commuting time is 40 min, the mode is mainly buses, and the cost is relatively low," whereas the work-home separation model presents a "differentiated commuting mode, where some commute toward the center (urban centers at all levels), some commute to nearby locations, and others commute to scattered locations." The overall urban public service facilities are "category-specific and non-homogeneous," while residential service facilities have the following characteristics: "education, commercial services, cultural and recreational, and community service facilities are different in terms of quality and quantity, post and telecommunications and municipal administration facilities are well equipped, and medical and healthcare facilities are insufficient."

(4) In terms of social integration characteristics, the overall social integration degree of resettlement areas in Nanjing is low and for the four areas of Yaolin Xianju, Xishan Huayuan, Tianjingshan Community, and Mufu Mountain Villa selected as samples. When analyzing by dimension, the degree of integration in each resettlement area differs considerably in each dimension, with the greatest differences seen in the psychological dimension, followed by the economic and spatial dimensions, and the least difference in the social and institutional dimensions.

(5) In terms of the mechanism of social integration, economic, social, psychological, institutional, and spatial factors have direct influence on social integration, while personal characteristics factors indirectly influence social integration through economic, social, and psychological factors. Additionally, there is mutual influence between economic, social, and psychological factors. In terms of the level of effect, psychological factors have the highest level of effect, followed by institutional (including "differences between resettlement areas") and spatial factors, while economic, social, and personal characteristics factors have less effect.

(6) In terms of spatial resettlement improvement strategies (macro-scope), first, it is necessary to consider integrating the planning of resettlement areas for land-lost peasants into the urban system, regulating the organization pattern of subsidized housing in the form of special planning at the level of overall planning, and proposing the layout and standards of land use in resettlement areas under guidance and control indicators at the level of zone planning and detailed planning. Second, it is necessary to optimize the location of resettlement areas. Within the main city, it is recommended to use part of the expropriated land, use land in a mixed way, renew partially old urban areas to construct resettlement areas, or simply designate certain commercial housings as resettlement areas. In industrial parks, part of the land or land in the vicinity can be used to construct resettlement areas. To construct resettlement areas at

the fringe of the city, it is proposed to consider adopting public transportation-oriented development models. In addition, it is also encouraged to construct mixed communities. The first strategy is to mix residential areas according to gradient classification, and it is recommended to mix resettlement housing with ordinary mid-scale commercial housings for middle-income residents to achieve residential integration through the shared use of public space and public service facilities. The second strategy is scattered mixed housing within residential areas based on a certain proportion, namely, building scattered subsidized housings within commercial housing area or building subsidized housings led by public rental housing.

(7) In terms of spatial resettlement improvement strategies (mid-scope), first, it is necessary to control the scale of resettlement areas. According to the organizational pattern of "residential community—grassroots community" in subsidized housing, the suggested scale is about 1500 households (about 5,000 people), which is roughly a land scale of about 10–20 ha. Second, it is recommended to improve the public transportation system in resettlement areas. Optimize the urban road system, strengthen the connection between the new and main city, improve the public transportation system, improve accessibility of stations, consider developing transit-centric communities, and build a transfer system that integrates traveling by foot, bicycles, and buses.

(8) In terms of spatial resettlement improvement strategies (micro-scope), there should be diversified allocation of service facilities and diversified configuration according to the characteristics of the facilities and their supporting standards: adopting a diversified supply approach based on the characteristics and attributes of the facilities and optimizing the configuration standards of residential facilities according to the diversified needs of residents. It is also necessary to provide diverse and flexible design of housing units. Taking the housing design experience of similar domestic and foreign groups as reference, it is proposed to adopt a housing design with reasonable function and suitable area for land-lost peasants in Nanjing to adapt to diversified usage patterns.

(9) In terms of improvement strategies for social integration (policy guidance), it is first necessary to reform the existing land expropriation system, establish reasonable compensation standards, explore differentiated compensation methods, establish a regulatory system, and establish a special independent agency to help resolve conflicts of interest between different stakeholders. Second, it is necessary to improve social security policies, establish a social security fund for land-lost peasants, and complete social security items for them. Furthermore, it is necessary to establish a sound re-employment mechanism. Based on the employment situation of land-lost peasants, develop employment channels to provide employment opportunities, strengthen professional training to increase employment opportunities, encourage land-lost peasants to start their own businesses, and guide them to choose their own careers.

(10) In terms of social integration improvement strategies (community work), it is (1) necessary to improve community services and explore community potential, (2) enrich community activities and promote residents' community participation, and(3) strengthen community education and cultivate the sense of ownership of land-lost peasants.

(11) In terms of social integration improvement strategies (social acceptance), first, we need to eliminate inherent social bias and call on mass media to guide citizens to evaluate peasants objectively and psychologically accept land-lost peasants. Second, it is proposed to advocate social care and call on all sectors of the society to care for, support, and help land-lost peasants improve their production and living conditions in the city.

(12) In terms of social integration improvement strategies (self-integration), first, land-lost peasants need to correctly understand the government's land expropriation actions and truly enhance their awareness of active integration from a psychological perspective. Second, they should change their traditional concepts in lifestyle and production and integrate into the urban society with practical actions.

References

Alba R, Nee V. Remaking the American mainstream: assimilation and contemporary immigration. Boston: Harvard University Press; 2005.

Bao H, Wu C. On the construction of social security system for land-lost farmers. Management World. 2002;18(10):38.

City of Chula Vista. Aita Planning and Design. Pedestrian Master Plan. Chula Vista: Cityworks. 2010.

Deng T. Active land-loss and passive land-loss of peasants. Rural Econ. 2004;22(11):30–2.

Ding M. Research on the residence space for land-lost peasants living in urban fringe area—taking Hangzhou city as example. Post-graduate Thesis. Xi'an: Xi'an University of Architecture and Technology; 2009a.

Ding M. Study on housing of land-lost Farmers in urban fringe. Xi'an: Xi'an University of Architecture Tech. 2009b:60–62.

Ga R, Huang K. Concept exploration of social integration in the west and its inspiration. Theoretical Horizon. 2008;16(1):20–25.

Gordon M. Assimilation in American life. Chicago: University of Chicago Press; 1964.

Guo D. Study on the planning and design of subsidized housing in the context of urbanization—with Nanjing as an example. Doctoral Thesis. Nanjing: Southeast University; 2010. p. 73

Guo D, Jin L. Study on the spatial layout characteristics and optimization strategies of Nanjing's Subsidized Housing. Mod Urban Res. 2011;26(3):87.

Guo D, Li J, Wang Z. Research on spatial distribution characteristics of indemnificatory housing in Nanjing and optimizing strategies. Modern Urban Res. 2011a;3:83–88. http://www.tuliu.com/read-24869.html

Guo D, Li J, Wang Z. Research on the spatial distribution characteristics and optimization strategy of affordable housing in Nanjing. Modern Urban Res. 2011b;26(3):83–88.

Hua F, Wang S. Public transport oriented planning and construction of new district—A case study of planning and construction of public transport community in Genbei new district, Hangzhou.

diversity and inclusion. Proceedings of 2012 China Urban Planning Annual Conference (05. Urban Road and Transportation Planning); 2012.
Hang L, Qian C. Analysis and suggestions on the construction of subsidized housing in Nanjing. Bus Econ. 2013;(13):21–24.
Huntington SP. Who are we? The challenges to America's national identity. Beijing: Xinhua Publishing House; 2005.
Jiang L. Urbanity and the urban adaptation of migrant workers. Sociol Res. 2003;18(9):92–6.
Jiang X, Zheng B. The inspiration of foreign and Hong Kong's subsidized housing on China. Urban Dev Stud. 2011;(3):21.
Jin L, Lu X. Agricultural land property rights system in the process of building a harmonious society. Macroecon Manage. 2007;12:51–54.
Jin J, Lin J, Yang C, Gao Z, Zhou Q. Research on effective Supply mechanism of public Service Facilities in urban residential areas: A Case study of Beijing. Urban Development Res. 2007;6:95–100.
Lang Y, Huang W, Wang Y. From social exclusion to social integration: localisation of migrant workers and the construction of harmonious communities within cities in China. Res Rep Polit Contemp China. 2007;6(5):115–42.
Lei H. Study on the interests of land-lost peasants in the urbanization process of Beijing suburbs. Doctoral Thesis. Beijing: China Agricultural University; 2005.
Li Z. Research on the planning strategy of Beijing's subsidized housing based on mixed housing model. Master's Thesis. Beijing University of Civil Engineering and Architecture, Beijing. 2015.
Long W. A study of mixed housing based on the perspective of residents in subsidized resettlement areas. Master's Thesis. Nanjing University, Nanjing. 2013.
Lu C, Pang, P. A study on the mechanism and countermeasure of residential segregation: large social housing districts in France and their enlightenment to the affordable housing construction in China. Urban Plann. 2013;37(6):52–56.
Lu S, Wei W. Structural equation modeling of social integration of urban immigrant population–a survey based on the pearl river delta region. Population Econ. 2011;32(05):17–23.
Ma X, Tong X. Social Integration of urban new Immigrants: A Case study of Y Community in Shanghai. University; 2008;2:15–22.
Mao D. Empowerment, interaction and identity: suburban farmers' urbanization under the perspective of role theory. Sociol Res. 2009;24(4):28–60+243.
Ni J. Study on the expropriation of urban and rural fringe area and land-lost peasants in Beijing—Taking Yizhuang development area as example. Post-graduate Thesis. Beijing: China Agricultural University; 2004.
Park, R, Burgess E. Introduction to the science of sociology. Chicago: University of Chicago Press; 1924.
Qiu H, Lin B. Principles and applications of structural equation modeling. Beijing: China Light Industry Press; 2009. p. 69–89.
Qiu H. Principles and applications of structural equation modeling. Beijing: China Light Industry Press; 2009.
Ren Y, Wu M. Social integration of urban migrants: literature review. Population Rev. 2006;30(3):87–94.
Su, Q. Social security of land-lost peasants in the process of urbanization: the case of Beisi Taizi Village in Huanggu District, Shenyang. Post-graduate Thesis. Guilin: Guangxi Normal University; 2005.
Tian K. Survey analysis and reflections on the urban adaptability of migrant workers. Sociol Res. 1995;10(5):90–5.
Wang F. A case study on rights and interests protection of farmers lost farmland and sustainable development of agriculture. Post-graduate Thesis. Beijing: China Agricultural University; 2005.
Wang M. Discrepancy and mean-variance deviation decision methods for determining weights in comprehensive multi-indicator evaluation. China Soft Sci. 1999;14(8):100–101+107

Wang T. Architectural pattern language—town—building—structure. Beijing: China Architecture & Building Press; 2007. p. 43.

Wang H. A study on land-lost peasants and policy exclusion issues in the process of urbanization. Agricult Econ. 2008;3:3–6. http://news.163.com/13/0314/06/8PTJOB8F00014AED.html

Wang C, Li M, Sun Q. Analysis on the planning of resettlement area for land-lost farmers in the process of urbanization—Taking Nanjing City as an example. Huazhong Archit. 2009;5:114–122.

Wen Z, Hau, Kit T, Herbert WM. Structural equation model testing: cutoff criteria for goodness of fit indices and Chi-Square test. Acta Psychologica Sinica. 2004;49(2):186–94.

Wu B. Preliminary study on structural equation model. Tianjin: Tianjin University; 2006. p. 9.

Wu M. Structural equation modeling: operation and application of AMOS. Chongqing: Chongqing University Publishing House; 2009.

Xin S, Chen Y, Zhang C. A review of research results on the application of structural equation modeling theory. J Indus Technol Econ. 2014;34(5):61–71.

Yang J, Qin Q, Sun LQ. Active land-lost peasants: an analysis from the perspective of social stratification. J Anhui Agricult Sci. 2009;37(8):3788–791.

Yi D. Tructural equation modeling methods and applications. Beijing: China Renmin University Press; 2008.

Yan B. Human urbanization and the elimination of urban adaptation barriers of land-lost peasants: the case of Jiangning District, Nanjing. People's Tribune. 2013;26:141.

Ye J. The concentrated residence of the landless farmers in the city suburbs and the cultural adaptation of migrants. The Ideol Front. 2010;36(2):61–65.

Yuan Q, Ma X. Supply of public service facilities in Guaranteed residential areas: A case study of Guangzhou. City Planning Rev. 2012;36(2):24–30.

Yue Z, Li S. Concept construction and empirical analysis of social integration for rural-urban migrants in China. Mod Econ Sci. 2012;34(1):1–11.

Zhang Y, Wang F. Studies on the employment of land-lost peasants. Xinjiang State Farms Econ. 2006;(12):1–5.

Zhang W, Lei K. A structural model of social identity of new urban immigrants. Sociol Res. 2009;24(4):61–87+243–44.

Zheng S, Zhang Y. Spatial location of subsidized housing: theoretical basis, international experience and reality in China. Mod Urban Res. 2010;9:20.

Zhang C. Research on the design of the resettlement for land-lost peasants living in suburbs in the context of urbanization. Post-graduate Thesis. Nanjing: Nanjing University of Science and Technology; 2013.

Zhao X, Li Z. Local identification and integration of new immigrants in China's big cities. Urban Plann. 2013;37(12):22–29.

Zhou J, Zhang K. The basic framework of optimizing the planning structure of urban residential districts. Urban Plann. 1999;43(6):63.

Zhou J, Jiang D, et al. The scale of land use and planning design of residential areas. Urban Plann. 1999;23(1):39.

Zhou HAO. Measurement and theoretical perspectives of immigrant assimilation in China. Popul Rev. 2012;36(3):27–37.

Zhou J, Yao P, Xu L, Chen Z. Land acquisition system reform in the context of urban-rural integration: Nanjing's compensation and resettlement policy reform and its performance. Mod Urban Res. 2014;8:25–30.

Zhu L. On the urban adaptability of the peasant-worker strata. Jianghai Acad J. 2002;8:82–88.

Chapter 4
Community Sample III Wave of Modernization and "Pangen Grass": Research on Changes in the Huis Settlement

Abstract Nowadays, with rapid development in science and technology, evolutionary changes in social structure, and continuous enhancement in people's living standards, the wave of modernization has changed the landscape of the world and impacted and challenged the traditional culture. Under this wave, although ethnic minority groups have made certain socio-economic achievements after the reform and opening up, they still have to face the grim reality of low economic aggregates, weak economic foundation, and increasing overall gap compared with the Han areas. Additionally, the specific characteristics of the ethnic minority, as one of the important components of traditional culture, are inevitably facing a new existential crisis. The traditional settlement space, which is closely related to this unique traditional culture, has also been affected by modernization to different degrees; hence, realistic problems such as disappearance of characteristic spatial patterns, difficulties in work–home commute, inadequate public facilities, decline in ethnic service, and difficulty in carrying forward traditional customs. As a typical representative of the ethnic minority group, the Huis community has not only "Jiaofang" (religious community) as an indispensable element in their traditional life but also intricate relationships and intertwined ties within the community, which is why people often use the term "Pangen Grass" (a type of plant with twisted roots and easy to grow) to describe the special structure within the community. However, under the continuous impact of modernization, their social structure and connections have changed differently, and the evolution of their social and spatial attributes has also shown a highly characteristic dynamic trajectory. To this end, this chapter focuses on the specific socio-cultural background and distinct ethnic identity of this ethnic group by taking the typical Qijiawan Huis Settlement in Nanjing as the research object to profoundly analyze and examine the spatial changes of ethnic minority settlements in China, reveal the existing problems in the evolution of Huis Settlement, and propose a series of renewal strategies and improvement recommendations around its social space.The selection and examination of this community sample is essentially a special exploration of the theme "integration-response" in the field of "ethnic minority—modernization."

X. Wu and L. Wang, *Settlement Spaces: Urban Survival Prospects of China's Special Communities*, Springer Geography,
https://doi.org/10.1007/978-981-16-4892-2_4

Keywords Ethnic minority settlement · Spatial attributes evolution · Social attributes evolution · Qijiawan Huis settlement · Renewal strategy

4.1 Definition of Research Object

4.1.1 Definition of Modernization

4.1.1.1 About Modernization

Due to the emergence of problems in the process of late modernization, Chinese scholars explicitly proposed the concept and theory of "modernization" as early as in the 1920s and 1930s to explore the challenges and impacts that external cultures bring to local traditional cultures. The rise of the concept and theory of "modernization" in Western theoretical circles in the first modern countries dates back to the golden period when developed Western countries witnessed sustained economic growth after World War II (1950s), and a large number of Western scholars studied the wave of industrialization-guided modernization that affected the world.

During more than half a century of urban development, scholars from different disciplines have made different interpretations of "modernization" from fields such as economics, sociology, political science and urban planning, which can be broadly divided into three types of perspectives:

The theory of civilizational transformation argues that modernization refers to the phased and revolutionary transformation of civilization forms in the history of human civilization. The theory of social transformation believes that modernization refers to the omni-bearing transformation of human society from traditional to modern in the fields of material skills, social systems, nation-state, organizational behavior, values, psychological attitudes, and lifestyles. The transformation of resource utilization defines modernization according to whether there is a major irreversible change in the way people use resources and their quantitative relation.

4.1.1.2 Modernization Studied in This Paper

The modernization studied in this chapter is a dynamic and processual concept with two dimensions, namely, established facts and future plans: the study on the background of modernization in the paper is based on the history of modernization, namely, the study of the current socio-economic, political, cultural, and psychological development of society as a whole. It is also a process that can be characterized using relevant indicators and divided into stages.

4.1.2 *Definition of Minority Settlements*

4.1.2.1 About Minority Settlements

In the global consensus, a "nation" is often a group that has stable territorial boundaries, who is given the symbolic meaning of the state, and has relatively clear political boundaries. It is what Anderson defined as "a political community that is imagined as inherently limited and sovereign". The term "ethnic minorities" refers to ethnic groups other than the main ethnic group (the most populous) in a multi-ethnic state. In the basic structure of China's Pluralist Unity, ethnic minorities refer to the 55 ethnic groups in addition to the Han Chinese identified and were recognized at the founding of the People's Republic of China (The PRC), including the Mongolian, Hui, Manchu, Tibetan, and Qiang people.

From the macroscopical level, the spatial distribution of ethnic minorities in China is characterized by "different ethnic groups living in a mixed way over vast areas, with some living in concentrated communities in small areas." Of them, "concentrated communities in small areas" refer to the ethnic groups living in autonomous regions, autonomous prefectures, and autonomous counties governed by local authorities, whereas "living in a mixed way over vast areas" refers to three scenarios, i.e., scattered settlements in urban area, rural areas, and ethnic townships.

From the microscopical level of the city, minority settlements belong to "urban scattered settlements" from the macroscopical perspective, which is a space formed by spatial units such as neighborhoods and communities in terms of spatial form, and people of the same ethnic culture live together at these places.

4.1.2.2 Minority Settlements Studied in This Paper

Combining the macro and micro meanings of minority settlements, this study considers minority settlements as spaces where minority populations of the same ethnic live together. They share common religious beliefs and values and are connected by similar lifestyles, customs, and habits.

The "minority settlements" studied in this chapter are the relationship-related settlements formed due to the spontaneous clustering of ethnic minorities in cities with predominantly Han Chinese populations, based on kinship and geographical and occupational ties. The focus of this study is the Nanjing Qijiawan Huis settlement (including the Huis populations that formerly lived and currently live here), which has a sizable population yet still distinct ethnicity. This area also has homogeneous, universal, and spontaneous settlement characteristics.

4.1.3 *Definition of Community Transformation*

4.1.3.1 About Community Transformation

Referring to sociologists' understanding of social transformations and the four models of social transformation summarized by Mr. Rongqu Luo (2013), community transformation refers to the partial or integrated changes in community structure due to change of time or related factors during the process of dynamic community development. This transformation is characterized by structural transformation, including hierarchical structure, occupational structure, organizational structure, values, etc. The impetus of community transformation can come from the external environment and the integration and differentiation within the community. Its basic forms can be classified as gradual micro-change, sudden micro-change, innovative macro-change, and conductive macro-change.

4.1.3.2 Community Transformation Studied in This Paper

The "community transformation" studied in this chapter mainly covers the evolution of social and spatial attributes of minority settlements and tries to fit and reflect the distinctive spatial characteristics and socio-cultural background of this particular ethnic group.

Social attributes contain the demographic composition, family relations, social organization, and lifestyle and customs of ethnic minorities, while spatial attributes contain community functions, residential space–time, employment space–time, and supporting facilities. The focus is on their evolution patterns and motivation mechanisms in the context of modernization, and the study period is mainly since the founding of the PRC in 1949.

4.2 Evolutionary Background of Huis Settlements

4.2.1 *Nanjing's Multi-Ethnic Profile Represented by the Hui*

Since ancient times, Nanjing has been an important city within the cultural economics and political development among regions south to the Yangtze River. Many literati, scholars, and people of different ethnic and cultural backgrounds have settled here, thus making Nanjing an important city along the southeast coast where a large population of multiple ethnic minority groups gather together (see Figs. 4.1 and 4.2).

The sixth population census shows that within the 11 districts of the city, the residential population is 8,004,680 people with 54 ethnic minorities (apart from Deang people) with a population of 99,100 people, of which there are more than 70,000 Hui

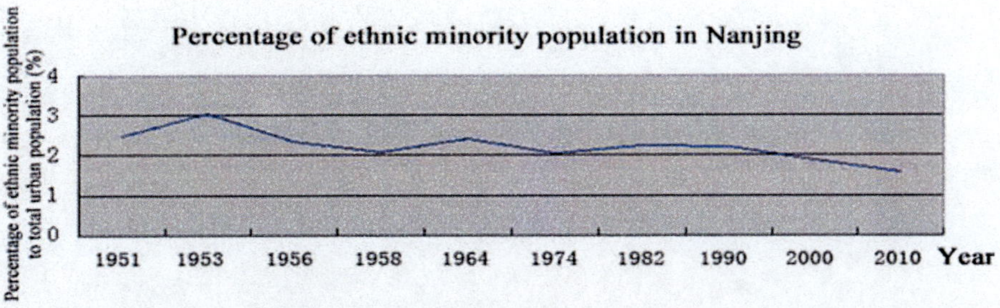

Fig. 4.1 Evolutionary chart of ethnic minority population in Nanjing. *Source* Own elaboration according to statistics of past population census

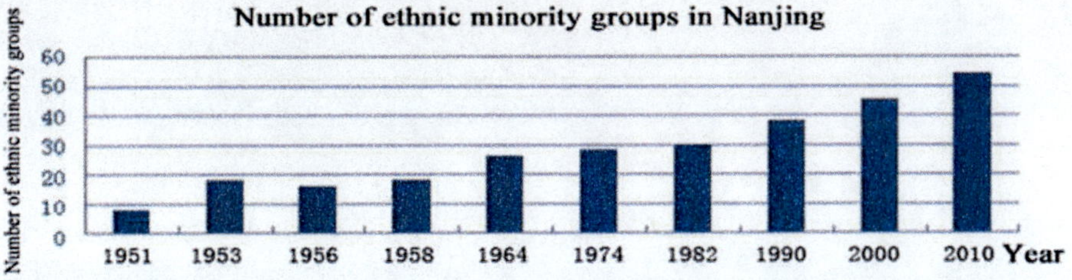

Fig. 4.2 Evolutionary chart of the number of ethnic minority groups in Nanjing. *Source* Own elaboration according to statistics of past population census

people, accounting for more than 70% of the total ethnic minorities. Additionally, there are 10 ethnic minority groups with a population larger than 1000, in addition to Hui. They are Manchu, Mongolia, Miao, Zhuang, Korean, Tujia, Uyghur, Buyi, and Yi people.

To outline the overall distribution of minority populations in Nanjing, three ethnic minorities with the most population, Hui (76.23%), Manchu (6.73%), and Mongolian (2.23%), are selected for comparative spatial analysis. The main urban area of Nanjing was studied, and the scope was fine-tuned at the "sub-district" level of the administrative division involving 44 sub-districts in 8 districts (Fig. 4.3). The population data were obtained from the sixth Nanjing population census (2010), and the land use data were obtained from the current land use map of Nanjing, collected by Arcgis.

4.2.1.1 Population Density

Population density is the most commonly used indicator of population distribution, which indicates the number of people per unit area of territory. The formula for measuring population density is.

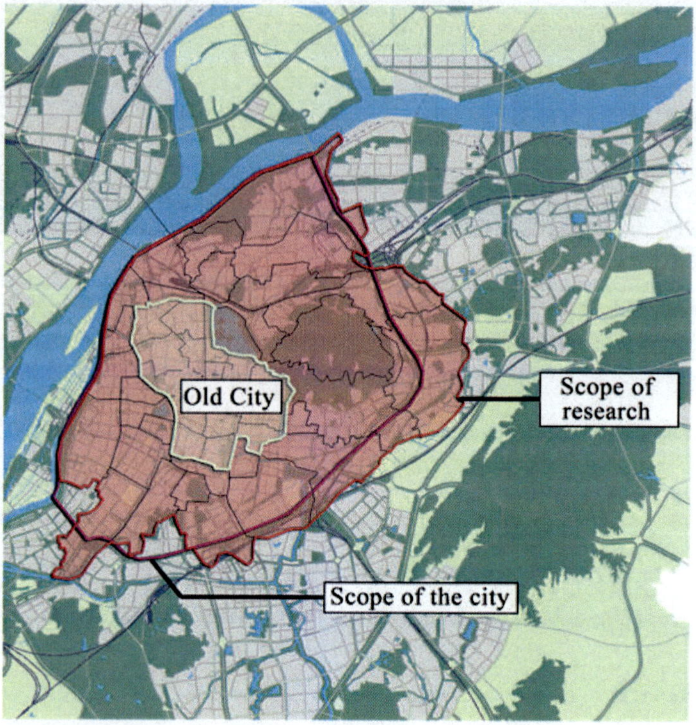

Fig. 4.3 Scope of research. *Source* Own elaboration

$$Di = \frac{\mathrm{P}i}{\mathrm{A}i}.$$

The specific research method is as follows: calculate the population density of ethnic minorities in each statistical unit, use the SPSS statistical software to cluster the 44 statistical units into five categories, and draw the population density distribution of the overall ethnic minority population and that of each community in Nanjing.

It can be seen that the density distribution of the overall and each ethnic minority in Nanjing has the following characteristics (see Table 4.1).

4.2.1.2 Geographical Ratio of Population

Geographical ratio is the proportion of the population in a part of the territory to the total population in the whole territory, and this indicator reflects the proportion of the ethnic minority population in each statistical unit to the ethnic minority population in the whole statistical unit. The formula for measuring the geographical ratio of the population is

Table 4.1 Population density distribution of ethnic minorities

Type	Population density distribution of ethnic minorities	Characteristics of population density distribution of ethnic minorities	
Overview of ethnic minorities	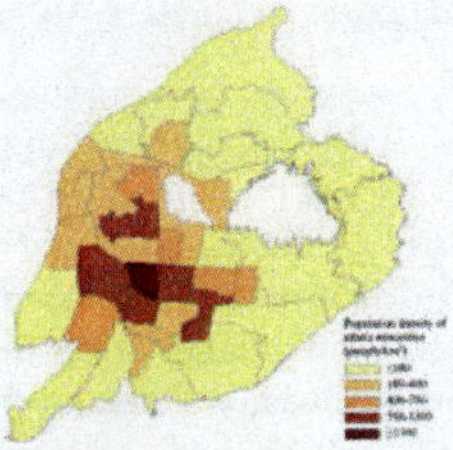	Distribution pattern	The first-level streets with the highest overall ethnic minority population density are Chaotiangong residential district in the southern part of the old city, and the fifth-level streets are distributed in a circular pattern along the southern, northern, and eastern fringes of the main city Overall in the city, the density is high in the west, low in the east, high in the south, low in the north, high in the inner city and low in outer areas, forming a circle-like trend
		Overall structure	Single core + circle-like
Hui people	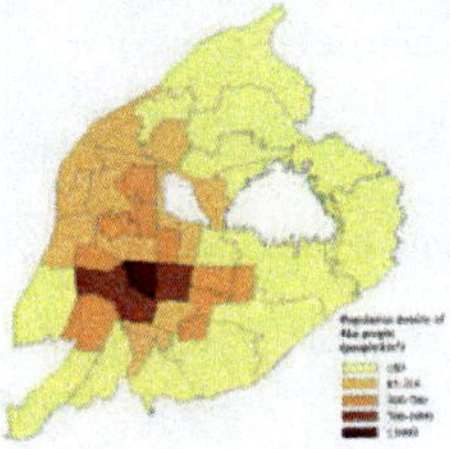	Distribution pattern	The first-level streets with the highest overall Huis population density are Chaotiangong residential district in the southern part of the old city, and the fifth-level streets are distributed in a circular pattern along the southern, northern, and eastern fringes of the main city Overall in the city, the density decreases successively from the southern part of the old city to the east-western and southern fringes of the city, forming a fan shape
		Overall structure	Single core + fan shape
Manchu people	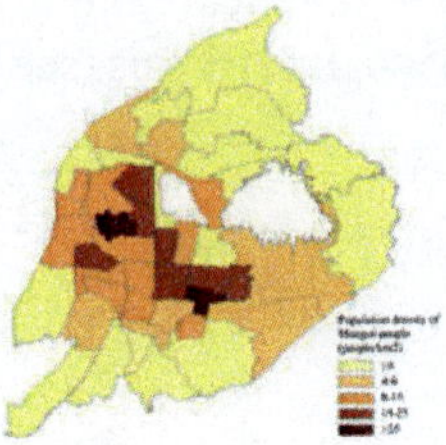	Distribution Pattern	The first-level streets with the highest overall Manchu population density are Ninghailu in the north of the main city and Daguanglu residential districts in the east of the main city, and the fifth-level streets are distributed by pieces on the north and south edges of the city Overall in the city, the second- and third-level streets in terms of densities are clustered and extended along the northwest to southeast direction, forming dual cores distributed in clusters

(continued)

Table 4.1 (continued)

Type	Population density distribution of ethnic minorities	Characteristics of population density distribution of ethnic minorities	
		Overall structure	Dual cores + clusters
Mongol people	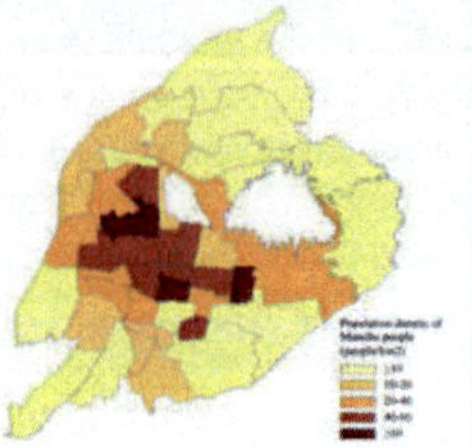	Distribution pattern	The first-level streets with the highest Mongolian population density distribution consist of four sub-districts, Ninghai Road and Hunan Road residential districts in the north, Hongwu Road residential district in the south, and Yueyahu residential districts in the east. The fifth-level streets are distributed in pieces on the north and south fringes of the city Overall in the city, the spatial distribution has a multi-point center with higher density areas and their contiguous areas expanding outward, forming a fan shape
		Overall structure	Multi cores + fan shape

Source Own elaboration according to statistics of the 6th population census

$$G_i = \frac{p_i}{\sum pi} \times 100\% .$$

The specific research method is as follows: calculate the geographical ratio of ethnic minority populations in each statistical unit, use the SPSS statistical software, Hierarchical Cluster, and Ward's method to group the 44 statistical units into five categories, draw the geographical ratio maps of the overall ethnic minority population and that of each community. It can be seen that the geographical ratio of the overall and each ethnic minority in Nanjing has the following characteristics (see Table 4.2).

4.2.1.3 Spatial Cluster Zoning of Ethnic Minority Population in Nanjing

The measurement of spatial clustering of the ethnic minority population is the necessary basis for analyzing the spatial clustering characteristics of ethnic minority population distribution.

In describing the spatial clustering of ethnic minority populations, population density and geographical ratios are used as two basic indicators, and the combination of "density index" and "ratio index" is used to quantitatively express the clustering of minority populations. The evaluation method is as follows.

Table 4.2 Geographical ratio of ethnic minorities

Type	Geographical ratio of ethnic minorities	Characteristics of geographical ratio of ethnic minorities	
Overview of ethnic minorities	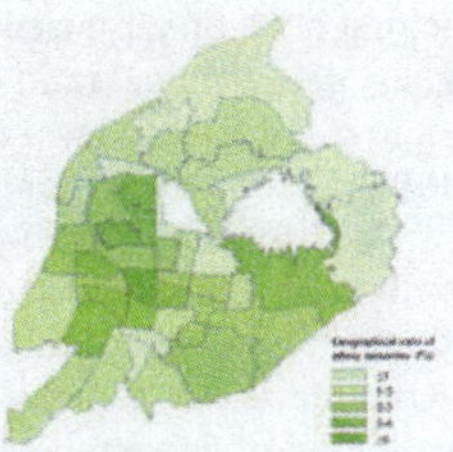	Distribution pattern	The first level of sub-districts with the highest overall geographical ratio of ethnic minorities is Chaotiangong residential district in the southern part of the old city, and the fifth level sub-districts are mainly distributed at the northern fringe of the city Overall in the city, the spatial distribution of the sub-districts in terms of geographical ratio is expanded in clusters
		Overall structure	Single core + clusters
Hui people	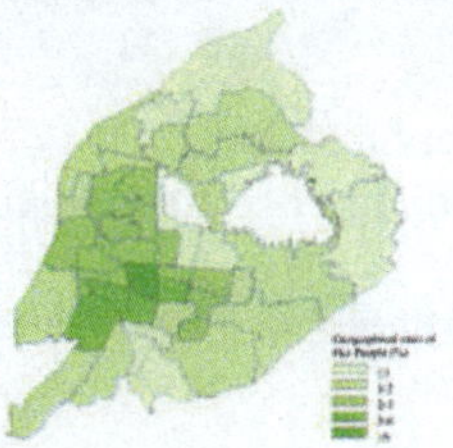	Distribution pattern	The first level of sub-districts with the highest geographical ratio of Huis people is Chaotiangong residential district in the southern part of the old city, and the fifth level sub-districts are mainly distributed at the southern and northern fringes of the city Overall in the city, the ratio decrease successively from the southern part of the old city to the fringe of the city, forming a fan shape
		Overall structure	Single core + fan shape
Manchu people	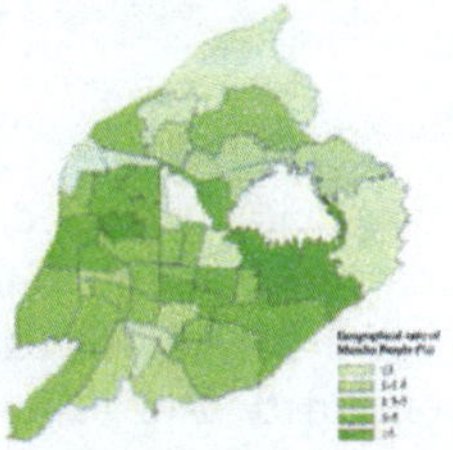	Distribution pattern	The first level of sub-districts with the highest overall geographical ratio of Manchu people is Ninghailu and Xiaolingwei residential districts in the northern part of the old city and the eastern fringe area, the fifth level sub-districts are distributed by pieces on the northern fringes of the city Overall in the city, the geographical ratio forms different clusters and double cores
		Overall structure	Double core + clusters

(continued)

Table 4.2 (continued)

Type	Geographical ratio of ethnic minorities	Characteristics of geographical ratio of ethnic minorities	
Mongol people	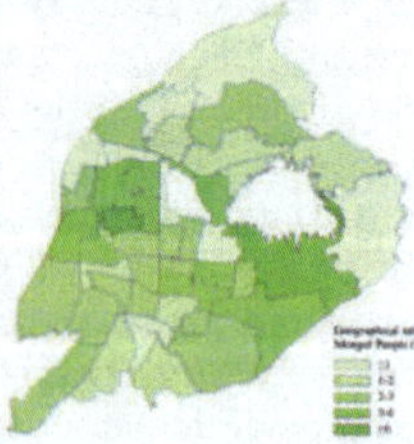	Distribution pattern	The first level of sub-districts with the highest geographical ratio of Mongol people is the Xiaolingwei residential districts in the eastern fringe of the city. The fifth level of sub-districts is distributed mainly in the southwest and northeast edges of the main city Overall in the city, the spatial distribution of geographical ratio forms relatively independent clusters
		Overall structure	Clusters

Source Own elaboration according to statistics of the 6th population census

【**STEP 1**】 The density index (Id) and geographical index (Ig) of the ethnic minority population are obtained by nondimensionalizing the density (D) and geographical ratio (G) of a statistical unit.

The density index (Idi) of statistical unit i is calculated as follows:

$$Idi = \frac{Di}{Avg(Di)(i = 1, n)}$$

The ratio index (Igi) of statistical unit i is calculated as follows:

$$Igi = \frac{Gi}{Avg(Gi)(i = 1, n)}$$

【**STEP 2**】 Construct a coordinate system with the standardized indices, where X axis is labeled density index and Y axis is labeled ratio index of the ethnic minority population respectively, and a slope of -1 is drawn with the intersection point of the two as the reference point. On this basis, the combination of "density index" *Id* and "ratio index" *Ig* can be used to classify the employment clustering space of migrant workers into high, medium (including two categories), and low levels (Fig. 4.4). Therefore, the types of population clustering areas for ethnic minorities can be divided as follows (see Table 4.3)

【**STEP 3**】 According to the measurement and classification of the ethnic minorities' population distribution in terms of spatial clusters, all statistical units were projected to the area, thus generating the graded distribution map of the overall ethnic minorities and that of each community. This study then calculated the number and proportion of each type of cluster and analyzed the characteristics of its spatial structure.

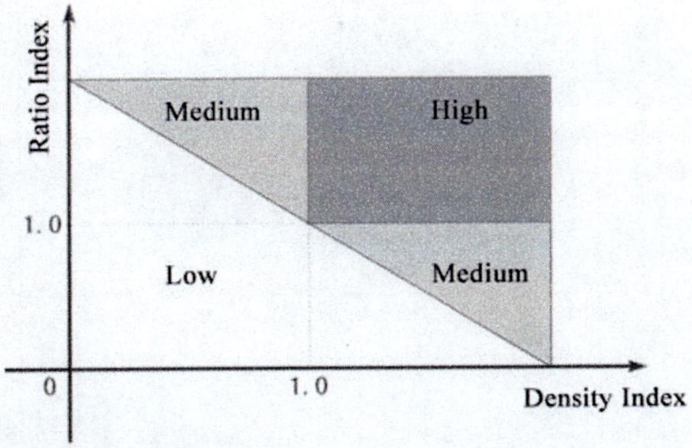

Fig. 4.4 Spatial cluster zoning of ethnic minority. *Source* Own elaboration

Table 4.3 Classification of ethnic minority population clusters

Classification	Clustering index characteristics	Characteristics
High population clustering area	Density index *Idi* ≥ 1; Ratio index Igi ≥ 1	High density, high ratio
Mid population clustering area (Type I)	Density index Idi ≥ 1; Ratio index Igi < 1; and Idi + Igi ≥ 2	High density, low ratio
Mid population clustering area (Type II)	Density index Idi < 1; Ratio index Igi ≥ 1; and Idi + Igi ≥ 2	Low density, high ratio
Low population clustering area	Idi + Igi < 2	Low density, low ratio

According to the above method, this study drew the distribution map of population clusters by ethnic groups and calculated the number and proportion of each type of cluster. It can be seen that the overall and individual population concentration zones of the ethnic minorities in Nanjing have the following characteristics (see Table 4.4):

Comparing and analyzing the spatial cluster areas of the overall and individual ethnic minorities in Nanjing, we list their commonalities and differences in Table 4.5.

4.2.1.4 Factors Influencing Spatial Distribution of Ethnic minority Populations in Nanjing

Historical spatial patterns of ethnic minorities

As a central city in Jiangnan region with a long history, Nanjing has been home to a large number of ethnic minorities for many years due to its superior natural conditions, convenient transportation facilities, intermingled ideology and culture, developed socio-economic conditions, and related political factors. Historical spatial information about Hui, Manchu, and Mongol people is provided in Table 4.6.

As can be seen from the table, the main historical settlement areas of the Hui people and the current high cluster areas of the population coincide well, indicating

Table 4.4 Commonalities and differences in spatial clusters of ethnic minority populations (Overall and by Ethnicity)

Type	Classification map of ethnic minorities' population cluster areas	Topology graph of ethnic minorities' population cluster areas	Proportion of different types of cluster areas
Overview of ethnic minorities	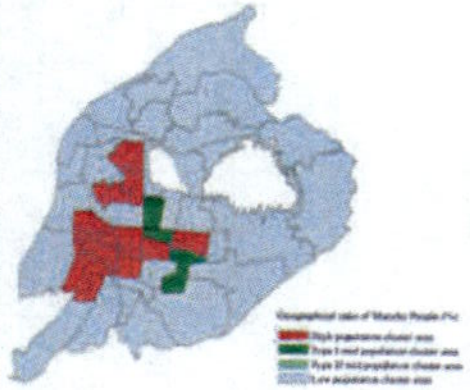	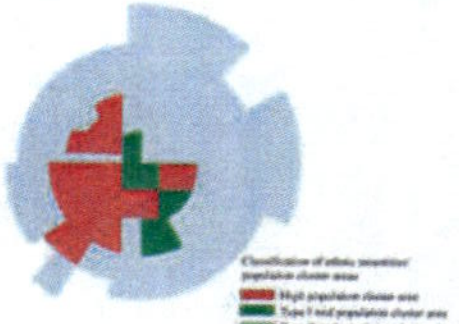	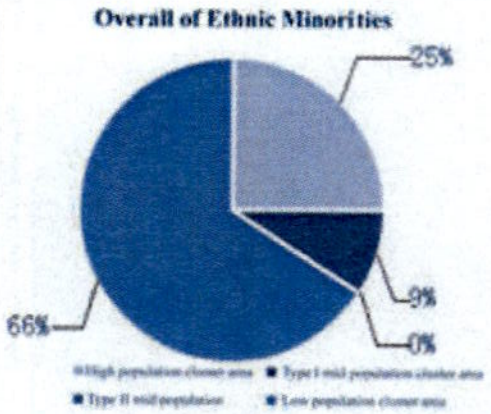
Hui people	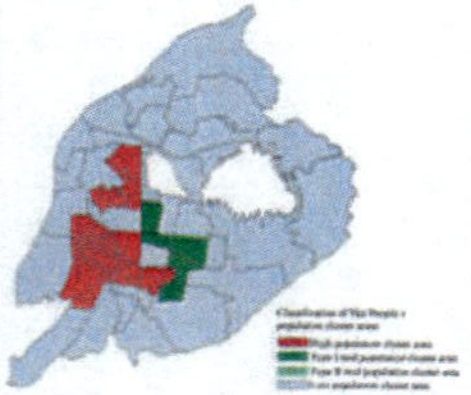	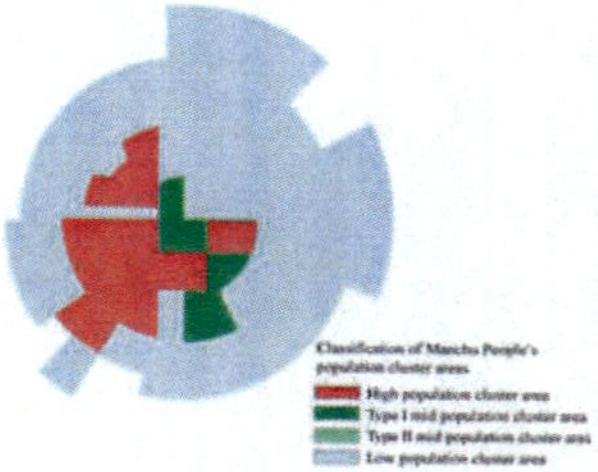	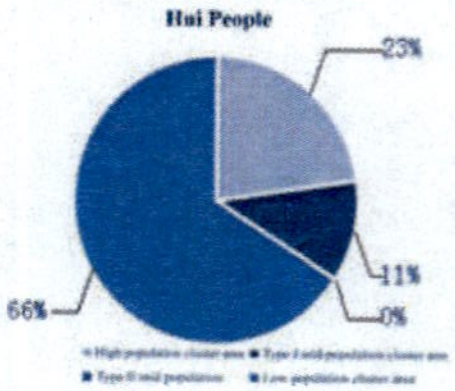
Manchu people	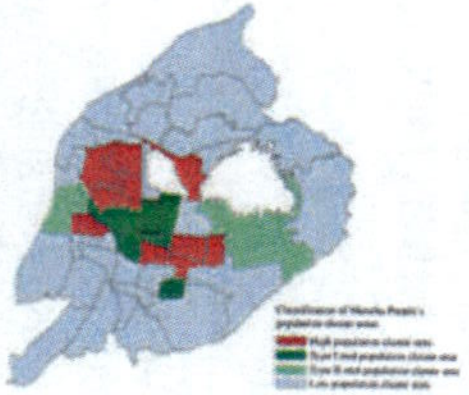	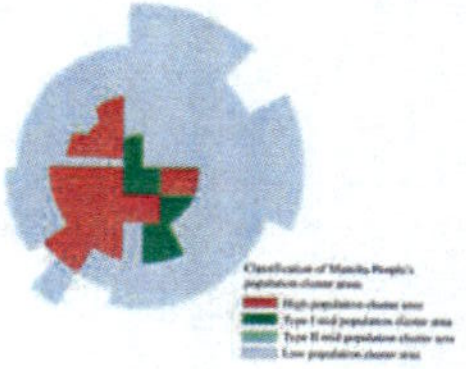	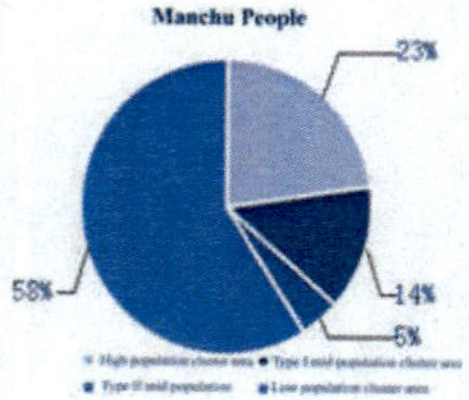

(continued)

Table 4.4 (continued)

Type	Classification map of ethnic minorities' population cluster areas	Topology graph of ethnic minorities' population cluster areas	Proportion of different types of cluster areas
Mongol people	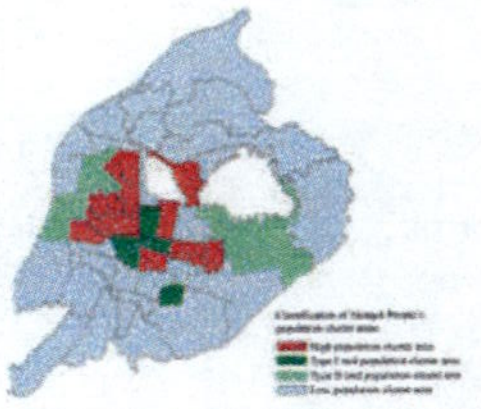	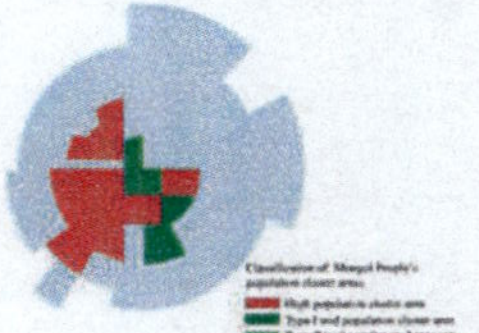	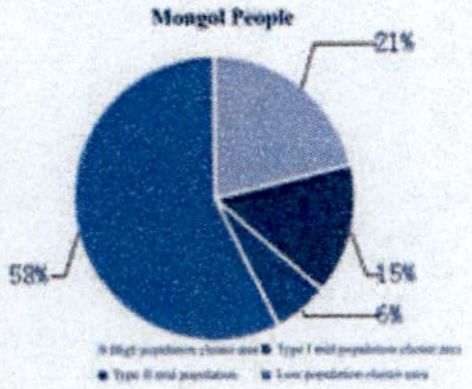
Legend	High population cluster area; Type I mid population cluster area; Type II mid population; Low population cluster area		

Source Own elaboration

Table 4.5 Commonalities and differences in spatial clusters of ethnic minority populations (Overall and by Ethnicity)

		Overall	Hui people	Manchu people	Mongol people
Commonalities	Number	The number and proportion of clusters are similar			
	Distribution	The distribution of clusters is almost isomorphic		High population cluster areas are mostly located at the northwest and south-east of the city Mid population cluster areas are mostly located in the middle of the old city and are connected to the high population cluster areas Low population cluster areas are mostly located at Jiangdong and Xiaolingwei sub-districts in the east and west of the city	
Differences	Distribution	The high population cluster areas are distributed on the west and east sides of the main city	The high population cluster areas are distributed on the west of the main city	The high population cluster areas are mainly distributed on the northwest and east-south of the main city	The high population cluster areas are mainly distributed at the northwest of the main city
		"Double-slice + fan-shaped radiation" structure	"Double-slice + fan-shaped radiation" structure	The "ring-shaped" structure	"Scattered radiation" structure

Source Own elaboration

Table 4.6 Historical spatial distribution of ethnic minorities in Nanjing

Ethnicity	Hui	Manchu	Mongol
Origin of ethnicity	At the end of the Tang Dynasty and the beginning of the Yuan Dynasty, the Islamic culture of the Western regions was introduced and fused with the Han culture	The formation of the ethnic community was driven by political factors at the end of the Ming Dynasty	Tribes engaged in hunting and nomadic activities along Reka Argun' at the beginning of Tang Dynasty
Age of settlement	Beginning of Yuan Dynasty	Beginning of Qing Dynasty	Beginning of Yuan Dynasty
Demographic composition	Local Muslims and Muslim merchants who came to the south from the northwest to engage in trades. "*Nowadays, the Hui people generally consider the Central Plains as home, especially in regions south of the Yangtze River.*" ①	Qing government's army was stationed there	Yuan government's army was stationed there. "*The imperial order stipulates that in Jiangnanzhou, the officials shall all be Mongol and Hui people*" ②
Occupational composition	Merchants	Officers and Soldiers	Officials
Spatial distribution graph	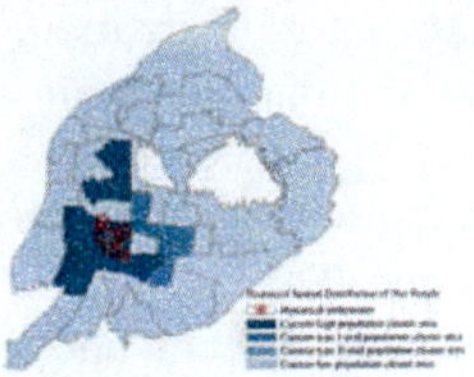	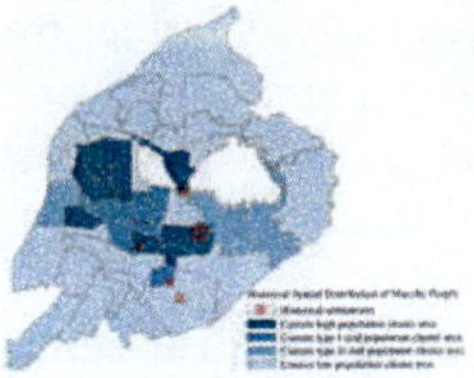	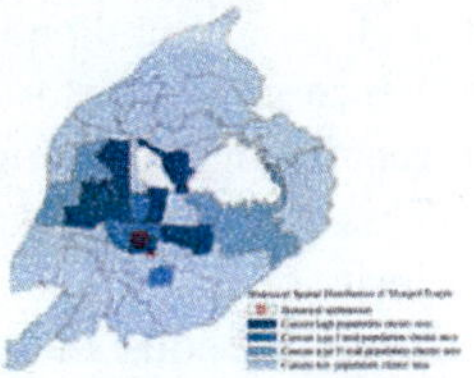
Historical location selection	The Hui population in Nanjing was at its peak in the Ming Dynasty With Tongji Gate and Dazhong Bridge as the center, it was mainly distributed in the area of Qingliang Gate, Shicheng Gate, Sanshan Gate, and Jubao Gate in a fan-shape, and the population was relatively concentrated in the area of Fuqiao where Jingjue Temple was located ③	The Manchu population in Nanjing was at its peak in the Qing Dynasty. They were mainly located in the Forbidden City of the Ming Dynasty and on the fringes of the urban areas where military defense was required④	The Mongol population in Nanjing was at its peak during the Yuan Dynasty According to the *Lushisi* (Institutions of urban civil administration) at that time, 86% of Nanjing's population included Mongol people located roughly in the center of the city and south-east of the city⑤

(continued)

Table 4.6 (continued)

Ethnicity	Hui	Manchu	Mongol
Current status comparison	The distribution of the Huis settlements in the history is similar to that of the current situation, showing a trend where multiple points form into a surface	Manchu people's settlements were scattered in a circular pattern, including the largest settlement around the Ming Forbidden City, and the settlements distribution in general is similar to the current situation in terms of geographical location	Mongol people's settlements in the history were concentrated in the core of the city and differed greatly from the current population's settlement areas

Source Own elaboration based on relevant historical materials
**Note* ①(Early Yuan) Zhou Mi. *Xinzashi*. Volume 1. Hui People's Moraine. ②(Yuan) An administrative rule promulgated in the third year of Dade (1299), see the "History of Yuan". ③④⑤ Nanjing Local History Editorial Committee. *Nanjing Ethnic and Religious Records* [M], Nanjing: Nanjing Publishing House, 2009:12–18

that the historically established pattern has a high impact on the distribution of the Huis population. The main historical settlement areas of the Manchu people and the cluster area around Yueya Lake sub-districts are similar in terms of geographical location, indicating that the historically established pattern also has some influence on the distribution of the Manchu population. However, the historical settlement area of the Mongol people is differs further from the current situation, the continual influence of the historical pattern is the weakest, and the clustering degree of the current Mongol settlements is also the weakest.

This difference is mainly related to the settlement motivation of each ethnic group. Majority of Hui people in Nanjing are businesspeople. Therefore, their settlement is more endogenous and spontaneous. The settlements of Manchu and Mongol populations are mainly driven by external forces (e.g., political and military factors), and their overall population distribution has undergone major changes passively along with the change of historical dynasties, evolution in military patterns, and policy concepts (e.g., many Manchu populations were politically discriminated against in the early years of the Republic of China and hid their ethnic identity when marrying Han Chinese (Nanjing Local History Compilation Committee 2009). **Minority cultural embeddedness.**

The study analyzed the innate differences in the social characteristics of the ethnic minorities in Nanjing and their evolution over the course of history, which is roughly shown in Table 4.7.

An analysis of the most typical and easily identifiable key social characteristics of the ethnic minorities in the table shows that the ethno-cultural identity of the urban Hui people is better maintained and sustained compared with the other ethnic groups, while those of the Manchu and Mongol ethnic groups are nearly extinct. This difference is fundamentally constrained by the superstructure from two perspectives.

Table 4.7 Social characteristics of ethnic minorities in Nanjing (▲ indicates that the characteristic exists, △ indicates that the characteristic has declined or disappeared completely)

Ethnicity	Religion	Religious architecture	Language	Lifestyle	Social networks	Customs and traditions	Costumes
Hui People	Islam▲	Mosque▲	Chinese, Arabic ▲	Live around a mosque △	Mosque Board of Directors▲ Industry Unions△	Mercantile focused ▲	Taqiyah or Hijab▲
Manchu people	Shamanism △	Temple△	Manchu language△	Family gathering△	Eight Banners System △	Fishing, hunting, riding and archery △	Qizhuang △
Mongol people	Shamanism △	Shrine △	Mongol language△	Live scatteredly△	Clan League and Banner regime △	Horse herding and hunting △	Long robe and boots △

Source Own elaboration based on relevant sources

Religious beliefs: The formation of the Hui ethnicity is closely related to the Islamic religious belief. Under the influence of the Islamic doctrine, the lifestyle of this ethnic community is subject to strict regulations. Additionally, when getting married, people from other ethnicities are converted to Islam, which also dissolves the potential impact of intermarriage on ethnic assimilation. The main religious belief of Manchu people is shamanism. However, as the relationship between Manchu people and other ethnicities such as Han people, Mongol people, and Tibetan people became deeper, the late shamanism was integrated with elements of Buddhism and Han culture. As for Mongol people, since the establishment of its unified regime in the Yuan Dynasty, the original belief in shamanism "*Tengri*" retreated to a secondary position, and the rulers of the Yuan Dynasty started accepting the Han culture and converting to Buddhism.

Social network: The mosque board unifies the management of the Huis community and allies with industry unions to form industry management institutions. The two organizations support each other and cooperate. Although Huis industry unions were abolished after the founding of the country, the mosque board of directors can still maintain the social network structure of the Huis community through religious activities and other means.

The militia and civilians of the Manchu people were unified under the Eight Banner administration, which is a form of social organization that integrates the functions of administration, military attack and defense, and the organization of production, and the "banner land" is delineated as the living area of the Manchu people. As the history of the Qing dynasty states: "When people leave the banner land, they serve as soldiers; when they come back, they serve as farmers; neither can be neglected (Qing)." Although the Qing government strived to maintain the traditional habits of ethnicity, when the "banner land" settlement was broken, the ethnic identity of the Manchu people began to decline. By the Yongzheng period, many Manchu officials "were not even able to speak the Manchu language (See (Qing))." Since the Yuan Dynasty, the Mongols have been living with the Han Chinese for a long time and have been continuously influenced by the advanced Han culture. "When the Northerners first arrived, we were still customary to shoot and hunt, but later we gradually learned to cultivate and sow like the Han (Yuan)." While gradually eliminating the economic and cultural differences with the Han, the increasing number of intermarriages also enabled them to further merge biologically with the Han, and thus, they became "not different from the Han (Ming)."

It can be seen that, in addition to the influence of historical distribution patterns, the unique social characteristics of each ethnic group constitute one of the important factors affecting the differences in the clustering of each ethnic group nowadays. The Hui's religious beliefs, social networks, lifestyles, and other national characteristics have formed a complete chain affecting the survival and development of the Hui ethnic group and have generated their mutually supportive ethnic identity, ensuring that the Hui's ethnic identity is relatively intact until now.

On the other hand, the settlements of Manchu and Mongol were built along with the establishment of ethnic regimes, and their ethnic identities have gradually dissolved due to the influence of reasons such as changes in the historical regime, collapse of social networks, and alternation of policies. This dissolution of ethnic identity becomes, in return, a great obstacle to their ethnic identification and clustering life.

Configuration of Public Service Facilities for Ethnic Minorities

To encourage the daily activities of various ethnic minorities and protect and develop the traditional culture of ethnic minorities, government at all levels in Nanjing have built and reconstructed several corresponding special service facilities as the basic space carriers for providing services to ethnic minorities and maintaining national solidarity. Of them, the distribution of major public service facilities for ethnic minorities at the municipal level is shown in Table 4.8.

The public service facilities for ethnic minorities in Nanjing can be divided into six categories covering all aspects of life, including education, culture, medical care, elderly care and funeral and interment service, etc. In terms of spatial distribution, they are mainly scattered in the southern part of the old city. In terms of service subject, as many as 67% of the municipal public service facilities for ethnic minorities are for the Huis population. In addition, there are 24 major halal food sites throughout Nanjing, all of which show that compared with the other two ethnic groups, the Huis population has a stronger demand and desire to maintain specific public service facilities. The consumption spaces for the Hui people, which are dominated by Islamic ideology, have become important places for Hui Muslims to realize their self-worth and identity in the current city (Xiaofei et al. 2014). Moreover, the close connection between the spatial layout of specific public service facilities and that of Huis Settlement further strengthens the spatial clustering of its population. Therefore, it can be said that the spatial configuration of specific public services in the city is another reason why the clustering of Hui people is much higher than that of Manchus and Mongols in Nanjing.

In summary, compared with Manchu and Mongol ethnic minorities, the Huis population has strong uniqueness and consistency in terms of historical continuity of spatial distribution, ethnic cultural roots, and spatial correlation of related services and can be used as a typical representative and sample of the "ethnic minority settlement" to study the dynamics of community evolution.

4.2.2 Modernization Context for Evolution of Huis Settlement

World modernization has spanned more than 250 years since the British Industrial Revolution started the process of global modernization in the 1760s,[1] and the study

[1] There are three main viewpoints on the starting point of world modernization. The author, referring to most of the current academic views, chooses the British Industrial Revolution of the 1760s, which had a profound impact on various aspects of society, such as productivity, political institutions and

Table 4.8 Distribution of public service facilities for major ethnic minorities in Nanjing

Facility type	Name of facility	Location	Key service objects
Education	Hui's Kindergarten in Jianyi District	Binhu sub-district	Hui people
	Ethnic Primary School in Chaotiangong	Chaotiangong sub-district	Hui people
Cultural activities	Nanjing Cultural Activity Center for Ethnic Minorities	Xinjiekou sub-district	All ethnic minorities
Medical	Hui's Hospital in Jianyi District	Chaotiangong sub-district	Hui people
Senior care facilities	Yahe Senior Residence for Ethnic Minorities	Honghua sub-district	All ethnic minorities
Funeral facilities	Nanjing Funeral Service for Hui People	Ningnan sub-district	Hui people
Others	Seven communities including the Pingshi Street community in Chaotiangong sub-districts, Qijiawan community, and others have established the current few ethnic minority service brands such as Home for Ethnic Minorities and Station for Ethnic Minorities	City-wide	All ethnic minorities

(continued)

Table 4.8 (continued)

Facility type	Name of facility	Location	Key service objects
Map of public service facilities for major ethnic minorities	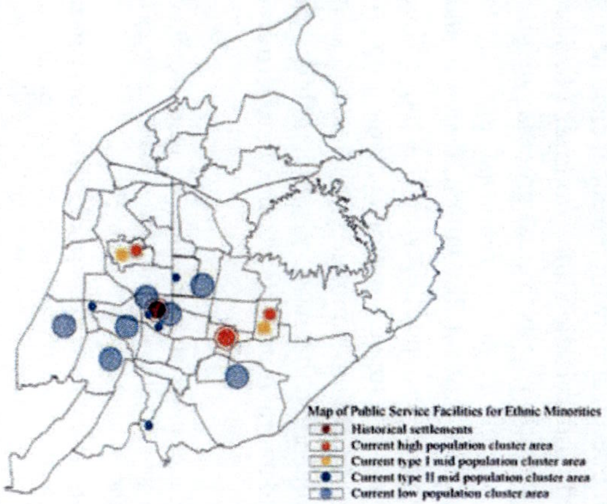		

Source Own elaboration

of world modernization has also had a long history of 60 years since the 1950s.[2] However, the broad meaning of modernization has made the subject an interdisciplinary and complex issue. Scholars from different fields of study and different perspectives also have different focus in their research. Therefore, topics such as modernization and how to measure the degree of modernization of a country or region have not had uniform definitions.

4.2.2.1 Definition and Characteristics of Modernization

After a comparative study of Eastern and Western religions, German sociologist Max Weber summarized the origins of modernization development in the West as a rationalist pursuit of science and technology under the influence of religion. Additionally, the transition from traditional to modern society is perceived a process of rationalization (Webber 2010). The American scholar Bendix regards modernization as a type of social transformation that exists in the economic and political progress of several pioneering societies and the process of social transformation in late-developed societies (Zapf and Lu 2000). According to Princeton Professor Black, modernization refers to the process of social transformation that has occurred and is occurring in economic, political, ideological, and other aspects under the influence of the scientific and technological revolution (Black 1992).

Luo Rongqu, a pioneer scholar of modernization research in China, synthesized various Western theories and proposed a "single core and multiple lines" view of development from the perspective of history (Luo 2013), indicating that each country should have a different path of modernization with the enhancement of productivity as the core. Mr. Luo also divided modernization into two levels. In a broader sense, he considers modernization as a worldwide historical process, which refers to the rapid transformation of human society from the traditional agricultural society to modern industrial society since the Industrial Revolution, which is also a transformation of psychological attitudes, values, and lifestyles. In a narrow sense, it refers to the development process of late-developed countries catching up with advanced ones and adapting to the modern world environment. He Chuanqi, a contemporary modernization researcher in China, noted that modernization includes the formation, development, transformation, and international interaction of modern civilization and

---

cultural structures, as the starting point of modernization; while the other two viewpoints define the starting point as the Scientific Revolution of the sixteenth-seventeenth centuries and the Age of Enlightenment of the seventeenth-eighteenth centuries, respectively.

[2] The mid-twentieth century was the period of the emergence of Western modernization theories, and the important events were: in June 1951, the symposium of the editorial board of the American journal Culture Change considered that the term "modernization" could be used to describe the transformation from an agricultural society to an industrial society. In 1958, Daniel Lerner published THE PASSING OF TRADITIONAL SOCIETY: Modernizing the Middle East, a book that explicitly introduced modernization theory and became one of the most important works in classical modernization theory.

the innovation, selection, dissemination, and withdrawal of civilization elements as well as the international competitions in catching up with advanced world levels. He also emphasized that the current world modernization can be divided into two stages: the first modernization (transition from agricultural to industrial society) and second modernization (transition from industrial to knowledge society).

Summarizing the views of several scholars of different eras, it can be seen that in essence, modernization reflects the dramatic changes and new characteristics that have emerged in society as opposed to the traditional ones since the Industrial Revolution in England. Meanwhile, domestic scholars focus more on the application of modernization in China under special national conditions of developing countries, and contemporary scholars pay increased attention to the new trend of modernization in the transition from an industrial to knowledge society.

4.2.2.2 The Relationship Between Modernity and Tradition

It is worth noting that although modernity exists in relation to tradition, it is not separate from or in opposition to tradition.

China's failure to develop modern consciousness was largely due to its semi-enclosed natural geographical pattern—located in the eastern part of Eurasia, bordered by the Pacific Ocean in the east, and with treacherous mountains and deserts in the west—which means that the emergence and development of traditional Chinese cultural thoughts had taken place in a relatively independent environment. Simultaneously, the economic structure centered on the small peasant economy and the cultural system centered on Confucian benevolence suppressed people's spirit of retrospection. In the political system, the "long division will lead to a long unity, and a long unity will then lead to a long division" pattern tended to be an internal structure with considerable internal continuity and stability. These characteristics have not only allowed our ancient traditional thoughts to continue for many years but have also resulted in the absence of challenges from others as well as the power of reflection to a certain extent. We should realize that modernization indeed originated from Europe—but it does not equate with westernization—and that there is a relationship between modernity and traditional Western culture that is connected and relatively independent as well as that modernity and traditional culture in any region should always constitute and maintain the tension of mutual criticism (Wu 2002).

As Mr. Ji Xianlin said: "Modernization in the East is inextricably related to the modernization in the West. However, the distinction between modernization in the East and West signifies that there are differences, and the biggest or most fundamental difference lies in the basic mode of thinking (Ji 2000)." Modernization is pluralistic, a social development and transformation based on the background of rational modern cultural knowledge; the modernization process of different countries and regions is bound to have different development trajectories, development speeds, and even forms because of their different internal and external environments as well as subjective and objective conditions (Qian 2012). Modernization is not the same as westernization. Although China, as a post-modernization country (exogenous modernization), has many commonalities with the original modernization countries (endogenous modernization) in the pursuit of modernization, it still needs to explore its own development path that is different from that of the developed countries.

4.2.2.3 Evaluation of the Degree of Modernization

Scholars from all domains have been trying to propose a reasonable indicator system of modernization for many years. However, there is no recognized absolute standard yet.

Among these indicators, the most influential qualitative models mainly include the Hakone model proposed in the Hakone Conference in Japan in 1960, the Levy model proposed by the American sociologist Levy in 1962, and the model of modern man proposed by the American sociologist Professor Inkeles in 1974. Quantitative models mainly include the following: the Blake Standard proposed by the American scholar Blake in 1966, the Inkeles Standard proposed by the American sociologist Professor Inkeles, the GNI per capita defined by the World Bank, the Human Development Index of the United Nations Development Programme, the indicator system in the United Nations Millennium Development Goals in 2000, etc. (see Table 4.9).

Based on the various indicator systems mentioned in the above table, it can be seen that in addition to the comprehensive consideration of social development, academia also pays more attention to "human modernization",[3] which is the reason for the five indicators (62.5%) related to individual development. Additionally, there are three flexible indicator systems for different regions and development stages, suggesting the concern for late-modernized countries.

In the current series of China Modernization Report, the basic elements of the study of the national modernization process are analyzed from six major fields (China Center for Modernization Research, CAS 2010): national economy, society, politics, culture, environment, and the individual. The quantitative evaluation of the first modernization (transition from an agricultural to industrial society) is based on the 10 modernization indicators of Inkeles. The quantitative evaluation of the second modernization (transformation from industrial society to knowledge society) has given more weight to knowledge innovation and dissemination indicators. It includes 16 specific indicators in 4 categories. The difference between the two evaluation indicators also shows the evolution of our country's focus on the new development

[3] As early as 1971, Dr. Salazar Bandi, a renowned Chilean scholar, gave a speech at a seminar on "Development Choices" held at the Vienna Institute for Development Studies to review the modernization quest of developing countries and put forward a view that had a profound impact on the academic community, that is, "backwardness and underdevelopment are not just a bunch of statistical indices that can delineate the socio-economic pattern but also can express a psychological state.".

Table 4.9 Main index system of modernization evaluation

Model	Indicator system	Number of indicators	Main content	Major characteristics
Qualitative models	Hokane model	8	Including energy applications, cultural diffusion, population distribution, social networks, institutional organization, etc	The standard of modernization was proposed for the first time, which reflects that modernization is not only socio-economic but also includes comprehensive human development
Quantitative models	Blake standard	10	Including income, employment structure, industrial structure, degree of urbanization, education, health, and level of communication	It tries to reveal the difference between "pre-modernized society" and "highly modernized society" in terms of the level of economic development and social mobility. Not highly operational and difficult to measure
	Inkeles standard	11	Including income, employment structure, industrial structure, degree of urbanization, culture and education, population longevity, and growth	It mainly focuses on economic indicators and human development indicators. Data is easily accessible, but there are problems such as low indicators, missing items, and absence of measurement methods
	GNI per capita defined by World Bank	1	Countries and regions of the world are divided into three categories according to GNP per capita	The indicators have been updated over time

(continued)

Table 4.9 (continued)

Model	Indicator system	Number of indicators	Main content	Major characteristics
	Human Development Index of the United Nations Development Programme	3	The three basic indicators include the following: life expectancy, adult literacy rate, and logs of GDP per capita	It focuses on human development, reveals national development priorities, and provides some basis for countries to formulate development policies. Data is easily accessible and the indicators are applicable to all populations
	Indicators system in the "United Nations Millennium Development Goals"	8	Including 8 general goals and 48 targets on poverty, family, women and children, environment, social equity, etc	It focuses on issues of poverty, disease, environment, etc., and assesses the government's responsibility in eliminating poverty and promoting equity

Source Own elaboration based on relevant sources

situation, which can provide a theoretical background and technical methodological reference for this study (see Table 4.10).[4]

4.2.2.4 Division of Modernization Stages

Stages of World Modernization

According to Mr. Luo Rongqu's review of the three waves of world modernization, world modernization can be divided into three basic stages—the first stage is the social transformation driven by political and economic revolutions, when the process of industrialization spread to Western Europe. The second stage is the process of modernization being spread to the surrounding heterogeneous cultural regions after

[4] Although there are still some controversies about Inkeles' indicators (see Table 2.10), as there is a lack of more authoritative modernization indicators to keep the consistency with the following analysis, this paper still refers to Inkeles' indicators and the data results of China Modernization Report as the main modernization indicators.

Table 4.10 Comparison of the composition of indicators of the two modernizations

	Indicators of the first modernization	Indicators of the second modernization
Knowledge innovation		Funding investment in knowledge and innovation
		Personnel input in knowledge innovation
		Patent output for knowledge innovation
Knowledge dissemination	Adult literacy rate	Popularizing rate of secondary school education
	Popularizing rate of university education	Popularizing rate of university education
		Popularizing rate of TV
		Popularizing rate of the internet
Life quality	Percentage of urban population	Percentage of urban population
	Proportion of doctors	Proportion of doctors
	Infant survival rate	Infant survival rate
	Life expectancy	Life expectancy
Economic quality	GNI per capita	Energy consumption per capita
	Proportion of agricultural labor force	GNI per capita
	Agricultural value-added ratio	PPP per capita
	Service industry Value-added Ratio	Percentage of value-added of physical industries
		Proportion of labor force in physical industries
		Industrial value-added ratio
		Industrial labor force ratio

Source Own elaboration based on relevant sources

the pioneer modernizing countries in Western Europe have achieved certain achievements. The third stage is the modernization process in which the underdeveloped countries worldwide entered modernization after undergoing the national liberation process since the two world wars (see Table 4.11).

Table 4.11 Historical stages of world modernization

Stage	Time	Major events	Major modernizing countries and regions	Characteristics
The first wave of modernization	1780–1860	The British Industrial Revolution, the French Political Revolution, and the American Revolutionary War	Western European countries such as England, France, and Germany	Politics: capitalist society Productivity: mass production by steam and machines Culture: emphasis on the emancipation of thoughts
The second wave of modernization	1860–early twentieth century	The colonies of the Western countries expanded rapidly	North America, Australia, Japan, while Latin America along with Egypt, Turkey, and China have also begun to be impacted by the wave of modernization	Politics: capitalist society and germinating socialism. Productivity: electricity and steel
First development crisis	Early twentieth century to second half of the twentieth century	Two world wars	Russia	Politics: capitalist society, colonialism and socialist society
Third wave of modernization	Second half of the twentieth century to the present	Post-war developments	Influencing countries worldwide	Politics: capitalist and socialist societies Productivity: oil and energy, synthetic materials, and microelectronics

Source Compiled from Luo, Rongqu. *A New Theory of Modernization—China's Road to Modernization* [M]. Wuhan: East China Normal University Press, 2003: 107–115

Division of China's Modernization Stages

The most common thought in academic circles at present is that China's modernization can be divided into three stages: 1. 1840–1911, from the Opium War in the late Qing Dynasty to the eve of the establishment of the Republic of China. 2. 1912–1949, from the beginning of the establishment of the Republic of China to the eve of the founding of the PRC. 3. 1949 to the present, that is, the development stage since the founding of the PRC. Among them, the 3rd stage can be further subdivided into the planning period from 1949–1977, the reform period from 1977–2001, and the catching-up period from 2002 to the present (see Table 4.12).

According to statistical analysis, as of 2008, 10 regions in China (excluding Hong Kong, Macao, and Taiwan) have completed or basically achieved the first modernization stage,[5] and Jiangsu Province is ranked sixth among them (China Center for Modernization Research,CAS 2010). As of 2007, seven regions in China have reached the level of medium developed countries in the second wave of modernization, and Jiangsu Province is also among them. Therefore, in the current period of catching up and unbalanced development of China's modernization, Nanjing has gone through a relatively complete stage of modernization development.

This section starts with the basic definition of "modernization," discusses the relationship between modernity and tradition, and summarizes the criteria for evaluating the degree of modernization and the stages of modernization. It can be seen that modernization is a huge and inevitable human social transformation that has greatly impacts people, society, ideology, and material space.

First, modernization includes not only all aspects of social development but also human development. Second, modernization and the development of traditional thoughts and culture are not completely opposed to each other, and people should reflect on traditional culture in the wave of modernization to seek the continuation of cultural connotations. Finally, Nanjing, as a region with a high level of modernization in China, provides relatively good focus and carrier for reflecting on the current urban construction by studying the changes of its urban space and traditional culture under the wave of modernization.

4.2.3 Historical Evolution of the Huis Settlement

As the ancient capital of Six Dynasties with a history of more than 2400 years, Nanjing has been a land flowing with milk and honey located to the south of the Yangtze River, a city prosperous in culture and economy since ancient times as well as a place of vital military importance.

Long before Nanjing was founded as a city in Wuyue (era), ancestors from the north and south gathered here to live together. By the time of the Southern Liang Dynasty, Jiankang already had a population of more than 280,000 households, making it the largest city in China at that time. Although it is difficult to precisely determine the ethnic composition of that population now, the policy of *Tuduan* (people and households will be listed in the register based on the current place of residence) and *Jianji* (validation and verification of household registration) for the minority groups from the north during the Southern Song Dynasty show that

[5] Complete the transition from an agricultural society to an industrial society, reaching the average level of the developed industrialized countries in 1960.

Table 4.12 Historical stages of China's modernization and their characteristics

Stages		Time	Major event	Economy	Society	Politics	Culture
Start of modernization	Late Qing dynasty	1840–1911	Losing in the Opium War	Agricultural economy, and start of industrialization and mechanization	Feudal society, agricultural society, modern education, and modern transportation	Reformation of feudalism	Enlightenment ideas and modern publications
Partial modernization	Republic of China	1912–1949	Founding of the Republic of China	Agricultural economy, partial industrialization, partial mechanization, and electrification	Agricultural society, modern education, modern health, and modern transportation	Republicanism, Manchu Restoration	Advocating democratic science and the New Culture Movement
Full Modernization	Planning period	1949–1977	Founding of the PRC	Planned economy, full industrialization, mechanization, and electrification	Agricultural society, urbanization, public health, and modern education	Republicanism, new democracy	People's literature, television literature, and cultural revolution
	Reform period	1978–2001	Reform and opening up	Industrial economy, marketization, industrialization, and informatization	Agricultural society, urbanization, information technology, and popularization of compulsory education	Republicanism, reform, and opening up	Popular literature, internet literature, modern technology, and modern philosophy

(continued)

Table 4.12 (continued)

Stages		Time	Major event	Economy	Society	Politics	Culture
	Catching-up period	2002 to present	Joining the World Trade Organization (WTO)	Industrial economy, new industrialization, globalization, and greenization	Industrial society, new urbanization, welfare, and greenization	Republicanism and political civilization building	Popular culture, internet culture, and cultural industrialization

Source Organized from the China Modernization Report published by China Center for Modernization Research, CAS, China Modernization Strategy Research Group. [M]. Beijing: Peking University Press, 2010: 282

Nanjing already had a large ethnic minority population at that time (Nanjing Local History Compilation Committee 2009b). As a result of alternations in dynasties, the population in Nanjing gradually developed a multi-ethnic pattern, where the Han population became the dominant group and various ethnic minorities co-existed at the same place. The ethnic and religious evolution constitutes an important vein in the ancient city of Nanjing. Therefore, the following section will focus on the Huis Settlement in Qijiayuan, Nanjing, and outline the history of the Huis Settlement at three levels: urban construction, development of the Hui people, and settlement evolution.

4.2.3.1 Before the Song Dynasty

[Urban construction] Before the Song Dynasty, although Nanjing was deliberately restrained and suppressed by the rulers of the north, the economy and culture of the city developed greatly. The capital of the Southern Tang Dynasty shifted southward from Jiankang City during the Six Dynasties and incorporated the economically wealthy areas on both sides of the Qinhuai River into the city. Its palace city was located in the area of present-day Hongwu Road and generally formed a central axis pattern along Hongwu Road and Zhonghua Road, which had been used during the Song and Yuan dynasties.

[Evolution of the Hui people] In the second year of Yonghui of the Tang Dynasty (year 651), an Islamic envoy from *Dashi* (Arab Empire) visited Chang'an and was received by Emperor Gaozong of the Tang Dynasty, which was the first diplomatic contact between an Arab country and China and the beginning of the introduction of Islam into China (Nanjing Local History Compilation Committee 2009c). During the Tang and Song dynasties, the Hui people of Nanjing were engaged in trade and commerce, and most of them were Muslim merchants who came to China by sea and land from Western Persia and other countries. As an important commercial city in the Song Dynasty, Jiankang absorbed many northern minorities who came to the South to live together with the local Han Chinese, thus further promoting ethnic interaction and integration.

[Qijiawan Settlement] During the Tang Dynasty, the Qijiawan area[6] was located on the west side of the central axis at a prosperous commercial and residential area in the southwest of the city, and this basic pattern continued to influence many subsequent dynasties.

[6] In the historical analysis of the various dynasties before the founding of the PRC, the term "Qijiawan area" is used in this paper instead of "Qijiawan Huis settlement" in the latter part of the paper.

4.2.3.2 Song and Yuan Dynasties

[Urban construction] Throughout the Song and Yuan dynasties, the center of Nanjing's urban area had been the local political center, and the basic pattern of the city had not changed much from the Tang and Song dynasties.

[Evolution of the Hui people] During the Yuan Dynasty, China became a country with an unprecedentedly vast territory englobing multiple ethnic minorities, and the Hui population was officially defined as an ethnic minority by inclusion in the *Zhizheng Jinling Xinzhi* (Nanjing local chronicles of Zhizheng). In terms of population size, Jiankang City had a population of 2919 Hui people according to the *Lushisi* of the city, accounting for 86% of the total Hui population. In terms of economic structure, the majority of the Hui population was officers, soldiers, and domestic servants, and the population mobility was high (Nanjing Local History Compilation Committee 2009c).

[Qijiawan Settlement] During the Song and Yuan dynasties, the Qijiawan area was adjacent to the local power center of Nanjing City to the east and close to the water transportation hub and the scenic Shuiximen to the west. Several powerful people settled in the area, including a large number of Hui population. In addition, during the Song and Yuan dynasties, numerous political and cultural facilities were built around the area, such as the Jiankang Governor's Office ((Southern Song Dynasty) Li 1992) in the east and the Jiangnandonglu Transportation Department's bureau in the north. During the Song Dynasty, the Anyuan Building and the Hexi Building were also built in the area (Southern Song Dynasty 1990).

4.2.3.3 Ming Dynasty

[Urban construction] In the late Yuan and early Ming dynasties, Zhu Yuanzhang set the Yuan Censorate in the Wangfu Garden that was east to Qijiawan as the government office (Qing) Bi (ed) 1999) and moved to the new palace after the construction of the Ming Forbidden City. After the expansion of Nanjing City, the roads could be classified into official streets, side streets, and alleys. The southern part of the city, the new district on the east of the Forbidden City, and the military defense area in the north of the city were relatively independent and formed their own system. The southern part of the old city was the location for the residents' marketplace at that time, and *Shibafang* (eighteen workshops), where handicrafts gathered, were built.

[Evolution of the Hui people] The Ming Dynasty was an important stage in the development of the Hui community in Nanjing as it was then that the population tended to stabilize. In terms of population size, according to *Kezuo Zhuiyu* written by Qiyuan Gu in the 20th year of Wanli of Ming Dynasty (1592), it can be estimated that the Hui population in Nanjing during the early Hongwu period was nearly 100,000. In terms of the structure of the migrant population, in addition to the Hui soldiers and officers stationed in the city, there was also a large Hui population among the 200,000 artisans dispatched for the large-scale construction in the city (Nanjing Local History Compilation Committee 2009d). In terms of religion and culture, the books written by Hui scholars in Nanjing at the end of Ming Dynasty pioneered the "interpreting of the Islamic Scripture with Confucianism" in Islamism in China, which strongly strengthened the spread and understanding of Islamic doctrine among the Hui people and enabled the training of the successors of Imam in the form of mosque education. This practice continues to this day.

[Qijiawan Settlement] In the Ming Dynasty, because of the eastward shift of the political center of the city, the Qijiawan area began to move away from the political power center and became an important part of the commercial center of the old southern part of the city and the Pingshi Street in the area also assumed the role of the north–south trunk road of the city. Among them, the "Eighteen Workshops," including the Cloth Workshop, Leather Workshop, Qinhua Workshop, and Brocade Workshop, were all located in the current Qijiawan neighborhood. Until the late Ming Dynasty, there were scenes where "from Dazhongqiao to its west, from Huaiqingqiao to Sanshajie, on the west of Doumeqiao, to Sanshanmen, and on the north from Cangxiang to Yecheng, till the east to Neqiao, and at last to Zhongzhenjie, merchants of all trades were gathered the capital city" (Ming) Gu 1997). The majority of the merchants in each area were the Hui people. Thus, in addition to the workshops and the old-style Chinese private banks, buildings embodied with the Hui characteristics, such as the mosques, were also scattered in the streets around the city.

4.2.3.4 Qing Dynasty

[Urban construction] The Qing government established the viceroy of Liangjiang in Jiangning Prefecture, which oversaw the present-day provinces of Jiangsu, Anhui, and Jiangxi, with Nanjing remaining a place of strategic importance in terms of cultural, political, and military aspects and a regional economic hub at the time. The urban space began to develop across the Yangtze River, with weaving as the dominating industry of their economic structure. The loss of the Opium War in 1840 can be considered as the beginning of modernization of the country, while the opening of Nanjing in 1898 and the resulting international trades and commerce started to impact the traditional social, cultural, and economic structure of Nanjing.

[Evolution of the Hui people] The Hui people in the Qing Dynasty had the same development trend as in the Ming Dynasty. In terms of population size, the Hui population in Nanjing in 1851 was approximately 40,000, accounting for 4.44% (Nanjing Local History Compilation Committee 2009e) of the total population at that time. In terms of economic structure, the Hui population was mostly engaged in the weaving industry in the early Qing Dynasty. Further, according to *Shangyuanjiangning Tuzhi* (Local Chronicles of Shangyuan and Jiangning), by the second year of Xuantong (1910), because of the rapid growth of the Hui population, industries with strong Hui ethnic characteristics, such as jade, felt, and pastry, witnessed a significant increase in the number of people engaged. However, with the start of the Opium War and the beginning of the modernization wave in China, weaving and small handicraft industries gradually declined. In terms of settlement distribution, the Hui population was primarily clustered in Xinqiao, to the west of Shangfuqiao, near Xiafuqiao Mosque, as well as in Tuqiao and Ruizifang areas. While in terms of religious culture, by 1850, there were 36 mosques in Nanjing (most of them were destroyed during the Taiping Heavenly Kingdom but were subsequently restored and gradually rebuilt) and 13 Hui trade unions.

[Qijiawan Settlement] During the Qing Dynasty, the Qijiawan area continued to have its commercial functions as in the late Ming Dynasty, and many residential houses of the distinguished men in Jinling who were successful in the imperial examinations were constructed here. The famous Former Residence of Gan Xi was built during the Jiaqing Dynasty.

4.2.3.5 The Republic of China Period

[Urban construction] In 1927, after the Kuomintang chose Nanjing as the capital, it started to prepare the city plan of Nanking, and the urban road network established at this time laid the basic framework of the current urban trunk road system. During this period, Nanjing was divided into six functional districts, including the central political, residential, industrial, and park district, among which the old southern part of the city was preserved intact as a traditional residential district to succeed the pattern of the Ming and Qing dynasties.

[Evolution of the Hui people] In the early years of the Republic of China, under the influence of the fights between the warlords, the Hui population in Nanjing gradually started to migrate out of the city. However, after the capital was set in Nanjing in 1927, the population in the city started to increase once again. In terms of population size, according to the Capital Police Profile, there were 37,443 Muslim believers by 1934 (only religious believers were registered in the national government census), accounting for 5.05% of the city's population. In terms of economic structure, in addition to the self-employment ventures in catering and handicrafts, the Hui population was also engaged in establishing numerous enterprises that had modern characteristics (later, because of the effects of war, these national enterprises either moved out of the city one after another or were forced to close). In terms of religion and culture, Islamic groups, such as the Nanjing Islamic Association, were established, 13 elementary schools for the Hui population were built to teach Islamic scriptures and doctrines, and the Hui people even created some important Islamic publishing houses to promote traditional national culture during the War of Resistance against Japan.

[Qijiawan Settlement] In the early years of the Republic of China, industrial business remained prosperous in the Qijiawan area; however, with the rise of the city's new commercial center, Xinjiekou, and the construction of several roads around Qijiawan, its status as the city's transportation and commercial center, which had been developed over the years, rapidly declined.

4.2.3.6 After the Founding of the PRC

1949–1958

[Urban construction] The First Five-Year Plan, which had been implemented since 1953, marked the beginning of large-scale, planned modernization and socialist construction in China, and the urban construction of Nanjing developed considerably during this period. Because of the relocation of the Jiangsu Provincial People's Government to Nanjing, the restructuring of colleges and universities and the development of industries, the urban constructions during the First Five-Year Plan were primarily focused on institutions, colleges, and universities, and industrial land (Zhou et al. 2004), as a result of which the city changed from a consumer-oriented city to a production-oriented one.

[Evolution of the Hui people] The development of the Hui population at the early stage of the founding of the country was protected and supported by the government. In terms of population size, according to the first census in 1953, the Hui population in Nanjing was 20,401, accounting for 2.95% of the total urban population. In terms of religion and culture, there were still 32 mosques during the initial post-liberation period (Taipinglu Temple, Neqiaowan Temple, and Caoqiao Temple were newly built). After the establishment of the Nanjing Mosque Administration Committee in 1958, the mosques in the city were merged and unified, and the number of mosques was reduced to seven; however, the construction of kindergartens and elementary schools for the Hui community was restored.

[Qijiawan Settlement] During this period, a certain number of guaranteeing policies in terms of the basic livelihood of ethnic minorities were implemented in Qijiawan area. Because of the great attention paid to the industrial construction and production-oriented investment at this stage of urban development, there was not much space for the development of self-employed home-based workshops in the area.

1959–1977

[Urban construction] During the Great Leap Forward period, because of the unrealistic construction goals and slogans, the urban construction was blindly adventurous, industrial advancement was put forward at an excessive speed, and the supply of urban housing was tight, thus greatly increasing the burden of the old city. Additionally, the uncontrolled and unordered urban construction and indiscriminate demolition of private buildings during the Cultural Revolution had caused irreparable damage to many urban green spaces as well as historical and cultural relics. Simultaneously, under the influence of multiple factors, such as the severe shortage of investment in non-production construction and the return of a large number of young people to Nanjing from the countryside, the housing supply in the city fell into a disorderly and uncontrollable mess.

[Evolution of the Hui people] In terms of population size, according to the second census in 1962, the Hui population in Nanjing was 38,454, accounting for 2.31% of the total urban population. In terms of economic structure, the development of traditional handicraft industry was seriously hampered under the guiding ideology because the national priority at the time was promoting large-scale industrial construction. In terms of religion and culture, the Cultural Revolution period interfered and blocked the normal life of the Hui population; mosques were occupied and Islamic associations were forced to suspend their activities.

[Qijiawan Settlement] During this period, the Qijiawan Huis Settlement and the entire Hui population in Nanjing had the same fate and suffered from identical changes. Consequently, the religious activities essentially came to a halt, the development of handicrafts was hampered, and the Hui population had a difficult life; the Caoqiao Mosque in the community was repurposed for the use of storage and resettlement.

1978–1989

[Urban construction] After the reform and opening up, the urban construction ideology of "emphasis on production but not living" began to change and the housing problems of citizens were granted more attention. According to the Nanjing Urban Master Plan (1980–2000), residential construction at that time primarily focused on altering the old city, and the construction of residential areas were mostly large-scale settlements and barracks-style courtyards invested by the government, which largely changed or even destroyed the traditional structure of Nanjing's old city.

[Evolution of the Hui people] According to the third census in 1982, the population of Hui in Nanjing was 46,124, accounting for 2.16% of the total urban population. In terms of settlement distribution, the main settlements of the Hui population in Nanjing in the 1980s were Qijiawan, Yeshan Daoyuan, and Shiliu Xincun (Nanjing Local History Compilation Committee 1986), and in terms of religious culture, the Islamic Association gradually resumed its activities after the Third Plenary Session of the Central Committee in 1978.

[Qijiawan Settlement] During this period, the Jingjue Mosque and Caoqiao Mosque in the Qijiawan area were repaired and opened one after another. The community also built the Lirongji pressed salted duck factories and the Chaotiangong roasted seeds and nuts factory, which solved the employment problem of many Hui people immediately and contributed to the economic and cultural development of the Huis Settlement.

1990 To Present

[Urban construction] After the 1990s, the amount of construction and development in cities grew rapidly (see Fig. 4.5). In 1998, the State Council issued the "Notice on Further Deepening the Reform of the Urban Housing System and Accelerating

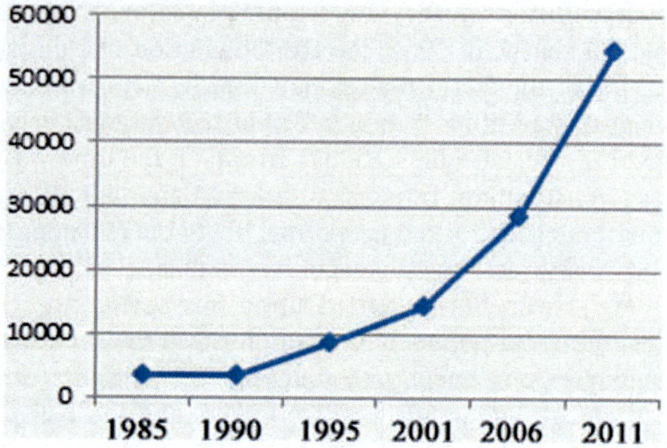

Fig. 4.5 Statistics of completed housing area in Jiangsu Province (Unit: 10,000 m^2). *Source* Yearbook of Jiangsu Provincial Bureau of Statistics

Housing Construction," which marked the entrance of housing construction into the privatization and marketization stage and brought increasingly significant impacts on urban construction. The central position and complex functions of Nanjing's old city have not been effectively alleviated, and the demand for urban construction because of continuous population growth led to the development of every bit of unused land and the scattered construction of high-rise buildings in the old city, which further affected and changed the traditional structure and spatial pattern of the old city.

[Evolution of the Hui people] In terms of population size, according to the three censuses since 1990, the urban Hui population was 54,408, 58,691, and 61,295, accounting for 2.03, 1.61, and 1.18%, respectively, of the city's total population. As more and more Hui populations migrate and disperse throughout the city, their economic structure, religious culture, and other traditional ethnic characteristics tend to decline. Presently, only three mosques are reserved in Nanjing, two of which (the Jingjue Mosque and Caoqiao Mosque) are located in the streets where the Qijiawan Huis Settlement is located.

[Qijiawan Settlement] During this period, in the mid-1990s, the Qijiawan Huis Settlement experienced large-scale land expropriation and the widening of Dingxinlu, Cangxiang, and Hongtuqiao, which broke the original cohesive pattern of living of the Huis around the mosque. The scattered construction of the multi-storey community Rongzhuangxincun, the high-rise neighborhoods Jindingwan, and Henglong Garden further altered the traditional pattern of the settlements that were dominated by low-rise dwellings. The renewal of the historical district in the southern part of the old city was often accompanied by demolition and reconstruction of varying scales, which also led to the forced migration of the local population and the decline of national commerce and religious activities (see Figs. 4.6 and 4.7). With the deterioration of living conditions and the continuous migration of the Hui population, the Qijiawan Huis Settlement became one of the major colliding points between modern development and traditional culture in old Nanjing. Consequently, the following analysis is a dynamic study of the socio-spatial evolution of the Qijiawan Huis Settlement, which inherits traditional culture and faces the challenges of modern lifestyles simultaneously (Table 4.13).

Fig. 4.6 Current status of Pingshijie. *Source* Own photograph

4.2.4 Research Strategies on the Evolution of Huis Settlements

4.2.4.1 Definition of the Scope of the Research

[Scope of the Research] The Qijiawan Huis Settlement is located at the Chaotiangong sub-districts, Qinhuai District, in Nanjing. The research area selected in this study is centered on the Caoqiao Mosque in Qinhuai District, and the scope is limited to the Huis Settlement area, where the community is relatively concentrated within Jianye Road to the north, Shengzhou Road to the south, Mochou Road to the east, and Zhongshan South Road to the west, comprises four communities: Qijiawan Community, Anpin Street Community, Pingshi Street Community, and Rongzhuangxincun (see Fig. 4.8).

The total area to be studied is approximately 52 hectares, and the existing household population is about 28,000 in total, among which about 2500 are Huis population. When distributing the questionnaires, we focused on visiting various communities within the research area as well as Nanyuan Community, Taoyuanju Community, Jingming Jiayuan, Yinlong Huyuan, Xishan Huayuan, etc., where the land expropriated Huis of Qijiawan are resettled. A total of 115 questionnaires were distributed, of which 104 were valid, with a male to female ratio of 1.2:1.

[Period of the Research] The evolution outline and transformation pattern of the Huis Settlement since the founding of the PRC in 1949.

[Subjects of the Research] The Huis population that used to and currently live in the Qijiawan Huis Settlement in Nanjing.

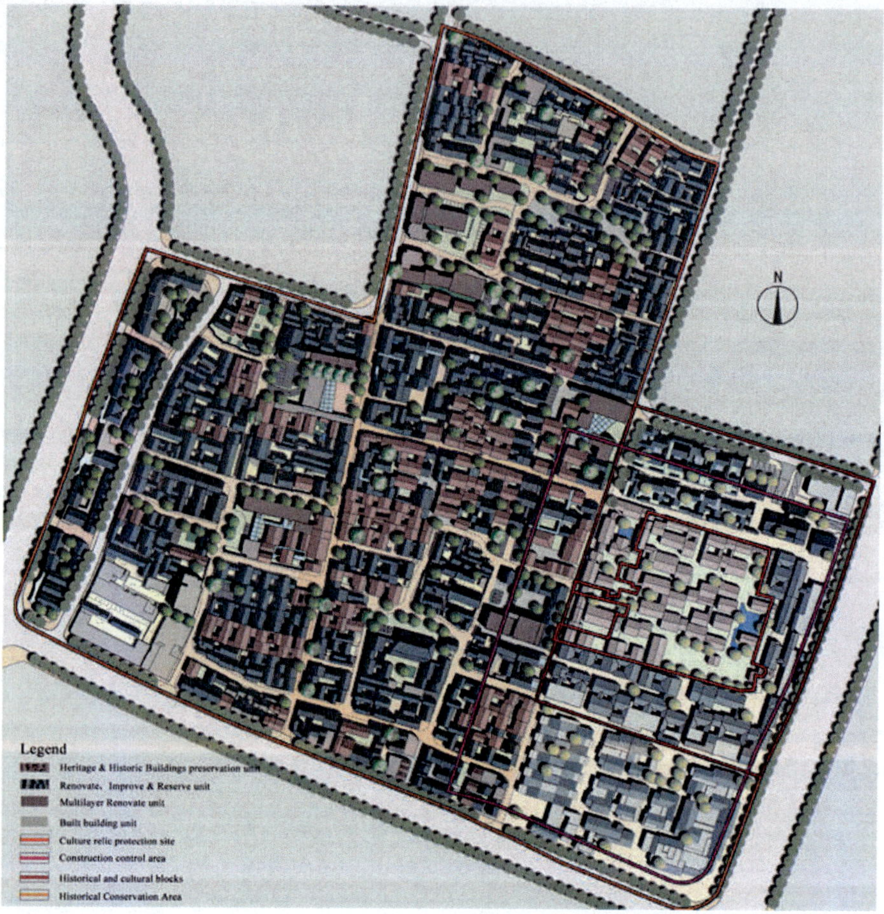

Fig. 4.7 Guidance Map of Nanbuting's Urban Design. *Source* South Historic District Conservation planning and urban design of Old Town in Nanjing. Nanjing Planning Bureau, 2012

The main reasons for choosing the sampling subject and research scope are as follows.

First, according to the analysis of relevant census statistics, it is found that the Hui is the most widely distributed ethnic minority group that has the seventh-highest urbanization rate in China, and the community has an obvious mixed spatial distribution pattern with the Han Chinese (Ma, 2002), while still maintaining clear religious beliefs and distinct ethnic customs and habits. Therefore, the Huis population can be selected as a typical group and characteristic sample for the study of the evolution of minority settlements.

Second, as the largest Hui community in Nanjing, the Qijiawan Huis Settlement has a long and important historical position; under the current wave of modernization, it is facing the same evolution and alternation of social and spatial attributes as

Table 4.13 Summary table of the development stages of the Qijiawan Huis settlement in Nanjing

Dynasties	Urban location of the Qijiawan area	Urban construction background	Background of Hui development	Events in Huis settlement
Before Song		Political suppression and economical and cultural development, including the economically prosperous areas on both sides of the Qinhuai River into the city—an urban pattern that influenced later generations was established	Western merchants came to China by the sea via the "Spice Road," built mosques and public cemeteries in their places of residence, and their descendants were the ancestors of the Hui people of Nanjing	–
Song and Yuan		Significant development in politics, economy, and culture. As a local political center, Jiankang was partially transformed because of the political situation; however, the overall pattern of the city did not change largely	During the Yuan Dynasty, the Hui population was compiled in the *Zhizheng Jinling Xinzhi* and became an officially recognized ethnic minority	Increased transportation status and proximity to political centers as well as growth of political and cultural facilities in the district

(continued)

Table 4.13 (continued)

Dynasties	Urban location of the Qijiawan area	Urban construction background	Background of Hui development	Events in Huis settlement
Ming		The eastward shift of the political center. The city was divided into three functional areas: the Forbidden City, residents' market, and the military defense area. Qijiawan was located among the residents' market area to the south of the city, thus laying the foundation for economic prosperity	Rapid growth of the Hui population was led by military personnel and artisans serving in the army. The first "Interpreting of the Islamic Scripture with Confucianism" was launched to spread in Islamism in China, and mosque education was introduced	The political status declined and the economic status increased, thus becoming an important part of the commercial center of the old southern part of the city
Qing		Establishment of the viceroy of Liangjiang in Jiangning Prefecture, where substantial political, economic, and cultural development took place Development of urban space across rivers The Opium War at the end of Qing Dynasty became the beginning of China's modernization	The population employed in the weaving industry increased, forming an employment structure with specific Hui characteristics Settlement areas were developed to some extent and widely distributed The number of mosques reached a historical peak	The golden age of economy and commerce in the settlement, where weaving and handicrafts industries nonetheless declined in the late Qing Dynasty

(continued)

Table 4.13 (continued)

Dynasties	Urban location of the Qijiawan area	Urban construction background	Background of Hui development	Events in Huis settlement
The Republic of China		A brief period of development against the background of social instability The city was divided into six functional areas, such as the central political, residential, industrial, and park district. The southern part of the city was preserved as a traditional residential area and the overall pattern did not change much	A significant rebound in the Hui population after Nanjing was chosen as the site of the capital The emergence of a group of Hui industrialists who founded modern factories The increase of religious groups and schools among the Hui social organizations and the improvement of their social life to some extent	Along with the construction of urban roads and the formation of a new city center, the area's transportation advantages and commercial status over the years began to decline
1949–1958		Dramatic expansion of urban space The main urban constructions were provincial and municipal institutions, universities, and colleges and industries, showing the alternating characteristics of leaping forward and filling in the vacant areas	Ethnic minority status was restored Nanjing Mosque Administration Committee was established	Ethnic minority status was restored Hui residents were engaged in socialist economic construction, and their occupational structure became diversified

(continued)

Table 4.13 (continued)

Dynasties	Urban location of the Qijiawan area	Urban construction background	Background of Hui development	Events in Huis settlement
1959–1977		Urban construction was dominated by productive investment and development, which was blindly adventurous and caused indiscriminate demolition and unordered constructions; however, the impact had not yet extended beyond the old city The situation of urban housing supply was serious	Forced closure or occupation of mosques The disintegration of the religious community system and forced cessation of the activities of Islamic associations	The cessation of religious activities at the Caoqiao Mosque and Jingjue Mosque. Problems such as ethnic discrimination, inadequate distribution of materials, and difficulties in the living of residents emerged
1978–1990		After the reform and opening up, a large number of people going and working in the countryside during the Cultural Revolution returned to the cities A "circle-type urban clustering layout" structure was proposed as the general layout of the city Urban construction was concerned with the public's housing and living needs	The size of the population decreased compared with the pre-reform period Islamic associations resumed religious activities	Restoration of activities at the Jingjue Mosque and reconstruction of the Caoqiao Mosque Establishment of a halal food factory and development of an economy with ethnic characteristics

(continued)

Table 4.13 (continued)

Dynasties	Urban location of the Qijiawan area	Urban construction background	Background of Hui development	Events in Huis settlement
1990 to present	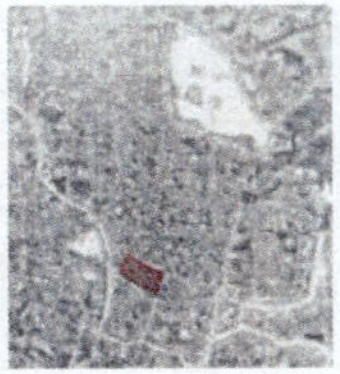	The spatial pattern of the metropolitan area with the main city as the core was basically drawn The modernization of the city has influenced and impacted the traditional structure and spatial pattern of the city	Only three mosques remained With the migration and dispersal of the Huis population, its economic structure, religious culture, and other traditional features started to diminish	Large-scale land expropriations began in the 1990s and resulted in the relocation of the original Hui households Several new multi-storey modern settlements were constructed The characteristic pattern of the settlement, which was dominated by traditional residential houses, has been broken

Source Yin (2005)

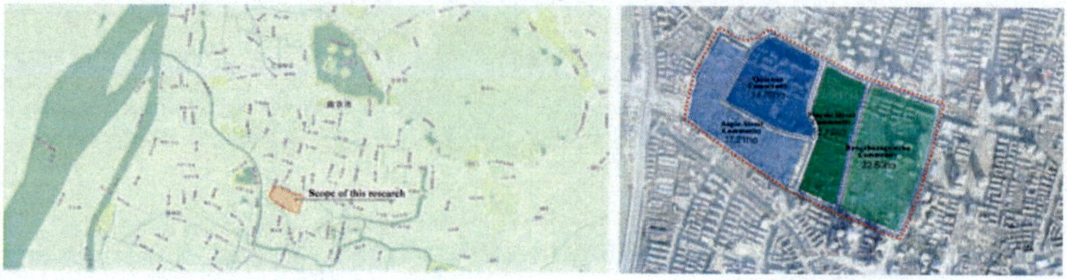

Fig. 4.8 Scope of research. *Source* Own elaboration

many other Huis settlements and thus can also represent and reflect the characteristic patterns and cause of problems of similar communities in their evolution (see Figs. 4.9 and 4.10).

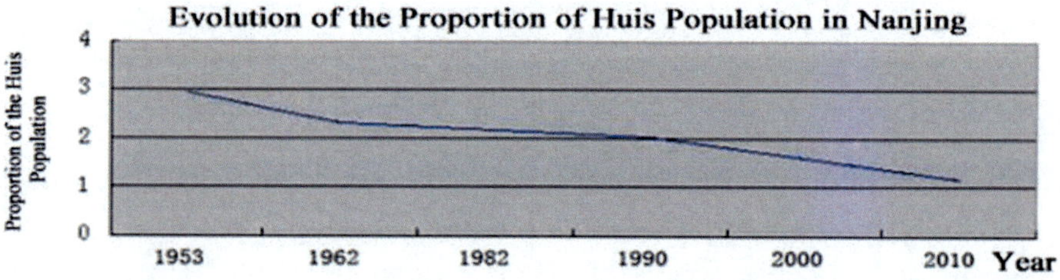

Fig. 4.9 Evolution of the proportion of Huis population in Nanjing (The proportion of the Hui population in Nanjing refers to the percentage of Hui population in Nanjing to the total population in the urban area (excluding Jiangning, Lihe, Lishui, and Gaochun)) *Source* Own elaboration according to statistics of the sixth population census

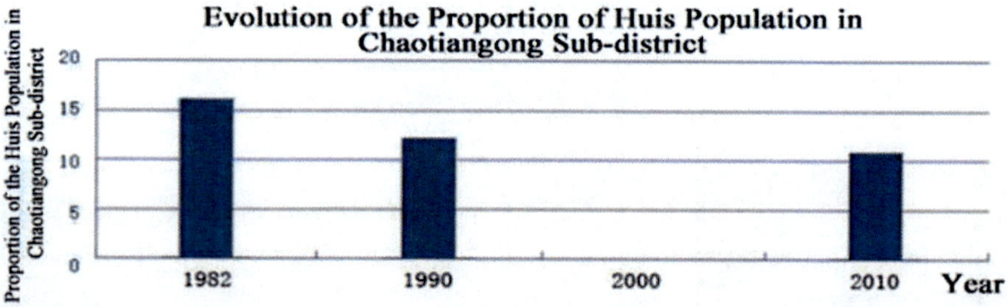

Fig. 4.10 Evolution of the proportion of Huis population in Chaotiangong Sub-district (Chaotiangong sub-district is where the Qijiawan Hui community is located. The proportion of the Hui population here refers to the proportion of Hui population in Chaotiangong sub-district to the total Hui population in the urban area, for the calculation of which the data of the 2000 Fifth National Population Census is missing) *Source* Own elaboration according to statistics of the sixth population census

4.2.4.2 Establishment of Research Strategy

[Division of Stages] The socio-spatial evolution of minority settlements under the wave of modernization can essentially reflect the realistic scenes where the conflict and fusion between tradition and modernity, history, and future in China's current urban construction can be seen. The evolution of the community is also closely related to the alternation of its multicultural background; thus, the division of the evolutionary stages of the Huis settlements is also inevitably dependent on changes in the stages of their background.

Considering this, this study will take into account the key evolutionary moments from the following four major aspects, namely, international modernization development stages, China's modernization development stages, Nanjing's urban construction situation, and changes in the Qijiawan Huis Settlement area, to consolidate and delineate the evolutionary stages of the Qijiawan Huis Settlement in Nanjing since the founding of the country (see Figs. 4.11 and 4.12).

[Specific Strategy] On the one hand, through the analysis of domestic and foreign literature and theoretical approaches, this study summarizes the multi-ethnic profile of Nanjing in general and distinguishes the modern background and historical context of the changes of the Huis settlements. On the other hand, the study combines the ethnic uniqueness of the research object—the Hui population and its settlement space—with primary data to conduct a historical comparison and dynamic analysis

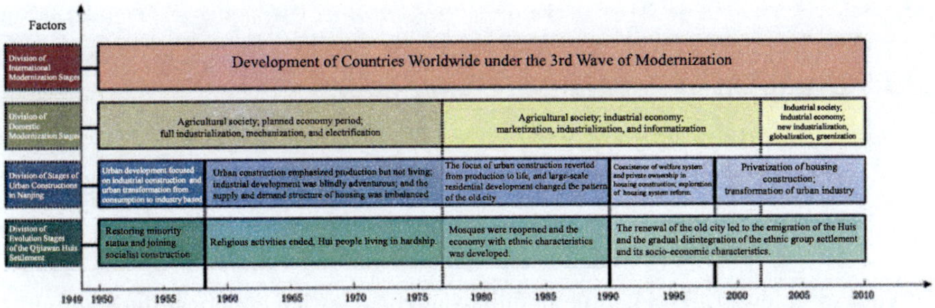

Fig. 4.11 Schematic diagram of the stages in the evolutionary background of the Qijiawan Huis settlement in Nanjing (since 1949). *Source* Own elaboration

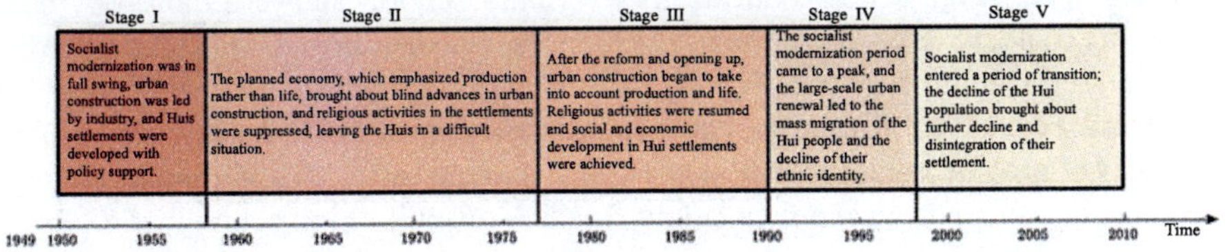

Fig. 4.12 Schematic diagram of the evolution stages of the Qijiawan Huis settlement in Nanjing (since 1949). *Source* Own elaboration

from the perspectives of social and spatial attributes through qualitative and quantitative approaches. The aim is to explore the multiple evolutionary characteristics of the social space of the Huis Settlement in the context of modernization as well as its problems and countermeasures.

[Data Source] The main methodologies used are a combination of relevant historical documents, field research, sample questionnaire statistics, and thematic interviews.

4.3 Evolution of Social Attributes in Huis Settlements Under the Wave of Modernization

Based on the aforementioned background of the Huis Settlement in Nanjing, the establishment of research strategy and division of evolutionary stages, this section analyzes the characteristics of the Hui people in each evolutionary stage in the Qijiawan Huis Settlement by selecting social attributes in terms of demographic composition, family relations, social organization, and lifestyle, and customs based on the characteristics of the community. Through a vertical comparison of the settlement itself across stages and a horizontal comparison of the modernization stages of the country in the same period, this study summarizes the general patterns and characteristics of the evolution of the social attributes of the urban Huis settlements.

4.3.1 Evolution of Population Composition

4.3.1.1 Evolution of the Education Level of the Hui Population

Evolution Characteristics of the Education Level

The education level of Hui people in Qijiawan Huis Settlement constitutes a highly educated trend of "**steady increase in the number of people with college education or above**." Specifically, there is a small decrease in the proportion of Hui people with elementary school education or below, and the proportion of Hui people with junior high school and high school education (secondary school) shows a more obvious decreasing trend, whereas the number of Hui people with college education or above has increased significantly (see Fig. 4.13).

During 1949–1958 (Stage I), the education levels of Hui people were more generally of middle and low education levels. Among them, 26.23% had an education level of elementary school or below, 34.43% of junior high school, 27.87% of high school (technical secondary school), and only 11.48% were highly educated and had a college education or above.

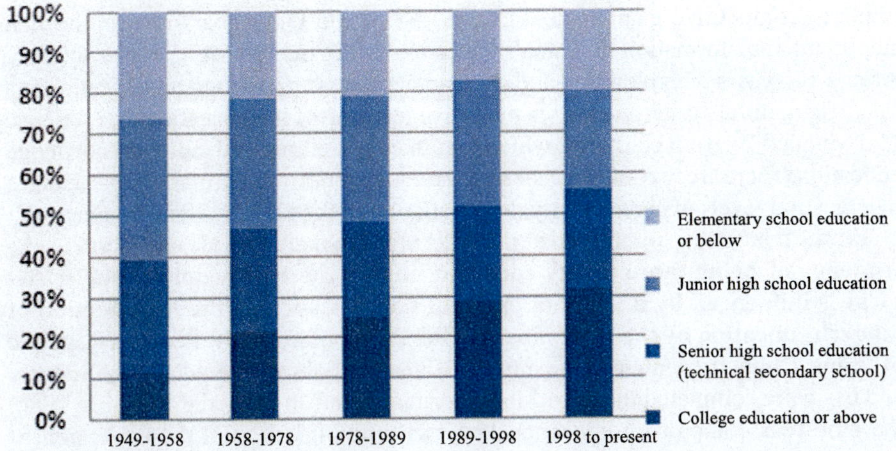

Fig. 4.13 Evolution of the education level of the Hui people in the Qijiawan Huis settlement. *Source* Data from the research group's sample study on the Qijiawan Hui Settlement in Nanjing (2014)

Between 1958 and 1998 (Phase II–IV), there was a relatively large increase in the proportion of highly educated people with college degrees or higher. This proportion increased from 11.48 to 28.85% across 40 years, whereas the proportion of people with the other three types of education level decreased slightly. Among them, the proportion of people with elementary school education and below decreased from 26.23% to 17.31%, and the proportion of people with junior high school and high school (technical secondary school) education level both decreased by around 4%, stopping at 30.77and 23.08%, respectively, by the end of the 1990s.

Since 2000 (Stage V), the education level of the Hui people tended to stabilize. The proportion of people with higher education above college level reached 31.71%, whereas the proportion of people with the other three types of education level only fluctuated slightly and remained basically the same as the previous stage (19.51, 24.39, and 24.39% from elementary to high school level, respectively).

In addition, only seven respondents attended Hui elementary school and received education related to traditional Hui culture. When asked whether they would like to send their children or third-generation children to schools with Hui characteristics, the respondents stated that such schools were not competitive under the pressure of further education and employment, so they did not want their children to get back to similar schools.

Analysis of the Evolution of Education Level

In ancient times, education in China was established to adapt to the ancient social structure that was based on manual labor, small peasant economy, and hierarchical society, and the main content was humanities education. The abolishment of the imperial competitive examinations at the end of the Qing Dynasty was the first step in the transformation of China's education under the impact of modernization. After a 100 years of exploration and development, the current modern education is a mature structural system formed in the active learning of the educational experience of developed Western countries, which focuses on scientific and cultural knowledge education. There are two main reasons for the educational transformation and changes in education levels experienced by the population in the Huis Settlement areas.

First, from the perspective of external environmental conditions, the tendency of being more highly educated among the Hui people in the settlement is influenced by a series of external policies, such as the introduction of modern education by the government. After the founding of the People's Republic of China, China implemented the modern education policy for eradicating illiteracy and basically completed the expansion of primary education by the 1980s. In 1986, the nine-year compulsory education law was promulgated, and the expansion of junior high school education was achieved at the beginning of the twenty-first century. Presently, the education in China has entered a new stage of modernization with the scaling up of high school education and the generalization of higher education and is currently facing new demands for improving the quality of education.

Second, from the perspective of the intrinsic needs of individuals, the Hui people in the settlement have also undergone a process of psychological acceptance from traditional ethnic education to modern education. According to the famous French educator Durkheim, the mission of education is to prepare individuals with the physical and mental conditions that they should be equipped with as members of society and of a specific group. Likewise, the education received by the Hui people in the Qijiawan Huis Settlement would change accordingly with the needs of society. The traditional education method of urban Huis population is the scripture education taught by the imam of the mosque, and the content of education is centered on Islamism that is limited to the religion-related ethnic culture. Under the impact of modern socio-economic and cultural structures, the Huis have started to leave the mosque and enter the general education system to acquire the necessary scientific and cultural knowledge through a modern education system, which is different from the traditional ethnic education (see Fig. 4.14).

The improvement in the education level of the Huis population is no exception; however, the proportion of Hui people with low education levels (elementary school and below) has not decreased significantly and is concentrated among the older people who do not have access to education previously, which may be related to the higher aging level of the Hui in the settlement.

[Example] Mr. Z, whom we met in Dafei Fuxiang, has a junior high school education and has spent six years in Elementary School for Hui people about 60 years ago, along with most of the Hui Min residents at that time. However, unlike most Hui people, Mr. Z believes that the merger of several elementary schools, including Elementary School for Hui people, is inevitable and has a

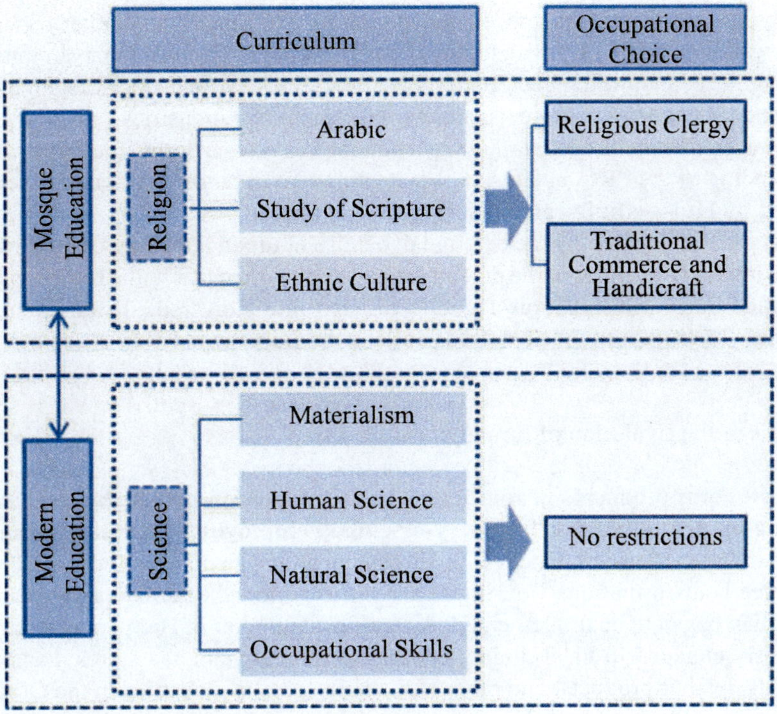

Fig. 4.14 Comparisons between mosque education and modern education. *Source* Data from the research group's sample study on the Qijiawan Hui settlement in Nanjing (2014)

promising outlook. "I heard that nowadays students in Elementary School for Hui people have to take a final exam on ethnic culture, so that they learn traditional culture without falling behind in mathematics, physics, and chemistry, so the current education is still much better than that of our time."

4.3.1.2 Evolution of the Employment Structure of the Hui Population

Evolution Characteristics of the Employment Structure

According to relevant historical documents, the Hui population in Nanjing before the founding of the PRC was mainly engaged in the service industry—such as jewelry, jade, and antique business—and food and beverage restaurants and bistros, while the residents of the Qijiawan Hui Settlement were mainly engaged in related industries supported by the cow and sheep butchery trade. The *Jianyewenshi* (Chronicle Records of Jianye) states that "since the Ming and Qing dynasties, eight out of ten households in Qijiawan have been engaged in the cow and lamb leather business." There was even a "jingle" made up by the Hui people of Qijiawan that summarized their ethnic economic characteristics—"selling chickens and ducks, slaughtering cows and sheep, acting as agents, and working in all five industries."

However, this traditional employment pattern was broken during the early years of the founding of the PRC, and an employment pattern of "**industry-based—supplemented by Hui—ethnic-related services**" was formed.

This pattern—driven by the external dynamics of urban industrial transformation and the internal dynamics of the pursuit for higher quality of life by the Hui people and diversified lines of occupation—reoriented itself into a diversified situation where "**all kinds of employment directions co-exist homogeneously**" (see Table 4.14).

Analysis of the Evolution of Employment Structure

First, the comprehensive promotion of industrialization before the reform and opening up promoted the "industry-oriented" employment pattern of the Hui people in the settlement. Under China's general guidance of "One Industrialization and Three Transformations[7]" during the transition period, the public–private partnership policy began to be implemented. A significant number of Hui people who were previously engaged in halal dining, beef and lamb butchers, tea stores, handmade garments, leather products, and other Hui ethnic-specific industries in the Qijiawan Hui Settlement began to work in factories, and the scale of Hui workers continued to grow as a result of the full-scale industrialization process. However, as urban Hui people had long depended on family workshops where residential and occupational functions were integrated together, their education level was generally low and their employment was mainly limited to positions as general workers with low technical threshold requirements.

Second, the industrial transformation of China's cities and the restructuring of state-owned enterprises brought about a diversified employment pattern for Huis population in urban settlements. The massive layoffs triggered by the restructuring of state-owned enterprises have led to a rapid decrease in the proportion of Hui people engaged in industrial production. In addition, the urban industrial transformation since 2000, which functioned as an extrinsic motivation—as well as the intrinsic incentive of the Hui people to improve their own education level and broaden their career choices—had resulted in the diversification of the types of employment of Hui.

Simultaneously, the urban renewal of Qijiawan area has disrupted the established form of the Hui Settlement. The scattering and mixed living of ethnic minority populations, while leading to an increased fragmentation of the specific demand for Hui ethnic-related products, had also hindered the regional clustering and development of ethnic-related services.

---

[7] That is, "socialist industrialization," "transformation of agriculture," "transformation of handicrafts," and "transformation of capitalist industry and commerce.".

Table 4.14 The evolution of the employment structure of Hui people in Qijiawan Huis settlement area

Stage	Employment by industry diagram	Brief analysis
Stage I (1949–1958)	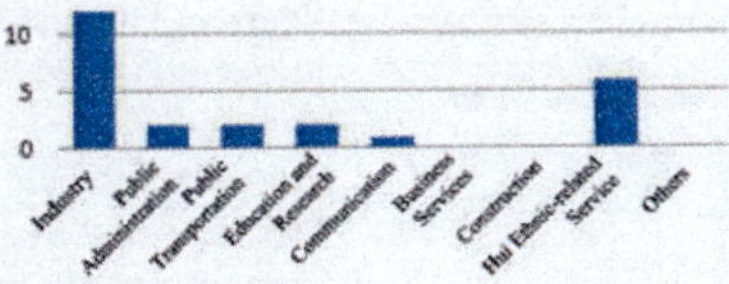	The industries engaged by the Hui people in Qijiawan mainly comprised industry (48%) and Hui ethnic-related services (24%) The Hui ethnic-related service industries in this period included beef and lamb stores, tea and beverage stores, tailor stores, leather products stores, etc. The business mode was mainly small-scale stores along the streets with families as the units
Stage II (1958–1978)	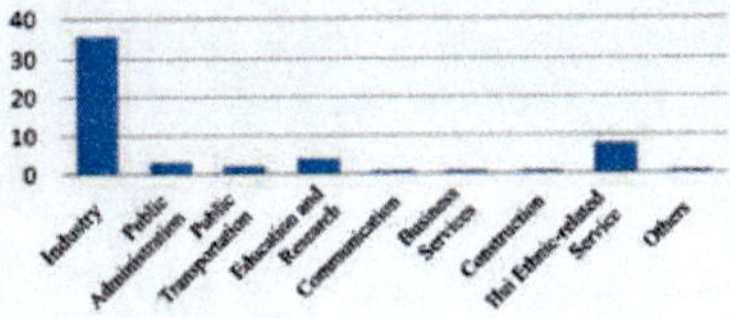	With the implementation of the public–private partnership policy, more Hui people entered factories (63%). However, limited by their education level, they were mostly general workers engaged in large-scale machinery manufacturing, pharmaceuticals, textiles and other second- and third-class industries. The Hui ethnic-related service industry was also restructured from the original family-run workshops to large state-run dining businesses
Stage III (1978–1990)	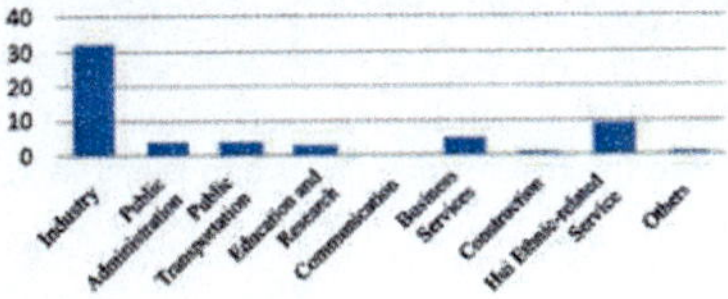	The sectors in which Hui people worked were still dominated by industry (54%). However, under the influence of the reform and opening up, the number of Hui people working in commercial services gradually increased (from 1.7% to 8.5%), and the types of industries they were employed were diversified
Stage IV (1990–1998)	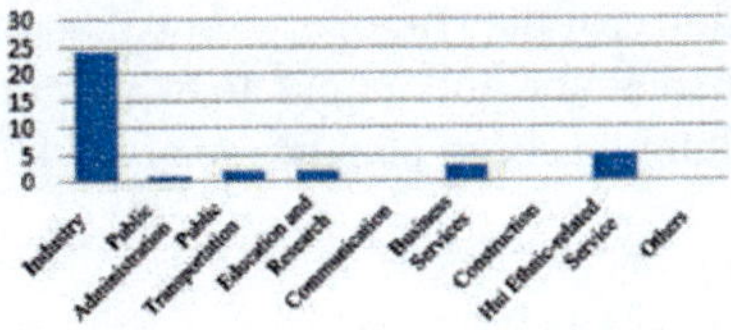	Although the direction of employment of Hui people changed slightly from the previous period, it was still mainly industrial and most Hui workers did not seek for other job opportunities after being laid off

(continued)

Table 4.14 (continued)

Stage	Employment by industry diagram	Brief analysis
Stage V (1998 to present)	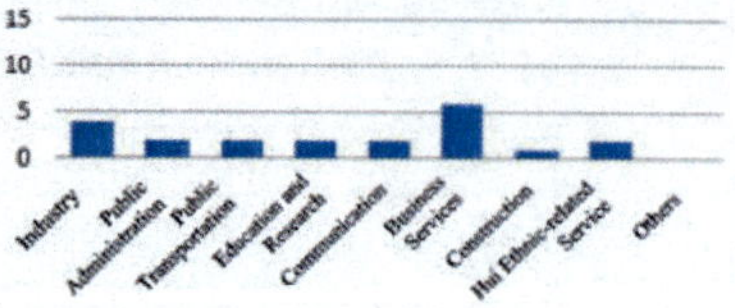	The employment direction of the current Hui residents is dominated by commercial services (28.6%), with 9.5% of the population returning to Hui ethnic-related services and a relatively balanced proportion of employees in the other industries

Source Data from the research group's sample study on the Qijiawan Hui Settlement in Nanjing (2014)

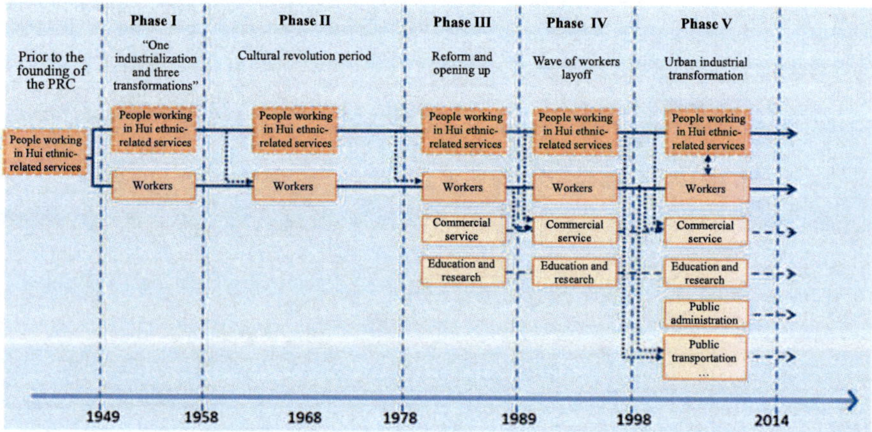

Fig. 4.15 Analysis chart of the evolution of the employment structure of the Hui people in the Qijiawan Huis settlement. *Source* Data from the research group's sample study on the Qijiawan Hui settlement in Nanjing (2014)

According to the research, nowadays, the staff of Hui restaurants in Qijiawan area is not limited to Hui people, and the proportion of Hui staff in two randomly sampled halal restaurants is less than 30%.[8] This phenomenon not only reflects the diversification of employment options for the Hui people but also implies the impact of modernization on ethnic industries and the extent to which the foundation of traditional ethnic customs and culture has been shaken (see Fig. 4.15).

[Example] Mr. Y is an enthusiastic man with a great sense of national pride. He is currently 75 years old; he spent his earlier years working at the factory

[8] On the day of research sampling, there were 19 staff in total in the two halal restaurants, of which only five were Hui.

in Yanziji, and his commuting time was around an hour's drive every day. His children are now relatively successful businessmen. Mr. Y said proudly, "I constantly taught them to do business when they were young. Hui people tend to act as agents and try to work in different fields. We had this tradition. Integrity was also an important quality. We should be honest but not untruthful, upright but not corrupted. It is the foundation of our ethnic group."

4.3.2 Evolution of Family Relationships

4.3.2.1 The Evolution of Inter-Ethnic Marriage

"Inter-ethnic marriage" is one of the most important indicators of changes in the family relations and life customs (and often religious beliefs) among ethnic minorities.

Evolutionary Features of Inter-Ethnic Marriages

The inter-ethnic marriage rate in Qijiawan Hui Settlement area shows a "linear increase" in general. To be specific, the inter-ethnic marriage rate from stage I to stage III increased only to a small extent, from 25.9 to 41.4% across the 40 years from the beginning of the founding of the PRC to 1990. However, since 1990, the increase of inter-ethnic marriage rate in the Qijiawan Huis Settlement has accelerated, rising to 82.5% in just 25 years from 1990 to the present (see Fig. 4.16).

Evolutionary Analysis of Inter-Ethnic Marriage

First, the change in the traditional concept of marriage among the Hui community is the critical reason. In the Islamic doctrine, common faith should be the primary requirement for the Hui when choosing a spouse, and this has been an important criterion that the Hui people have consistently followed since the formation of the ethnic minority group (see Fig. 4.17). However, with the impact of comprehensive modernization after the founding of the PRC, inter-ethnic communication in the cities has become an unstoppable trend (Rong 2004). The social relations of urban Hui youths in choosing their spouse have changed from kinship to local connections, and the criteria for choosing spouses were no longer limited to ethnicity and beliefs but focused instead on more practical factors, such as personal characteristics and economic resources.

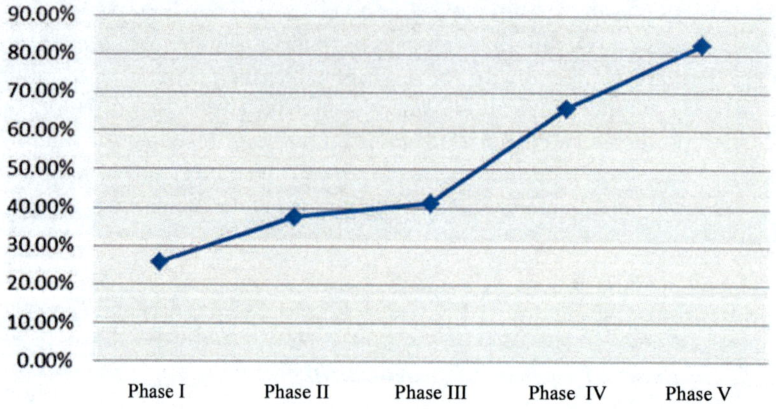

Fig. 4.16 Evolution of inter-ethnic intermarriage among the Hui people in the Qijiawan Hui settlement area. *Source* Data from the research group's sample study on the Qijiawan Hui settlement in Nanjing (2014)

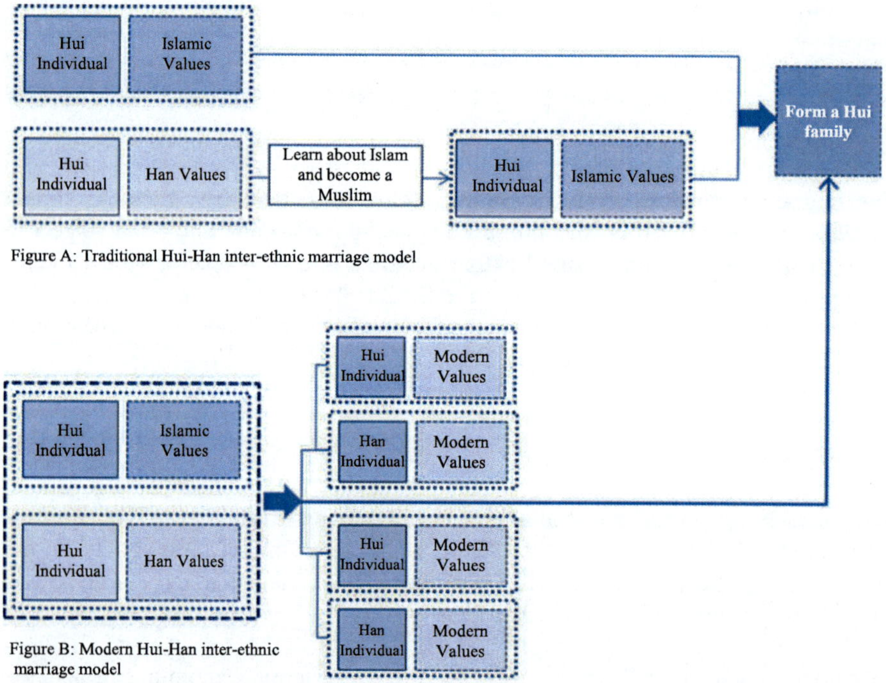

Fig. 4.17 Schematic diagram of the evolution of Hui-Han inter-ethnic marriage patterns. *Source* Own elaboration

Second, the breakdown of the traditional settlement pattern is an important external factor contributing to modern inter-ethnic interaction. As the largest traditional Huis Settlement in Nanjing, the traditional ethnic settlement pattern was broken after the urban renewal in the late 1990s, and the massive outflow of the original Hui population brought about a stagnation in intra-ethnic interactions, which made it difficult for Hui youth to meet prospective suitable native spouses on the one hand and created more opportunities for inter-ethnic interactions than before through mixed living between different ethnic groups on the other.

To solve this problem, the Caoqiao Mosque initiated the establishment of the "Hui Marriage Introduction Center" for the city's Hui people in 1995, providing consultation and introduction services for those in need. However, the average number of consultations per week was 3–5 individuals and the success rate was 3–5 couples per year, which reflects the irreversible increase in inter-ethnic marriages among the Hui people today.

[Example] Mr. L (pseudonym), 56, who lives in Hongtuqiao, is a native of Qijiawan area. Since the founding of the country, he had moved many times because of job changes, expropriation of old houses, and other issues. He has now moved back to settle in Hongtuqiao of Qijiawan area with his father.

Mr. L said that although the large family had long been separated in the early years, the traditional religious beliefs and living habits were still well maintained. Mr. M from next door "has a 31 year-old niece working in a bank and has not got married yet as she insisted on finding a Hui spouse." The traditional cultural perspective cannot reverse the trend of decreasing interactions with Huis outside the family, whose effect has been highlighted in the marriage of the younger generation.

4.3.2.2 Evolution of Family structure

Evolution Characteristics of Family Structure

The family structure of the Hui people in the Qijiawan Hui Settlement has changed a great deal since the founding of the PRC and generally follows the evolutionary trajectory of the family structure in China, showing the characteristics of "**diversified structure, small size, and aging population**" (see Table 4.15).

Analysis of the Evolution of Family Structure

Within the traditional cultural concept of China, which was based on the small peasant economy, the family is the most basic fundamental unit that provides comprehensive

Table 4.15 Evolution of the family structure of Hui people in Qijiawan Huis settlement[9]

Stages	Diagram of family structure	Brief analysis
Stage I (1949–1958)	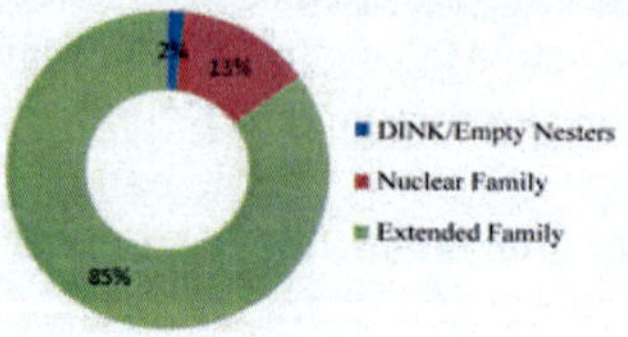	The structure of Hui families in the Qijiawan Huis Settlement was dominated by extended families with multiple children living together (85%), nuclear families living with unmarried children, accounting for 13%, and there were only 2% of DINK couples or empty nesters
Stage II (1958–1978)	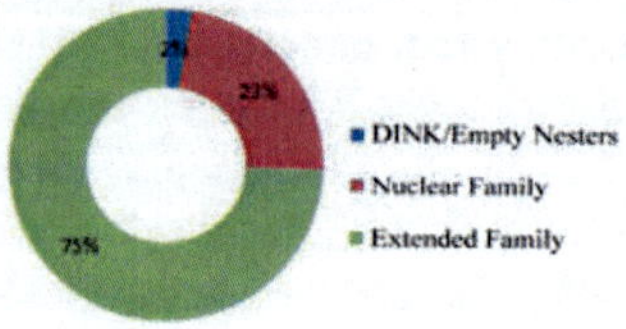	The structure of Hui families in the Qijiawan Huis Settlement has relatively small changes compared with the previous period: the proportion of extended families living together with multiple children decreased slightly (75%), the proportion of nuclear families living with unmarried children increased to 13%, while the proportion of DINK couples or empty-nester families remained low at 4%
Stage III (1978–1990)	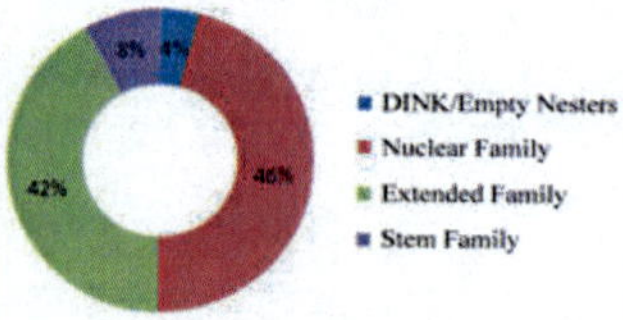	The overall structure of the Qijiawan Huis Settlement, which was absolutely dominated by extended families, was broken, forming a situation where extended and nuclear families were of a similar proportion (42% and 46%, respectively), and the proportion of nuclear families exceeded that of extended families for the first time. Stem families began to account for a certain proportion (8%), while the proportion of DINK couples or empty-nester families remained at 4%
Stage IV (1990–1998)	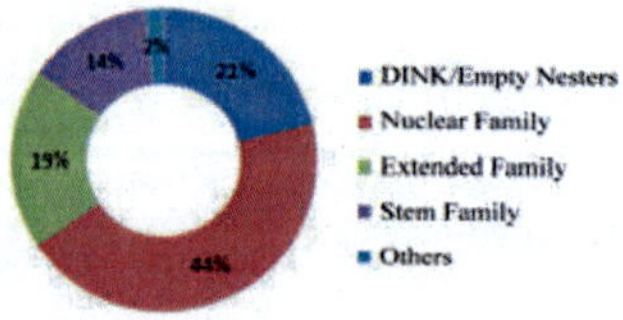	The proportion of nuclear families tended to stabilize (44%), while the proportion of extended families shrank further (19%). The corresponding proportion of stem families increased to 14%, the proportion of DINK couples or empty-nester families increased to 21%, while single or divorced families accounted for 2%

(continued)

[9] Nuclear family: family model comprising a couple and unmarried children; Stem family: family model comprising a couple and married children; Extended family: family model comprising parents and more than two pairs of married children, or multiple nuclear families with married siblings; DINK/empty nester family: family model of a couple without children or living separately from their children.

Table 4.15 (continued)

Stages	Diagram of family structure	Brief analysis
Stage V (1998 to present)	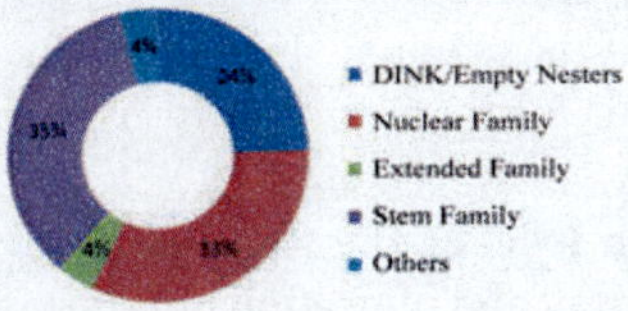	The proportion of stem families in the Qijiawan Huis Settlement increased significantly (from 14 to 35%). The proportion of extended families decreased from 19% to only 4%, and the proportion of nuclear families also decreased slightly (from 44 to 33%). The proportion of DINK or empty-nester families and other type of families did not fluctuate much, stabilizing at 24 and 4%, respectively

Source Data from the research group's sample study on the Qijiawan Hui Settlement in Nanjing (2014)

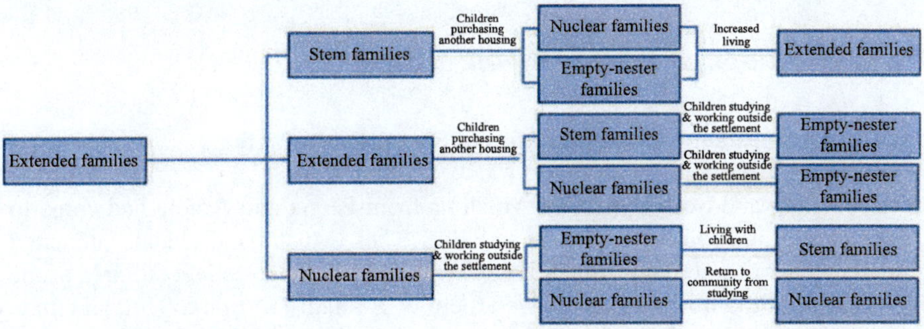

Fig. 4.18 Analysis of the evolution of the family structure of the Hui people in the Qijiawan Hui settlement. *Source* Data from the research group's sample study on the Qijiawan Hui settlement in Nanjing (2014)

protection for family members and serves as the platform for their social life. In the urban Hui people's concept, family is a guarantee for the integration of study, work, and the life of family members, and its structure has become a typical model that carries and reflects the changes of Hui family relations. Since the reform and opening up, the traditional family structure of the Hui, which is "mainly multigenerational," has become "nucleated" for the following reasons (see Fig. 4.18).

First, it was the result of the socialist modernization or transformation of "One Industrialization and Three Transformations." This not only transformed the traditional handicraft industry of the Qijiawan Hui people but also fundamentally shook their integrated and family-based organizational mode of production and living. In addition, the occupational differentiation and commuting considerations drove many Hui workers to live away from home.

Second, the comprehensive one-child policy and the increasing cost of raising children in the 1980s. The result is that post-1980s Hui families often chose to have

1–2 children, and the departure of their children after turning 18 directly led to the transformation of the stem/nuclear family into an empty-nester family.

Third, the reform of the housing system and urban renewal in the late 1990s. This changed the housing conditions of the original Huis and reduced their housing size per capita, making it more difficult for them to live as large multigenerational families.

Fourth, the augmenting education level of the Hui people in the settlement. Along with this factor, population mobility and migration rates have increased; more and more Hui youth preferred to choose the modern small family patterns, and the phenomenon of late childbearing among young Hui couples has led to an increase in the proportion of single-person and nuclear families.

In addition, the aging tendency of Hui in the settlement is one of the important reasons for the increasing proportion of empty nesters and stem families among the current Hui population.

4.3.3 Evolution of Social Organization

4.3.3.1 Evolutionary Characteristics of Social Organizations

Since the Tang and Song dynasties, Muslims from Persia and Arabia had come to China for trade and political exchange activities. These Muslim individuals interacted with each other and actively absorbed and learned from the excellent cultures of the Han Chinese at the time, which allowed them to gradually form into Hui, an ethnic group with common religious beliefs and customs. From the historical process of the formation and development of the Hui over the centuries, it can be seen that the Islamic culture has become the most prominent spiritual foundation in Hui's ethnic culture and a connection that cohered the "different ethnic groups living in a mixed way over vast areas while some lived in the concentrated communities in small areas" pattern. This culture has also penetrated all aspects of Hui's traditional philosophy and lifestyle. The mosques thus established have become the spiritual and functional hub of the Hui settlements and are the fundamental reason why the Hui has been able to gain a foothold in the predominantly Han Chinese society of China.

In the Qijiawan Huis community, the Hui social organizations represented by the religious community system also assume the characteristic central function of community cohesion and operation.

However, this system was basically disbanded after the religious reform in 1958, and with the advancement of urban modernization and the establishment of new organizations, the Hui social organizations in the settlement have shown a trend of "**unification of institutional functions and generalization of inter-ethnic interactions.**" More specifically, it is manifested in the following aspects.

I. **During the "One Industrialization and Three Transformation," the cattle industry union and the leather and bone industry union in Qijiawan Huis**

Settlement that collaborated with the Caoqiao Mosque were dissolved. In 1958, during the religious reform period, the Nanjing Islamic Association and other religious organizations established by the government directly replaced the board of directors of the original mosques. The original 32 mosques, private schools for Huis, and girls' schools for Hui were merged into 8 mosques and handed over to the Housing Authority for "renting or management." At this point, the social organizations of the Qijiawan Hui Settlement "**transformed from the embodiment of a compound functions of religious, work, and life customs to serving a single function based only on religion**" (see Table 4.16).

II. Hui residents are actively involved in socialist construction. The modern society with dialectical materialism as the core of its ideology has shaken the traditional Islamic cultural foundation of the Hui. In the early years of the founding of the PRC, the socialist transformation activities of the "One Industrialization and Three Transformations" were in full swing, and the Hui people in the Qijiawan Hui Settlement, like other residents, were subjected to the management of the new dual urban community organizations of the "institutional system" and the "residents' committee" and had a high desire to join the new regime. According to relevant data, the number of Hui cadres in Nanjing's municipal bureaus and departments had reached 713 by early 1959, including 391 CPC Party members, where more than half of them were Hui residents in the Qijiawan Settlement (see Table 4.17).

III. The fracture of social connections and the weakening of the clustering of living space have led to increasing difficulties in intra-ethnic interactions. The research found that the social interactions of the Hui residents in Qijiawan have gradually evolved from "a pattern where the Hui residents predominantly socialize within their neighborhood to a pattern where they interact with their Han neighbors or co-workers" since the founding of the country, which shows that inter-ethnic interactions have become the main trend in their interpersonal relationships (see Table 4.18).

Analysis of the Evolution of Community Organizations

First, from the perspective of the external environmental conditions, the wrong ethnic and religious policies implemented in the late 1950s seriously interfered with and destroyed the established religious activities. During this period, all the mosques were closed or occupied, resulting in the Hui people in the Qijiawan Huis Settlement having no place to worship as well as having to hide their ethnic identity, which shook the social organization based on the mosques. It was not until the implementation of ethnic and religious policies and the restoration of religious activity sites in the early 1980s that the number of mosques in the main city of Nanjing was restored to three; however, it was still far from the 32 that existed at the beginning of the founding of the country. The decrease in the number of mosques and the continuous migration of Hui residents resulted in such a consequence that the

Table 4.16 Evolution and status of Hui social organizations in the Qijiawan Huis settlement

Item	Muslim community	Industry union	Mosque board of directors	Islamic association	Community residents committee
Main functions	Religious affairs, rituals of life (slaughter of animals on wedding and funeral occasions)	Defend the interests of the industry and the rules of the trade, and finance the mosque within the workshops	Support the education of Hui people, provide charity to poor compatriots, and propagate the doctrine	Assist the people's government in conducting religious affairs and propagating doctrines	Manage the public affairs of the community and maintain normal order
Start and end date	Yuan Dynasty-1958	Qing Dynasty -1958	Qing Dynasty -1958	1959 to present	1954 to present
Nature of organization	Community-based grassroot organizations with an Islamic cultural identity	Community-based grassroots organizations	Community-based grassroot organizations with an Islamic cultural identity	Religious activity groups managed by the State. Administration for Religious Affairs	Basic organization of urban grassroots authority in China
Consequence	Dissolution of the religious community system has led to the loss of the function of the joint management service in terms of religious affairs and daily life of the Hui people	Disintegration of Hui trade unions, resulting in the loss of intra-Hui trade connections and mutual support functions	Dissolution of Mosque Board, resulting in the loss of guarantee for the education and public welfare of the Hui people	It is mainly responsible for religious affairs at the municipal level and does not get involved in community management matters, and it has little interference with other related elements of ethnic traditions	It serves to manage the daily life of all residents in the district, but there is a lack of a specific department for minority customs

Source compiled from Nanjing Islamic Association. *Nanjing Hui Islamic History Manuscript* [M]. Nanjing Islamic Association, 1999

Table 4.17 Statistics on the distribution of Hui CPC carty members in Nanjing in 1959

Type	City	Suburban counties	Municipal enterprises	Finance and Trade System	Colleges and Universities	Secondary Technical Schools	Hospital	Transportation, Post and Telecommunications	Party and government organs	Railway	Urban Construction	Total
Number of people	137	52	114	6	6	2	3	17	26	20	8	391

Source Nanjing Hui Islamic History Manuscript [M]. Nanjing Islamic Association, 1999: 195

Table 4.18 Evolution of Daily Socialization Objects of Hui People in Qijiawan Huis settlements

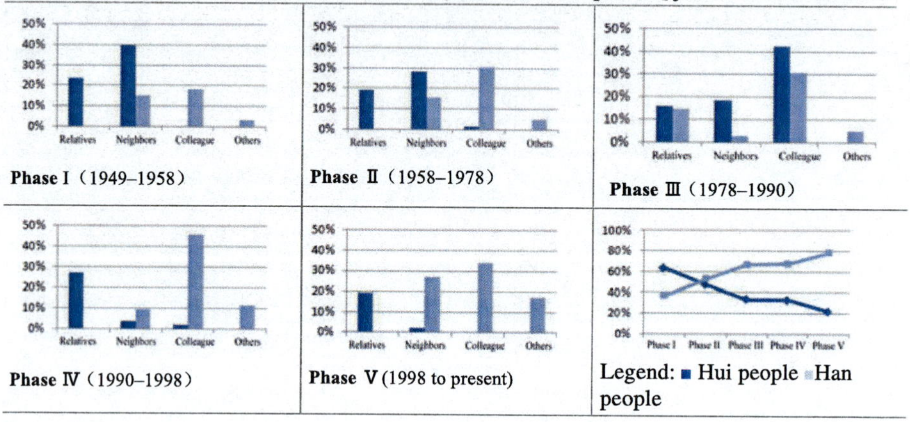

Source Data from the research group's sample study on the Qijiawan Hui Settlement in Nanjing (2014)

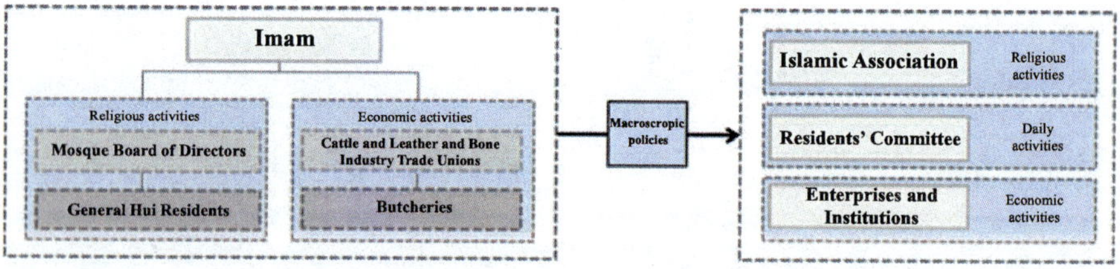

Fig. 4.19 Analysis of the evolution of Hui social organizations in the Qijiawan Huis settlement. *Source* Own elaboration based on relevant information

group was unable to maintain the original community organization and management function of the religious community system centered around the mosque, which was the external reason for the change of social organization in the Qijiawan Huis Settlement (see Fig. 4.19).

Second, in terms of the subjective motivation of the Hui people, their lack of identification with their own cultural origins and ethnic identity had also contributed to the decline and dissolution of social organizations that had ethnic characteristics. Since the 1980s, although the mosques and relevant religious activities have been gradually restored, the majority of the Hui population has been influenced by the long-term modernization process of the country and has accepted the modern education system affected by the Han culture. These people started to engage in employment activities unrelated to their own beliefs and customs, and their identification with the Islamic culture itself was no longer high. This was because the Hui people were more concerned about their economic interests and the improvement of their material environment, which has become the internal reason for the disintegration of the distinctive social organizations and the loss of their spiritual functions in the Qijiawan Huis Settlement.

> **[Example]** Y, 25, is the youngest Hui in Qijiawan that the team interviewed during the research. After graduating from university, Y worked as a white-collar worker for one year and then joined the mosque to help manage the clerical affairs. Because of her work and a good ethnic and cultural atmosphere in her family, Y deeply appreciates the difficulty of communicating and interacting with her fellow Hui people nowadays. In less than four months, she has already helped the mosque to organize many Hui fellowship activities and lectures on Hui culture, which has greatly promoted exchange among the Hui community around Qijiawan and Nanjing as a whole.

After the PRC was founded, the ethnic characteristics of the Qijiawan Huis Settlement are mainly manifested in the continuation of religious beliefs and adherence to dietary taboos. An analysis of the changes in the Halal diet and religious beliefs of the Hui population by stages reveals an "**increasing ignorance for dietary and religious traditions among Hui after the reform and opening up**," while the participation in religious activities has shown the characteristics of "**polarization and serious ageing**" (see Table 4.19).

In addition, in terms of traditional language, only a few Hui imams are familiar with the language (e.g., "teaching language in the mosque" and "Xiao'erjing") and use it in their sermons, while the Hui people in general only greet each other with the Arabic word "Salem" when they meet, and the language has long ceased to play a role in intra-community communication. In terms of traditional costume, the typical costume tradition of "Taqiyah for males and Hijab for females" is noted only when the Hui people worship, and the costume they wear when not in mosques is no different from that of the Han Chinese, making it difficult to identify their ethnic identity.

4.3.3.2 Analysis of Evolution of Living Customs

First, the influence of the national ethnic and religious policy and its related policies is a macro factor in the evolution of the traditional living customs of the Hui people in the Qijiawan Huis settlement. With the initial establishment of the socialist public ownership economy, the reforms in the superstructure also came along, which inevitably involved changes in ethnic religion, customs and practices, as well as the evolution of the old system of rules and regulations. Although the religious reforms in the early period of the founding of the country were moderate, the closure of mosques and suspension of religious activities during the Cultural Revolution fundamentally undermined the traditional social structure of the Hui people and affected their cultural confidence and cohesiveness. In addition, during the planned economy period, the government strictly forbade the "slaughtering of cattle," leading to inadequate supply of beef and mutton. As a result, some halal beef stores were only able to plan for a daily ration of 1.4.2 kg of beef, which was not enough to sustain the business and meet the basic dietary needs of the Hui. This was another reason why some Hui had to change their dietary habits.

Table 4.19 Evolution of living customs of Hui people in Qijiawan Huis settlement area

Stages	Diet and religion	Frequency of worship	Brief analysis
Stage I (1949–1958)	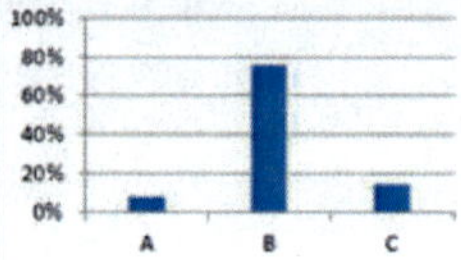	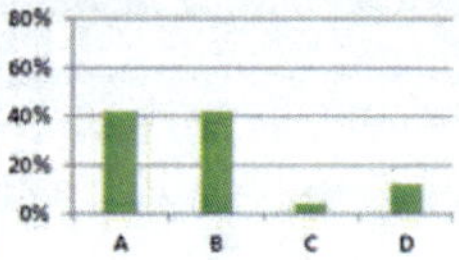	Most Hui in the Qijiawan Huis Settlement could observe the Hui dietary taboos, and only 8.3% of the non-Muslim Hui did not abide by the dietary taboos at all During this period, the Hui people worshipped more frequently, and more than 80% of them went to the mosque weekly
Stage II (1958–1978)	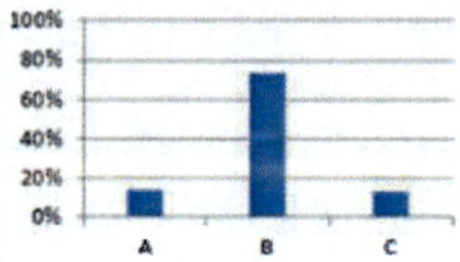	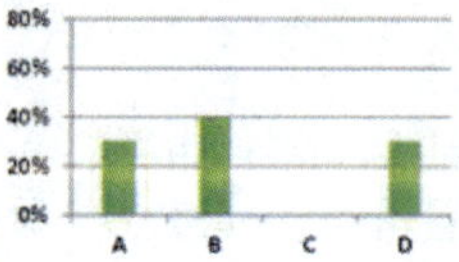	There was little change in dietary and religious customs among Hui, but the proportion of non-Muslim Hui who do not observe dietary taboos increased slightly (up to 13.9%) The frequency of worship by Hui people during this period decreased slightly compared with the previous period, and the percentage of Hui who worship less than once a month on average increased to 30%
Stage III (1978–1990)	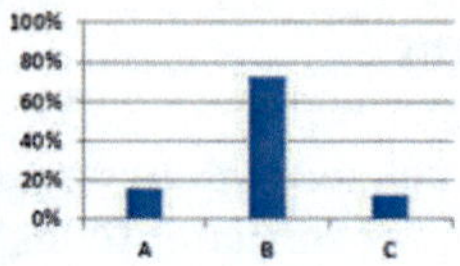	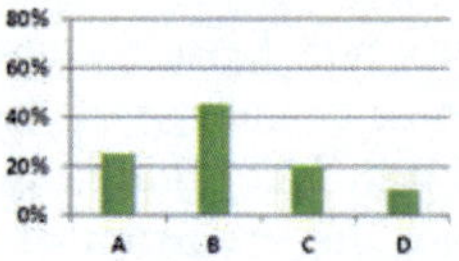	The dietary habits of the Hui retained the characteristics as the previous phase. However, in terms of religious customs, the frequency of worship changed significantly, and it is dominated by more than once a week (70% in total), and the low frequency of once a month or less is increasing (30% in total)

(continued)

Table 4.19 (continued)

Stages	Diet and religion	Frequency of worship	Brief analysis
Stage IV (1990–1998)	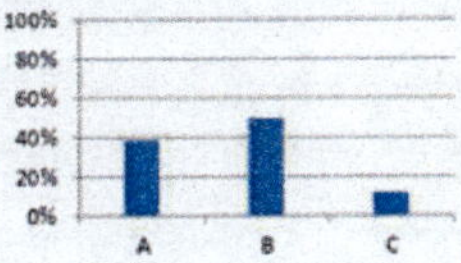	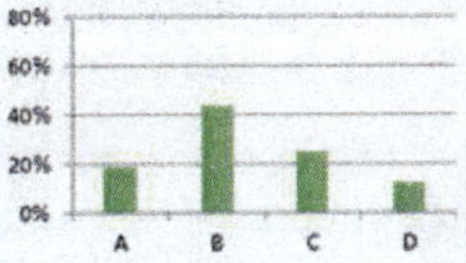	The proportion of Hui who do not abide by traditional taboos in the Qijiawan Huis settlement has increased sharply (36.6%), breaking the original living customs dominated by the observance of dietary taboos The frequency of worship in this period continued the characteristics of the previous phase, with a slight decrease in the proportion of daily worshipers (18.8%)
Stage V (1998 to present)	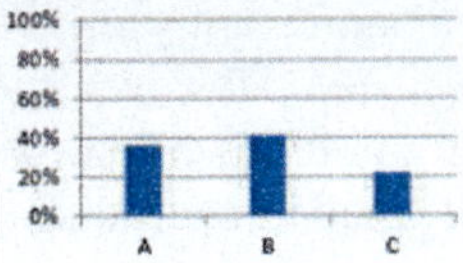	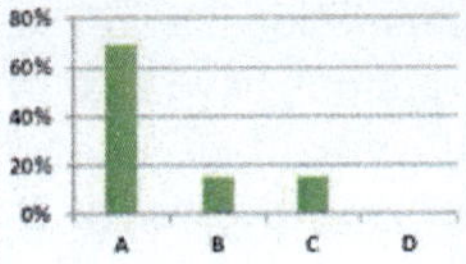	The percentage of Hui who both observe dietary taboos and attend worships has increased (21.9%) in the Qijiawan Huis Settlement The frequency of worship among the Hui in this period increased rather than decreased, with the proportion of those who worshiped daily reaching 69.2%
Note	A does not observe; B only observes dietary taboos; C observes religious and dietary taboos	A 1–5 times a day; B 1–5 times a week; C 1–3 times a month; D fewer than 1 time a month (*Note: The above statistics are only for Hui residents who attend religious services.)	

Source Data from the research group's sample study on the Qijiawan Huis Settlement in Nanjing (2014)

Second, changes in the basic family structure and compact community lifestyle are the mid-scope factors in the evolution of the traditional living customs of the Hui people in the Qijiawan Huis settlement. The tendency of miniaturization and nucleation of the family structure, as the tiniest unit constituting social life, has led to more independent lifestyles of family members. Family structure is the tiniest unit constituting social life, and its tendency of miniaturization and nucleation leads to more independent lifestyles of the family members. At the same time, urban renewal has broken the traditional settlement pattern of the Hui community, impacted Halal food stores with Hui ethnic characteristics, and weakened the interactions and networks within the Hui group. Traditional cultural customs have changed due to the impact on the family and community.

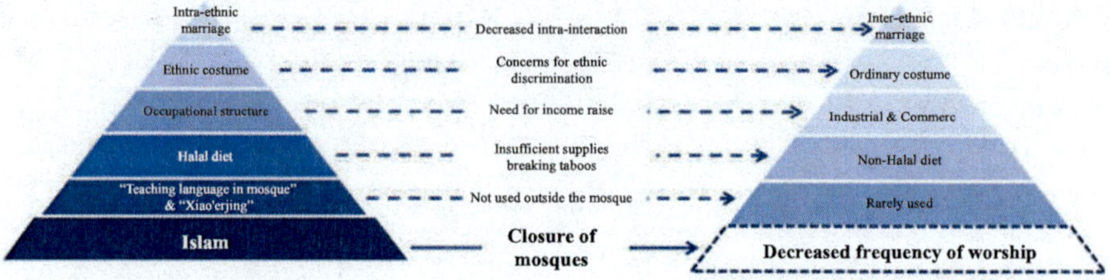

Fig. 4.20 Analysis of the evolution of Hui living customs in the Qijiawan Huis settlement. *Source* Author's own drafting

Finally, the weakening of the Hui people's own identification with their traditional ethnic culture is a microscopic factor in the evolution of the traditional living customs of the Hui people in the Qijiawan Huis settlement. While the education level of the Hui population has increased significantly since the founding of PRC, their employment options have also diversified. However, they have also inevitably lost their original sense of ethnic identity in terms of individual thought and cultural awareness while being involved in a fully modernized educational system and employment structure. Several surveys of the mosques in the main city of Nanjing found that majority of the local Hui people that worship in Nanjing are retired Hui workers (as they have relatively plenty of time and had been religious for a long time). In one of the mosques, there were more than 20 staff members, although they were engaged in ethnic-religious related work, fewer than 10 of them actually practiced Islam and worshipped on a weekly basis. These factors reflect the potential influence of individual Hui people on the inheritance of traditional living practices in terms of ethnic and cultural identity (see Fig. 4.20).

[Example] Mr. M, who is now 96 years old, moved from Qijiawan to live near Mochou Lake after 1997 and is hale and hearty. Every Friday, Mr. M, accompanied by his eldest son, would go to Jingjue Mosque to attend the Jumu'ah, which he has been doing for nearly 20 years without any hindrance. However, when talking about his life before retirement, Mr. M. said, "I was very busy working at the telegraph office in the 1970s and 1980s and often came home after 11:00 p.m., so I didn't have time to worship . . . at work, I often ate at the unit restaurant, so I wasn't able to be particular about it." There are quite of lot of Hui people like Mr. M. who only started attending religious activities after retirement, and the regular Jumu'ah every Friday became an important part of social interaction in their later years.

4.4 Evolution of Spatial Attributes of Huis Settlements in Nanjing

Based on the above analysis of the evolutionary stages of the Qijiawan Huis Settlement and the evolutionary characteristics of its social attributes, this section continues to focus on the unique spatial settlement form of the Hui people and the spatial utilization methods that reflect their ethnic characteristics and analyzes the general characteristics of the evolution in spatial attributes and their influencing factors in the Qijiawan Huis settlement area from the aspects of community function, residential space–time, employment space–time, and service facilities.

4.4.1 Evolution of Community Functions

4.4.1.1 Evolution of Land Use Functions

Evolutionary Characteristics of Land Use Functions

The research found that in terms of land use functions, the Qijiawan Huis settlement has undergone an evolutionary process from "**single function of mainly residential use of land to a mix of multifunctional use of land,**" and the spatial pattern also evolved "**from a traditional pattern of living around a mosque to an open layout with mixed functions**" (see Table 4.20).

Evolutionary Analysis of Land Use Functions

Modernization wave enhanced the urban location of the Qijiawan Huis settlement: Located in the south of Nanjing's old city and close to the city center of Xinjiekou, Qijiawan Huis Settlement is a typical segment of the traditional city with a rich cultural and historical heritage. Influenced by the accelerated urban construction in Nanjing since the 1980s and the agglomeration effect of public service center in Xinjiekou, the upgrade of the location conditions of the area around Qijiawan has become a new development point of the city. The role transformation of the district has brought about the rise of its commercial development value and receding of its original residential function. Furthermore, improved traffic and road conditions (transformation of Jianye Road on the north side, Zhongshan South Road on the east side, Shuiximen Street on the west side, and the widening of Dingxin Road and Cangxiang Lane inside the area) broke the original spatial pattern of the Qijiawan settlement while facilitating urban traffic and travel.

Influence of modernization wave on functional layout of residential areas: The functions of urban residential areas in China have transformed from mixed to single and then back to the mixed function. Traditionally, urban residential areas

Table 4.20 Evolution of land use function in the Qijiawan Huis settlement

Stages	Land use map	Pattern diagram	Brief analysis
Stage I (1949–1958)		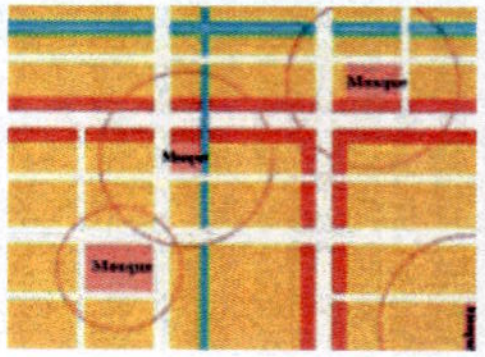	The function of the settlement is mainly residential, with only some Hui stores open along the streets aligned in the east—west direction in Qijiawan and in the north—south direction in Pingshi Street, with a single function and a dense and regular road network. The function is relatively homogeneous and the road network is dense and regular During this period, there were many religious places such as Caoqiao Mosque, Dahui Fuxiang Girls' School, DenglongxianMosque, and Jingjue Mosque adjacent to the settlement, which is a continuation of the traditional spatial pattern of living around the mosque

(continued)

Table 4.20 (continued)

Stages	Land use map	Pattern diagram	Brief analysis
Stage II (1958–1978)	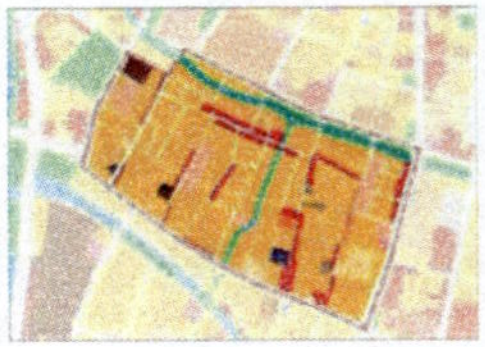	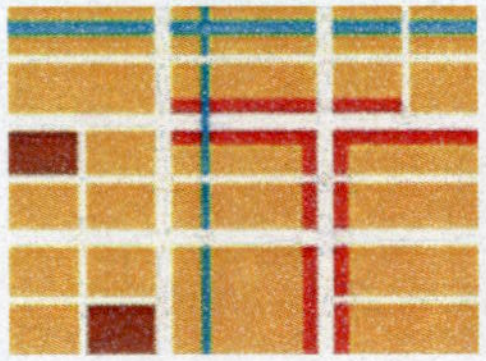	Under the influence of the planned economy, industrial sites such as food factories and radio equipment factories started to appear in the settlement, and its residential function remained, yet the traditional family-operated stores of the Hui people gradually decreased During this period, the spatial pattern of "living around the mosque" no longer existed due to the occupation of the mosque and the cessation of religious activities

(continued)

Table 4.20 (continued)

Stages	Land use map	Pattern diagram	Brief analysis
Stage III (1978–1990)	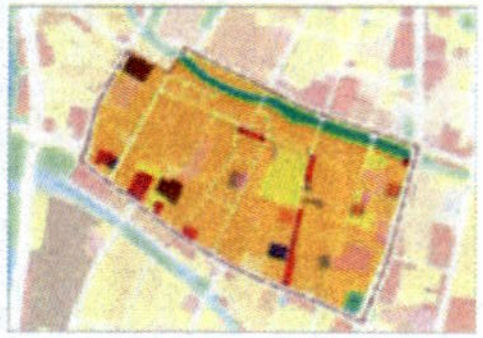	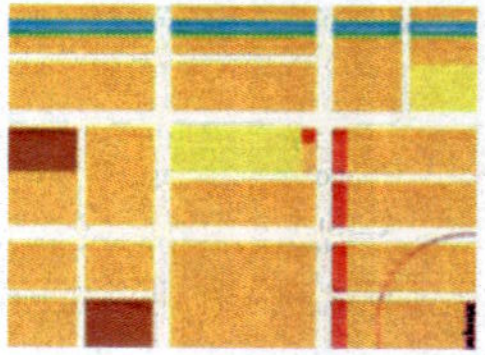	With the acceleration of urban construction after the reform and opening up, new residential buildings gradually appeared in the settlement, the number of traditional stores open along the streets decreased further, the density of the road network decreased, and the scale of the neighborhood increased Only the Jingjue Mosque was reopened during this period, and there was lack of mosques and religious activities within the settlement

(continued)

Table 4.20 (continued)

Stages	Land use map	Pattern diagram	Brief analysis
Stage IV (1990–1998)	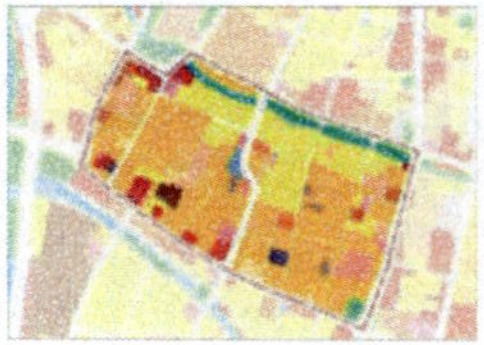		As the period with the most dramatic spatial changes in the Qijiawan Huis settlement, it was marked by the obvious adjustment and transformation of the road network, which had a significant impact on the overall pattern of the settlement: construction of Dingxin Road, demolition of Dahui Fuxiang Girls' School, and relocation of more than 100 Hui families marked the complete disappearance of the core location features of Qijiawan

(continued)

Table 4.20 (continued)

Stages	Land use map	Pattern diagram	Brief analysis
Stage IV (1998 to present)	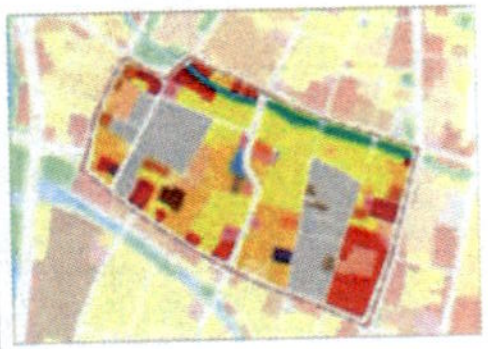		The display of traditional ethnic minorities who lived in the south of the city in Nanbuting neighbourhood, which is located in the southeastern part of the city currently, largely drove the development of this area, while other areas face further expropriation and renovation Even with the new Caoqiao Mosque, it is impossible to restore the spatial pattern of the old Hui community of living around the mosque, since most new residents are Han Chinese
Legend	Mosque; Public facilities; Class II residential land; Administrative facilities; Class III residen-tial land; Water; Industrial land; Public green space		

Source Data from the research group's sample study on the Qijiawan Huis Settlement in Nanjing (2014)

not only had residential functions but also had stores formed spontaneously to meet the needs of nearby residents for life services, which was especially evident in Huis settlements, a population good at business. Additionally, the construction of working-unit courtyards after the PRC was founded led to the self-sufficiency of residential areas and isolation and separation of public service functions from these areas. With the increasingly diversified needs of today's residents, the planning and construction of commercial housings have often been closely related to education, medical care, commercial services, and other types of facility, and the new trend of "marketization of business and socialization of services" has led to the mixed functions of the Huis settlement again.

Impact of modernization wave on traditional cultural concepts of the Hui people, especially their religious outlook: The evolution of the religious architecture in the Huis settlement of Qijiawan—from the mosques being closed or demolished under the impact of the religious reform in 1958—to the Caoqiao Mosque being restored and reconstructed in 2005, ethnic religious activities have been missing in Qijiawan for nearly half a century. The stagnation of traditional religious activities and withdrawal of related organizations have not only destroyed the traditional pattern of the original mosque but also shaken the foundation of the traditional Hui culture. From having a strong belief to being indifferent, the gradual change in the religious outlook of several generations of Hui people is the fundamental reason for the low sense of self-identity and weak demand for spontaneous clustering settlement among contemporary Hui people.

4.4.1.2 Evolution of Architectural Texture

Evolutionary Characteristics of Architectural Texture

With the function of land use changing from single to diverse, the texture of the Qijiawan Huis settlement also transformed from the "**traditional texture of small blocks, dense road networks division, low-rise, and small volume buildings to the modern texture of high-rise and large volume buildings**" since the PRC was founded (see Table 4.21).

Evolutionary Analysis of Architectural Texture

Demand for compound functions of land use due to urban development under the wave of modernization: The evolution of the architectural texture of the Qijiawan Huis settlement is directly influenced by the mixed use of land, and the demand for functional compounding of modern office and commercial facilities often translates into the multi-component of space. Therefore, the composite evolution of the functions of land use is one of the important reasons why the architectural texture of the Qijiawan Huis settlement tends to be high-rise and large volume.

Table 4.21 Evolution of Architectural Texture in the Qijiawan Huis Settlement

Stages	Architectural texture analysis diagram	Brief analysis
Stage I (1949–1958)		The Qijiawan Huis Settlement is dominated by traditional residential buildings that are small in scale and high in density; the buildings are mainly 1–2 stories in height, with a dense and homogeneous distribution pattern in space
Stage II (1958–1978)		Influenced by the national policy of comprehensive industrialization, large-volume buildings such as schools and factories began to appear in the Qijiawan Huis settlement, but the overall texture is still dominated by small blocks, dense road networks, and small-scale residential building characteristics
Stage III (1978–1990)		As urban residents had increasing demand for life quality under the wave of modernization, the newly built Pingshi Street community in the Qijiawan Huis settlement contrasts with the size of the surrounding traditional buildings

(continued)

Table 4.21 (continued)

Stages	Architectural texture analysis diagram	Brief analysis
Stage IV (1990–1998)		With the accelerated modernization of the city, the Qijiawan Huis settlement experienced the most drastic spatial changes. During this period, the widening of many roads, especially the construction of Dingxin Road, broke the traditional road pattern of small blocks and dense road networks. The construction of many multi-storey and high-rise residential communities and office buildings also broke the small-scale traditional residential architectural texture
Stage IV (1998 to Present)	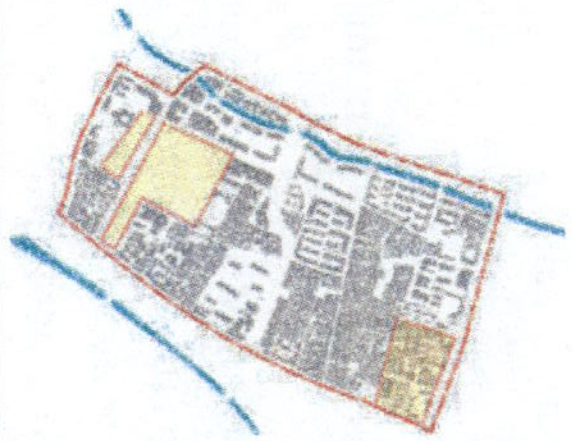	The architectural scale of the Qijiawan Huis settlement was further segmented, except the area around Gan Xi's former residence in the southeast corner, where the traditional scale and pattern are still maintained and restored. Most housings in the area were expropriated and the residents were relocated. The traditional architectural texture of the old southern part of the city has been gradually mixed with other urban areas in terms of scale and road network

Source Data from the research group's sample study on the Qijiawan Huis Settlement in Nanjing (2014)

Demand for urban space from the growing population under the wave of modernization: Since the PRC was founded, the population of Nanjing has increased from 1.35 million to 8 million (according to the sixth national census in 2010), with a population density of 1215 people/km^2. It is ranked right after Beijing, Shanghai, and Guangzhou in China. Within the city, the average population density of the eight sub-districts in the old city, including Chaotiangong sub-district where Qijiayuan is located, has exceeded 40,000 people/km^2.[10] Under the pressure of such a high population density, the emergence of a high-intensity land development model is inevitable, which led to the construction of high-rise buildings and large-scale land expropriation and renovation in the Qijiawan Huis settlements.

Negligence of urban developers to the traditional urban texture under the wave of modernization: As a typical traditional residential area in the southern part of Nanjing with a long history and ethnic characteristic settlement of the Hui people in Nanjing, Qijiawan did not receive enough attention at the beginning of urban renewal. While urban developers focused on improving the physical environment quality and accessibility of supporting facilities, they neglected the importance to preserve traditional styles and characteristic patterns, resulting in the gradual loss of traditional features in the architectural texture and scale of the Qijiawan Huis settlement, which has become an unidentifiable or fractured spatial fragment in the modern city.

4.4.2 Space–time Evolution of Settlement

4.4.2.1 Evolution of Time–space Structure of Settlement

This section draws on the relevant technical methods of time geography, collect, map and superimpose the residential "time" (by temporal stage) and "space" (by distance to Qijiawan Huis settlements) data at the city level, and then attempted to advance and deepen the discussion at different levels.

Characteristics of Time–space Structure of Settlement

First, this study considers the Caoqiao Mosque, the core of life having an obvious ethnic characteristic in the Qijiawan Huis settlement, as the reference point, calculates the time required to reach the mosque by various types of travel modes to divide the 44 sub-districts in the main city into four categories: core areas, basic areas, peripheral areas, and fringe areas (see Table 4.22 for the basis and manner of division).

---

[10] These eight main city sub-districts are: Chaotiangong, Xinjiekou, Shuangtang, Hunan Road, Hongwu Road, Wulao Village, Yijiangmen and Central Gate sub-districts. Source: Xinhuanet Jiangsu: http://www.js.xinhuanet.com/2012-08/12/c_112697335.htm.

Table 4.22 Sub-district division in the main Urban Area of Nanjing based on travel distance to Qijiayuan Huis settlement

Area division	Distance	Transportation mode	Transportation Time	Sub-districts
Core Area	≤1 km	Walking	≤15 min	Chaotiangong Sub-District
Basic Area	1–3 km	Bicycle/Public Transportation	10–30 min	Fenghuang, Huaqiao Road, Xinjiekou, Meiyuan New Village, Binhu, Nanhu, Wulao Village, Hongwu Road, Shuangtang, Fuzimiao Sub-districts
Peripheral Area	3–5 km	Bicycle/Public Transportation	30–45 min	Yijiangmen, Central Gate, Xuanwumen, Hunan Road, Ninghai Road, Jiangdong, Xinglong, Nanyuan, Zhonghua Gate, Yuhua Xincun, Houzai Gate, Ruijin Road, Daguang Road, Yueya Lake, Qinhong Sub-districts
Fringe Area	≥5 km	Bicycle/Public Transportation	≥45 min	Yanziji, Mufu Mountain, Maigao Bridge, Baota Bridge, Xiaoshi, Hongshan, Suojin Village, Xuanwu Lake, Yuejianglou, Jianning Road, Rehe South Road, Maqun, Xiaolingwei, Guanghua Road, Honghua, Ningnan, Sai Hongqiao, Shazhou Sub-districts
Sub-district Division Map	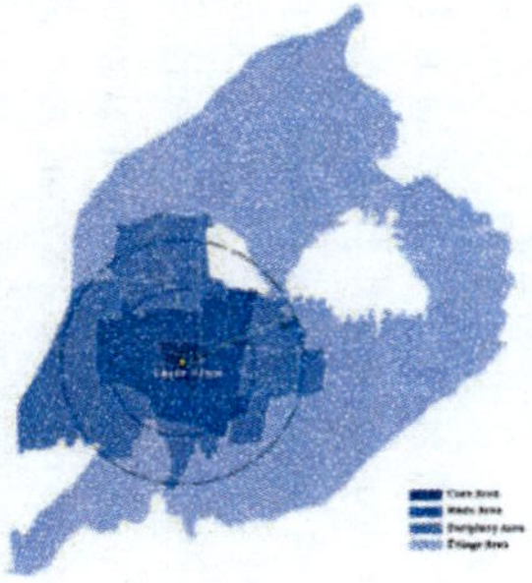			

Source Author's own drafting

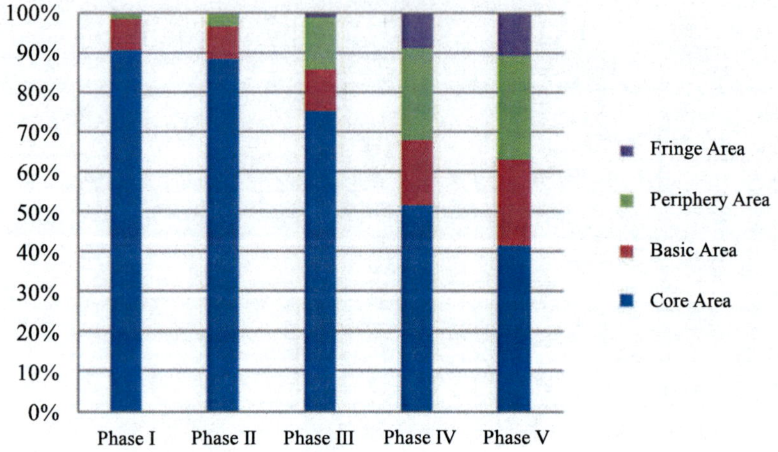

Fig. 4.21 Time–space structure of settlements in the Qijiawan Huis settlement. *Source* Data from the research group's sample study on the Qijiawan Huis Settlement in Nanjing (2014)

Based on the zoning of streets in the main city of Nanjing, the time–space structure of the living places for the former and current Hui residents in Qijiawan was mapped, and the differences in the time–space distribution of Hui living places were analyzed. The time–space structure of the Huis Settlement in Qijiawan shows that "there is a gradual change in the spatial distribution of settlement over time, from an "**absolute dominance of the core area" to a "balanced coexistence of all categories of areas.**"

When PRC was founded (Stage I), majority of the Hui people still lived in the core and basic areas (98.5%), and this proportion only decreased slightly in the next 30 years (Stage II 92.3%). In the 1980s (Stages III and IV), some Hui began to move out to the peripheral areas and broke the original residential structure. Not only was the proportion of Hui living in the peripheral areas higher than that of the basic areas but also the proportion of Hui in the core and basic areas dropped to 75.7%. Since 1998 (stage V), the four-external scattered settlements of Hui have formed an increasingly balanced spatial layout in each zone (see Fig. 4.21). Accordingly, the proportion of each spatial zone at each time stage also shows a trend of a gradual decline in the core zone and steady increase in the other zones (see Fig. 4.22).

Analysis of Time–Space Structure of the Settlement

Traditional economic system of the city has been transformed by the wave of modernization: The employment allocation system and the housing allocation system have been broken one after another, and the employment of the Hui people was no longer limited to traditional ethnic handicraft and specialty restaurants but had more choices of occupation like other citizens.

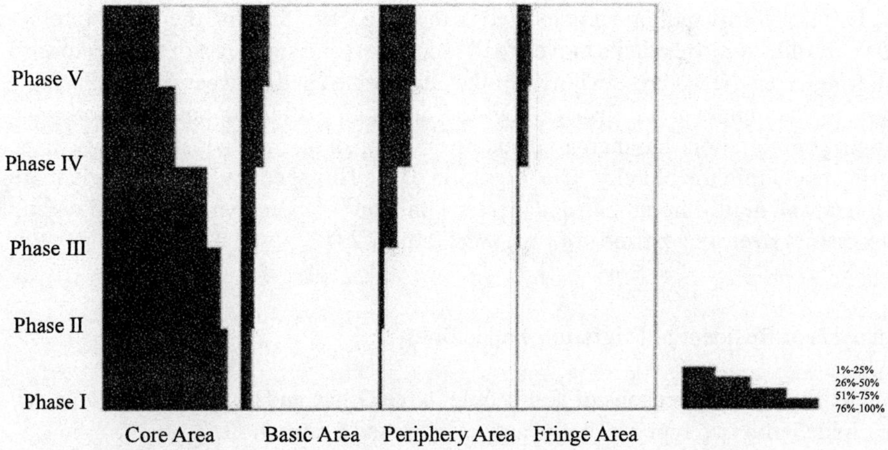

Fig. 4.22 Spatial structure of settlements in the Qijiawan Huis settlement. *Source* Data from the research group's sample study on the Qijiawan Huis Settlement in Nanjing (2014)

The renewal and reconfiguration of urban space under the wave of modernization: The reform of the land system has stimulated the migration and allocation of urban land resources and promoted the reconstruction of modern urban space. As part of the traditional urban area, Chaotiangong sub-district, where Qijiawan is located, is also one of the most important areas for the renewal of the old city, and as a result, a series of old city renovation and building demolition activities will inevitably contribute to the migration and dispersal of indigenous people from the settlement area to the main entire city.

4.4.2.2 Residential Migration Trajectory

Characteristics of Residential Migration Trajectories

Based on the data of change in residence of individual Huis at each stage, this paper uses the main urban area of Nanjing as the spatial projection plane and base map (with sub-district space as the unit) and the five major time stages divided in the previous paper as the z-directional vertical axis to superimpose the planar spatial trajectory and three-dimensional time–space trajectory of their residential migration, which can visually reflect the macro-trajectory of different individuals' residence during their life cycle. In general, the migration frequency and scope of the Hui people in the Qijiawan Hui settlement area have shown the overall characteristic of "**varies by age and mainly in the old city**".

A further analysis by age shows the following: the three-dimensional trajectory of the residences of Hui people over 55 years old is mostly vertical and linear, i.e., most of the Hui population insists on living in the Chaotiangong sub-district where the Qijiawan Huis settlement is located in the long term. Starting the 1990s (around 30 years old), some residents moved to the southwest and southeast of the city toward the fringe area. Nowadays, majority of the Hui people aged between 35 and 55 years live in Chaotiangong sub-districts, but the migration pattern is not significant, the number of migrations has increased but the direction of migration is still mainly in the basic area within the old city. The migration of the Hui aged under 35 years is more frequent and multi-directional, and there are only few of them living in Chaotiangong sub-district over long periods of time (see Table 4.23).

Analysis of Residential Migration Trajectories

The main migration reasons of Hui people in the Qijiawan Huis settlement area can be divided into two types, namely, active and passive migration.

From the perspective of a passive migration behavior, the renewal and expropriation of the existing housings of residents is undoubtedly the most important causal factor: This factor is characterized in the migration trajectory of individual Hui people by that fact that there is no age difference in the large-scale migration behavior of the group between the 1980s to 1990s.

From the perspective of an active migration behavior, the life course of individual Hui people is possibly the most important motivation: Cross-referencing the age at migration time and overall residential migration rate[11] can help identify the relationship between the two. It can be deduced that the residential migration rate increases gradually until the age of 25, continues to scramble after a small decline at the age of 30, reaches a peak at the age of 55, and then falls sharply to a stable level. It is also clear from the sample survey that the high migration rates in the 20–30 and 50–60 age groups are due to the need of getting married and having children, people would want to have an independent living space and take care of each other and their children when they get older. This shows that the regularity of life courses (such as marriage and divorce, childbirth, retirement and house purchase) is also one of the important factors influencing the migration of individual Hui residents (see Fig. 4.23).

[11] Residential migration rate = the number of people of a given age who have moved residentially as a percentage of the total population of that age.

Table 4.23 Residential migration trajectories of individual Hui people in the Qijiawan Hui settlement

Age	Plane migration trajectory	Three-dimensional space–time trajectory
55 y/o and above		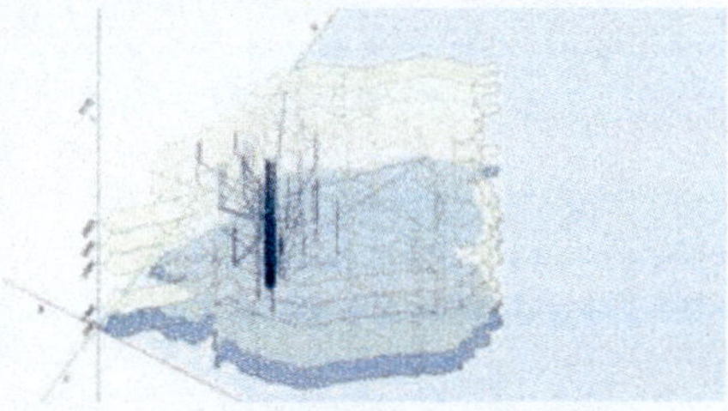
35–55y/o	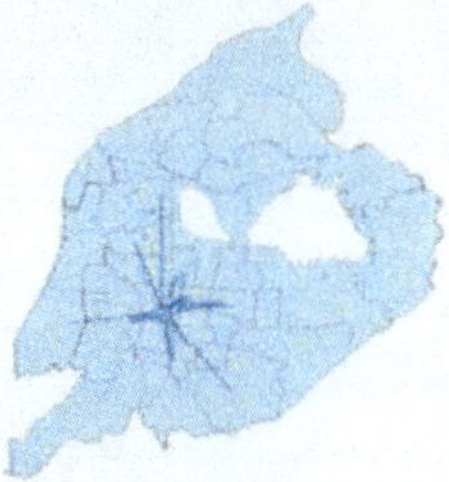	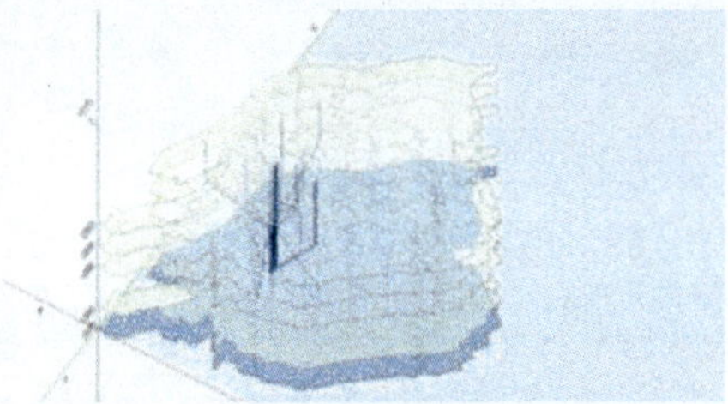
25–35y/o		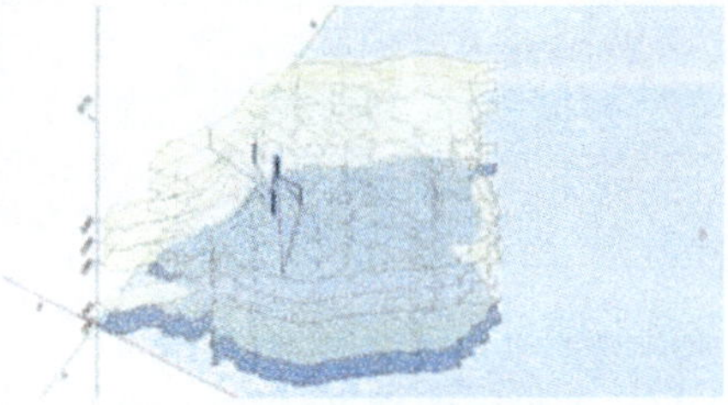
25 and below		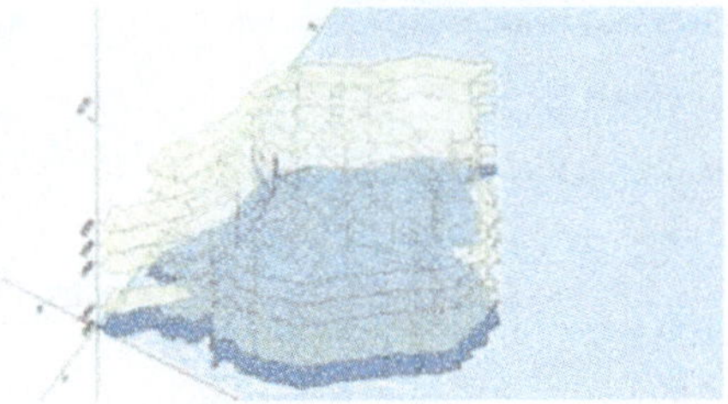

(continued)

Table 4.23 (continued)

Age	Plane migration trajectory	Three-dimensional space–time trajectory
Overall	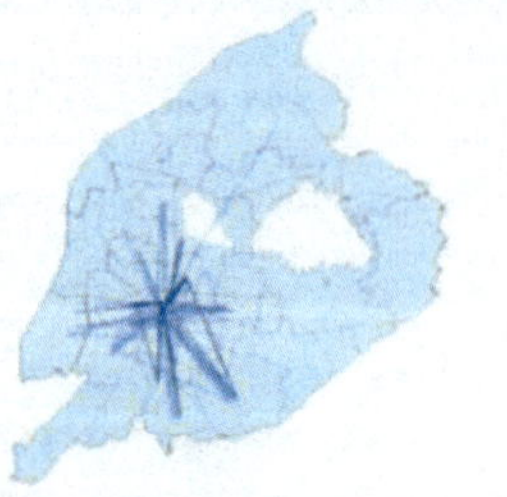	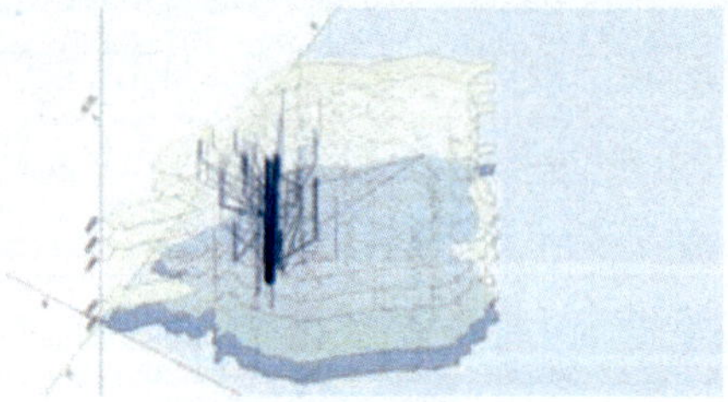

Source Data from the research group's sample study on the Qijiawan Huis settlement in Nanjing (2014)

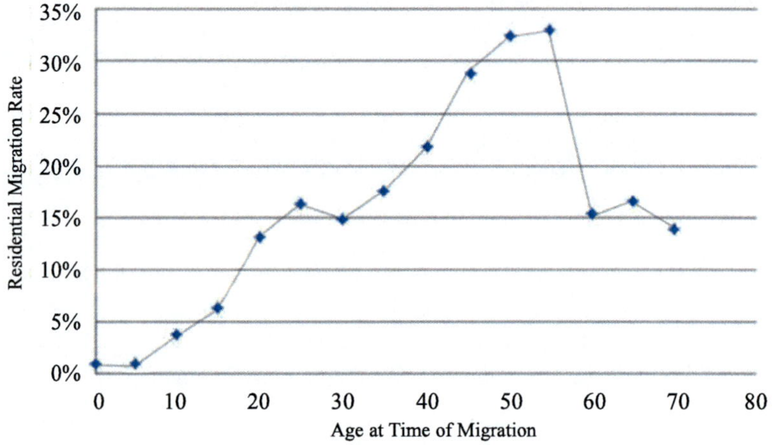

Fig. 4.23 Analysis of the relationship between the residential migration rate of the Hui People and the age at the time of migration in Qijiawan Huis settlement. *Source* Data from the research group's sample study on the Qijiawan Huis settlement in Nanjing (2014)

4.4.3 Evolution of Residential Attributes

4.4.3.1 Evolution of Housing Types

Evolutionary Characteristics of Housing Types

As an important area within the old southern part of Nanjing, the traditional residential buildings in the Qijiawan Huis settlement are mostly represented by Jiangnan courtyard-style dwellings, while the modernization of the city after the reform and opening up has brought about the construction of many high-rise modern neighborhoods. Although this change has largely impacted the traditional pattern and spatial

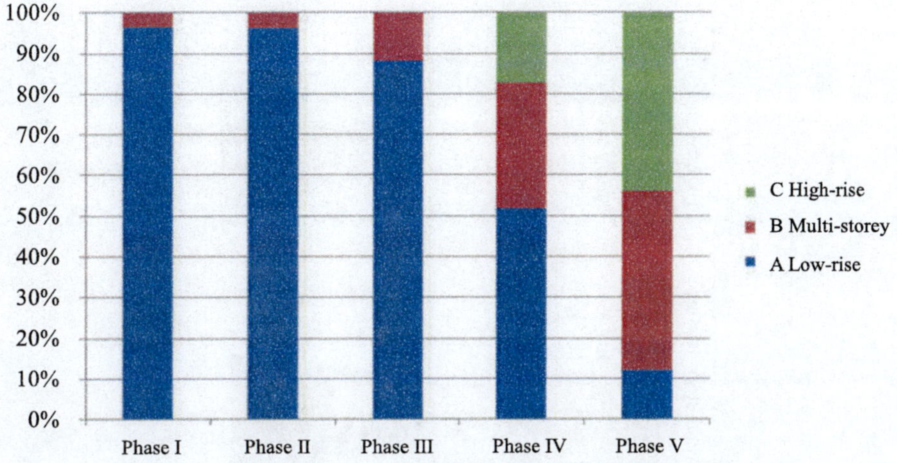

Fig. 4.24 Evolution of the number of floors of Hui residents' housings. *Source* Data from the research group's sample study on the Qijiawan Huis settlement in Nanjing (2014)

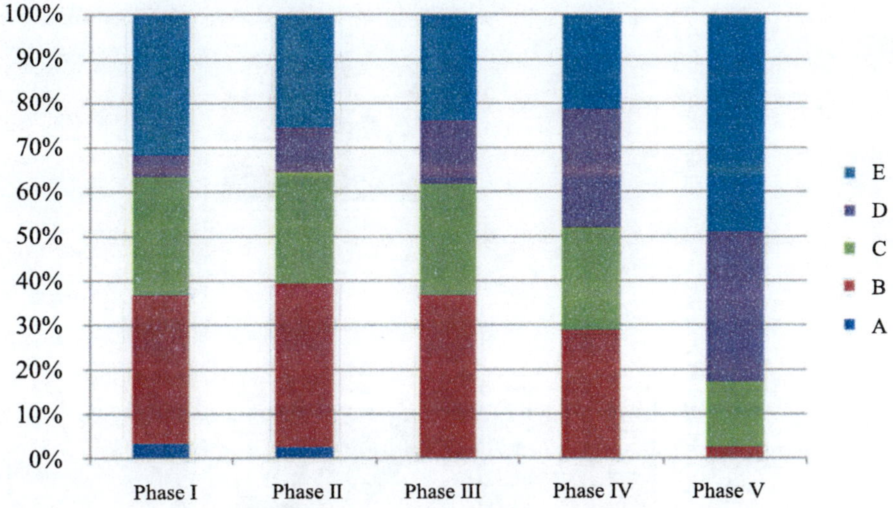

Fig. 4.25 Evolution of housing types in Qijiawan Huis settlement. *Source* Data from the research group's sample study on the Qijiawan Huis settlement in Nanjing (2014)

texture of the traditional area of Qijiawan, it is an objective fact that the living conditions of this area have gradually improved since the PRC was founded, and the trend of "**changing from traditional low-rise courtyard-type dwellings**" to "**modern multi-rise residential communities with complete facilities**" has been observed in the housing types (see Figs. 4.24 and 4.25 and Table 4.24).

Table 4.24 Typical housing types in the Qijiawan Huis settlement

Type A bungalow housing	Type B bungalow housing	Type C bungalow housing	Multi-story housing	High-rise housing
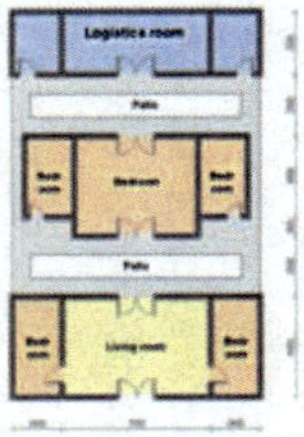	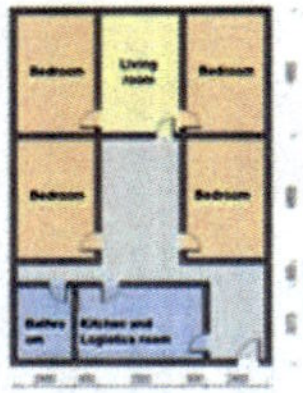	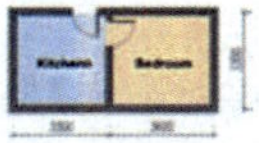	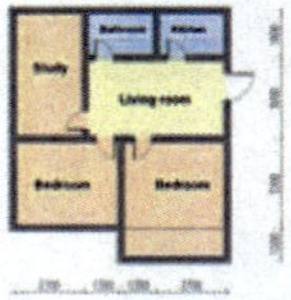	

Source Data from the research group's sample study on the Qijiawan Huis Settlement in Nanjing (2014)

Before the PRC was founded, most residences in the Qijiawan Huis settlement were built in the Qing Dynasty or the period of the Republic of China, and Hui people mostly lived in closed traditional courtyards comprised multiple housings, where each family usually lives in one unit (e.g., Type A Bungalow Housing in the table). These housings are usually of wooden structure with sloping roofs, with the typical features of "black bricks and small tiles forming crow-stepped gables, and lattice windows hanging over the cloisters," the number of floors of these residences are below three, and the floor area (including courtyards) ranges from 50–300m^2. At the same time, there is also a smaller housing type (e.g., Type B Bungalow Housing in the table), which has mainly an area of 20–30 m^2.

After the founding of PRC to the end of 1980s (Stage I–Stage III), the housing situation of majority of the Hui people is not different from that before the founding of the country, which still had the traditional wooden structure of low-rise housings, and the house types are mostly studio type, 1 bedroom + 1 living room type, and larger than 2 bedrooms + 1 living room type, the housing area varies greatly among different types.

In the 1990s (Stage IV), high-rise buildings (17.3%) began to appear among Hui people's residence, the proportion of multi-story buildings increased from 3.3% at the beginning of the founding of the country to 30.8%, and the proportion of 2 bedrooms + 1 living room housing increased from 5% at the beginning of the founding of the country to 26.9%.

Since 1998 (stage V), only 12.2% of the Qijiawan Hui people still live in traditional low-rise brick dwellings, and multi-story and high-rise buildings have become dominant at this stage (both 43.9%), of which large area housings (larger than two bedrooms + one living room) account for 82.9%.

Analysis of the Evolution of Housing Types

Renewal of traditional urban residential areas under the wave of modernization has changed the existing living conditions: The living conditions of Hui people living in the Qijiawan Huis settlement have changed significantly since the 1990s, which coincides with the modernization and demolition of traditional residential buildings in the area. After the expropriation and resettlement, the Hui people who had the opportunities for in-situ resettlement moved back to the newly built high-rise houses, which improved their living conditions, but affected the traditional spatial pattern and texture of the Qijiawan area.

Wave of modernization has also enhanced the willingness and ability of the Hui people to choose their residence: From the previous relevant analysis, it can be seen that with the diversification of employment options and steady increase in education and income levels since the founding of the country, the Hui in the Qijiawan Huis settlement have gradually become more competent in improving their living conditions and enhancing their living environment on their own.

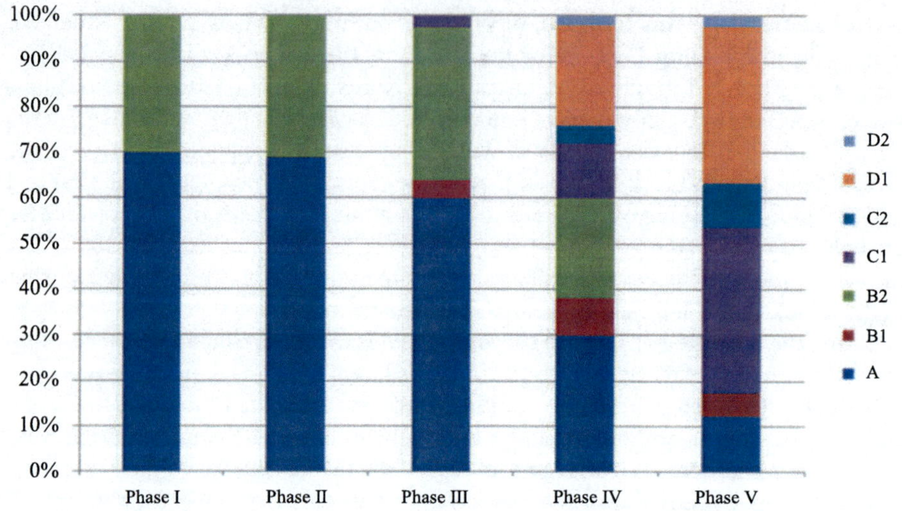

Fig. 4.26 Evolution of Hui people's residential ownership in Qijiawan Huis settlement. Legend: A self-built housing; B1 purchased housing from work units; B2 rented housing from work units; C1 purchased commercial housing; C2 rented commercial housing; D1 purchased compensation resettlement housing for land expropriated households; D2 rented compensation resettlement housing for land expropriated households. *Source* Data from the research group's sample study on the Qijiawan Huis settlement in Nanjing (2014)

4.4.3.2 Evolution of Residential Ownership

Evolutionary Characteristics of Residential Ownership

The residential ownership of Hui people in the Qijiawan Huis settlement has evolved from mainly **self-built and rented public** housing in the early years of the founding of PRC to "**diversified housing types and privatized property rights**" (see Fig. 4.26).

In the early years of the founding of the country, the state applied socialist transformation on privately owned housings, where the housings were rented, allocated, and managed by the government; it is commonly referred to as "renting or management" by the Housing Authority. During the Cultural Revolution, many households lost their original property rights and the ownership of their homes was changed from "private" to "public rental housing."

1978–1990 (Stage III): The state attempted to reform the privatization of public housing, which made the privatization of housing property rights possible again; additionally, some Hui residents in Qijiawan began to consciously purchase housings from work units.

In 1998 (Stage IV), in the process of deepening the housing reform, the State Council made clear the goal to stop housing allocation and gradually implement the monetization of housing allocation. At the same time, the Qijiawan Huis settlement experienced a large-scale urban renewal and construction transformation. Hui residents, who were affected by both effects, started all at once to have multiple types of housing ownership and housing types, the majority of which were self-built housing, rented housing from work units and purchased compensation resettlement housing due to expropriation.

Since 1998 (Stage V), the proportion of Hui people buying commercial housing has further increased. These people, together with those buying resettlement housing for land expropriated households, constitute the main body of the current residential ownership distribution of Hui. The proportion of Hui living in self-built housing has also gradually decreased from 70% at the beginning of the founding of PRC to 12.2% at present, and Hui people who used to rent housing from their work units no longer exist.

Analysis of Evolution of Residential Ownership

The housing policy reform played a decisive role: From the "renting and managing" of self-built houses since the founding of the country to the allocation of social houses by work units and purchase of commercial housing and compensation resettlement housing for land expropriated households, the evolution of the ownership of residential buildings in the Qijiawan Hui settlement is closely related to the adjustment and reform of the state's housing economic policies at the stage of each evolution.

Stage of family life cycle in which Hui residents are also has an influence. The change of housing and residential ownership due to the purchase of a house, establishment of a small family, or change of occupation are other reasons for the evolution of the residential ownership of the Hui population.

4.4.4 Evolution of Employment Time–Space

4.4.4.1 Evolution of Time–space Structure of Employment

Evolutionary Characteristics of Time–Space Structure of Work Locations

Based on the sub-district zoning of the main city of Nanjing in the previous section, this study analyzes the differences in the temporal and spatial distribution of Hui people's work locations by mapping the time–space structure of the work locations of the former and current Hui residents in the main urban area of Nanjing. It can be summarized that the temporal structure of the employment of Hui in the Qijiawan Huis settlement shows a pattern of "**sharp decline in the proportion of industrial workers after 1998, while a steady increase in the proportion of commercial service workers, and a stable proportion of workers in other industries.**" At the same time, the spatial structure shows a distribution trend of "**the core area is**

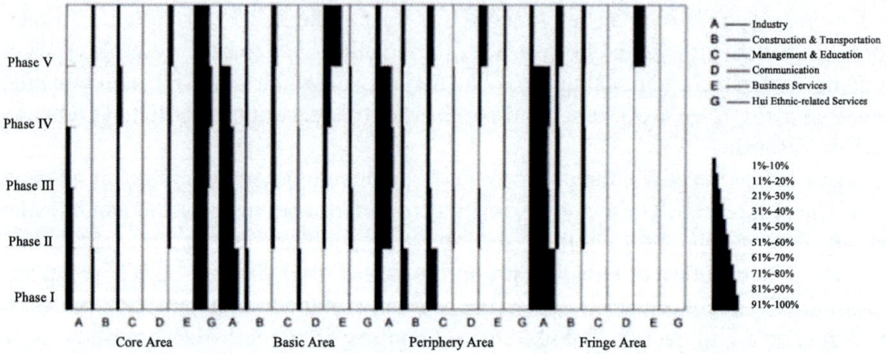

Fig. 4.27 Time–space structure of Hui's employment locations in the Qijiawan Huis settlement. *Source* Data from the research group's sample study on the Qijiawan Huis settlement in Nanjing (2014)

dominated by Hui ethnic-related service workers, while the other three types of areas are dominated by industrial workers" (see Fig. 4.27).

In the early years of the founding of PRC (Stage I), the work locations of Hui people were mostly concentrated in Qijiawan, and most of the population worked locally in Hui ethnic-related service industries (88.5%). After 1958 (Stage II), a large number of Hui people migrated to the peripheral (37%) and fringe (52.5%) areas and became workers. This proportion continued to decline steeply until the late 1990s, which was accompanied by a rapid increase in the proportion of commercial service workers (73.5% in Stage V).

Analysis of Time–Space Structure of Work Locations

The relatively stable employment structure of the Hui people over time originates from the limitations of their own employment types: The educational level of the Hui population in the Qijiawan Settlement has been steadily increasing in general; however, due to the cumulative effects of traditional Hui culture and lack of vocational education, employment was limited to handicraft and in industries that require lower knowledge and skills; thus, the frequency of job change was also limited. This contributed to a more stable employment structure in Huis settlement areas over a long period of time (basically through Stages II–IV).

The limited spatial variation in the employment location of Hui people reflects the spatial projection of staged adjustments in related policies: The relatively solidified employment structure of the Hui in the settlement also faced the recurring of phases and changes, which were mainly reflected in the alternation of employment direction and location between Phase I and Phase V. The main reason for this is the policy of "One Industrialization and Three Transformations" in the early years of the founding of PRC and the "Suppress the second industry and develop the third industry" adjustment of industries in old cities at the end of the twentieth century, which led to the return of some of the population (9.5%) to Qijiawan and Hui ethnic-related services. For these Hui people who returned, although the social space of Qijiawan has been declining under the impact of modernization, it is still the most representative and hometown of the Hui population and Hui culture in Nanjing and is still the preferred place of certain Hui people that attracted them to adjust their employment option and return to the unique Hui economy.

4.4.4.2 Migration Trajectory of Employment Location

Characteristics of Migration Trajectory of Employment Location

Following the analysis technique used for "residential migration trajectory" in the previous paper, the planar spatial trajectories and three-dimensional time–space trajectories of Hui's employment migration can be superimposed on different temporal stages, which aims to visually reflect the macro trajectories of employment locations of different individuals throughout their life cycle. In general, the employment locations of the Hui in the Qijiawan Huis settlement have shown the characteristics of "**age-specific and long-distance migration**" against the background of low overall migration rate.

An analysis by age shows the following: Hui people aged 75 years and above are mainly employed in the old city, and the migration direction is from Chaotiangong Sub-District to the basic areas, with the migration period mainly concentrated in Stage II. For those aged 55–75 years, many of them worked in the peripheral sub-districts at first, and the migration direction of those whose employment location changed was mostly from Chaotiangong Sub-District to the basic areas, with the migration period in Stage III. The place of employment and the direction of migration of the Hui people aged between 45 and 55 years are more random, the distance is also longer, there are only few Hui people employed locally in Chaotiangong Sub-District. The migration direction of employment locations of the Hui people aged 35–45 years is no longer limited to the old city, but began to shift and move between the peripheral and fringe sub-districts (see Table 4.25).

Analysis of Migration Trajectory of Employment Location

To analyze the influencing factors of employment location migration, this paper compares the age of Hui at the time of employment migration with the overall employment location migration rate to define the relationship between the two. The scatter plot and its fitted curve show that the employment migration rate of the Hui people is not high at all ages (not exceeding 14%), it continues to rise starting from the age of 20 years, and reaches the highest value at the age of 40–45 years and then drops steeply. In addition, from the sample survey, it can be concluded that the higher migration rate of employment locations of Hui people aged 40–45 years is mostly related to the layoff wave of workers in the 1990s. These workers were forced to

Table 4.25 Migration trajectory of Hui people's employment location

Age	Plane migration trajectory	Three-dimensional space–time trajectory
75 y/o and above		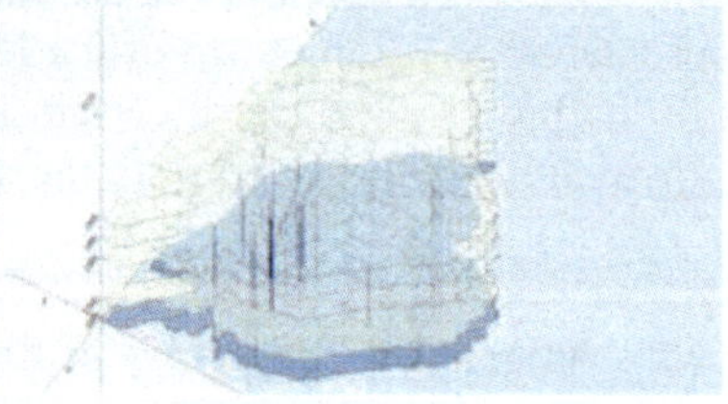
55–75y/o		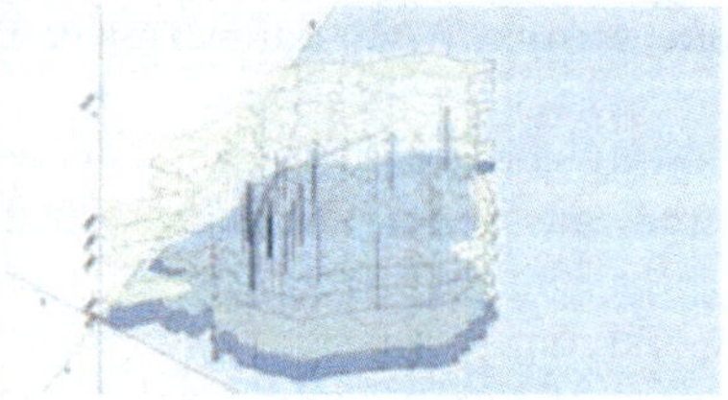
45–55y/o		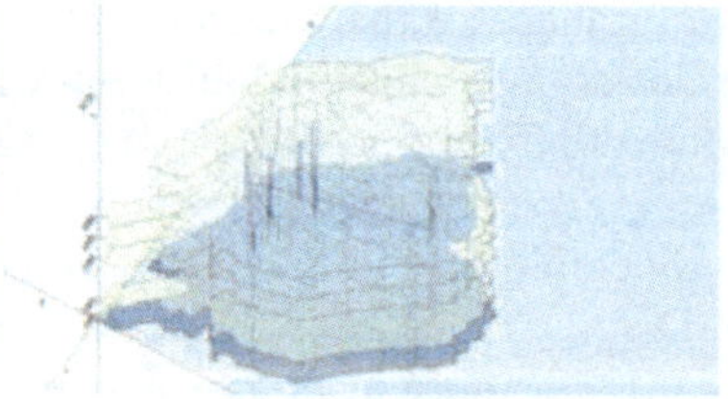
35–45y/o	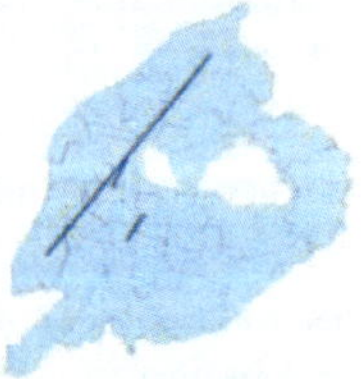	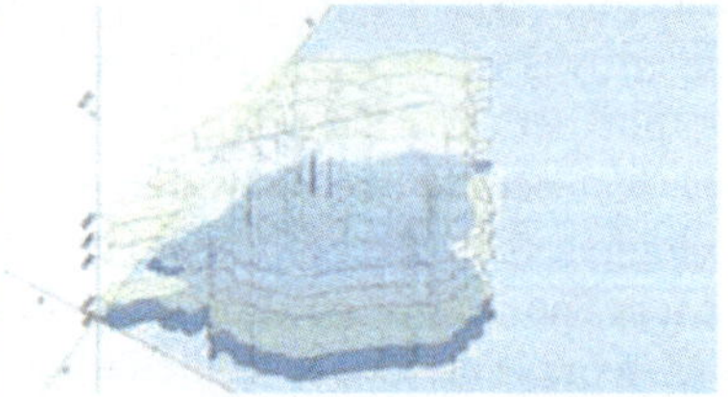
35 and below		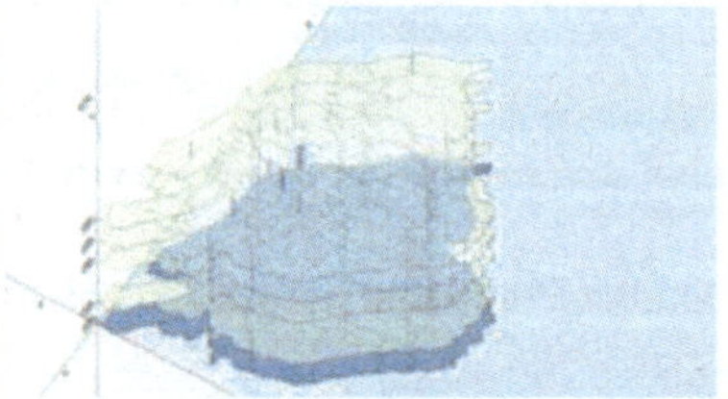

(continued)

Table 4.25 (continued)

Age	Plane migration trajectory	Three-dimensional space–time trajectory
Overall Comparison		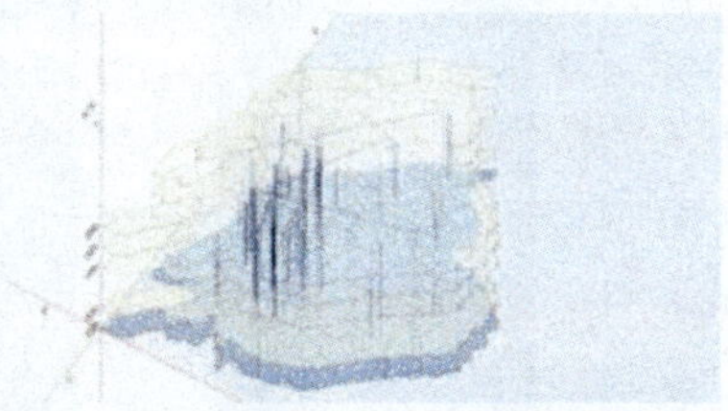

Source Data from the research group's sample study on the Qijiawan Huis settlement in Nanjing (2014)

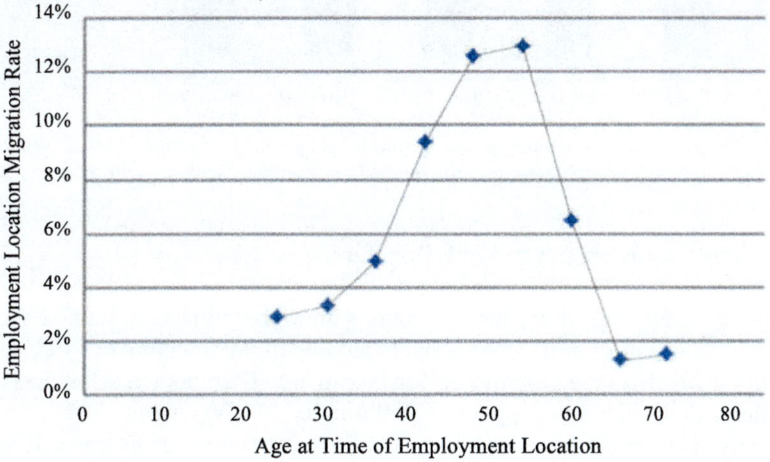

Fig. 4.28 Analysis of the association between employment migration rate and age at the time of migration in Qijiawan Huis settlement. *Source* Data from the research group's sample study on the Qijiawan Huis settlement in Nanjing (2014)

find alternative jobs and employment after they lost their jobs before retirement age, resulting in a higher migration rate for this age group (see Fig. 4.28).

4.4.5 Evolution of Work-Home Commuting

4.4.5.1 Evolution of Work Commuting

The commuting characteristics of Hui people in the Qijiawan Huis settlement have evolved over the years into a low-cost and stable work-commuting pattern of "**half an hour in time, mainly on foot and by bicycle.**"

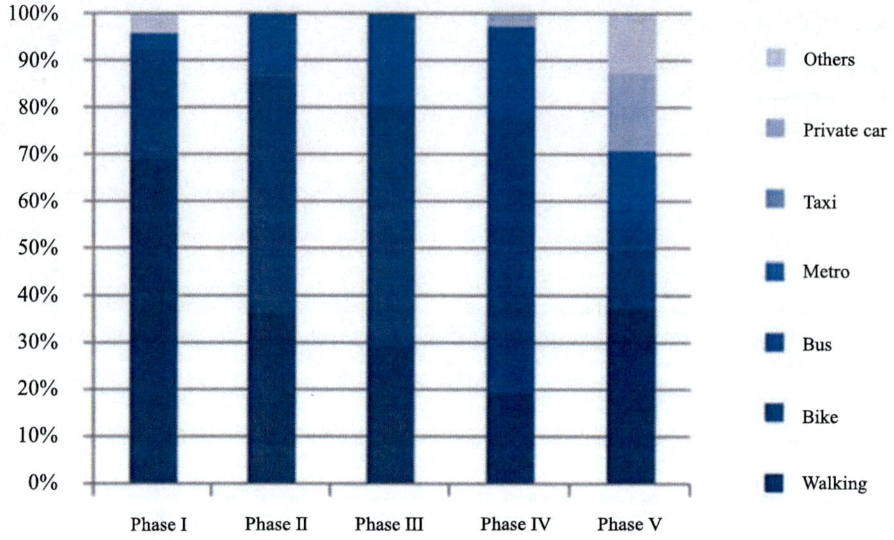

Fig. 4.29 The evolution of commuting time of the Hui people in Qijiawan Huis settlement. *Source* Data from the research group's sample study on the Qijiawan Huis settlement in Nanjing (2014)

Evolutionary Characteristics of Work Commuting

In terms of commuting time, the commuting time for more than half of the Hui people was less than 15 min (68%) at the beginning of the founding of PRC, and the proportion of Hui people commuting for less than half an hour reached 88%; this proportion stabilized after dropping to 71.9% in the 1960s (see Fig. 4.29).

In terms of commuting modes, the vast majority of Hui residents traveled mainly on foot and bicycle (92.3%) at the early stage of the founding of PRC; since then, the proportion of this mode gradually decreased throughout all stages. Until at the current stage, the walking mode accounts for 50% of the overall commuting modes, which indicates the coexistence of multiple commuting modes (see Fig. 4.30).

In terms of commuting distance, based on the average speed of walking, cycling, and traveling by public transportation, the spatial distances reachable within half anhour are 2, 3, and 5 km, respectively. As can be seen from the table below, the commuting distance of Hui people in Qijiawan Huis settlement at the beginning of the founding of PRC and in the current period was mostly concentrated within 5 km, but in Stage II–IV, there is still a certain amount of Hui workers whose commuting distance exceeded 10 km (see Table 4.26).

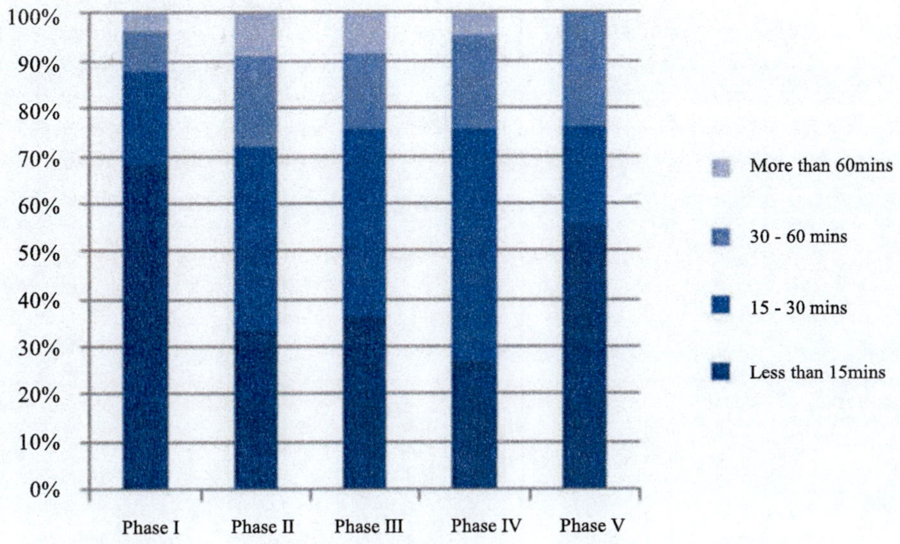

Fig. 4.30 Evolution of commuting patterns of Hui people in Qijiawan Huis settlement. *Source* Data from the research group's sample study on the Qijiawan Huis settlement in Nanjing (2014)

Analysis of Evolution of Work Commuting

The commuting time of Hui people in the Qijiawan Huis settlement is approximately half an hour, their work locations have evolved from "**in vicinity—scattered—in vicinity**," and their commuting modes have experienced the evolution of "**on foot—by bicycle—on foot.**" There are two reasons for this evolution:

The industrial landscape of the city determines the employment locations of the Hui people in the settlement: As mentioned above, the employment structure of the Hui population in the settlement is relatively stable, and some of them have not even changed their employment until retirement. Social transformation was only carried out at partial stages, and the employment locations were the influence and guidance of policies such as "suppress the second industry and develop the third industry," which is manifested as follows: with the adjustment, filtering, and attraction of the city's industrial landscape and the objective location of enterprises in the context of different stages, the employment locations of Hui people with different employment orientations have changed in phases and industries.

The economic income level affects the choice of commuting mode of the Hui in the settlement: According to the research interviews and data cross-analysis, before the 1990s, Hui residents would choose to commute by bicycle even if their commuting time exceeded 30 min or even one hour, while at the present stage, with the overall increase of income level, the commuting distance has been the main consideration for Hui people when choosing commuting mode, and multiple modes of commuting coexist. At the same time, the dual choice of employment and residence has driven

Table 4.26 Distribution of employment locations of Hui people in Qijiawan Huis settlement area

Phase I (1949–1958)	Phase II (1958–1978)	Phase III (1978–1990)	Phase IV (1990–1998)	Phase V (1998 to present)
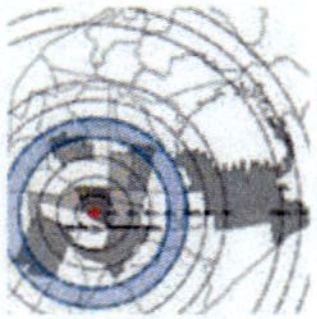	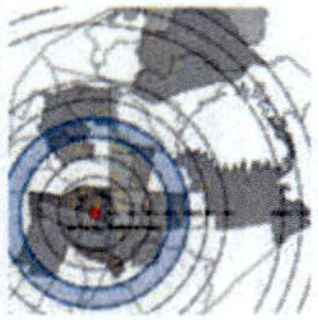	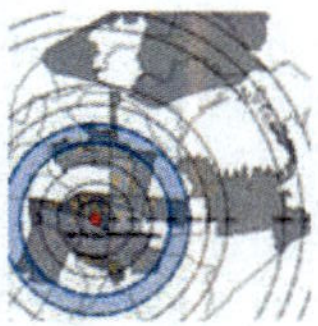	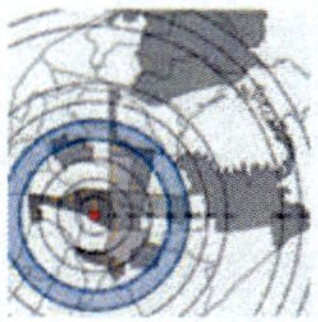	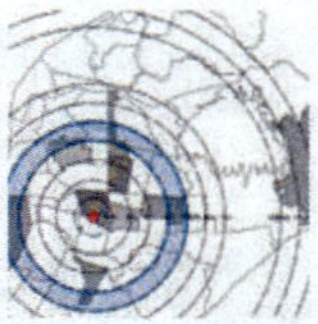

Source Data from the research group's sample study on the Qijiawan Huis settlement in Nanjing (2014)

some Hui to move back to Qijiawan for commuting reasons, resulting in an increase in the proportion of people commuting on foot in less than 15 min.

4.4.5.2 Evolution of Commuting Frequency

Evolutionary Characteristics of Commuting Frequency

Based on the sample survey data of Hui workers in Qijiawan Huis settlements, this study mapped out the commuting directions between the sub-districts of the settlement and the work place of Hui's residents and calculated its frequency to reveal the distribution characteristics of each street associated with the settlement in terms of commuting. Assuming that the number of Hui residents living in the Chaotiangong sub-district, where Qijiawan is located, and going to X sub-district for employment (i.e., same starting point and commuting to the same direction) is Ai, and the total number of work-home commuting samples is Bi, the commuting frequency Ci in that direction is obtained and the formula is as follows:

$$\mathrm{Ci} = \frac{\mathrm{Ai}}{\mathrm{Bi}}$$

The commuting-associated sub-districts can be classified into three categories: high commuting frequency sub-districts (Ci > 10%), medium commuting frequency sub-districts (5% < Ci ≤ 10%), and low commuting frequency sub-districts (Ci ≤ 5%).

By substituting the employment commuting data of Hui workers in each stage into the above formula, it can be seen that the commuting frequency of Hui people in Qijiawan Huis Settlement has shown the characteristic of "**gradually changing from the equilibrium state at the beginning of the founding of PRC to the low equilibrium state.**"

The work-home separation pattern of Hui workers in Qijiawan Huis settlement during Stage I shows a centripetal orientation with commuting in vicinity as the main focus. The work-home separation patterns of Hui workers in Stage II–IV show decentralized commuting within the main urban area, while the separation patterns of Hui workers in Stage V are significantly different, showing the coexistence of commuting in vicinity and decentralized commuting. Based on the separation patterns of Hui workers in each stage since the founding of PRC, a graphical analysis of the "work-home" connection was carried out in terms of residential opportunities, employment opportunities, and commuting patterns in the settlement (see Table 4.27).

Analysis of Evolution of Commuting Frequency

Change of commuting frequency of the Hui in the settlement is directly related to the change of their occupational structure: In the early period of the founding of the

Table 4.27 Diagrammatic representation of the work-home commuting of Hui people in the Qijiawan Huis settlement

Stages	Commuting map	Brief analysis
Stage I (1949–1958)	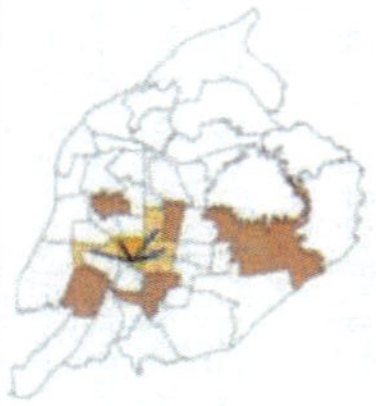	[Housing Opportunities] The majority are low-rise traditional dwellings from the Ming and Qing dynasties, and the public supporting facilities are not complete [Employment Opportunities] Small stores and enterprises such as food factories and radio factories [Overview of Work-home Commuting] Work-home are relatively balanced. The sub-district with the highest commuting frequency is Chaotiangong Sub-District where the Qijiawan Huis settlement is located, while the medium commuting frequency sub-districts are Binhu Sub-District, Xinjiekou Sub-District, Wulao Village Sub-District and Hongwu Road Sub-District, which are adjacent to Chaotiangong Sub-District
Stage II (1958–1978)	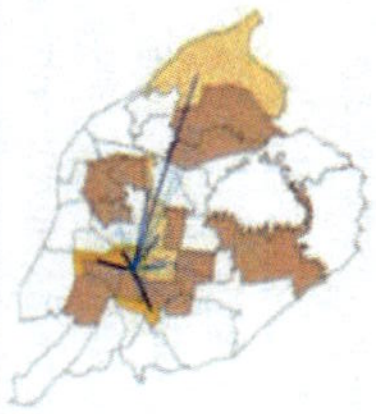	[Housing Opportunities] The majority are low-rise traditional dwellings from the Ming and Qing dynasties, and the public supporting facilities are not complete [Employment Opportunities] Small stores, food factories, radio factories, etc [Overview of Work-home Commuting] Relatively low work-home balance. The sub-districts with the highest commuting frequencies are Chaotiangong Sub-District where the Qijiawan Huis settlement is located, Hubin Sub-District, and Zhonghua Gate Sub-District. Medium commuting sub-streets are spread to the north of the main city, which reaches as far as Yanziji Street, where the commuting time is more than one hour
Stage III (1978–1990)	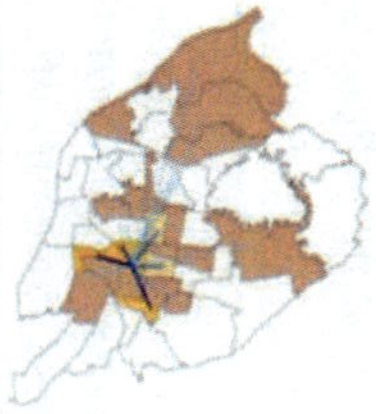	[Housing Opportunities] The majority are low-rise traditional dwellings from the Ming and Qing dynasties, and the public supporting facilities are not complete. At the same time, some new multi-story communities are built with complete supporting facilities [Employment Opportunities] Enterprises and institutions such as food factories, radio factories, primary and secondary schools, and the Planning and Research Center of the Provincial Department of Transportation [Overview of Work-home Commuting] Relatively low work-home balance. The sub-district with the highest commuting frequency is still Chaotiangong Sub-District where the Qijiawan Huis settlement is located; medium commuting sub-streets are concentrated in Xinjiekou Sub-District and Hongwu Road Sub-District

(continued)

Table 4.27 (continued)

Stages	Commuting map	Brief analysis
Stage IV (1990–1998)	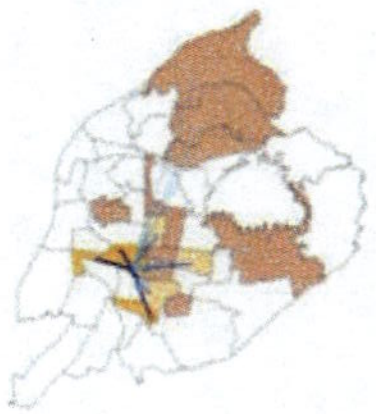	[Housing Opportunities] Low-rise traditional residential houses with incomplete facilities coexist with new multi-story and high-rise residential communities with complete facilities [Employment Opportunities] Primary and secondary schools, the Planning and Research Center of the Provincial Department of Transportation, supermarkets, and other enterprises[Overview of Work-home Commuting] Relatively low work-home balance. Compared with the previous period, the range of commuting sub-districts has been reduced, and there is little change in the high-commuting frequency sub-districts and low commuting frequency sub-districts, but there is a decrease in commuting to the southwestern part of the main city
Stage IV (1998 to present)	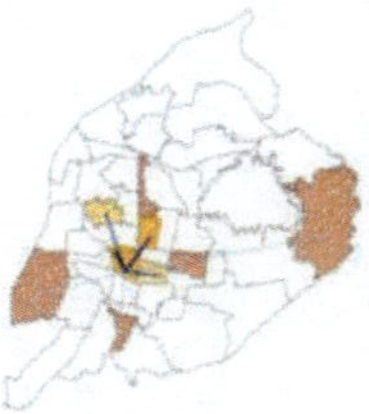	[Housing Opportunities] Multi-story, high-rise residential communities with more complete service facilities [Employment Opportunities] Elementary and middle schools, Planning and Research Center of the Provincial Department of Transportation, high-rise commercial office buildings, supermarkets, and other enterprises [Overview of Work-home Commuting] Relatively low work-home balance. The sub-districts with the highest commuting frequencies are Chaotiangong Sub-District where the Qijiawan Huis settlement is located, and Xinjiekou Sub-District, while low commuting frequency streets are scattered in all directions in the main city
Legend	Ci>10% 5%<Ci≤10% Ci≤5%	

Source Data from the research group's sample study on the Qijiawan Huis settlement in Nanjing (2014)

PRC, Hui people were mainly employed in the service industry with ethnic characteristics or in factories, the locations of which are concentrated in the neighborhoods of the main urban areas; however, the subsequent "One Industrialization and Three Transformations" policy drove a large number of Hui people to work in factories, and with the city's "Suppress the second industry and develop the third industry" reform and adjustments of its industrial landscape, the corresponding changes in employment location and commuting frequency were brought about. In the current stage of urban transformation, more and more Hui people tend to diversify their employment in terms of industry and location and thus changing their work-home commuting frequencies.

Change of Hui people's commuting frequency in the settlement is also related to the floating of their economic income: The continuous improvement of the income level of the Hui population at this stage has brought about the improvement of the quality of life and the diversification of commuting methods, so when choosing employment locations, they are no longer limited by the length of spatial distance. This results in a higher degree of work-home separation and lower degree of balance in commuting frequency and coverage area.

4.4.6 Evolution of Supporting Facilities

4.4.6.1 Evolution of Supporting Facilities

Evolutionary Characteristics of Supporting Facilities

By analyzing the layout of supporting service facilities in the Qijiawan Huis settlement since the PRC was founded, we can see the following trends (see Table 4.28): **in terms of spatial structure, the nested commercial layout of the settlement with the mosque as the center is declining. In terms of spatial form, the linear layout along the inner streets was transformed into a piecewise layout along the main urban roads, and in terms of facility types, Hui ethnic-specific stores are gradually disappearing and facility types tend to diversify**.

Analysis of Evolutionary Characteristics of Supporting Facilities

Connected influence of the urban environment: Under the impact of modernization and urban construction, the interconnection and mutual influence within the city got closer and tighter. Similarly, the Qijiawan Huis settlement is bound to be affected by the changes in the surrounding environment, while its traditional cohesive structure is moving from closed to open, the configuration of its service facilities is inevitably becoming more complex and diversified (see Figs. 4.31, 4.32, 4.33, and 4.34).

Change in the occupational structure of the Hui people: In the 1960s, after the three major socialist reforms, many Hui people who were previously engaged

Table 4.28 Evolution of supporting facilities in Qijiawan Huis settlement

Stages	Diagram of the evolution of supporting facilities	Brief analysis
Stage I (1949–1958)	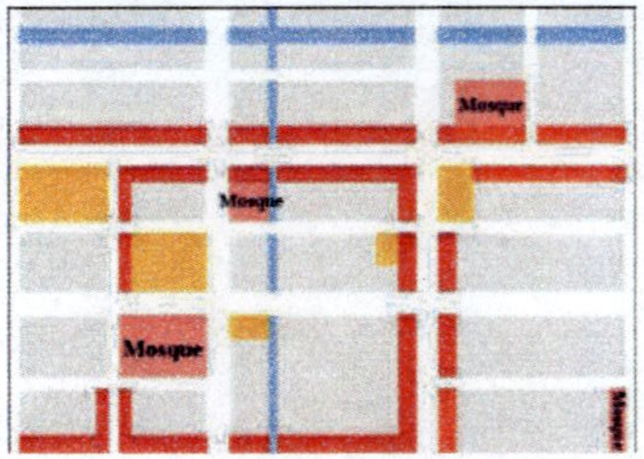	There are 8 mosques and girls' schools in Qijiawan Huis Settlement, and many primary and secondary schools supported by the mosques Commercial facilities mainly include halal food stores, beef and mutton and its product stores, department stores, thrift stores, etc., and most of them consist of small family-owned stores along the street (with residence on top and stores on bottom), which are mainly distributed linearly along Shengzhou Road, Qijiawan, and Panshi Street More than 80% of the service facilities in this period were operated by Hui people and were closely related to their special needs
Stage II (1958–1978)	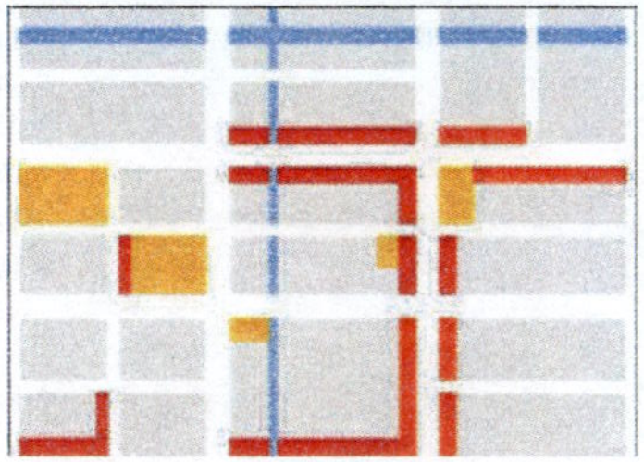	Religious activities were suspended and the mosques within the Qijiawan Hui settlements disappeared, while some schools continued to operate. Commercial facilities are still dominated by halal food stores, and beef and mutton stores, etc., but a part of the stores went through collectivization, and a part of the stores were closed, resulting in a decline in the total number of commercial facilities The proportion of general commerce increased during this period, but the proportion of the halal industry decreased rather than increasing

(continued)

Table 4.28 (continued)

Stages	Diagram of the evolution of supporting facilities	Brief analysis
Stage III (1978–1990)	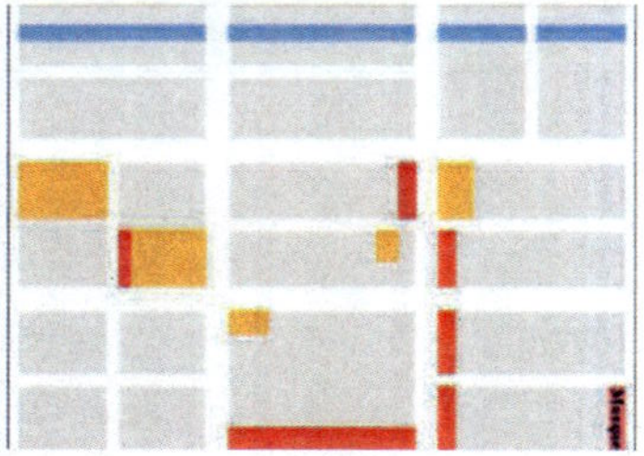	After the reform and opening up, the Jingjue Mosque in the southeast of the Qijiawan Huis settlement gradually resumed religious activities, and schools were opened gradually Commercial facilities were mainly restaurants and antique stores, and the number of stores decreased sharply and were mostly distributed along the Pingshi Street and Shengzhou Road, which are south to the settlement During this period, commercial services with Hui characteristics were no longer dominant in the settlement
Stage IV (1990–1998)	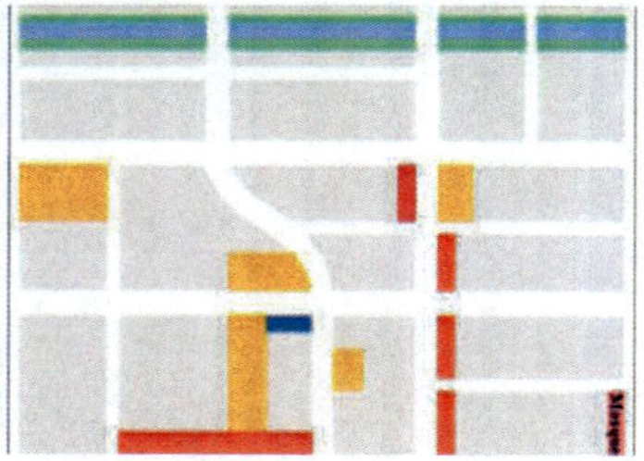	Only one mosque was preserved in the vicinity of the Qijiawan Huis Settlement The evolution of commercial facilities continues the trend of the 1980s, while the number of Hui ethnic-related commercial services continued to decrease In terms of medical facilities, a Hui hospital that offers regular medical examinations was established, and it focused on the health of Hui people In terms of public activity space, the green space along both sides of the river in the north was renovated to provide residents space for recreational and sports activities The widening of Dingxin Road and establishment of several bus stations also improved the accessibility of the area

(continued)

Table 4.28 (continued)

Stages	Diagram of the evolution of supporting facilities	Brief analysis
Stage IV (1998 to present)	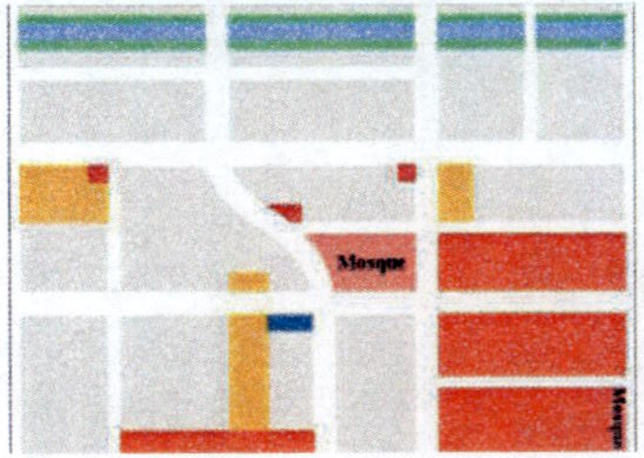	The Caoqiao Mosque in the Qijiawan Hui Settlement was renovated and opened, while several elementary schools were merged In terms of medical facilities, the Hui Hospital also added a Hui nursing home In terms of commercial facilities, Xinanli in the southeast corner of the settlement is an important urban connecting point to enhance the vitality of the area, integrating museum exhibitions, culture and education, dining, and leisure functions. Moreover, there are only three small-scale halal restaurants left, while most other stores along the street are thrift stores or repair stores operated by Han people With the construction of the subway station in the northeast corner of the settlement, the accessibility of the area has been further enhanced
Legend	Mosque; General Service Facilities; Medical Facilities; Public Green Space; Water Area; Primary & Middle School; Hui Ethnic-related Service Facilities	

Source Data from the research group's sample study on the Qijiawan Huis settlement in Nanjing (2014)

Fig. 4.31 Shengzhou Road and Pingshi Street at the beginning of the founding of PRC. *Source* http://blog.sina.com.cn/s/blog_7c3326200101ct2a.html

Fig. 4.32 Shengzhou Road and Pingshi Street at present. *Source* Author's own photograph

in small handicraft and halal food and beverage industries switched careers and became collective workers, thus causing the services they used to operate to weaken and shrink in terms of content and scale.

Change of Hui's residence locations: The emigration of Hui and migration of Han residents not only reduced the size of the Hui ethnic-specific services industry but also decreased the number of consumers who have a demand for it, aggravating the decline of the Hui specialty economy and ethnic-related service industry.

Shift in the ethnic consciousness of the Hui people: The impact of modernization on the lives of Hui people is not only limited to employment and housing but has also infiltrated into traditional culture and ethnic consciousness. As a result, the core position of mosques has been degrading, the Hui people have a weaker ethnic

Fig. 4.33 Anleyuan Halal restaurant at the beginning of the founding of PRC. *Source* http://blog.sina.com.cn/s/blog_7c3326200101ct2a.htm.l

Fig. 4.34 Anleyuan Halal restaurant at present (Wangfu Street, 2014). *Source* Author's own photograph

conscience, religion has become part of their leisure activities rather than a spiritual basis, and the former characteristic lifestyle and its related services have become dispensable.

4.4.6.2 Evolution of Supporting Facilities with Ethnic Characteristics

Evolutionary Characteristics of Supporting Facilities with Ethnic Characteristics

By analyzing the ethnic characteristics of supporting facilities in the Qijiawan Huis settlement since the PRC was founded, the changes in the layout tend to be as follows (see Table 4.29). **The overall structure has changed from a piecewise band-like distribution to a point-like distribution, and the types of facilities changed from diversified to homogenous.**

Analysis of Evolution of Supporting Facilities with Ethnic Characteristics

The phased adjustment of state policies is the direct external factor affecting the supporting facilities with ethnic characteristics: The religious reform in 1958 and suspension of religious activities hindered the development of religious beliefs not only spatially but also ideologically. After 1978, with the gradual recovery of religious movements, the state began to pay attention to the customary consciousness and special needs of ethnic minorities and, therefore, started to restore and establish some medical and educational facilities in a targeted manner, but the overall declining trend of supporting facilities with ethnic characteristics in the settlement areas is difficult to be reversed in the short term.

The indifference of the ethnic consciousness of the Hui people is an important internal cause of the decline of supporting facilities with ethnic characteristics: Although religious activities have been restored after the reform and opening up, when looking from the perspectives of individual thoughts and cultural consciousness, Hui people implicitly lost their sense of ethnic identity and self-identification after being integrated in the comprehensive modern education system for a long time, which will undoubtedly objectively reduce their demand for Hui-specific facilities in the settlement and lead to the gradual decline of the ethnic-related economy in the area.

4.5 Renewal Strategy for Huis Settlement

Culture is an important characteristic of a nation. Tylor, a famous British ethnographer, once defined culture as "that complex whole which includes knowledge, belief, art, morals, law, custom, and any other capabilities and habits acquired by man as a member of society.[12]" The adoption of "Convention on the Protection and Promotion of the Diversity of Cultural Expressions" at the 33rd General Conference

[12] Taylor, E.B. Primitive Culture [M]. Shanghai: Shanghai Literature and Art Publishing House, 1992:1.

Table 4.29 Evolution of supporting facilities with ethnic characteristics

Stages	Diagram of evolution of supporting facilities with ethnic characteristics	Brief analysis
Stage I (1949–1958)	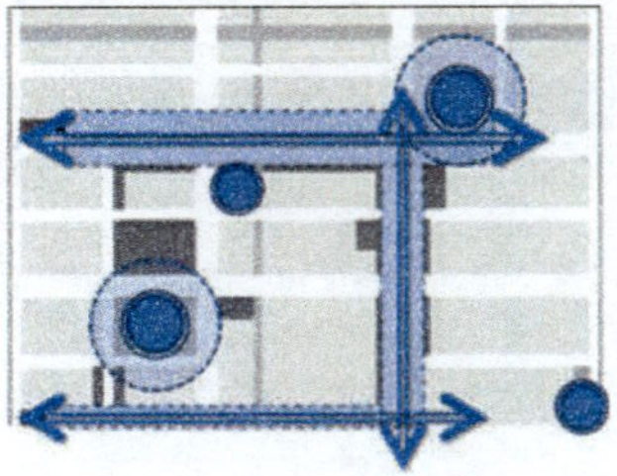	The facilities that had ethnic characteristics in the Qijiawan Huis settlement were of various types and evenly distributed, and the overall structure was distributed in a band-like manner The main types include mosques and girls' schools, Hui primary and secondary schools supported by the mosques, halal food stores, beef and mutton stores, department stores, thrift stores operated by Hui people, etc Its spatial distribution was aligned in a linear way along Shengzhou Road, Qijiawan and Pingshi Street, and around the mosque in a planar pattern
Stage II (1958–1978)	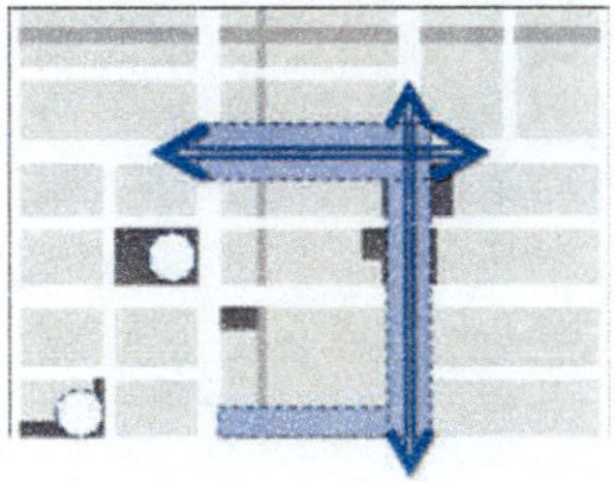	The religious activities were suspended in this stage, the mosque within the Qijiawan Huis settlements disappeared, ethnic-related facilities were greatly impacted, and the overall structure was distributed in a band-like manner The main types are Hui elementary school, halal restaurants operated by Hui people, beef and mutton stores, etc Its spatial distribution is aligned in a linear way along Shengzhou Road, Qijiawan, and Pingshi Street, but their number has declined significantly compared with the previous stage

(continued)

Table 4.29 (continued)

Stages	Diagram of evolution of supporting facilities with ethnic characteristics	Brief analysis
Stage III (1978–1990)	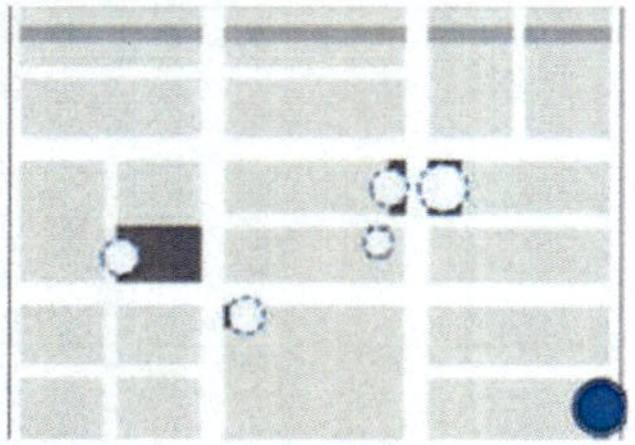	After the reform and opening up, religious activities were resumed at the Jingjue Mosque, which is located on the southeast side of the Qijiawan Huis Settlement, the mosque only retained its religious significance but lost its socio-economic and cultural influence on the settlement The supporting facilities with ethnic characteristics in the settlement consist mainly of a few halal food stores, which were distributed in a scattered point-like manner
Stage IV (1990–1998)	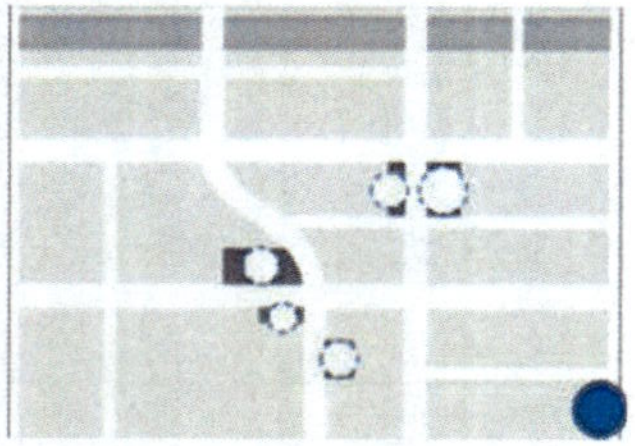	There is only one remaining mosque around the Qijiawan Huis Settlement The supporting facilities with ethnic characteristic in the settlement were still mostly distributed in a point-like pattern, but a hospital for Hui people has been constructed in addition, which showed respect to the ethnic customs of the Hui people and paid attention to their health
Stage IV (1998 to present)	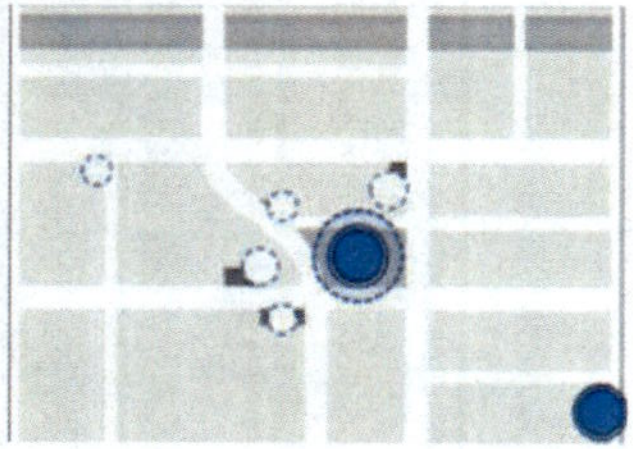	The Caoqiao Mosque in the Qijiawan Huis Settlement was renovated and opened, and several halal restaurants were built next to the mosque. In addition, a nursing home for Hui people was added to the Hospital. However, this did not reverse the decline in the number of ethnic-related facilities in the settlement, and the overall distribution pattern has still been mainly point-like
Legend	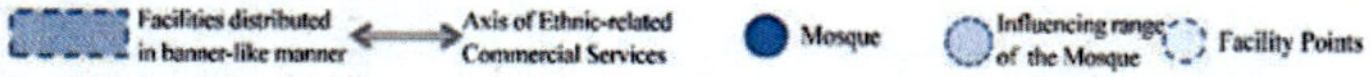	

Source Data from the research group's sample study on the Qijiawan Huis settlement in Nanjing (2014)

of UNESCO in 2005 marked that the principle of cultural diversity has been raised to an ethical and moral level that the international community should observe and recognize; in the meantime, it acquired the nature of an international legal instrument.

4.5.1 *Existing Problems in Huis Settlement*

Cultural identity consists of two dialectical aspects: identification of ethnic minorities and Han people with their own traditional culture and identification of ethnic minorities with other cultures brought about by the wave of modernization that has occurred and will continue to exist. The sustainable development of national culture can only be achieved by individuals loving their own traditional culture and actively accepting foreign cultures.

To build a renewal strategy for minority settlements, we should establish values that sustain the coexistence of multiculturalism in our country and examine the representation of culture and its evolution based on cultural identity in terms of living customs, residential space–time, and employment space–time. The basic idea is as follows: based on the previous study on the evolutionary pattern of the Qijiawan Huis settlement, this section sorted out and selected the main problems faced in the evolution and their negative causes (not all of the aforementioned evolution are negative and passive) to provide a basis for the following strategic discussion (see Table 4.30).

4.5.2 *Renewal Strategies for Huis Settlement*

Based on the above summary, it can be seen that throughout the evolution of the social space of the Qijiawan Huis Settlement, development and problems were present hand in hand, and opportunities and challenges coexist. The above problems and their causes are often compounded and intertwined with each other. By further consolidating and summarizing the multiple causes of the problems, this study further sorted out five basic renewal strategies as follows (see Fig. 4.35).

4.5.2.1 Development Positioning Level

Interpretation of City Master Plan

According to the Nanjing Master Plan (2007–2020), the Qijiawan Huis settlement is located in the center of the old city of Nanjing, which is densely populated and in urgent need of commercial services. It is also the core area for shaping the historical features of the old city, which is important for displaying the urban landscape of Nanjing.

Table 4.30 Summary of issues in evolution of the Qijiawan Huis settlement

	Segmented category		Major problems	Major causes
Social attributes	Demographic composition	Education level	–	–
		Occupational level	Lower proportion of employees in Hui ethnic-related services	Urban industrial transformation and weakening of ethnic consciousness
	Family relations	Inter-ethnic marriage	Increased interethnic marriage rate	Weakness of the agglomeration Hui people
		Family structure	–	–
	Social organizations		Institutional homogenization	National phased policies, weakened sense of ethnic identity
	Living customs		Traditional ethnic customs are in decline	National Phase Policy, Settlement Life Patterns and Cultural Identity
Spatial attributes	Community functions	Land use functions	–	–
		Architectural texture	Disappearance of traditional texture	The demand of compounding urban functions, designers' negligence of urban texture
	Residential space–time	Residential space–time	–	–
		Residential migration trajectory	–	–
		Spatial association of residential migration and settlement areas	Low association	Renewal of old urban areas, weakening of ethnic consciousness
		Residential types	–	–
		Residential ownership	Privatization of property types	National housing policy orientation, household life cycle stages

(continued)

Table 4.30 (continued)

Segmented category		Major problems	Major causes
Employment space–time	Employment space–time structure	–	–
	Employment migration trajectory	–	–
	Spatial association between employment migration and settlement areas	–	–
	Work commuting	Commute by walking and by bicycle	Occupational structure and income level
	Commuting frequency	High work-home separation	Occupational structure and income level
Supporting facilities	General supporting facilities	Incomplete supporting facilities at all levels	Urban positioning and the changing social attributes of Hui people
	Supporting facilities with ethnic characteristics	Ethnic-related service industry in decline	National phased policy adjustment, weakening of national consciousness

Source Own elaboration

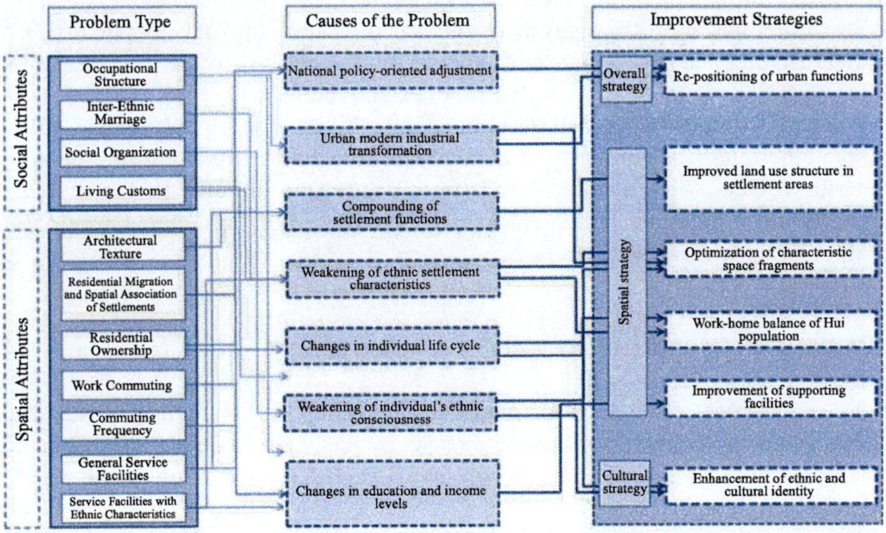

Fig. 4.35 Derivation of the renewal strategy of the Qijiawan Huis settlement. *Source* Author's own drafting

In terms of land use functions, commercial land and historical preservation land co-exist. As for industrial development, modern commercial and cultural creativity co-exist. In terms of spatial appearance, modern high-rise commercial offices with large volume and traditional low-rise residential buildings with small volume are equally important. It can be seen that the Qijiawan Huis settlement faces the dual challenges of the impact of the modernization wave and the continuation of traditional history and culture.

Interpretation of District Master Plan

Qinhuai District, where Qijiayuan is located, has been the social, economic, and cultural center of Nanjing since ancient times and has a rich cultural and historical heritage. The "Nanjing Qinhuai District Master Plan" (2013–2030), which was compiled at the end of 2013, integrates the advantages of historical and cultural as well as urban commercial aspects of the district, forming a double growth pole of urban culture and economy. Regarding the future urban functions and spatial development of the Qijiawan Huis settlement, the compilation of this plan provides a more detailed consideration and interpretation compared with the master plan (see Table 4.31).

By summarizing the various positioning guidelines of the Nanjing Qinhuai District Master Plan (2013–2030) for the Qijiawan Huis settlement, it can be seen that the plan can basically reflect the dual advantages of the rich historical and cultural resources of the area; however, it does not make targeted interpretation and emphasis on the ethnic and cultural characteristics of the Hui ethnic group in the settlement.

Therefore, based on the analysis of the urban master plan and the planning of Qinhuai District and combining the field research and analysis of the Hui people in Qijiawan Huis settlements, this study further extends and optimizes its existing positioning and proposes the overall development positioning of Qijiawan Huis settlements in the future as a minority settlement that **integrates urban commerce and tourism services, with historical culture and urban commerce as the driving force for development, Hui's culture as the highlight, and has traditional small-scale spatial appearance.**

4.5.2.2 Land Use Structure Level

Under the impact of the wave of modernization after the founding of PRC, the traditional pattern of living around the mosque in the Qijiawan Settlement has been transformed into a trend of "independent living groups, scattered Huis settlements, and increasingly open communities."

Therefore, with reference to the structural pattern of traditional Huis settlements and the functional positioning of Qijiawan Huis settlements in the previous section, this study proposes an optimization strategy for the land use layout model (see Fig. 4.36): First, set up mosques or grassroots ethnic service centers in the residential

Table 4.31 Interpretation of master plan of Qinhuai district

Relevant systems	Relevant blueprints	Brief analysis
Functional positioning	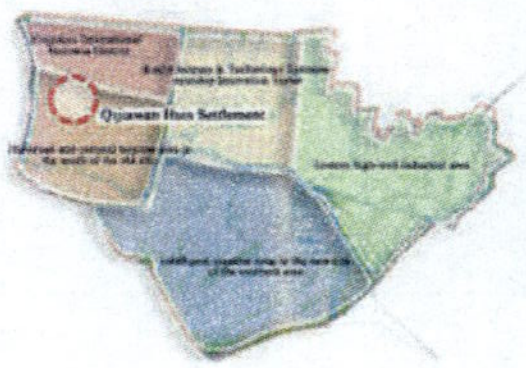	Located in the northern part of the historical and cultural tourism area in the south of the old city and close to the international business district of Xinjiekou, its dual driving factors of history and culture, and urban commerce are an advantage and a challenge for the traditional Huis settlement of Qijiawan
Land use planning		The planned land use of the Qijiawan Huis settlement is mainly for residential and historical and cultural preservation purposes; this land use layout emphasizes that the need for the future of the Qijiawan area not only consists of assuming residential functions but also developing the role of cultural exhibitions
Spatial layout structure		The Qijiawan Huis settlement is spatially located in the areas where commercial and historical land zones overlap and is an important connecting point linking the Fuzimiao commercial center and the Xinjiekou international business service area

(continued)

Table 4.31 (continued)

Relevant systems	Relevant blueprints	Brief analysis
Industry development and layout guidance	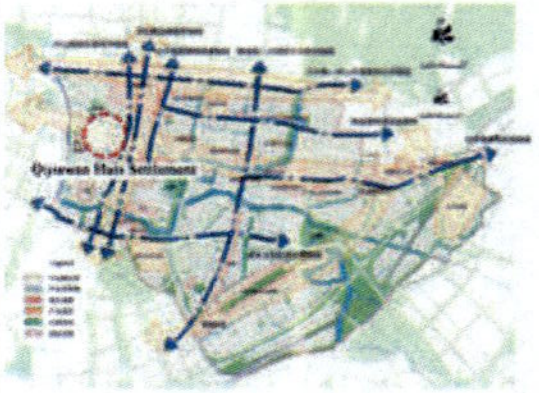	Qijiawan Huis settlement is located in the district-level industrial integration zone between Shengzhou Road and Guanghua Road, and its eastern side is adjacent to the major economic axis of Zhongshan South Road, which is a key location for industrial "transformation, upgrade and optimization" in the Qinhuai District, of which the cultural and creative industries with ethnic characteristics are important development points in driving the industrial transformation of the old southern part of the city
Historical and cultural heritage conservation planning	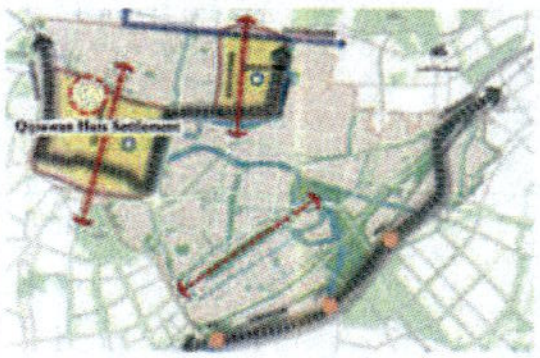	In terms of the historic resource demonstration structure, the Qijiawan Huis settlement is located in the historic district of the south part of the city and on the axis of Zhongshan East Road. In terms of the protection of historical sites, Nanbuting historical and cultural district, Pingshijie historical landscape areas and Dahui Fuxiang General historical site are all located within the settlement. In terms of the protection of cultural relics and monuments, there is the former residence of Gan Xi, which is a site protected at national level, multiple immovable cultural relics protected at municipal level, the settlement is also adjacent to many national and provincial protected sites such as Chaotian Palace and Jingjue Mosque. Therefore, the settlement contains abundant historical and cultural heritage resources

(continued)

Table 4.31 (continued)

Relevant systems	Relevant blueprints	Brief analysis
Urban design guidance		In terms of the spatial guidance of urban design, the Qijiawan Hui settlement is located in the historic district of the old city in the south, adjacent to the urban landscape axis of Zhongshan East Road. In terms of the landscape line of sight, it is within the important line of sight of Xinjiekou in the direction toward Zhonghua Gate; the building height is ≤ 18 m, and the local plot is ≤ 12 m
Public service facilities Planning		In the commercial supporting facilities in the Qijiawan Huis Settlement, there are three distinct commercial streets planned by Qinhuai District to meet modern needs, namely, the Cangxiang Cultural and Commercial Pedestrian Street, the Chaotiangong Antique Street and the Nanbuting Handicraft Workshop Street. Additionally, one new community-level sports facility is planned in the settlement, yet there are no other new facilities
Adequate tourism resource and development planning enhancement		The Qijiawan Huis settlement is located in the tourism area between Chaotiangong and the Former Residence of Ganxi, which is a key development area in the future. It is planned to be positioned as an ethnic culture area, aiming to attract all kinds of tourists by tapping into Qinhuai folk culture and gathering industrial businesses featuring folk culture

Source Compiled by the author based on the Nanjing Qinhuai district master plan (2013–2030)

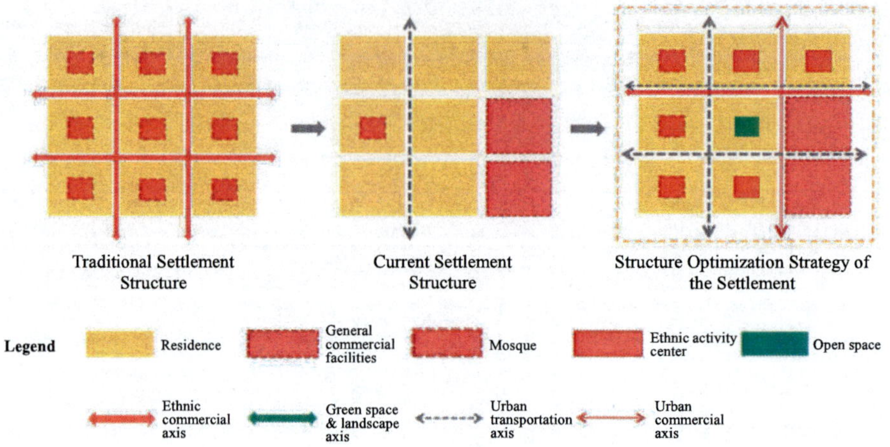

Fig. 4.36 The land layout improvement model of the Qijiawan Huis settlement. *Source* Author's own drafting

in need to establish a new pattern of Huis settlements around the mosques. Second, introduce street-level commercial facilities with ethnic characteristics in combination with modern commercial layout models to strengthen the relationship between the residential clusters and their relationship with the city. Finally, establish open spaces within the settlement to form a vibrant settlement core together with the surrounding supporting facilities.

The future land use structure of the settlement can provide guidance to the land use function at the general level, coordinate the relationship between the settlement and the city, and emphasize the spatial layout pattern of the settlement with ethnic service points as the core, which may help improve the work-home balance of Hui people in the settlement, strengthen the intra-ethnic interaction of the community, and prosper the culture of ethnic characteristics.

4.5.2.3 Characteristic Space Level

In conjunction with the above-mentioned overall functional positioning and land use structure recommendations, this study further selected representative spatial segments in the Qijiawan Huis settlement for analysis and made strategic recommendations for the improvement of functional and spatial models (see Table 4.32).

Table 4.32 Optimization strategies for characteristic spatial segments

Typical space	Space optimization schematic	Analysis and strategy
Traditional residence fragment		**[Analysis of the problem]** A number of traditional residential fragments with high historical value, high architectural quality, and representative significance still exist in the Qijiawan Huis Settlement, which jointly form the spatial spiritual core of the settlement The optimization strategy should focus on preserving the traditional spatial pattern and texture based on the idea of ethnic cultural identity and provide the possibility of sustainable development
Optimization objectives	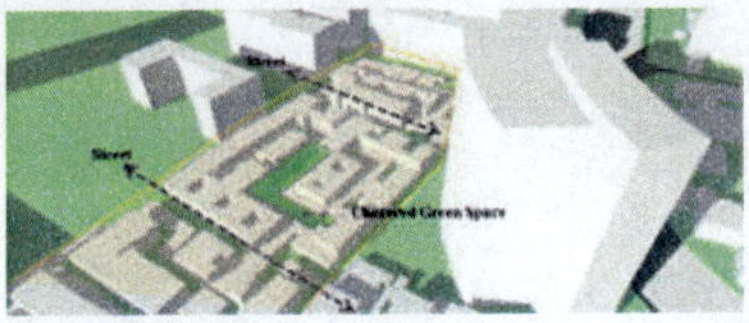	**[Specific Strategies]** Protect and preserve as much as possible, demolish the poor quality self-built houses, restore the traditional courtyard space and scale texture, clear the street structure, build open space for residential groups, and form a positive interaction between the traditional way of living and modern needs
Residential waterfront fragments		**[Problem Analysis]** The Inner Qinhuai River on the north side of the Qijiawan Huis Settlement is an excellent open space and landscape resource, and the current buildings in the settlement are adjacent to the street and facing back to the river, so this featured resource has not been used effectively
Optimization objectives	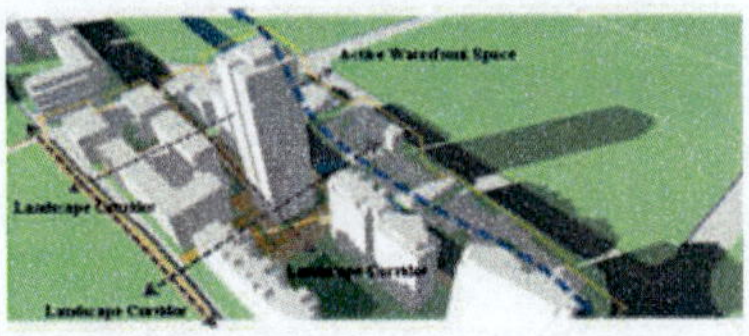	**[Specific Strategies]** Actively use the waterfront resource to improve the quality of the spatial landscape, and open up the connections between space, traffic, and line of sight within the settlement It will decentralize the motorized traffic to the periphery area, create open space joints in the residential groups along the river, and provide a pleasant living space environment and a place for communication activities for the residents of the settlement

(continued)

Table 4.32 (continued)

Typical space	Space optimization schematic	Analysis and strategy
Urban commercial fragments		**[Analysis of the problem]** Xinanli in the southeast of the Qijiawan Settlement is a city-level commercial district, but it is not closely connected to the internal functions and space of the settlement. It is possible to improve the supporting facilities of the settlement and provide employment opportunities for Hui residents by strengthening the connection
Renewal objectives	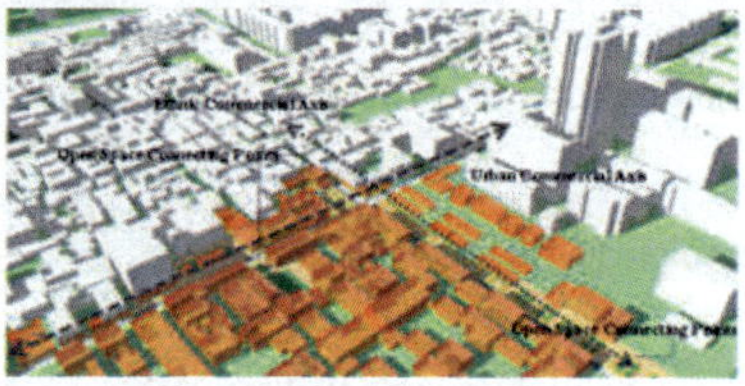	**[Specific Strategies]** In terms of function, the Xinanli neighborhood will be used to form a joint for the interaction between the settlement and the city, further develop urban and community-level ethnic industries, and improve related supporting facilities. In terms of space, the traditional space and scale texture characteristic of the settlement should be preserved and the open space should be enhanced

Source Author's own drafting

4.5.2.4 Work-Home Balance Level

Historically, the gradual settlement and growth of ethnic minorities in cities has not only relied on the socio-cultural ties derived from traditional bonds such as kinship and geography but also on the overlapping support of community-based economy and ethnic industrial system. Therefore, encouraging the restoration and development of ethnic-specific economies and constructing integrated work-home communities in Huis settlements is not only a guarantee for achieving balanced work-home development of Hui people but also one of the necessary conditions for the settlement to embark on a sustainable development path through the systematic integration of life and production.

Strengthen macro-controlling role of the government departments on market concept education and overall coordination of interests: Relevant local departments should protect and support the economy with ethnic minority characteristics through appropriate policies and support, plan the industrial space with ethnic characteristics and its industrial chain, and tap the market potential of the ethnic minority economy to maintain their relatively clustered living conditions from the work-home integration perspective.

Encourage Hui residents to take the initiative in gaining modern knowledge and expertise outside of their settlements to meet the challenges of modernization: Hui people should be motivated to transform their inherent traditional mindset, take advantage of the current opportunities of the changing industrial landscape of the city, and enhance their employability to cope with the ever-changing challenges of modernization. The government should increase the possibility of diversified local employment for residents.

Combine specific ethnic markets with urban demand and restore or strengthen popular ethnic-specific economies (such as halal restaurants and ethnic handicraft products): Plan and build related industrial sites and special neighborhoods, shape and expand the influence of ethnic brands, seek new sources of sustainable development for ethnic economies on the basis of the self-sufficiency model. The ethnic identity and sense of belonging of ethnic minority populations should be strengthened through multiple channels, combined with the ethnic-specific economy, and they will become an inexhaustible positive development driving force for the settlement (see Fig. 4.37).

In terms of the specific construction of the work-home community, the following strategies can be adopted at the micro level: First, considering that the current situation and tendency of the Qijiawan Huis settlement is developing from a closed status

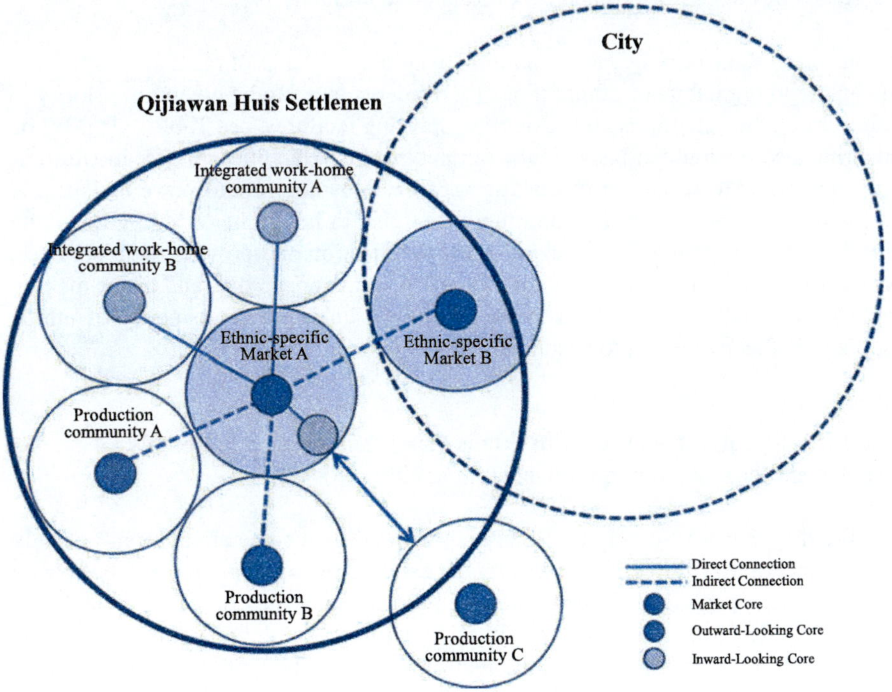

Fig. 4.37 Structure of ethnic-specific market in Qijiawan Huis settlement. *Source* Author's own drafting

to an open status, the construction of the work-home community may be based on the principle of "stores on bottom and residence on top." Second, according to the change of spatial scale, the spatial carriers of work-home communities in Qijiawan can be divided into three levels and modes: unit, neighborhoods, and clusters. Third, in terms of the specific application of the work-home community model, it is suggested that it should be adapted to local conditions and categories. Of them, the unit improvement model is applicable to the traditional residential houses with certain historical characteristics that have "stores on bottom and residence on top," which are mainly distributed on both sides of Pingshi Street. The neighborhood improvement model is applicable to multi-story residential neighborhoods built around the 1990s, such as the residents of the Pingshi Street neighborhood renovating street-level storefronts. The cluster improvement model is mostly applicable to high-rise buildings constructed around the 2000s and used for commercial and residential purposes such as the ground floor commercial stores in Tongkun Building on the west side of Zhongshan East Road (see Table 4.33).

4.5.2.5 Supporting Facilities Level

Establish a Sound System of Supporting Facilities, Particularly Considering the Ratios of Facility Types and Sizes Related to ethnicity

At the initial stage of policy and planning, the practical needs of residents, especially the ethnic living habits of ethnic minority residents, need to be given top priority in determining the categories and scale of supporting facilities (see Table 4.34). While ensuring the allocation of basic non-profit public ethnic facilities, private institutions are encouraged to set up profit-making organizations not only to serve the specific ethnic minority groups in the community but also to help other ethnic groups and a wider range of people in achieving self-identification and provide them services, which will form a situation in which non-profit, quasi-non-profit, and profit-making facilities coexist in a pluralistic manner, closely linking the prosperity of ethnic facilities in the settlement to sustainable development dynamics.

Prioritize Layout of Facilities with Ethnic Characteristics and Preferably Supplement the Construction of General Facilities

Considering that well-configured facilities with ethnic characteristics can not only serve Hui people within this community but can also be extended to meet the special needs of Hui people in the surrounding areas and even the entire city, which can further help strengthen the sense of cultural identity and belonging within and outside the ethnic group, the layout of ethnic facilities is often the first test for the sustainability of traditional culture in the settlement. The specific spatial strategies are as follows: first, rely on the original historical commercial interfaces such as Qijiawan and Cangxiang, reshape the commercial axis with ethnic characteristics and build a

Table 4.33 Suggested spatial patterns of work-home community

Level	Unit	Neighbourhood	Cluster
Scale	10–50 m	50–150 m	150–500 m
Model construction	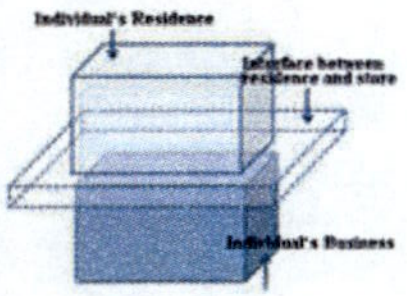	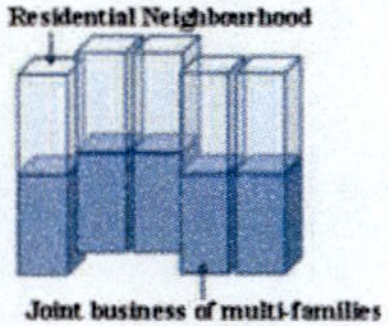	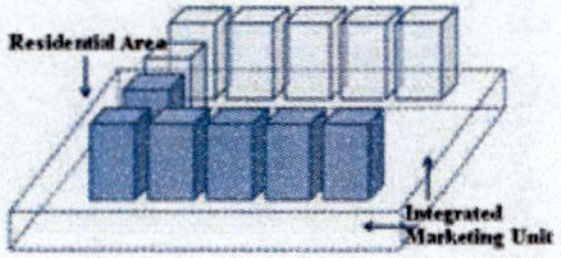
Renewal example			

Source Author's own drafting

Table 4.34 Proposed supporting facility for minority settlements

Facility category	General facilities		Facilities with ethnic characteristics	
	Facilities	Recommended size (m^2)	Facilities	Recommended size (m^2)
Education	Including kindergarten and elementary school	Consider based on 70 students per 1000 people and a 500-m service radius	Including kindergarten and elementary school for ethnic minorities	Consider based on 80 students per 1000 people and a 1000–1500-m service radius
Healthcare	Including clinics, health posts, hospitals, etc	2500–3000	Including clinics, health posts, hospitals, etc. for ethnic minorities	1000–1500
Culture and sports	Including cultural activity stations, libraries, and clubhouses	4000–5000	Including ethnic minority activity stations, marriage agencies, cultural stations, etc	1000–2000
Commercial services	Including various commercial facilities, meat and vegetable markets, etc	2000–4000	Including halal meat and vegetable markets and restaurants	1000–1500
Financial post and telecommunications	Banks, savings banks, post offices, etc	16,000–20,000	–	–
Municipal utilities	Including parking lots, bus stops, etc	Bicycle parking lot of 1000–1500 m^2; at the same time, set up 200–300 motor vehicle parking lots and 2 bus lines in conjunction with the community center	–	–
Administration	Including residents' committees, sub-district offices, police stations, etc	3000–5000	Including ethnic houses, etc	1000–2000

(continued)

Table 4.34 (continued)

Facility category	General facilities		Facilities with ethnic characteristics	
	Facilities	Recommended size (m^2)	Facilities	Recommended size (m^2)
Others	Including human defense facilities, street non-profit facilities, etc	1500–2500	Including religious construction facilities such as mosques	Allocate a mosque for every 300 Muslim residents; set the size of the mosque with due consideration to the service radius

Source Compiled by the author with reference to relevant norms (based on a settlement size of about 30,000 people)

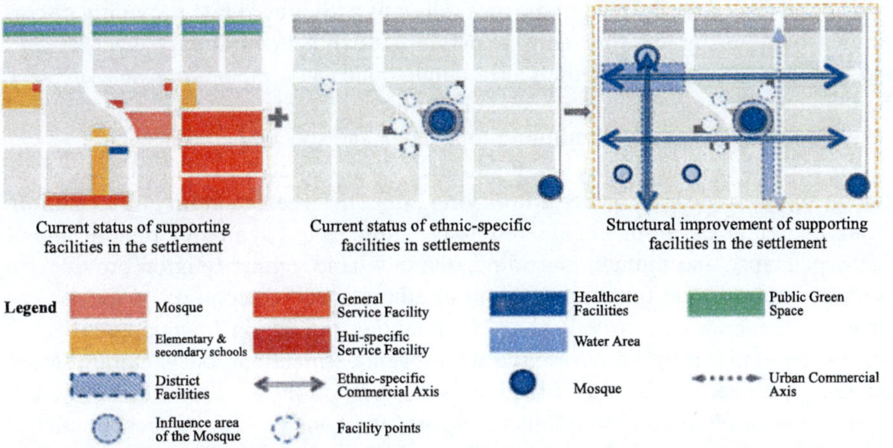

Fig. 4.38 Improvement of supporting facilities in Qijiawan Huis settlement. *Source* Data from the research group's sample study on the Qijiawan Huis settlement in Nanjing (2014)

network of ethnic services; second, strengthen the deployment of ethnic services at important spatial connection points, such as urban road intersections, to enhance the urban radiation of ethnic services in the settlement; finally, allocate grassroots ethnic service centers in the residential clusters within the settlement to meet the necessary service needs at the grassroots community level (see Fig. 4.38).

4.5.2.6 Ethnic Culture Level

Strategies for Religious Beliefs

There is an inseparable endogenous mechanism between religion and culture (Zhuo 1988), and the history of human culture is that of religion during a long period of historical development. The ethnicity and religion of the minority culture represented by the Hui people show a high degree of long-term integration. Religion is not only faith but also a code of conduct that guides and influences all aspects of the life of Hui people and serves as a social network and instills values in the social structure of Hui and becomes even the main criterion for shaping and dividing the ethnicity (Zhou 2007).

On one hand, combine policy orientation with ethnic identity, develop the dual drive from state-level and ethnic-level, and guide the free and solid development of religious beliefs on the basis of ethnic and cultural identity. Improve and supplement the basic places and facilities for religious activities (e.g., mosques, free private schools, ethnic-specific elementary schools, etc.) in the settlement areas to encourage the orderly development of religious activities, while providing opportunities for ethnic minority residents to participate in other community-level services and facilities to further strengthen intra-ethnic bonds and interactions, and enhance the religious and cultural consciousness of ethnic minorities and the social capital of the community by means of cultural propaganda and special activities.

On the other hand, develop vigorously ethnic-specific cultural industries and strengthen the interactive amplifying effect between economy and religion. Ethnic economic development and religious culture have a symbiotic relationship of interdependence and mutual constraints. On one hand, ethnic religion provides an important cornerstone for the formation of ethnic-specific economy, while, on the other hand, economic prosperity lays the material foundation for maintaining and developing ethnic religion. While the wave of modernization boosts cultural globalization, we need to also pay attention to the promotion of cultural diversity, full exploitation of ethnic and religious resources, and construction of special cultural industries in the context of modernization, such as religious and cultural tourism, ethnic commercial streets, and medium and large-scale halal food enterprises.

Strategies for Cultural Education

The educational level of Hui people in the settlements is directly related to the competitiveness of the new generation of Hui people in the labor market and their relative position and space for development in the political, economic, and cultural aspects of the city. Reasonable guidance of the development direction of ethnocultural education in the settlement can provide a guarantee strategy from social and spatial aspects.

From the mentality perspective, establish ethnic cultural confidence: The establishment of ethnic cultural identity is the ideological basis for developing ethnic

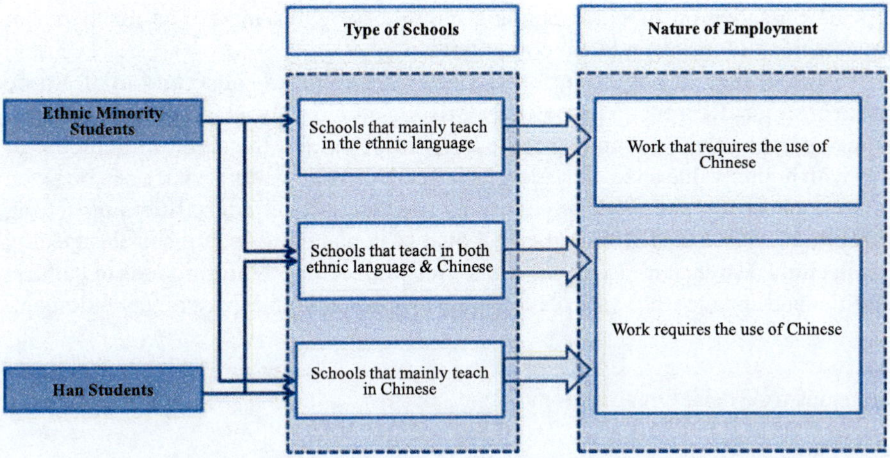

Fig. 4.39 Patterns of students choosing school types by ethnicity.. *Source* Drawn from the following literature—Ma, R. Some Thoughts on Education for Ethnic Minorities in China [J]. Journal of Xinjiang Normal University (Philosophy and Social Science Edition) 2010, 31 (1): 15

minority culture education, which includes the self-identification of ethnic minority culture and external identification from other ethnic cultures. Only ethnic cultures recognized in the larger society can be developed in all aspects of social life.

From the educational content perspective, incorporate the content of ethnic characteristics into the modern education system: The pragmatic concept of modern education that emphasizes higher level education and employment makes us fully aware that only by ensuring the competitiveness of ethnic minority students in higher education and employment can we break through the stagnation of Hui people in terms of ethnic cultural awareness and customs and integrate them into the modern education system spontaneously. At the same time, considering the need for the inheritance and identification of ethnic culture, it is advisable to incorporate ethnic culture, traditional customs, ethnic-specific economy, and other special contents in conjunction with the transmission system of scientific and cultural knowledge in modern education. This will not only guarantee the future career prospect and modern adaptability of ethnic minorities but also strive for the inheritance and renewal of traditional culture (see Fig. 4.39).

From the educational structure perspective, the educational concept of ethnic culture should be deeply implemented: To build a comprehensive and three-dimensional structure of ethnic education covering all levels, it is possible to rely on rich and flexible educational paths, including lay the foundation of ethnic cultural identity in the daily family education, set up ethnic culture courses at basic education level, such as primary and secondary schools, strengthen training related to ethnic-specific industries in vocational education, conducting religious study courses in religious activity sites such as mosques, focus on the promotion of ethnic culture and the organization of corresponding cultural activities in service management organizations such as residents' committees, etc.

It is necessary to pay attention to the protection and construction of ethnic cultural space: Respect and protect the existing ethnic cultural buildings (mosques, ethnic-specific schools, ethnic-specific hospitals, etc.). For the ethnic cultural buildings with higher value characteristics, its historical ethnic cultural space can be given a new connotation and modern vitality by upgrading its urban positioning, setting protection levels, and improving spatial quality in planning. In terms of the specific architectural design, it is also necessary to fully reflect the ethnic attributes of cultural facilities and enhance the sense of belonging to the culture and space of the settlement.

Strategies for Social Organizations

The traditional social organizations in the Huis settlement can provide more targeted services for the Hui than the government administration and play an important role in the process of social formation and spatial clustering of the Hui. These traditional social organizations were once disintegrated with the religious reform in the early years of the founding of PRC. Subsequently, due to the socio-economic transformation under the wave of modernization, they faced the challenges of stratification of social interests and logic of market supply and demand in their organizational reconstruction, coupled with the multiple resistance brought by the weakened ethnic identity of the Hui people themselves and the diversification of individual needs, their optimization can be considered from the following aspects:

From the external guidance perspective, it is necessary to change the mindset and identify with minority cultures and combine government-led management with autonomous management in ethnic minority settlements: Government departments should fully understand the intrinsic needs of ethnic minorities to establish their own social organizations, change the control-oriented management in the past to the nurturing-oriented management that moderately guides and encourages ethnic social organizations, incorporate the development of social organizations in minority settlements into the construction planning of basic systems, and provide certain support in terms of the system construction and financial investment.

From the internal dynamics perspective, it is necessary to encourage the ethnic minority population to initiate modern democratic and legal transformation on traditional ethnic social organizations and guarantee the effective participation of residents of the settlement in social organizations: As a grassroots institution that integrates the material, cultural, human, and organizational resources of the settlement, it not only influences the spatial construction of the settlement but also plays a crucial role in the social and cultural construction. The effective participation of minority representatives can not only guarantee their material needs but also effectively promote the preservation and renewal of traditional ethnic culture in modern urban communities.

Take the spatial-economic transformation of modern urban renewal or settlement areas project construction as an opportunity to re-examine the reconfiguration of ethnic minority social organizations during the transition period: A networked structure of social organizations in the settlement should be guided and established in due time to realize the necessary interaction for their participation in the modernization of the settlement (see Table 4.35).

4.6 Summary of Chapter

This chapter is essentially a special exploration of the theme "**Integration-Response**" in the field of "ethnic minorities—modernization". It firstly defined the concepts of modernization and Minority settlements and determined the reasonable division of community evolutionary stages through a multilinear analysis of the evolutionary background of Huis settlements. Next, focusing on the typical minority settlements in Nanjing, the Qijiawan Huis settlement, this research investigated and retraced the changes of this "Pangen Grass" under the impact of modernization, focused on the specific socio-cultural background and distinct ethnic identity of the community, which includes the evolution of social attributes (such as demographic composition, family relations, community organization, and living customs) and spatial attributes (such as land use function, living space, employment work-space, and supporting facilities) of the settlement, and analyzed the motives and mechanisms of changes in the Huis settlement under the wave of modernization. Finally, on the basis of revealing the existing problems of the settlement, we discussed the renewal strategies of the Huis settlement based on cultural identity.

The main conclusions of this chapter include:

(1) In terms of evolutionary stages, this research took the evolutionary key moments from four major aspects, modernization stages, China's modernization stages, Nanjing's urban construction, and changes of the Qijiawan Huis settlement, into account to consolidate and classify the evolutionary stages of the Qijiawan Huis Settlement in Nanjing since the founding of the People's Republic of China as follows: Stage I (1949–1958); Stage II (1958–1978); Stage III (1978–1990); Stage IV (1990–1998); and Stage V (1998-present).

(2) In terms of social attributes, the education level of Hui people in the Qijiawan Huis Settlement constitutes a trend of "steady increase in the number of people with college education or above," the employment structure gradually evolved from the pattern of "industry-based, supplemented by Hui ethnic-related services" in the early years of the country into a diversified situation where "all kinds of employment directions co-exist equally." The interethnic marriage rates in Qijiawan Huis settlement area shows a "linear increase" in general and it family structure shows the characteristics of "diversified structure, small size, and ageing population." The social organization of Hui people

Table 4.35 Recommended social organization configurations in minority settlements

Type	Organization		Type	Functions
General social organizations	Residents' committee		Political organizations	A grassroots-based mass self-governance organization
	Community home service center		Mass organizations	Provide residents with assistance in managing their family affairs
Ethnic-specific social organizations	Modern religious community	Ethnic support organizations	Religious organizations	Organize the orderly deployment of religious activities, promote religion and ethnic culture, and provide material and spiritual assistance to ethnic compatriots in need
		Mosque educational institution		
	Ethnic medical team		Healthcare organizations	Respect the customs of ethnic minorities and take care of the health of the disadvantaged people in the ethnic groups
	Ethnic-specific industry association		Economic organizations	Promote the development of ethnic minority-specific industries and help members find employment
	Ethnic marriage agency		Fellowship organizations	Strengthen intra-ethnic contacts and interactions and to facilitate intra-ethnic marriages

Source Author's own drafting

in the Hui settlement presents a characteristic of "unification of institution functions and generalization of inter-ethnic interactions." The living customs have an "increasing ignorance for dietary and religious traditions among Hui after the reform and opening up" since the reform and opening up, particularly the frequency of religious activities has shown a "polarization and serious ageing" pattern.

(3) In terms of the evolution of spatial attributes, the land use function of Hui settlements has undergone an evolutionary process from "single function of mainly residential use of land to a mix of multifunctional use of land," and the architectural texture shows a trend of "traditional texture of small blocks, dense road networks division, low-rise and small volume buildings to the modern texture of high-rise and large volume buildings." The spatial and temporal evolution of living in Huis settlements is characterized by the evolution from an "absolute dominance of the core area" to a "balanced coexistence of all categories of areas." The migration trajectory of residences has the characteristics of "varies by age, and mainly in the old city." The housing types in the settlement have the trend of "changing from traditional low-rise courtyard-type dwellings" to "modern multi-rise residential communities with complete facilities," and the residential ownership of the Hui people has evolved from mainly "self-built and rented public" to "diversified housing types and privatized property rights. The employment time in the Huis settlement shows a general pattern of sharp decline in the proportion of industrial workers after 1998, a steady increase in the proportion of commercial service workers, and a stable proportion of workers in other industries." Its spatial structure shows a distribution trend of "the core area is dominated by Hui ethnic-related service workers, while the other three types of areas are dominated by industrial workers." Their employment space has shown the characteristics of "age-specific and long-distance migration" and their commuting characteristics have evolved into a pattern of "half an hour in time, mainly on foot and by bicycle," which has also shown increasingly significant work-home separation characteristics. In terms of spatial structure, the nested commercial layout of the settlement with the mosque as the center is declining. In terms of spatial form, the "linear layout along the inner streets" was transformed into a "piecewise layout along the main urban roads." In terms of facility types, Hui ethnic-specific stores are gradually disappearing and facility types tend to diversify.

(4) On the level of development positioning of the renewal strategy, the overall development of Qijiawan Huis settlements in the future is positioned as a minority settlement that integrates urban commerce and tourism services, with historical culture and urban commerce as the driving force for development, Hui culture as the highlight, and has traditional small-scale spatial appearance.

(5) On the land use structure level of the renewal strategy, we propose to update the strategy of land layout pattern based on the overall functional orientation of the Huis settlement to combine the traditional residence pattern of Hui people with modern urban life.

(6) On the spatial level of the renewal strategy, this paper combines the above recommendations on the overall functional positioning and land use structure of the Huis settlement and further selects representative typical spatial segments of the Qijiawan Huis settlement for analysis and proposes strategic recommendations on functional and spatial improvement patterns.
(7) On work-home balance level, there should be integrated work-home communities built in the Huis settlement by enhancing the macro-control role of governmental departments, encouraging Hui people to take the initiative to learn modern knowledge and expertise outside the settlements, supporting the recovery and development of ethnic-specific economy in the settlements, and planning ethnic-specific industrial space and its industrial chain.
(8) On supporting facilities level of the renewal strategy, the government needs to establish a sound system of supporting facilities, particularly consider the ratios of facility types and sizes related to ethnicity. It is proposed to prioritize the layout of facilities with ethnic characteristics and preferably supplement the construction of general facilities.
(9) On the ethnic culture level of the strategy, in terms of religious beliefs, the Huis settlement needs to combine policy orientation with ethnic identity, develop the dual drive from state-level and ethnic-level, guide the free and solid development of religious beliefs on the basis of ethnic and cultural identity, develop vigorously ethnic-specific cultural industries, and strengthen the interactive amplifying effect between economy and religion. It is essential to establish ethnic cultural confidence, incorporate the content of ethnic characteristics into the modern education system, implement deeply the educational concept of ethnic culture, and pay attention to the protection and construction of ethnic cultural space. From social organization perspectives, it is necessary to combine government-led management with autonomous management in the ethnic minority settlement, encourage the ethnic minority population to initiate modern democratic and legal transformation on traditional ethnic social organizations, utilize the spatial-economic transformation of modern urban renewal or settlement areas project construction as an opportunity to re-examine the reconfiguration of ethnic minority social organizations during the transition period.

References

Black CE. The modernization in Japan and Russia. Beijing: Commercial Press; 1992.
China Center for Modernization Research, CAS, China Modernization Strategy Research Group. China Modernization Report. Beijing: Peking University Press; 2010.
China Center for Modernization Research, CAS, China Modernization Strategy Research Group. China Modernization Report. Beijing: Peking University Press; 2010. P. 282.
Ji X. East Asia and India: Preface to two modernization models in Asia. In: Chen F, editors. East Asia and India: two modernization patterns in Asia. Economic Science Press; 2000. pp. 1–2.

Luo R. A new theory of modernization—China's road to modernization. Wuhan: East China Normal University Press; 2013. p. 58.
(Ming) Factual Record of Taizu of the Ming Dynasty, vol. 109. Sixteen years of Ming Yongle; 1418.
(Ming) Gu Q. Kezuozhuiyu, vol. 1. Beijing: Zhonghua Book Company; 1997.
Nanjing Local History Compilation Committee. A Brief Chronicle of Nanjing. Nanjing: Jiangsu Ancient Books Publishing House; 1986. P. 16.
Nanjing Local History Compilation Committee. Nanjing ethnic and religious records. Nanjing: Nanjing Press; 2009a. p. 15.
Nanjing Local History Compilation Committee. Nanjing ethnic and religious records. Nanjing: Nanjing Press; 2009b. p. 1.
Nanjing Local History Compilation Committee. Nanjing ethnic and religious records. Nanjing: Nanjing Press; 2009c. p. 9.
Nanjing Local History Compilation Committee. Nanjing ethnic and religious records. Nanjing: Nanjing Press; 2009d. p. 10.
Nanjing Local History Compilation Committee. Nanjing ethnic and religious records. Nanjing: Nanjing Press; 2009e. p. 176.
Qian X. World modernization history (East Asia volume). In: Qian X, editor. World modernization history (East Asia volume). Nanjing: Jiangsu People's Publishing; 2012.
(Qing) History School of Liaoning University edited. Factual Record of Taizong of the Qing Dynasty, vol. 7; 1978.
(Qing) Bi Y, editors. Xu Zizhi Tongjian, vol. 213. Changsha: Yuelu Press; 1999.
Ma, Rong. Ethnic sociology: a study of race relations in sociology. Beijing: Peking University Press; 2004.
Rongqu R. A new theory of modernization–China's road to modernization. Wuhan: East China Normal University Press; 2013. p. 95–7.
(Qing) Wang Y. Shangyubaqi (Imperial Edict for the Eight Banners). Ninth year of Qing Yongzheng; 1731.
(Southern Song Dynasty) Compiled by Zuguang M, Yinghe Z, editors. Records of Jingding Jiankang, vol. 21. Beijing: Zhonghua Book Company; 1990.
(Southern Song Dynasty) Li X. The main records of the years since Jianyan, vol, 20. Shanghai: Shanghai Classics Publishing House; 1992.
Webber M. The protestant ethic and the spirit of capitalism. Guilin: Guangxi Normal University Press; 2010.
Wu G. Pluralistic modernity: from the September 11 disaster to Wang Hui's "Modernity in China." Shanghai: Shanghai Sanlian Bookstore Press; 2002. p. 169.
Xiaofei C, Shaowei A, et al. A study on the spatial location of business around urban mosques under informal system. Econ Geogr. 2014;34(6):112.
(Yuan) Shi L. Damingfu Journals of Zhengfu Emperor, vol. 10, Literary. Boyan-Zongdao Biography. Shanghai: Shanghai Editorial Office of Zhonghua Book Company; 1966.
Yin C. Research on the Protection of spatial Pattern of Nanjing Ancient City. Nanjing: Southeast University; 2005.
Zapf CL, Lu C. Modernization and Social Transformation, 2nd ed. Beijing: Social Science Literature Publishing; 2000.
Zhou D. Introduction to anthropology. Kunming: Yunnan University Press; 2007. p. 35.
Zhou L, et al. Preservation and renewal of Nanjing's Old City in the process of rapid modernization. Nanjing: Southeast University Press; 2004. p. 20.
Zhuo X. Religion and culture. Beijing: People's Press; 1988. p. 57.

Chapter 5
Community Sample IV: Service Supply and Facility Utilization: Research on the Level of Public Services in Migrant Worker Settlements

Abstract In the wake of the reform and opening up, China saw reforms in the rural system, changes in farmers' perceptions, and removal of urban–rural barriers. Alongside these transformations, the surplus rural labor force that had long been suppressed by the urban–rural divide policy was liberated. Under the influence of the push and pull factors of "urban–rural" migration, an unstoppable, large-scale "migrant worker boom" has gradually formed in urban areas. This type of active urbanization group differs from the land-lost peasant (passive urbanization) group described in Chap. 3. Active urbanization groups mostly form multiple types of distinctively characteristic residential patterns in cities through differentiated selections of residence and diversified places of residence. In addition to dispersed rentals and settlements, a typical phenomenon pertains to the spontaneous agglomeration of migrant workers in more economically developed areas and widespread formation of sundry settlement spaces. Some apt examples include the "Zhejiang Village" in Beijing, the "Henan Village" in Nanjing, and the mixed settlements of floating populations in Shenzhen and other regions. The scale of such communities ranges from hundreds to tens of thousands of residents.Regarding such migrant worker settlements, the research group has long focused on their structural characteristics, having systematic discussions on regional distribution (primarily the urban–rural fringe), demographic composition (relationship-related and mixed settlements), employment structure (mainly employment in the tertiary sector, and informal and localized employment), space and environments, etc. Xiao (Analysis of the employment space for metropolitan floating population: an empirical study for migrant workers. Southeast University Press, Nanjing, pp. 144–155, 2010). However, research on areas in relation to the level of public services enjoyed by migrant workers is inadequate, which serves as one of the quintessential manifestations of the "citizenization of

---

The majority of research finds that migrant workers from different regions often engage in different occupations, whereas migrant workers from the same region tend to engage in similar or related traditional occupations (a manifestation of industry relationships). For example, Anhui people often work in the fruit and vegetable business, Henan people deal in used item collection, Xinjiang people enter the catering sector, Zhejiang (Wenzhou) people go into the garment processing business, and Jiangsu people pursue work in home remodeling and interior design. Such instances are referred to as "regionalized employment".

X. Wu and L. Wang, *Settlement Spaces: Urban Survival Prospects of China's Special Communities*, Springer Geography,
https://doi.org/10.1007/978-981-16-4892-2_5

migrant agricultural population." For example, what is the state of public services in these settlements? What regulations and issues do members face in terms of the supply and utilization of facilities? How can their public services be improved and enhanced? On this ground, this chapter conducts a specialized study on the level of public services from the two angles of "objective supply and subjective utilization" based on first-hand data collected in migrant worker settlements. First, this study assesses and compares the level of supply of public services and facilities in migrant worker settlements and explores the underlying reasons behind the characteristics of facility supply. Next, this study investigates and analyzes the characteristics of the utilization of public services and facilities among migrant workers in these settlements and reveals the law governing group differentiation in the use of facilities. Finally, this study presents the match patterns of facility "supply-utilization" and discusses the approaches to enhancing the level of public services in migrant worker settlements based upon different match types. In essence, the selection and investigation of this type of community sample entail a specialized study on the sphere of "the inflow of the migrant agricultural population into cities" under the topic of "integration-response."

Keywords Migrant worker settlement · Public service and facilities · Supply · Utilization · Match pattern · Nanjing

5.1 Definition of Research Objects

5.1.1 Definition of Migrant Workers

Although perceived as a prime representative and significant branch of the floating population, migrant workers do not have a unified and precise definition or interpretation. A great number of scholars have discussed and provided different meanings for the floating population from the perspectives of different disciplines, such as economics (Rongshi 1996), population geography (Ruijun 1990; Shanyu 1999), and administrative management (Chengrong 1999).

Based on the characteristics by which the floating population varies relative to the geographical space, Xiao (2001) defined the floating population in the broad and narrow senses. According to the length of stay in the place of inflow, the floating population in the broad sense may be generally classified into three types, i.e., permanent migratory population, travel-based passing population, and temporarily residing population. The floating population in the narrow sense is limited to the travel-based passing population who stay in a certain region for a short span only.

In addition, based on the motivation (or nature) of the floating population, Jianmin et al. (1995) categorized the floating population into labor, management and service, public service, cultural, and social types. These five types of floating population can be further grouped into two categories: economic and non-economic. Of the two, the

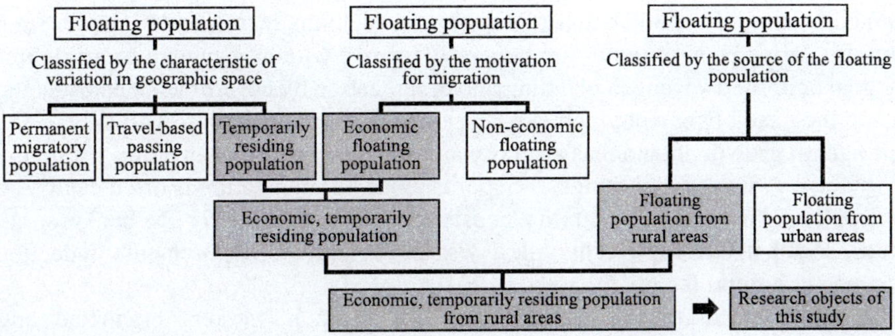

Fig. 5.1 Classification of migrant workers and definition of research objects. **Source* Own elaboration

economic floating population directly engages in economic activities of all sorts in cities to obtain income. Migrant workers belong to this category.

Finally, based on the source of the floating population, Qiang (2012) divided the floating population into two groups, namely, "migrant peasants" (i.e., the floating population from rural areas) and "migrant citizens" (i.e., the floating population from urban areas).

With reference to the different classification standards stated above, the "migrant workers" investigated in this chapter refer to the economic, temporarily residing population from rural areas, i.e., the surplus rural labor force entering urban employment primarily to earn a living (referred to as "migrant workers" in short). The migrant workers in this sense are distinguished from literary and art workers, new hires, and middle and senior managers in terms of employment status and positions (see Fig. 5.1).

5.1.2 Definition of Migrant Worker Settlements

Of the various types of settlement spaces, that spontaneously formed by migrant workers through active selection of residence, group rental, and group residence shows a greater inherent sense of rationality and inevitability in light of the naturalness and universality of its development. As a type of unique product of urbanization that straddles the divide between urban and rural areas, such settlement space displays marginality and heterogeneity at variance with ordinary urban and rural communities in terms of its social attributes, regional distribution, demographic composition, and social influence. Accordingly, this type of settlement space bears exclusive sample significance and typical value, which are not rendered by passive settlement spaces such as collective dormitories and construction sheds.

In view of the above-mentioned, the "migrant worker settlements" to be discussed in this chapter will focus on this type of active settlement space, particularly the spontaneously formed concentrated residential area where migrant workers make up the main body of residents and house rentals serve as a dominant residential mode. Such naturally derived communities are generally formed without having a priori definitive goal or unified advanced planning and organization by government departments. Rather, their establishment is achieved through the continuous coordination between their natural growth potential as an entity and spontaneous harmonization within the community. Among other things, the spontaneous harmonization is often catalyzed by deep-lying intrinsic adaptability consisted by factors such as life background, values, social psychology, economic laws, and functional requirements under the influence of natural factors (Xiao et al. 2010).

As for concentrated residential areas for migrant workers organized and constructed by the government, collectives, or corporations (e.g., collective dormitories for migrant workers, construction sheds, resettlement areas), they actually belong to a type of settlement space generated through passive selection of residence. In light of this, such settlement space is fundamentally different from the settlements naturally derived through the active selection of residence as described previously, thereby going beyond the scope of our discussion herein.

5.1.3 Classification of Migrant Worker Settlements

Examined in light of community composition and organization, migrant worker settlements formed through active selection of residence may be classified into two types. Notably, these two types of settlements are characterized by dissimilar formation mechanisms (discussed further below).

The first type of settlement gathers groups from the same township, village, industry, or ethnicity. With kinship, geography and industry-based relationships as their fundamental bonds, these settlements are referred to as "relationship-related settlements." Illustrative examples include the "Zhejiang Village" and "Xinjiang Village" in Beijing, "Anhui Village" in Nanjing, and "Dishi Village (Taxi Village)" in Shenzhen. The other type of settlement pertains to "mixed settlements," which do not constitute a distinct dominant bond or industrial system in their composition. The residents come from mixed sources and lack extensive connection and necessary communication with one another. Mixed settlements cannot compare with relationship-related settlements in terms of the cohesion of communities and representativeness of settlements. Multitudes of communities of this type have emerged in urban villages situated in cities such as Guangzhou and Shenzhen.

5.1.4 *Definition of Public Services and Facilities*

5.1.4.1 Introduction to Public Services and Facilities

Foreign scholars made an early start on research in relation to public services and facilities. For example, Samuelson (1954) established the classic concept of public goods, arguing that these are products characterized by non-excludability and non-rivalry in consumption. The most commonly used terminology in subsequent related studies includes "public services" and "public facilities." These two terms basically share the same meaning and refer to services and facilities provided directly or indirectly by the government and shared by all further categorized public services and facilities into two types. One denotes the facilities necessary for urban life, including transportation facilities, water supply and drainage, public order, and security. The other indicates facilities that improve the quality of life and level of development among residents, including medical care, libraries, parks, and green spaces (Junbo et al. 2010).

Chinese researchers conducted extensive studies on public services and facilities. The academia defines public services and facilities as "services or facilities that are directly or indirectly provided by government departments for all citizens to enjoy in order to fulfill their daily requirements for education, medical care, shopping, culture and entertainment, socialization, and other activities (Dawei et al. 2006)" In terms of the industry norms and standards, the Code of Urban Residential Areas Planning and Design (GB50180-93) defines public services and facilities as the sum of all facilities that can meet the basic needs of all classes and groups and provide corresponding software and hardware services in eight areas, including medical care, education, recreation and sports, business services, community services, financial telecommunications, municipal public utilities, and administrative management. The Standard for Basic Terminology of Urban Planning (GB/T50280-98) does not accord a definition of "public services and facilities" but provides descriptions of related concepts. In particular, the standard classifies urban infrastructure into two types: engineering and social. The former refers to engineering facilities such as energy supply, water supply and drainage, transportation, post and telecommunications, environmental conservation, and disaster prevention and protection. The latter pertains to cultural, education, medical, and health facilities. Of regional standards, the Planning Guidelines for the Supporting Standards of Basic Public Services and Facilities in the Rural Areas of Nanjing City, which was promulgated in 2011, defines public services and facilities as facilities that serve the function of basic public services, including education, health, inquiry, society, administrative management, business services, and municipal facilities. In the Supporting Planning Standard for Public Facilities in Nanjing City, which was promulgated in 2015, public services and facilities are classified into seven categories: education, medical care and health, culture and entertainment, sports, social welfare and security facilities, administrative management and community service facilities, and commercial and financial service facilities. In addition, the standard specifies the necessity of including parks and green spaces, the first and last bus terminals, and post and telecommunications in the scope of public services and facilities (see Fig. 5.2).

5.1.4.2 Public Services and Facilities under Study

The "public services and facilities" to be studied in this chapter refer to public goods or equipment of all classes and types directly or indirectly provided by government departments for public use in fixed places, including the four types of social infrastructure selected in the Standard for Basic Terminology of Urban Planning that are closely related to the daily lives of residents in migrant worker settlements as well as transportation facilities in the category of engineering facilities.

5.1.4.3 Classification of Public Services and Facilities

As an essential part of urban planning and construction, public services and facilities have gradually formed a multi-level, multi-class network of public service supply. Under such a spatial architecture, different classes of public facilities may display distinct discrepancies in scale, quantity, and service scope. In this light, a definite class division within the system should be established to facilitate the research targeted on the level of public services in each class.

However, currently no unified standard has been established for the classification of public services and facilities. The majority of scholars categorize public services and facilities into two classes: city and residential-area. Such class division may be inadequate for some major cities, wherein regional standards will be formulated and promulgated accordingly. For example, the Supporting Planning Standard for Public Facilities in Nanjing City (2015) divides public facilities into four classes, namely, city, district, residential community, and grassroots community. Of them, city-class and district-class facilities originate from the unified constructions primarily supplied by the government, whereas residential community class and grassroots community class facilities mostly originate from cooperative residential construction under the regulation of the market economy. Together, they constitute a public service system for residents.

In view of the different dominant factors in resource allocation at different levels (the difference between governments and markets) and discrepancies in service scopes, this study categorizes public services and facilities into three classes, i.e., city, district, and community. These classes respectively correspond to the four classes defined in the Supporting Planning Standard for Public Facilities in Nanjing City (2015). Their specific relationship is shown in Fig. 5.3.

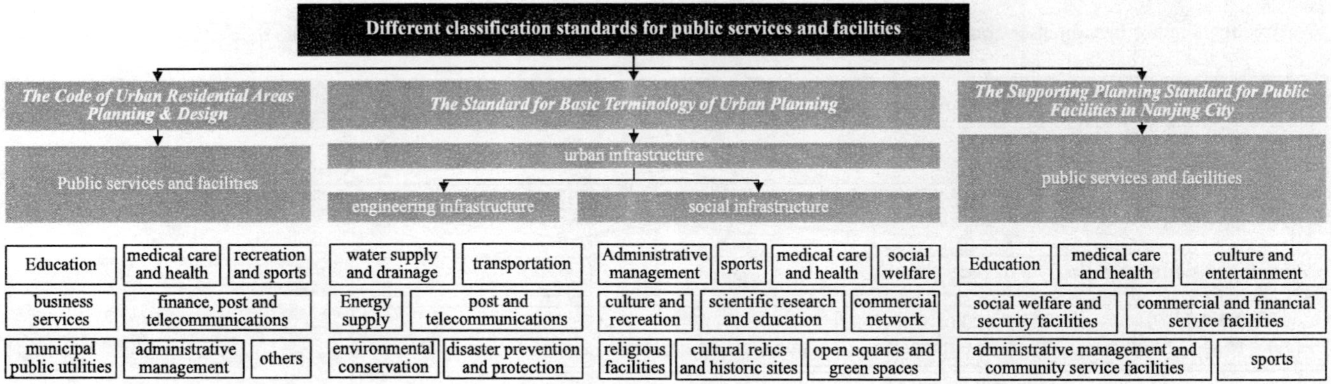

Fig. 5.2 Different classification standards for public services and facilities. **Source* The Code of Urban Residential Areas Planning and Design (GB50180-93); The Standard for Basic Terminology of Urban Planning (GB/T50280-98); Supporting Planning Standard for Public Facilities in Nanjing City (2015)

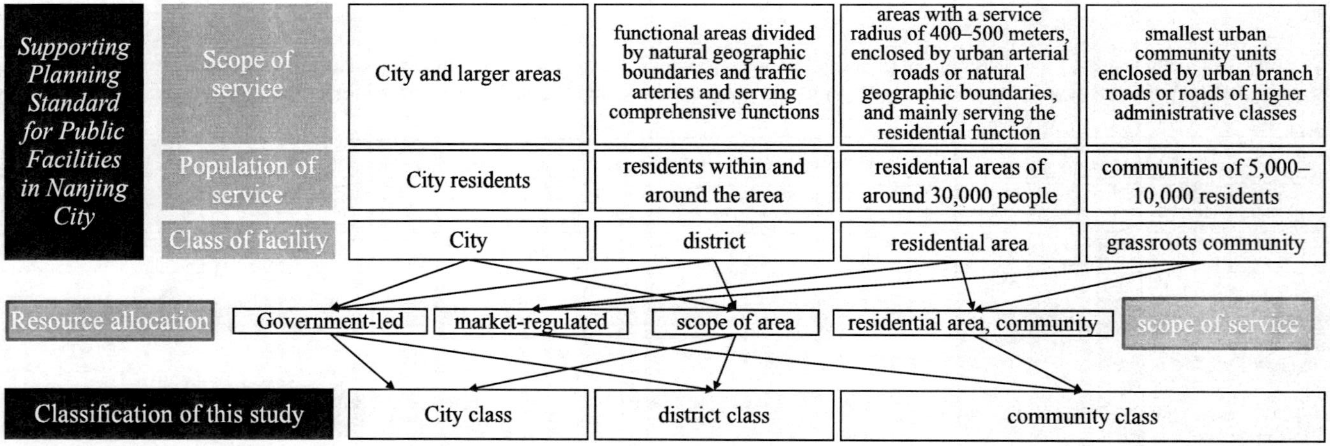

Fig. 5.3 Classification of research on public services and facilities. **Source* Supporting Planning Standard for Public Facilities in Nanjing City (2015)

5.2 Overview of Migrant Workers, Migrant Worker Settlements, and Public Services and Facilities

5.2.1 Relevant Background for the Formation of Migrant Workers

In the context of a planned economy, China established an urban–rural dual structure. Characterized by household registration management, this structure fundamentally adopts a static strategy of enclosure that focuses on restraining and controlling the spontaneous flow of populations of all types. In the 1980s, with the accelerated pace of reform and opening up and the continuously improving urbanization, China rapidly transformed from a leading country in traditional agriculture to a country of industrialization and urbanization. It was not until then that a dramatic change came about in the structure of labor force and employment, which, in turn, triggered a large-scale "migrant worker boom," wherein a substantial number of surplus rural workers migrated to cities. The reasons for such migration may be explicated through the four categories of factors proposed in *A Theory of Migration* by the American scholar Everett S. Lee, as detailed below.

5.2.1.1 Positive and Negative Factors

The large-scale migration of surplus rural labor to cities is essentially the result of comparison between positive and negative factors or the combined effect of push factors of the place of population outflow (rural areas) and pull factors of the place of population inflow (urban areas). This comparison is manifested mainly in three aspects:

Disparity in employment opportunities: On one hand, China's rural areas have long seen a disproportionate ratio between their population and land. The issue of "surplus" rural labor has become increasingly critical. On the other hand, the vigorous development of emerging industries and tertiary industries in urban areas may provide more employment opportunities and solutions for the large number of surplus rural workers mandatorily confined to their land by the urban–rural dual structure.

Income disparity between urban and rural areas: Since the reform and opening up, the gap in per capita income between urban and rural residents in China (including various subsidies, social welfare benefits, housing grants, and medical insurance for urban residents) has been increasing year by year. Conversely, secondary and tertiary industries have a comparative advantage in terms of benefits over the primary industry, which the agricultural population is engaged in. The advantage directly pulls agricultural workers into cities wherein they may enter secondary and tertiary industries for higher income, thereby creating a more distinct effect of push and pull factors in rural labor migration.

Regional economic disparities: With rapid economic development along coastal regions in China, labor-intensive, export-oriented processing and service industries have emerged at a startling speed. While the economic gap between these regions and the central and western regions has widened accordingly, a gigantic demand for labor resources has come about in the process. The large number of employment opportunities created thereby provides a pull force and direction for the gradient transfer of surplus rural labor.

5.2.1.2 Intervening Obstacles

Reforms in rural system: Due to the widespread implementation of the rural household responsibility contract system, the allocation of rural economic resources has broken away from the traditional model of centralized management and unified distribution adopted by people's communes. The system reform has transformed the "covert surplus" of rural labor force into "overt surplus" and endowed farmers the power to allocate their own labor force, thereby rendering the possibility for the spontaneous flow and transfer of farmers.

Social and economic transformation: The traditional urban–rural dual structure characterized by identity management through household registration has gradually slackened. For example, the supply of food and housing has become market-oriented, and employment opportunities have opened up. To some extent, such transformations have removed possible institutional obstacles and barriers that may confront farmers in their production and livelihood after they move to cities.

5.2.1.3 Personal Factors

Transformation of values: Due to the popularization of education in rural areas, cultural changes, and unblocked social media channels, traditional xiangtu concepts, attachment to land, and unwillingness to move prevalent among farmers have been shaken and changed in a profound and swift manner. Such transformation has further provided the farmers with the courage and confidence to flow "out of their homes and land."

5.2.1.4 Other Factors

Transformation of rural enterprises: With improved living standards, the evolution of market demand structures and transformation in the operating mechanisms of state-owned enterprises, rural enterprises have been driven onto the path of transformation. In response to the fierce competition in domestic and overseas markets, rural enterprises have begun using capital strength and replaced intensive labor with capital and technology. Such transformation has inevitably shrunk the labor absorption capacity of enterprises. The once-alleviated employment predicament of rural surplus labor has aggravated again. This issue also serves as an underlying reason for the formation of the "migrant worker boom."

Through the combined effects of the various factors stated above, a unique "migrant worker boom" has come about in China. Massive numbers of migrant workers transfer from rural to urban areas and from inland to coastal regions. This phenomenon denotes an inexorable law of urbanization as well as an inevitable trend of urban–rural integration in China. According to the 2000 national population census, China's floating population rose to 121.17 million, of which migrant workers accounted for 73%, reaching an astronomical number of 87 million. As per the National Monitoring Report on Migrant Workers issued by China National Bureau of Statistics, migrant workers in China totaled 286.52 million in 2017, surpassing that in the previous year by 4.81 million. Such a drastic surge has created objective conditions for the emergence of living space for the floating population of migrant workers.

5.2.2 Formation Mechanism of Migrant Worker Settlements

The millions of surplus rural laborers that migrate to cities at all levels are also faced with multiple choices of residence. Many migrant workers chose to rent houses at first but eventually formed settlements of various types through spontaneous agglomeration rather than dwelling in passive settlements such as collective dormitories and construction sites. Their selection is mainly influenced by a combination of sundry internal and external conditions.

5.2.2.1 External Conditions for the Formation of Migrant Worker Settlements

Scale distribution of migrant workers associated with class or regional distribution of a city

Class distribution: According to statistics, 27.8% of the migrant workers generally flow and stay in large cities, 45.1% stay in small and medium-sized cities, while the remaining population constantly moves between cities (Deyu et al. 1998). A comprehensive comparison of the number of migrant workers in China's cities at all levels reveals that, on average, China's major cities have significantly greater scales of migrant workers than its small and medium-sized cities.

Regional distribution: While in situ urbanization and intra-province employment have become a new trend among migrant workers in recent years, the National Monitoring Report on Migrant Workers shows that 54.8% of the migrant workers still enter employment in eastern regions. To be more specific, 21.88 million migrant workers have been absorbed into the Beijing-Tianjin-Hebei region, 54.52 million into the Yangtze River Delta region, and 45.36 million into the Pearl River Delta region. In other words, an overwhelming majority of China's migrant workers have settled in the relatively developed eastern regions (Weilue 1994).

Agricultural populations that migrate to economically developed regions and major cities on a massive scale have advantageous conditions and solid foundations to choose migrant worker settlements as an instinctive and universal spontaneous residence option. Such choice will be subject to specific restrictions in underdeveloped regions or small and medium-sized cities. In light of this, a larger population of migrant workers becomes an essential prerequisite for the formation of a migrant worker settlement.

Employment distribution of migrant workers: Determining the residential arrangements of migrant workers to some extent

Since the reform and opening up, secondary and tertiary sectors have played a significant role in resolving the employment issue for migrant workers in urban areas. In particular, the construction, business service, and manufacturing industries rank as the top three mainstream industries that migrant workers choose to enter. However, further analysis reveals that migrant workers engaged in different industries have made fundamentally distinct choices of living space, as detailed below.

Construction workers typically reside at renovation sites or in makeshift sheds built on construction sites. Alongside changes in construction projects, they move from one construction site to another and are less likely to dwell permanently in one fixed settlement of the floating population. For workers in the manufacturing industry, employment in state-owned enterprises, three types of foreign-invested enterprises, or large and medium-sized private enterprises normally entails unified accommodation arrangements. In such cases, workers are not required to form settlements spontaneously. By contrast, migrant workers engaged in the business service industry often need to find accommodations for themselves. To fulfill such living needs, they are more likely to rent houses and, hence, stand a stronger chance of spontaneously forming various settlements.

In light of this, a sizable population of migrant workers employed in the business service industry and informal sectors (e.g., private, individual, and collective enterprises) is another crucial requirement for the formation of migrant worker settlements.

Housing distribution in urban rental housing market: Providing space and possibility for migrant worker settlements

House rental serves as a dominant residential mode among migrant workers. In view of this and the community scale of migrant worker settlements, the distribution of housing in the urban rental housing market is a crucial factor in the formation of migrant worker settlements. At present, the scope of rental housing for migrant workers in China mainly includes the private houses owned by villagers in the urban–rural fringe and the vacant houses owned by residents in urban areas. Due to their scattered distribution, urban vacant houses are more suitable for dispersed rentals and settlements. Migrant workers renting such houses are less likely to agglomerate and form a "village" pattern. By contrast, rental housing in the urban–rural fringe has a wide range of advantages such as substantial numbers of housing listings, concentrated distributions, availability for clustered rentals, lower rents, convenient transportation, and better manageability. As such, the urban–rural fringe has become the first choice for the spontaneous settlements of migrant workers, and the main district that hosts settlement spaces of all sorts.

Based on the above discussion, whether the urban rental housing market has a clustered, concentrated supply of housing is also a fundamental requirement for the formation of migrant worker settlements. In the future, with enormous amounts of public housing flowing into housing supply channels, migrant workers settlements are expected to include one more option. Nevertheless, such settlement (relocation) areas will naturally be characterized by government cultivation and organization.

In more economically developed areas of China, the above requirements have been more universally fulfilled and covered in reality. Therefore, in cities that see relatively concentrated agglomerations of migrant workers, such as Beijing, Nanjing, Guangzhou, and Shenzhen, the formation and development of migrant worker settlements have become ubiquitous and natural. Small business vendors, waste collectors, service personnel, and temporary workers in non-state-owned enterprises may all spontaneously form settlements via relationship-related or mixed communities. However, these two types of settlements display significant differences in terms of the internal mechanism of settlement formation.

5.2.2.2 Internal Mechanism of the Formation of Relationship-related Settlements

Essentially, relationship-related settlements stem from the tenacious extension of the traditional concept of "xiangtu" under contemporary urban conditions. When breaking out of the previous pattern of agricultural production and establishing various economic ties with cities, these settlements retain inextricable linkages with China's traditional agricultural society. In addition to the continuity of cultural backgrounds, concepts and ideologies, and lifestyle habits, residents have maintained and continued the organizational mechanisms and social networks of the traditional agricultural society under the conditions of modern cities.

Such settlements provide migrant workers a relatively familiar small society. Within the society, residents engage with and assist one another. Gradually, they reach homogeneity and relevance of some sort in livelihood backgrounds, interests and demands, concepts and ideologies, and various other aspects. Externally, the residents may swiftly adapt to the cities they dwell in and establish their foothold therein while reducing possible friction with the main society as well as their dependence on the government. Among other things, it is the kinship and geographic views in the traditional concept of "xiangtu" that exert a cohesive effect and potential influence on the individual behaviors of migrant workers.

Kinship and geography in the traditional concept of "xiangtu"

Many sociologists attach foremost importance to kinship and geographic ties in their discussions on the primary ties and bonds of traditional agricultural society and their influence on the development orientation of farmers. They argue that these two major ties have incomparably profound and decisive influence on the spiritual world and social behavior of farmers and that both ties are deeply rooted in the soil of the traditional concept of "xiangtu relationship."

In the traditional agricultural society, the xiangtu relationship involves the social relationship between people and the fundamental relationship between people and nature (i.e., farmers and farmland). Due to the immobility of farmland and other means of production, traditional farmers have gradually come to the realization that "settlement for generations is normal, whereas migration is abnormal."[1] Out of such attachment and respect for the land, generations of traditional farmers have stoutly believed that farmers without land cannot be regarded genuine, having the ability to expand one's own farmland is a symbol of family prosperity and paramount ideal of farmers, and selling land for a living is an unforgivable wastrel behavior. The support and encouragement from the ruling class of feudalism have reinforced the "xiangtu complex," thereby forming a strong sense of agriculture as the foundation of society and land attachment as well as unwillingness to move in rural areas.

Because of an agricultural economy featuring long-term settlement in the same place; dependence on arable land; lack of mobility and change; and the lifestyle of "being born, bred, and buried in the same place," farmers have lived on the same land for countless generations. In the process, they gradually developed a xiangtu relationship that includes the relationship between farmers and between farmers and the land. From this relationship, Chinese farmers derived an enormous significance that they attach to kinship and geography. Generations of farmers have multiplied and maintained a multitude of expanded family societies that relate to each other through kinship ties. Moreover, they constructed a slew of neighborhood societies in the same region based on geographic ties. Owing to the attention and effort devoted to the continuation of such kinship and geographic ties, people have formed organizational mechanisms, social networks, concepts, and customs and passed them down from generation to generation while living together in the same settlement for a long time. In return, these organizational mechanisms, social networks, concepts, and customs generated comprehensive, far-reaching influence and restriction on people's spiritual world and social practices (Xiao-hong 1998).

On this basis, it is understandable "why land reforms and one radical political movement after another since 1949 have not changed the conventionality of Chinese farmers despite their powerful impact on class consciousness." Kinship and geographic ties have undergone tenacious extension under modern conditions,

[1] Specifically, "xiang" (countryside) refers to the location where farmers reside for generations, and "tu" (soil) denotes the foundation or means of livelihood for farmers. Due to the immobility of "soil", farmers sustaining on farmland have no alternative but to settle in fixed residence in relation to their farmland and remain therein for generations. On this basis, "xiang" and "tu" are inseparable. Source: Fei Xiaotong. Xiangtu China & Childbirth System [M]. Beijing: Peking University Press, 1998: 7.

regardless of whether they are perceived as a biological and more social form of relationship, a structural form, or a symbolic system. In addition, they continue to play a direct or potential role in all aspects of life. After the reform and opening up, kinship and geographic ties have further constituted an indispensable social foundation and organizational system for relationship-related settlements.

Choice of relationship-related settlements under the urban–rural dual structure

China has long maintained an institutionalized dual social structure. While creating obstacles for the spontaneous migration of rural surplus labor, this structure also provides objective background conditions for the tenacious extension of the traditional concept of "xiangtu" under contemporary urban conditions. Governed by a closed urban–rural policy characterized by strict control over household registration, migrant workers and urban residents showcase marked discrepancies in various aspects, including their identity status, cultural background, concepts, and ideologies. Moreover, a serious imbalance in social security has existed across diverse areas such as food supply, public housing distribution, employment and retirement, medicine, and health care. Such issues have presented multiple forms of stress and impediments for farmers who "leave their township and soil" in search of development in cities. Migrant workers in cities have constantly found themselves in dire predicaments. On one hand, they may not easily receive prompt and full acceptance by urban residents. Due to various reasons, these migrant workers may be labeled "outsiders" and treated with discrimination, defense, or alienation in their social life at the place of population inflow. Common reasons include having no farmland, having no relatives or friends, infringing on the potential interests of local residents, or having speech or other habits that urban residents find irritating. On the other hand, they are deprived of a sense of identity with rural residents as the latter regard them as "non-genuine" and "ungrounded" farmers who have lost their land.

Against this backdrop, the accumulation and continuation of kinship and geographic ties as a form of social capital has borne a certain kind of unconventional value and significance. Migrant workers are provided a marginalized small society to rely on under a "quasi-xiangtu relationship," whereby they may engage in activities such as production, business management, procreation, and life entertainment through the maintenance and preservation of the community as well as interaction and collaboration among residents. Due to the inveterate, long-standing concept of "xiangtu" and the potential influence of urban–rural isolation, majority of the migrant workers intuitively choose kinship and geographic relationships as the bond and basis for their settlements during the process of spontaneous agglomeration. This choice not only corresponds to the organizational structure, humanistic background, and historical culture of Chinese traditional agricultural society but also meets the basic needs of migrant workers for survival, adaptation, and development in cities. In light of this, such choice bears inherent plausibility, inevitability, and paradigmatic significance.

5.2.2.3 Formation Mechanismof Mixed Settlements

The formation of mixed settlements mainly arises from the passive selection of common locations by mixed agglomerations of migrant workers. As previously discussed, the urban–rural fringe is frequently the first choice for migrant workers who are housing tenants and, hence, inevitably attracts a horde of migrant workers to gather therein. If migrant workers who come from different places, serve in different occupations, and lack universal commonality and extensive contact with one another temporarily gather and reside in a specific geographical space, a mixed settlement is formed accordingly.

Objectively speaking, mixed settlements are not a product of the overall choice made by migrant workers with active reference to a certain standard but rather a collective embodiment of a series of individualized and unrelated choices. However, considering the formation of mixed settlements also stems from the spontaneous choice of migrant workers, such settlements bear inherent plausibility and inevitability, thereby serving as an essential supplement to relationship-related settlements against the background of massive migration of surplus rural labor.

This argument raises the following question: How can migrant workers break away from the two-fold influence of the traditional concept of "xiangtu" and the institutionalized urban–rural dual structure while choosing mixed settlements instead of relationship-related settlements? This may be explicated from the following two perspectives at least.

Possibility: Whether migrant workers satisfy the overall requirements for the formation of relationship-related settlements

Housing supply in the urban rental housing market as a factor: A substantial number of migrant workers in China resolve their accommodation issues via the urban rental housing market each year. Accordingly, the urban–rural fringe that has multiple advantages in this regard has become the key region for these types of migrant workers (primarily those engaged in business service and small industries) to seek short-term rentals. Moreover, many villages face new issues such as tight housing supply, surging housing pressure, and deteriorating environmental conditions due to the agglomeration of hundreds or thousands of migrant workers within a short time. The various types of houses in villages may no longer fulfill the diverse rental needs of migrant workers, who may have difficulty finding sufficient space in other places for establishing relationship-related settlements. When these issues arise, their instinct for relationship-related agglomeration could only give way to individual housing requirements. Alternatively, they may temporarily reside alongside migrant workers from different places and industries to form patches of mixed settlements or adopt the model of dispersed rentals and settlements in the vacant houses of urban residents or hotels.

Multiple sources in the "chain migration" of migrant workers as a factor: The formation of floating population settlements often begins with a small number of migrant workers dwelling in the urban–rural fringe. Once a settlement is formed, an increasing number of migrant workers from the same township, village, or ethnicity will go off in a "chain migration" manner to these pioneers. Such transfer lays the foundation for the formation of various types of settlements and may come about in one of the following two scenarios:

Scenario 1: When a relationship-related group organization expands to a specific scale and occupies a dominant position in the area, it transforms into the embryonic form and pronucleus of relationship-related settlements. For instance, the pioneers and founders of the "Zhejiang Village" in Beijing were reportedly Lu Bize and Lu Biliang, two brothers who restarted their business in Beijing after losing money in the clothing business in Inner Mongolia. Subsequently, a "chain migration" pivoting around the Lu brothers gave rise to a community size of 110,000 people (Yude et al. 1998). This was also the case with the "Henan Village" in Suojie Village, Nanjing.

Scenario 2: If the first migrant workers display a relatively mixed composition of sources, they are likely to bring more relationship-related groups of various types to agglomerate therein due to the respective "chain migrations" they occasion. If these groups cannot be integrated through a clear dominant bond, they will transform into a de facto mixed settlement. In light of this, such multi-source and individualized relationship-related agglomeration cannot easily alter the mixed settlement nature of the entire community from a partial perspective.

Necessity: Whether migrant workers need to rely on relationship-related communities

Relative scales of migrant workers: In general, compared with the original residents in the place of population inflow, migrant workers who are classified as a "disadvantaged group" in terms of scale are more inclined to seek asylum and assistance from relationship-related settlements. Once they exceed a specific scale or surpass the number of local residents to become an "advantaged group" (such as Shenzhen), migrant workers are no longer subject to the psychology of relationship-related agglomeration under the long-term influence of the "xiangtu" concept in selecting their location of residence.

Social structure of the place of population inflow: Based on the theory of "structuralist constructivism" proposed by French sociologist P. Bourdieu, social structures are characterized by a dual dimension of objectivity and subjectivity. These two dimensions exist respectively in the institutional form of things and the physical state of the "persistent dispositional system," i.e., Bourdieu's two pivotal concepts: field and habitus.[2] Of the two, the former endows the social system with the features of self-sustainment, harmony, and restraint, whereas the latter ingrains the natural disposition deeply in the people within the structure and induces a tendency toward resisting changes. From this argument, two types of scenarios may be derived:

[2] According to Bourdieu, "field" is understood as the relationship structure between an individual and other collective actors in the society while habitus is "a system of lasting transposable dispositions". The "disposition" here refers to a tendency to perceive, feel act and think in a certain way, and "lasting" means that although these dispositions could be changed as people's experiences accumulate, they are ingrained and resistant to changes.

Scenario 1: Large cities that have a long history and relatively stable social structure as a result of long-term evolution (e.g., Beijing, Nanjing) are more likely to display a subjective consciousness of self-sustainment and resistance to changes in the face of intervention of external forces. Their original dispositional system is also more likely to maintain continuation of some sort rather than undergoing transformation. This may well be another underlying reason why massive amounts of migrant workers have difficulty receiving full acceptance from local residents. In comparison with mixed settlements, relationship-related settlements are apparently better suited to the survival needs of migrant workers in the said cities.

Scenario 2: The social structure of emerging cities (e.g., Shenzhen, Dongguan) or frontier cities in the reform and opening up is in an open state of rapid growth and constant changes. Under the impact of external forces, their self-sustaining function and continuity of resistance to changes have been contained to a certain extent. Accordingly, external forces face relatively low levels of difficulty and resistance in their entry. In such a relaxed environment, migrant workers do not actually need to achieve self-protection and mutual support through relationship-related agglomeration. Rather, they form several mixed settlements through instinctive location selection.

Against the historical background of the "xiangtu" concept and the dual social structure, the interplay of the above factors has potentially changed the inherent dependence of Chinese migrant workers on relationship-related agglomeration and enabled them to display more diversified choices in the form of settlements. These settlement forms derive from a spontaneous and inevitable choice of migrant workers, hence bearing an inherent sense of representativeness and plausibility.

5.2.3 *Overview of Migrant Workers in Nanjing*

Based on the results of the sampling survey conducted on migrant workers in downtown Nanjing in 2015 by the research team as well as data collected from 2101 valid questionnaires, the overall characteristics of migrant workers in Nanjing City may be roughly inferred and described as follows.

5.2.3.1 Social Attributes of Migrant Workers in Nanjing

The investigation and analysis of the social attributes of migrant workers in Nanjing City mainly focus on indicators, including age composition, gender ratio, education level, place of origin, and marital status.

In terms of gender composition, male migrant workers in Nanjing take up a slightly higher percentage of the total migrant population than their female counterparts, with the former accounting for 55.78% and the latter 44.22% (see Fig. 5.4).

Regarding marital status, majority of the migrant workers in Nanjing are married, accounting for 83.65% of the total migrant population (see Fig. 5.5).

Fig. 5.4 Gender composition. *Source* Research team's survey sampling on migrant workers in Nanjing (2015)

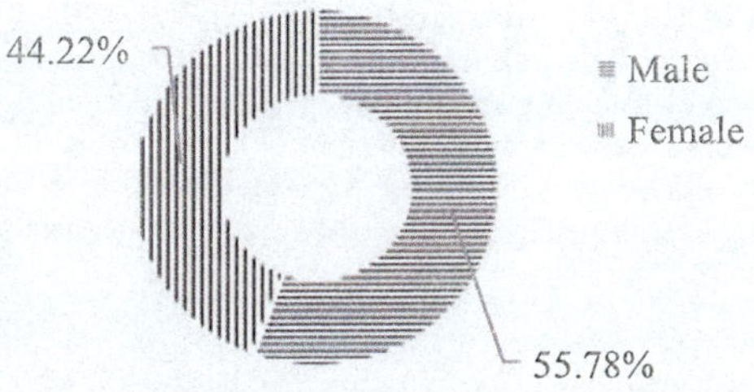

Fig. 5.5 Marital status. *Source* Research team's survey sampling on migrant workers in Nanjing (2015)

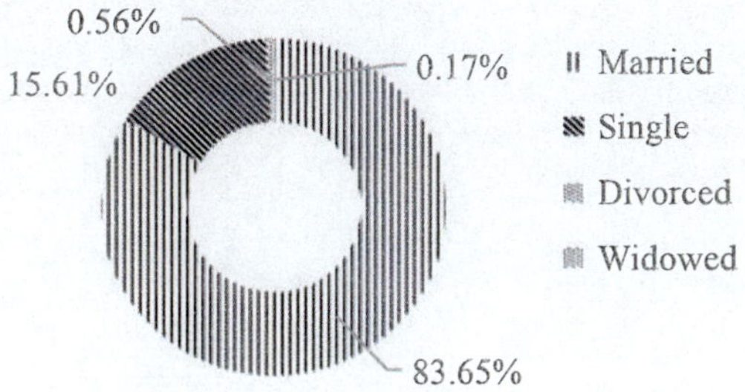

With regard to age composition, the bulk of migrant workers in Nanjing fall in the age groups of 20–39 and 40–59, accounting for 45.40% and 47.10%, respectively, with a high combined percentage of 92.50%. Adolescents under the age of 19 and senior citizens over 60 years are relatively few in number, accounting for 1.80% and 5.70%, respectively (see Fig. 5.6).

With regard to education level, migrant workers in Nanjing manifest a relatively low overall level of education. Migrant workers with an education level of middle school or below account for 74.58%, with migrant workers of middle school education totaling the highest percentage (41.27%). Migrant workers with education level of primary school account for 24.46%, illiterate migrant workers take up 8.85%, and migrant workers with a university degree or above occupy only 8.14% (see Fig. 5.7).

With regard to the place of origin, the mass of the migrant workers in Nanjing come from Jiangsu and Anhui Provinces, accounting for 36.19% and 40.63%, respectively. Such percentages showcase the characteristic of proximity-based migration. Barring Henan Province (5.91%), Jiangxi Province (3.05%), and Shandong Province

Fig. 5.6 Age composition. *Source* Research team's survey sampling on migrant workers in Nanjing (2015)

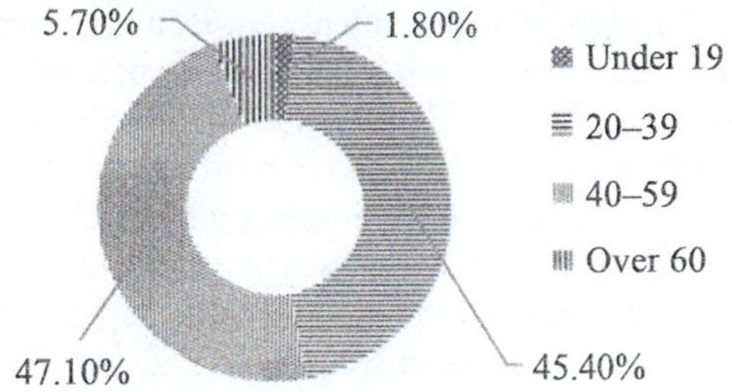

Fig. 5.7 Education level. *Source* Research team's survey sampling on migrant workers in Nanjing (2015)

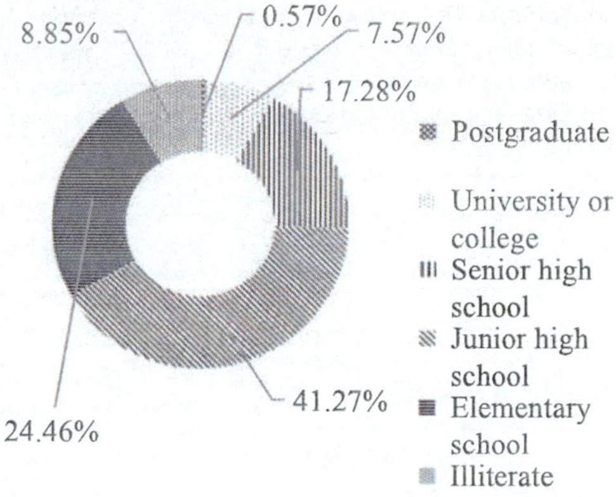

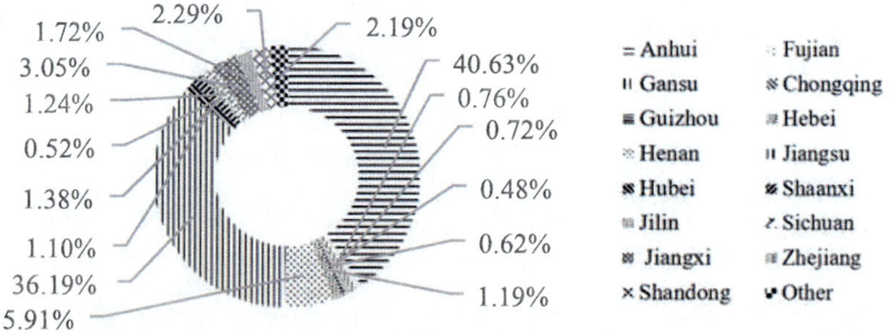

Fig. 5.8 Composition of place of origin. *Source* Sample data collected by the research team on migrant workers in Nanjing (2015)

(2.29%), migrant workers from other provinces account for less than 2% (see Fig. 5.8).

5.2.3.2 Economic Attributes of Migrant Workers in Nanjing

The investigation and analysis of the economic attributes of migrant workers in Nanjing mainly focus on indicators, including type of occupation, nature of work unit, and income level.

In terms of occupational composition, migrant workers serving in business service industries account for the highest percentage of the total migrant population in Nanjing, at 69.92%, followed by those working as production and transport equipment operators (15.47%) and professional and technical personnel (7.61%). These three categories together occupy as high as 93.00% of the total (see Fig. 5.9).

As for the nature of business unit, individual industrial and commercial households account for the highest percentage of the total migrant population in Nanjing, at 60.01%, followed by private and collective enterprises, which occupy 21.30%

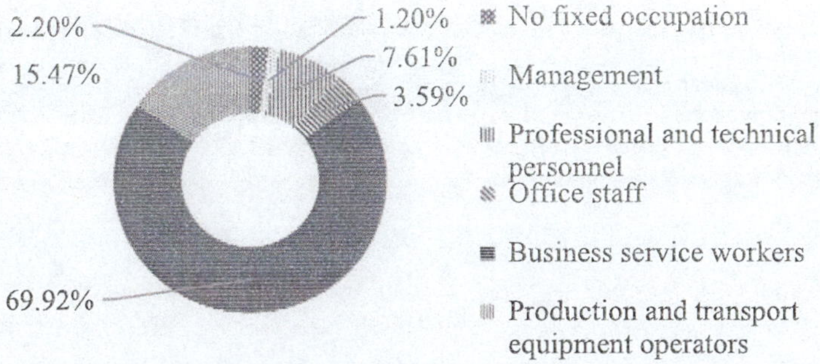

Fig. 5.9 Occupational composition. **Source* Research team's survey sampling on migrant workers in Nanjing (2015)

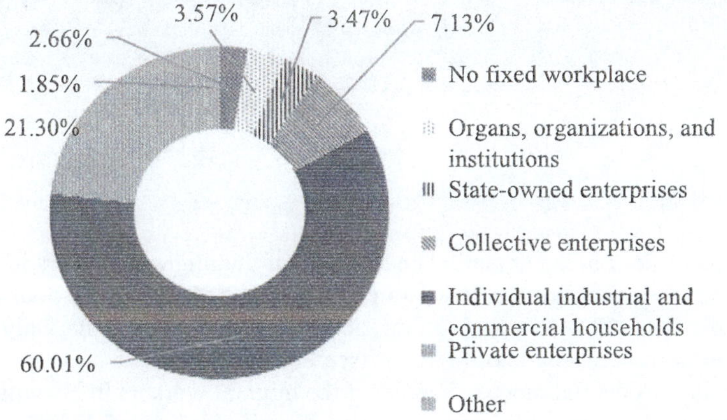

Fig. 5.10 Business unit. **Source* Research team's survey sampling on migrant workers in Nanjing (2015)

and 7.13%, respectively. The three combined account for 88.44% of the total (see Fig. 5.10).

Regarding the income level, majority of the migrant workers in Nanjing earn a household monthly income above CNY2,500. Among them, households earning CNY2,500–5,000 account for 75.60%, those earning more than CNT5,000 account for 19.75%, whereas those earning less than CNY500 account for merely 0.58% of the total (see Fig. 5.11).

5.2.3.3 Spatial Attributes of Migrant Workers in Nanjing

The investigation and analysis of the spatial attributes of migrant workers in Nanjing mainly focus on the micro characteristics of their living space, including the indicators of floor area per capita, residential modes, and housing sources.

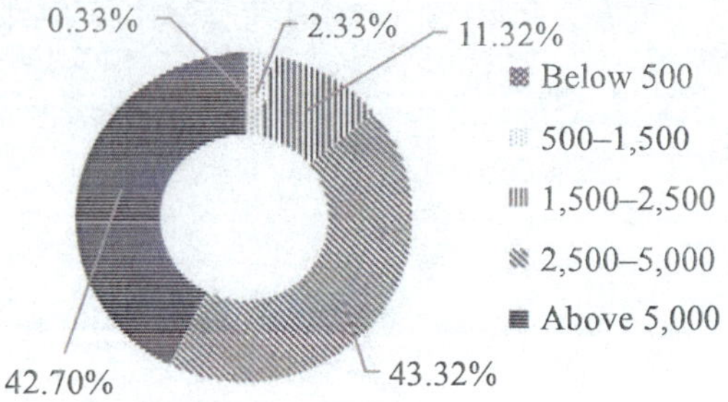

Fig. 5.11 Income level. **Source* Sample data collected by the research team on migrant workers in Nanjing (2015)

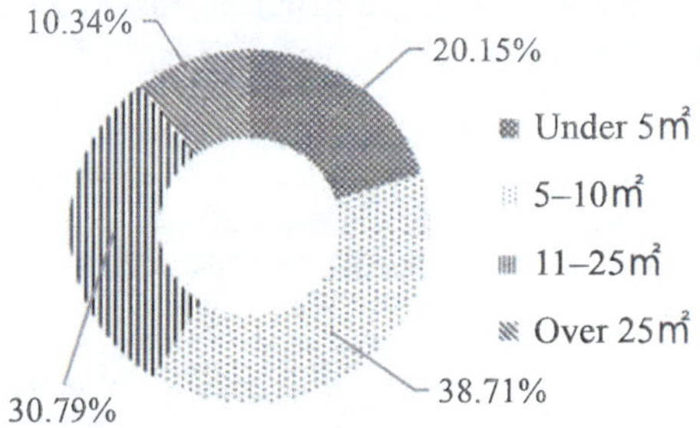

Fig. 5.12 Floor area per capita. **Source* Sample data collected by the research team on migrant workers in Nanjing (2015)

In terms of floor area per capita, nearly 60% of the migrant workers in Nanjing have a floor area per capita below 10 m^2. Of them, 20.15% have a floor area per capita under 5 m^2, 38.71% have 5–10 m^2, and 30.79% have 11–25 m^2. Only 10.34% have a floor area per capital above 25 m^2 (see Fig. 5.12).

Regarding residential modes, 76.64% of the migrant workers in Nanjing choose rental housing, 11.65% reside in collective dormitories, and 6.43% live on the construction site. Only around 5% choose to stay at the places of relatives or friends or find residence by other means (see Fig. 5.13).

With regard to housing sources, 46.37% of the migrant workers in Nanjing rent commercial housing, accounting for the highest percentage. Those renting public housing and living in dormitories account for the second and third highest percentages at 21.58% and 7.80%, respectively. The three types combined achieve a total of 75.75% (see Fig. 5.14).

5.2.3.4 Settlement Space for Migrant Workers in Nanjing

Using streets as the research unit and the size, density, and ratio of population as the research indicators, the authors conduct an extended analysis of the spatial attributes of migrant workers, i.e., a macro analysis of the settlement space of migrant workers in the downtown area. In the analysis, population size refers to the number of migrant workers within the designated street area, population density indicates the number of migrant workers per unit area within the designated street area, and population ratio denotes the ratio of migrant workers within the designated street area to the total

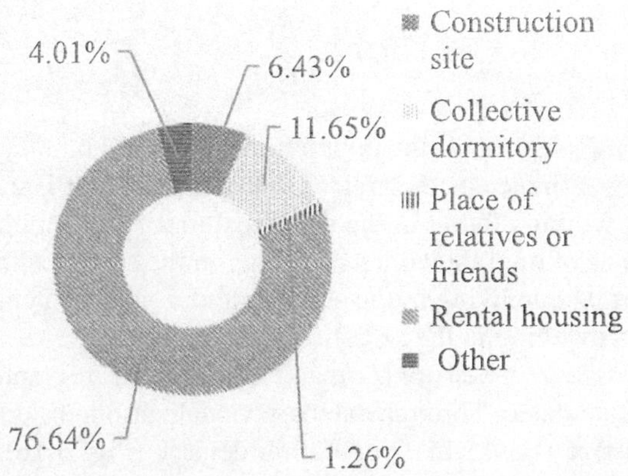

Fig. 5.13 Residential mode. **Source* Sample data collected by the research team on migrant workers in Nanjing (2015)

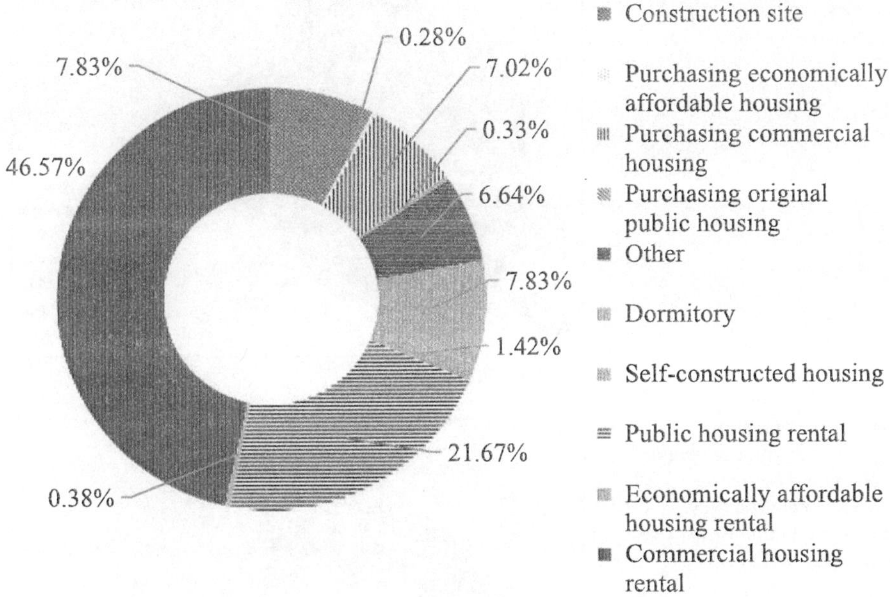

Fig. 5.14 Housing source. **Source* Sample data collected by the research team on migrant workers in Nanjing (2015)

population therein. The formulas for measuring population density and population ratio are as follows:

$$Di = \frac{P_i}{Q_i} \tag{5.1}$$

$$Gi = \frac{P_i}{Q_i} \tag{5.2}$$

In formula (5.1), P_i represents the population size of migrant workers residing in the statistical unit i, A_i indicates the area of the statistical unit i, and D_i denotes the density of migrant workers living in the statistical unit i In formula (5.2), P_i refers to the population size of migrant workers residing in the statistical unit i, q_i signifies the total number of people living in the statistical unit, and G_i symbolizes the ratio of migrant workers dwelling in the statistical unit i.

The population size (P_i), density (D_i), and ratio (G_i) of migrant workers in each statistical unit are calculated. Through database visualization, the distribution graphs for the population size (Fig. 5.15), population density (Fig. 5.16), and population

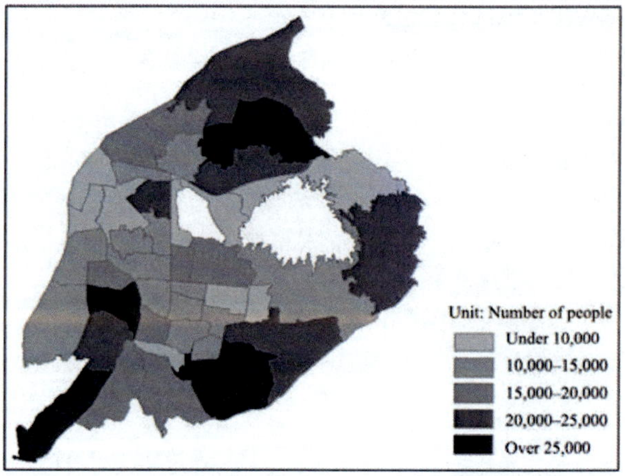

Fig. 5.15 Population size of migrant workers in Nanjing. **Source* Research team's survey sampling on migrant workers in Nanjing (2015)

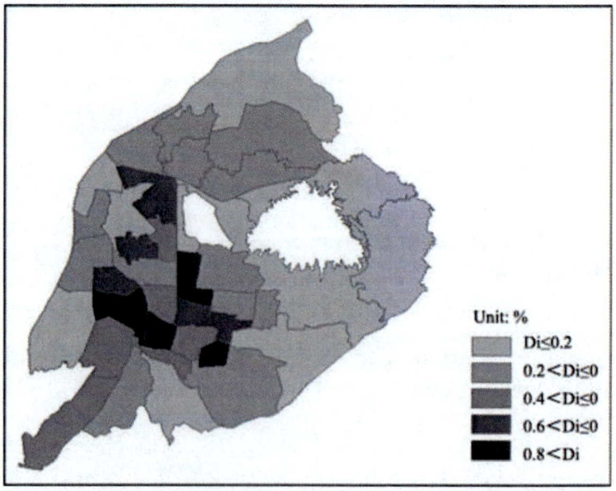

Fig. 5.16 Population density of migrant workers in Nanjing. **Source* Research team's survey sampling on migrant workers in Nanjing (2015)

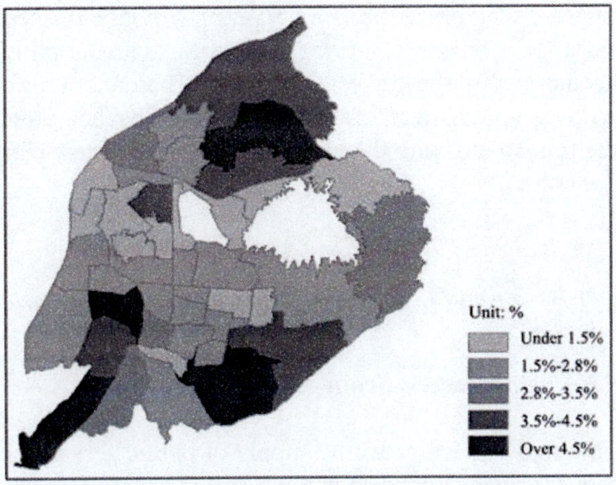

Fig. 5.17 Population ratio of migrant workers in Nanjing. **Source* Research team's survey sampling on migrant workers in Nanjing (2015)

ratio (Fig. 5.17) of migrant workers in Nanjing are output. The graphs indicate the following:

In terms of the population size of the migrant worker settlement space, its overall distribution displays a "multi-core and sector-shaped" pattern that involves greater numbers in outer areas and smaller numbers in inner areas. Of all areas, the areas around Xuanwu Lake and Yijiangmen Street have the smallest population sizes, whereas the four corners of the downtown area (e.g., Honghua Street, Maigaoqiao Street, Mochou Lake Street, and Shazhou Street) have the greatest population sizes. Areas with medium and high population sizes extend toward the old town district.

Regarding the population density of the migrant worker settlement space, its overall distribution manifests a "sector-shaped pattern with the southern area of the old town district as its peak." Of all the areas, three have the highest population density, i.e., Xinjiekou and Wulaocun Streets, Qinhong Streets, and Shuangtang and Mochouhu Streets, leading areas with medium and high population densities toward outer areas that have lower population densities.

Regarding the population ratio of the migrant worker settlement space, its overall distribution presents a "multi-core and block-shaped" pattern that shows higher ratios in outer areas and lower ratios in inner areas. Among all the street areas, three of them have the highest ratios, i.e., Honghua Street outside the downtown area, Shazhou and Mochouhu Streets, and Maigaoqiao Street. Taking these three areas as the nucleus, three vast stretches of medium to high population ratios are formed in the shape of blocks in the surrounding area, located in the outer corner of the old town district.

5.2.3.5 Conclusion

Majority of the migrant workers in Nanjing are married individuals aged 20–59, with slightly more males than females and a relatively low overall education level. Most migrant workers come from Jiangsu Province or the neighboring Anhui Province, serving in business service industries, with monthly salary of CNY 2500–5000 accounting for the highest percentage. Their major residential mode is rental housing, with Honghua Street, Maigaoqiao Street, Shazhou Street, and Mochouhu Street being the four streets with the highest ratios and greatest numbers of migrant workers (see Table 5.1).

5.2.4 *Overview of Public Services and Facilities in Nanjing*

5.2.4.1 Research Scope

This study investigates the supply of public services and facilities with its research scope focusing on the downtown area of Nanjing (see Fig. 5.18). There are three main reasons for this delineation. First, to maintain consistency with the spatial scope of the previous research on the attributes of migrant workers, it is preferable to select the same research scope for investigating public services and facilities. Second, the downtown area has seen the fastest economic development and most robust urban construction in Nanjing. It is also an area where facility constructions of all types demonstrate equally salient strengths and problems. On this account, the downtown area bears practical significance in research. Third, Nanjing is characterized by a massive city area. To avoid an overwhelming workload that ensues complete coverage research, the downtown area that bears more representative significance is selected as the research object, i.e., using the administrative division of the "street" class as the spatial statistical unit involving 41 streets in six districts.

As there are myriad types and complex classes of public services and facilities, this study adopts a classification method to examine the level of public services in the downtown area of Nanjing. To facilitate practical operation and enhance feasibility, the study selects several types of facilities that have relatively intimate association with the lives of migrant workers. The selection principles include ① selecting designated services and facilities related to the equity of urban public facility allocation; ② selecting the type of facilities with traceable and reliable sources of data related to scale, location, etc.; ③ selecting the major types of facilities with relatively close association with the daily life of the research subject. Through questionnaire distribution (see the section of "Category Selection" below for further details) and literature review, the authors ultimately select five categories and 14 types of public services and facilities (covering three classes of facilities).

The data required for this study include the population data of the streets in Nanjing and basic data of all types and grades of public services and facilities. Of them, population-related data come from the sixth population census data of Nanjing City; data on public services and facilities contain the number, spatial distribution, and scale of all types and grades of facilities, mainly derived from government bulletins, government websites, web maps, etc. The specific sources are detailed as follows (see Table 5.2):

Table 5.1 Changes in land acquisition and compensation system in Nanjing over the years

Attribute	Category of indicator	Major attributes			Cumulative percentage of top three attributes (%)
		No. 1	No. 2	No. 3	
Social attribute	Gender composition	Male	Female	–	100.00
		55.80%	44.20%	–	
	Marital status	Married	Single	Divorced	99.83
		83.65%	16.61%	0.56%	
	Age composition	40–59	20–39	Over 60	98.20
		47.10%	45.40%	5.70%	
	Education level	Middle school	Primary school	High school or technical secondary school	83.01
		41.27%	24.46%	17.28%	
	Place of origin	Anhui Province	Jiangsu Province	Henan Province	82.73
		40.63%	36.19%	5.91%	
Economic attribute	Occupational composition	Business service workers	Production and transport equipment operators	Professional and technical personnel	88.98
		69.92%	15.47%	3.59%	
	Nature of business unit	Individual industrial and commercial households	Private enterprises	Collective enterprises	88.44
		60.01%	21.30%	7.13%	
	Income level	CNY 2500–5000/month	Over CNY 5000/month	CNY 500–1500/month	99.42
		75.60%	19.75%	4.07%	

(continued)

Table 5.1 (continued)

Attribute	Category of indicator	Major attributes			Cumulative percentage of top three attributes (%)
		No. 1	No. 2	No. 3	
Spatial attribute	Floor area per capita	5–10 m^2	11–25 m^2	Under 5 m^2	89.65
		38.71%	30.79%	20.15%	
	Residential mode	Rental housing	Collective dormitory	Construction site	94.72
		76.64%	11.65%	6.43%	
	Housing source	Commercial housing rental	Public housing rental	Dormitory	75.75
		46.37%	21.58%	7.80%	
Settlement space	Population size	Streets with large population size	Honghua, Maigaoqiao, Mochouhu, Shazhou		
		Streets with small population size	Yuejianglou, Rehe South Road, Yijiangmen, Zhonghuamen, Ruijin Road, Yueyahu, Xuanwumen, Suojincun, Xuanwuhu		
	Population density	Streets with high population density	Xinjiekou, Wulaocun, Qinhong, Shuangtang, Mochouhu		
		Streets with low population density	Yuejianglou, Yijiangmen, Xinglong, Yuhua, Guanghua Road, Xuanwumen, Suojincun, Xuanwuhu, Maqun, Yanziji, Xiaolingwei		
	Population ratio	Streets with high population ratio	Honghua, Maigaoqiao, Mochouhu, Shazhou		
		Streets with low population ratio	Yuejianglou, Rehe South Road, Yijiangmen, Ninghai Road, Hunan Road, Zhonghuamen, Ruijin Road, Yueyahu, Xuanwumen, Suojincun, Xuanwuhu		

* *Source* Sample data collected by the research team on migrant workers in Nanjing (2015)

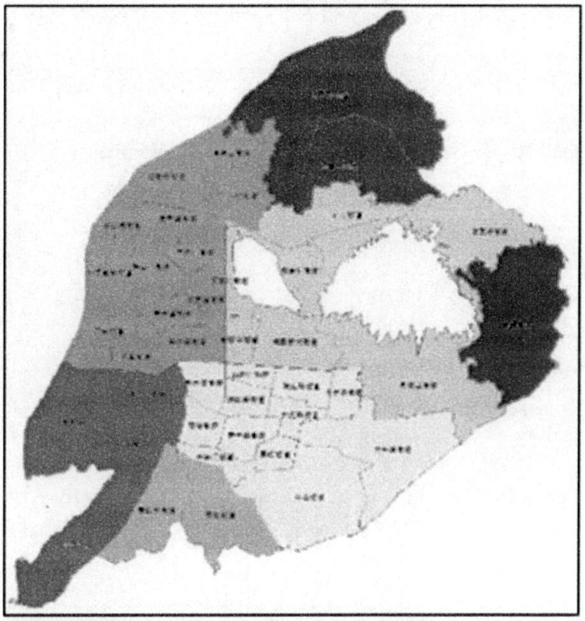

Fig. 5.18 Research units in the downtown area of Nanjing. **Source* Own elaboration

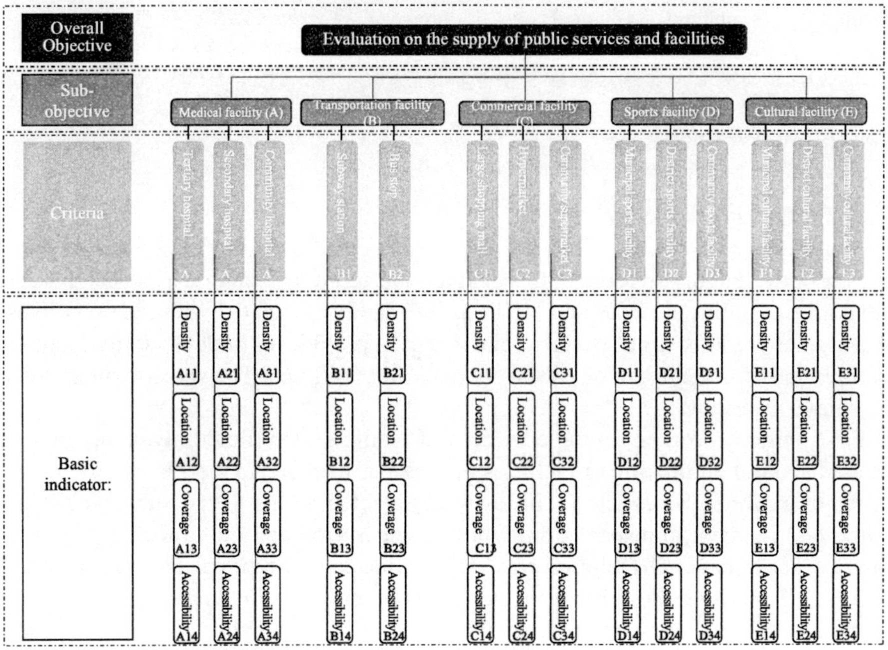

Fig. 5.19 Evaluation index system of public services and facilities for migrant workers in the core area of Nanjing. **Source* Own Elaboration

Table 5.2 Facility types and data sources

Classification of facility	Grade			Source of data
	Municipal	District	Community	
Medical facility	Tertiary hospital	Secondary hospital	Community hospital	*Nanjing Yearbook of Hygiene 2009* [M]; the website of the Nanjing Municipal Bureau of Public Health; Google Maps; official websites of hospitals; POI data of Nanjing public services and facilities
Commercial facility	Large shopping mall	Hypermarket	Community supermarket	CAD map of Nanjing City; Google Maps; official websites of all units concerned; POI data of Nanjing public services and facilities
Transportation facility	Subway station	Bus stop	-	Google Maps; Baidu Map; POI data of Nanjing public services and facilities
Sports facility	Municipal sports facility	District sports facility	Community sports facility	Website of Nanjing Municipal Sports Bureau; Google Maps; official websites of all units concerned; POI data of Nanjing public services and facilities
Cultural facility	Municipal cultural facility	District cultural facility	Community cultural facility	*Jiangsu Cultural Statistical Yearbook 2013* [M]; Google Maps; official websites of all units concerned; POI data of Nanjing public services and facilities

**Source* Own elaboration

5.2.4.2 Category selection

In view of the length and focus of this chapter, the authors meticulously select the key categories of public services and facilities intimately associated with the daily lives of "migrant workers" as a special social group to provide an in-depth study on their requirements for the usage of public services and facilities. The specific operations are shown as follows:

First, migrant workers in the core area of Nanjing were randomly sampled and queried about their ranking of facilities most closely related to their lives. Five categories of public services and facilities were selected as the actual subjects of this study, i.e., medical, transportation, commercial, sports and cultural facilities (the absence of educational facilities was largely due to institutional barriers in practical use). Subsequently, in pursuance of the 1–9 ratio scale in the analytic hierarchy process (AHP), a materiality assessment survey on various types of public services and facilities was developed and distributed to the migrant worker group and 25 questionnaires were recovered. Finally, based on the collected data, the materiality or weight of the five categories of facilities for this group was calculated (see Table

Table 5.3 Materiality or weight of the five categories of facilities for migrant workers in Nanjing

Weight of facility by category		Weight of facility by grade		Overall weight
Category of facility	Weight	Type of facility	Weight	
Medical facility	0.17335	Tertiary hospital	0.50404	0.08738
		Secondary hospital	0.21011	0.03642
		Community hospital	0.28586	0.04955
Transportation facility	0.25301	Subway station	0.42616	0.10782
		Bus stop	0.57384	0.14519
Commercial facility	0.33009	Large shopping mall	0.27203	0.08979
		Hypermarket	0.38902	0.12841
		Community supermarket	0.33895	0.11188
Sports facility	0.10986	Municipal sports facility	0.28314	0.03111
		District sports facility	0.23182	0.02547
		Community sports facility	0.48504	0.05329
Cultural facility	0.13369	Municipal cultural facility	0.41138	0.05500
		District cultural facility	0.36406	0.04867
		Community cultural facility	0.22456	0.03002

**Source* Own elaboration

5.3; for specific details on methodology, please refer to the section of "Assessment Methodology" below).

5.2.4.3 General Overview

This study mainly focuses on the core area of Nanjing, which has a total area of approximately 274.78 km^2. Based on the research objectives and subjects, 3168 public services and facilities were selected, covering 14 types and five categories. These include 152 municipal, 1382 district, and 1634 community facilities. According to the division of street units, the number of facilities in each unit was calculated and their distribution characteristics were described. Further, the statistical units were classified using the Jenks natural breaks classification method (NBC), and visualization output was realized and shown as below (see Table 5.4):

5.2.4.4 Assessment Methodology

This section offers a grade- and category-based discussion on the supply of public services and facilities in the core area of Nanjing, primarily using the single-factor measurement and multi-factor comprehensive evaluation methods. The assessment methodology is illustrated below.

Single-factor Measurement

Table 5.4 Overall distribution of the five categories of facilities in the core area of Nanjing

Facility distribution on the streets in the core area of Nanjing	
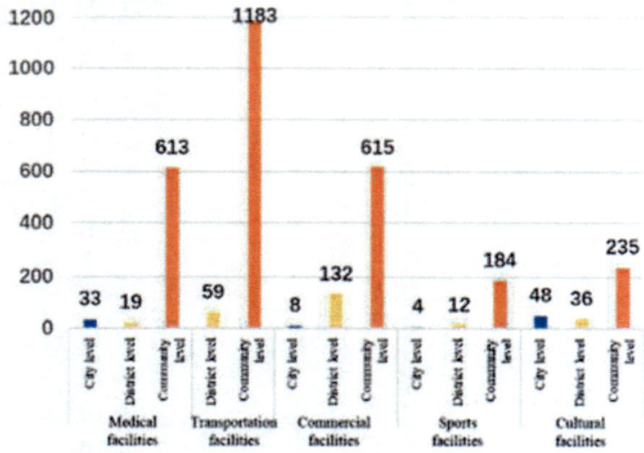	
Medical facility ·There are 653 medical facilities in the core area of Nanjing, including 33 tertiary hospitals, 19 secondary hospitals, and 613 community medical facilities (e.g., general hospitals, health stations, clinics) ·The streets with the largest number of medical facilities in the core area include Maigaoqiao Street, Yijiangmen Street, and Mochouhu Street	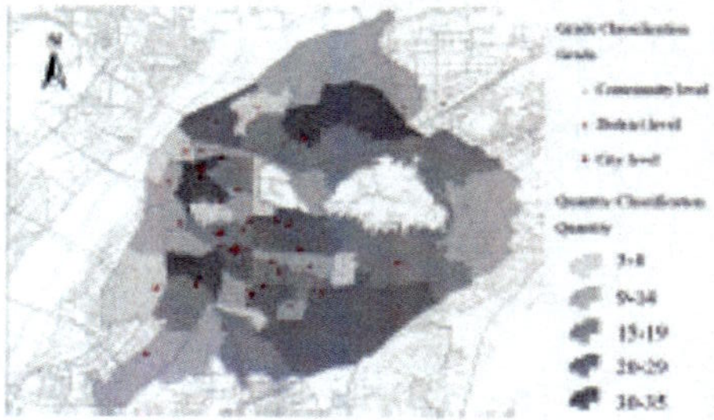

(continued)

Table 5.4 (continued)

Facility distribution on the streets in the core area of Nanjing	
Transportation facility ·There are 1242 transportation facilities in the core area of Nanjing, including 59 subway stations and 1183 bus stops ·The streets with the most transportation facilities in the core area include Maigaoqiao Street, Mochouhu Street, and Nanyuan Street	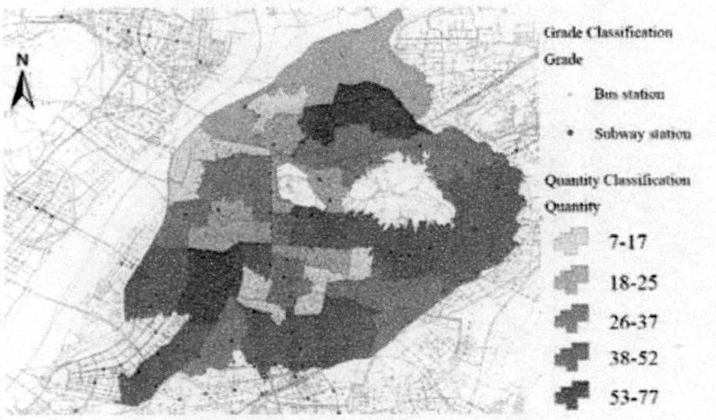
Commercial facility ·There are 755 commercial facilities in the core area of Nanjing, including 8 large shopping malls, 132 hypermarkets, and 615 community supermarkets (e.g., Baijia Supermarket, Suguo Supermarket, Trust-Mart) ·The streets with the largest number of commercial facilities in the core area include Maigaoqiao Street, Hunan Road Street, and Chaotiangong Street	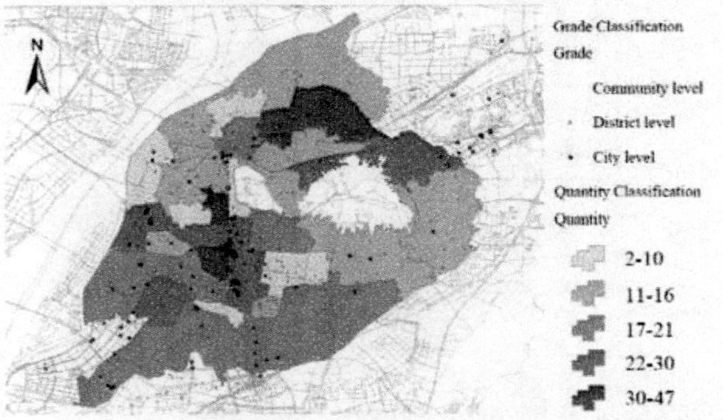

(continued)

Table 5.4 (continued)

Facility distribution on the streets in the core area of Nanjing	
Sports facility	
·There are 200 sports facilities in the core area of Nanjing, including 4 municipal, 12 district, and 184 community sports facilities ·The streets with the largest number of sports facilities in Nanjing include Mochouhu Street, Huaqiao Road Street, and Xinjiekou Street	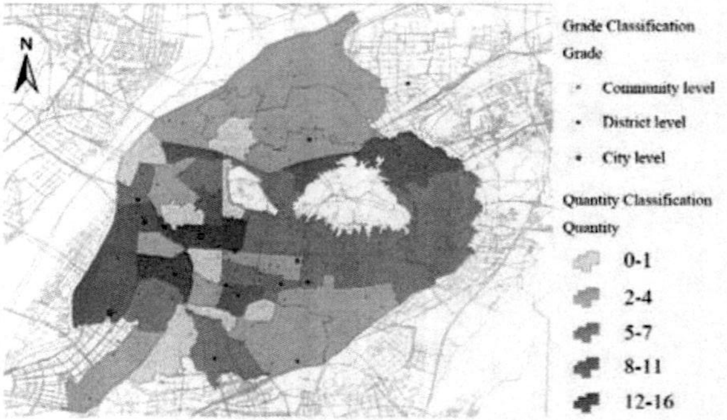
Cultural facility	
·There are 319 cultural facilities in the core area of Nanjing, including 48 municipal, 36 district, and 235 community cultural facilities ·The street with the largest number of cultural facilities in the core area is Fuzimiao Pedestrian Street	

* *Source* Own elaboration

In the single-factor measurement of the level of public service and facility supply, quantitative evaluation and analysis are conducted using a series of indicators and firsthand data (including data on population, facility distribution, and scale). Through the measurement, the objective conditions of facility supply are revealed. The quantitative indicators introduced to assess the level of public service and facility supply from different dimensions and angles include four spatial factors: **density, location, coverage, and accessibility**. The calculation method and implication of each factor are described as follows (see Table 5.5):

Multi-factor Comprehensive Evaluation—Selecting Evaluation Methodology

At present, the commonly employed comprehensive factor evaluation methods in urban research primarily include the following: principal component analysis (PCA), the Delphi method (expert investigation method), the analytic hierarchy process (AHP), and R-type factor analysis. Considering urban public services and facilities involve a multi-level and multi-grade complex integrated system and the supply level of each type of facility entails evaluation factors in multiple dimensions, this study opts for the analytic hierarchy process (AHP), a method that provides an uncomplicated system and a quantitative factor analysis. By constructing a hierarchical

Table 5.5 Single-factor measurement

Factor	Formula	Description	Implication
Density	$W_i = \frac{\bar{W}_i}{\sum_{i=1}^{n} \bar{W}_i}$ (5.3)	n_{ij}: Number of category-*j* facility in statistical unit *i*; A_i: Area of statistical unit *i* (km^2)	This measurement reveals the distribution density of facilities in the street, reflecting the quantity and scale of public services and facilities The greater the value, the more dense is the facility distribution and the higher is the level of supply
Location	$LQ_{ij} = \frac{\frac{m_{ij}}{p_i}}{\frac{M_j}{P}}$ (5.4)	m_{ij}: Scale of a certain category of public service/facility in statistical unit *i*; M_j: Overall scale of category-*j* facility in the core area; p_i: Population size of statistical unit *i*; *P*: Overall population size of the core area	This measurement reveals the degree of dominance of on-street facilities in the entire core area, reflecting the status and role of public service and facility supply The higher the location score, the higher is the status and role; 1 > LQij ≥ 0 indicates that a certain facility in the street has a below-average status or level, whereas LQij ≥ 1 suggests an above-average status or level

(continued)

Table 5.5 (continued)

Factor	Formula	Description	Implication
Coverage	$C_{ij} = \frac{a_{ij}}{A_i}$ (5.5)	a_{ij}: Area of the service district covered by category-j facility in statistical unit i (the road network-based polygon of the service district covered by facilities acquired through the ArcGIS Network Analyst); A_i: Total area of statistical unit i (km^2)	This measurement reveals the coverage of the service area of the facilities in the street, reflecting the coverage of the public services and facilities to the population The value of the factor is between 0 and 1. The lower the value, the higher is the proportion of service blind spots in the street (areas not covered by the service of facilities) and the lower is the corresponding level of supply[a]
Accessibility	$E_{ij(k)} = W_{j(k)} * S_{ij}^{-\partial}$ (5.6) (Jun-bo 2010)	$E_{ij(k)}$: Level of accessibility of public service/ facility $j(k)$ to street unit i; $W_{j(k)}$: Attractiveness index of facility, indicated by scale (number of hospital beds, floor area, site area, number of stations/stops, etc.); S_{ij}: Spatial isolation from the geometric center of street unit i to facility $j_{(k)}$ (indicated by street network distance extracted through the ArcGIS Network Analyst); □: Spatial isolation parameter, with a value ranging between 1 and 2, wherein 2 is taken more frequently.(Talen and Anselin , 1998)	This measurement reveals the level of accessibility of the facilities in the street, reflecting the degree of ease with which the population accesses public services and facilities The higher the factor value, the higher is the accessibility of the facility and the higher is the corresponding level of supply
	$G_{ij(k)} = \sum E_{ij}$ (5.7)	G_{ij}: Overall level of accessibility of category-j public service/facility to street unit i	

Source Own elaboration

[a]The radiation radii of various types of facilities are as follows: tertiary hospitals-3 km; secondary hospitals-2 km; community hospitals-1 km; subway stations-1.5 km; bus stations-0.5 km; large shopping malls-2.5 km; hypermarkets-2 km; community supermarkets-0.5 km; municipal sports facilities-4.67 km, district sports facilities-3 km; community sports facilities-1 km; municipal cultural facilities-3 km; district cultural facilities-2 km; community cultural facilities-1 km.

structure, the complex issue of "public service and facility supply" is broken down into several ordered levels. Next, the relative materiality of factors at each level is compared and analyzed to determine the weight of each factor. Finally, a comprehensive evaluation is conducted on the actual condition of public service and facility supply by means of multi-factor weighted accumulation (Shu-bai 1995).

Multi-factor Comprehensive Evaluation—Constructing Evaluation System

As public services and facilities involve multiple types and levels, this study constructs a three-level indicator system as shown below. Through multi-level and multi-factor overlay, the supply level of public services and facilities of all types in all grades is evaluated (and uniformly encoded at the same time; see Fig. 2.17).

Level-1 indicators: Level-1 indicators include the five categories of medical, transportation, commercial, sports, and cultural facilities, i.e., the types of urban public services and facilities that are relatively closely associated with the daily lives of migrant workers.

Level-2 indicators: The above five categories of facilities are further divided into 14 sub-categories, namely, tertiary hospitals, secondary hospitals, community hospitals, large shopping malls, hypermarkets, community supermarkets, subway stations, bus stops, municipal cultural facilities, district cultural facilities, community cultural facilities, municipal sports facilities, district sports facilities, and community sports facilities.

Level-3 indicators: Based on the 4*14 basic indicator matrix formed by a combination of the aforementioned four single factors (i.e., density, location, coverage, and accessibility) and the fourteen level-2 indicators, 56 basic indicators are derived for this study.

Multi-factor Comprehensive Evaluation—Establishing Weights for Evaluation

This study adopts the 1–9 ratio scale in the AHP (see Table 5.6) in combination with the weights for the 14 types of facilities in the five categories established in the aforementioned survey on the migrant worker group. Furthermore, through the expert scoring method, 25 copies of the weighted scoring table for the four types of single factors (i.e., density, location, coverage, and accessibility) are distributed to urban planning specialists. Subsequently, by means of data collection and statistical analysis, the relative materiality of the four types of single factors are determined via pairwise comparisons whereby the different scores of "√" are established.

In statistical analysis, the number of people in all the items marked "□" in the table is multiplied by the value of the corresponding grade, all the amounts are added, and finally, the total sum is divided by the total number of people surveyed to derive a ratio, which is the average score value of criterion A to criterion B.

$$Z = \frac{\sum (a * n)}{N} \tag{5.8}$$

In the formula, Z represents the average score value of criterion A to criterion B; a represents the score value;

Table 5.6 Ratio scale

No	Level of materiality	Value assignment
1	Both indicators bear equal levels of materiality	1
2	Indicator A bears a slightly higher level of materiality than indicator B	3
3	Indicator A bears a significantly higher level of materiality than indicator B	5
4	Indicator A bears an intensely higher level materiality than indicator B	7
5	Indicator A bears an extremely higher level of materiality than indicator B	9
6	Indicator A bears a slightly lower level of materiality than indicator B	1/3
7	Indicator A bears a significantly lower level of materiality than indicator B	1/5
8	Indicator A bears an intensely lower level of materiality than indicator B	1/7
9	Indicator A bears an extremely lower level of materiality than indicator B	1/9

**Note* The level of materiality indicated by No. 2, 4, 6, and 8 is the median value of the adjacent judgments shown above

n represents the number of people marked "□" in each evaluation;
N represents the number of valid questionnaires.

The AHP method is employed to calculate the relative weight of each indicator to the general objective. First, a judgment matrix is developed, following which the eigenvector, or the relative weight, of the matrix is obtained through the extraction of square roots, whereby the qualitative judgment of the scorer is converted into a quantitative description. Finally, the relative weight of each single factor indicator to the sub-objectives and the general objective is derived.

In the following, this method is demonstrated through the weight calculation procedure of the four basic indicators, i.e., density (A11), location (A12), coverage (A13), and accessibility (A14), under the sub-objective of tertiary hospitals (A1).

Step 1: Establish a judgment matrix, i.e., compare the relative materiality of all relevant factors at this level. In this step, the weight of each indicator to the higher-level objective is calculated using the expert weighted scoring survey. The judgment matrix is shown as follows:

A1	A11	A12	A13	A14
A11	1	0.76	0.34	0.54
A12	1.32	1	0.46	0.79
A13	2.94	2.17	1	1.23
A14	1.85	1.27	0.81	1

Step 2: Find the eigenvector.
Calculate the product M_i of the values in each row of the judgment matrix:

$$M_1 = 1 * 0.76 * 0.34 * 0.54 = 0.139536$$
$$M_2 = 1.32 * 1 * 0.46 * 0.79 = 0.4781579$$
$$M_3 = 2.94 * 2.17 * 1 * 1.23 = 7.8644501$$
$$M_4 = 1.85 * 1.27 * 0.81 * 1 = 1.9057856$$

Calculate the nth root W_i of M_i:

$$W_1 = \sqrt[4]{M_1} = \sqrt[4]{0.139536} = 0.61$$
$$W_2 = \sqrt[4]{M_2} = \sqrt[4]{0.4781579} = 0.83$$
$$W_3 = \sqrt[4]{M_3} = \sqrt[4]{7.8644501} = 1.67$$
$$W_4 = \sqrt[4]{M_4} = \sqrt[4]{1.9057856} = 1.17$$

Normalize the vector W_i:

$$W_i = \frac{\bar{W_i}}{\sum_{i=1}^{n} \bar{W_i}}$$

Obtain W_1= 0.14, W_2= 0.19, W_3= 0.39, W_4= 0.27.
Hence, the eigenvector is derived as follows:

$$W = (W_1, W_2, W_3, W_4)^T = (0.14, 0.19, 0.39, 0.27)^T$$

Step 3: Find the characteristic root AW

$$\begin{bmatrix} 1 & 0.76 & 0.34 & 0.54 \\ 1.32 & 1 & 0.46 & 0.79 \\ 2.94 & 2.17 & 1 & 1.23 \\ 1.85 & 1.27 & 0.81 & 1 \end{bmatrix} \begin{bmatrix} 0.14 \\ 0.19 \\ 0.39 \\ 0.27 \end{bmatrix}$$

$$(AW)_1 = 1 * 0.14 + 0.76 * 0.19 + 0.34 * 0.39 + 0.54 * 0.27 = 0.5700916$$
$$(AW)_2 = 1.32 * 0.14 + 1 * 0.19 + 0.46 * 0.39 + 0.79 * 0.27 = 0.7768033$$
$$(AW)_3 = 2.94 * 0.14 + 2.17 * 0.19 + 1 * 0.39 + 1.23 * 0.27 = 1.5667877$$
$$(AW)_4 = 1.85 * 0.14 + 1.27 * 0.19 + 0.81 * 0.39 + 1 * 0.27 = 1.0998399$$

The characteristic root is derived as follows:

$$\lambda_{\max} = \sum_{i=1}^{n} \frac{(AW)_i}{nW_i} = \frac{0.4934}{4 * 0.12} + \frac{0.8864}{4 * 0.22} + \frac{1.0388}{4 * 0.26} + \frac{1.548}{4 * 0.40} = 4.0118123$$

Step 4: Check and validate consistency.

(A) Consistency index of the judgment matrix (CI)

$$CI = \frac{\lambda_{\max} - n}{n - 1} = 0.0039374$$

(B) **Mean random consistency index of the same order (RI)**

The mean random consistency index (RI) refers to the arithmetic mean obtained for 500 eigenvalues of randomly generated judgment matrices of the same order (Huan-chen et al. 1996).

Order	1	2	3	4	5	6	7	8	9
RI	0.00	0.00	0.58	0.90	1.12	1.24	1.32	1.41	1.45

Random Consistency Index (CR):

$$CR = \frac{CI}{RI} = \frac{0.0039374}{0.9} = 0.0043749 < 0.1$$

The results show that the judgment matrix has satisfactory consistency.

On this basis, judgment matrices are developed according to the scoring of the survey. The weight of each factor at the levels of sub-objectives, criteria, and basic indicators are calculated and obtained. Finally, the relative weight of all indicators at all levels to the general objective is derived (see Table 5.7).

In addition, given that various indicators for the evaluation of public service and facility supply are gauged using disparate measurement units and orders of magnitude, nondimensionalization is required before calculation to avoid the exaggeration of weights due to different value ranges. This study adopts the linear scale transformation method to nondimensionalize the indicators, and the formula is

$$\boldsymbol{X_i} = \frac{\boldsymbol{x_i}}{\boldsymbol{x_{\max}}} \tag{5.9}$$

In the formula, $\boldsymbol{x_i}$ indicates the value of raw data, $\boldsymbol{x_{\max}}$ represents the maximum value, and $\boldsymbol{X_i}$ signifies the value of standardized data.

Table 5.7 Index system and weights for the evaluation of public service facility supply

General Objective A	Sub-objective B		Criteria C		Basic indicator D		Weight to sub-objective	Weight to general objective
	Item	Weight	Item	Weight	Item	Weight		
Evaluation of the supply level of public service facilities for migrant workers	Medical facility (A)	0.17335	Tertiary hospital (A1)	0.50404	Density (A11)	0.14239	0.07177	0.01244
					Location (A12)	0.19373	0.09765	0.01693
					Coverage (A13)	0.39015	0.19665	0.03409
					Accessibility (A14)	0.27373	0.13797	0.02392
			Secondary hospital (A2)	0.21010	Density (A21)	0.14239	0.02992	0.00519
					Location (A22)	0.19373	0.04070	0.00706
					Coverage (A23)	0.39015	0.08197	0.01421
					Accessibility (A24)	0.27373	0.05751	0.00997
			Community hospital (A3)	0.28586	Density (A31)	0.22091	0.06315	0.01095
					Location (A32)	0.18358	0.05248	0.00910
					Coverage (A33)	0.18404	0.05261	0.00912
					Accessibility (A33)	0.41147	0.11762	0.02039
	Transportation facility (B)	0.25301	Subway station (B1)	0.42616	Density (B11)	0.14239	0.06068	0.01535
					Location (B12)	0.19373	0.08256	0.02089
					Coverage (B13)	0.39015	0.16627	0.04207
					Accessibility (B14)	0.27373	0.11666	0.02951
			Bus stop (B2)	0.57384	Density (B21)	0.22091	0.12677	0.03207
					Location (B22)	0.18358	0.10534	0.02665
					Coverage (B23)	0.18404	0.10561	0.02672
					Accessibility (B24)	0.41147	0.23612	0.05974
	Commercial facility (C)	0.33009	Large shopping mall (C1)	0.27203	Density (C11)	0.14239	0.03873	0.01279
					Location (C12)	0.19373	0.05270	0.01740
					Coverage (C13)	0.39015	0.10613	0.03503
					Accessibility (C14)	0.27373	0.07446	0.02458
			Hypermarket (C2)	0.38902	Density (C21)	0.14239	0.05539	0.01828
					Location (C22)	0.19373	0.07537	0.02488
					Coverage (C23)	0.39015	0.15177	0.05010
					Accessibility (C24)	0.27373	0.10649	0.03515
			Community supermarket (C3)	0.33895	Density (C31)	0.22091	0.07488	0.02472
					Location (C32)	0.18358	0.06222	0.02054

(continued)

Table 5.7 (continued)

General Objective A	Sub-objective B		Criteria C		Basic indicator D		Weight to sub-objective	Weight to general objective
	Item	Weight	Item	Weight	Item	Weight		
					Coverage (C33)	0.18404	0.06238	0.02059
					Accessibility (C34)	0.41147	0.13947	0.04604
	Sports facility (D)	0.10986	Municipal sport facility (D1)	0.28314	Density (D11)	0.14239	0.04032	0.00443
					Location (D12)	0.19373	0.05485	0.00603
					Coverage (D13)	0.39015	0.11047	0.01214
					Accessibility (D14)	0.27373	0.07751	0.00852
			District sports facility (D2)	0.23182	Density (D21)	0.14239	0.03301	0.00363
					Location (D22)	0.19373	0.04491	0.00493
					Coverage (D23)	0.39015	0.09044	0.00994
					Accessibility (D24)	0.27373	0.06346	0.00697
			Community sports facility (D3)	0.48504	Density (D31)	0.22091	0.10715	0.01177
					Location (D32)	0.18358	0.08904	0.00978
					Coverage (D33)	0.18404	0.08927	0.00981
					Accessibility (D34)	0.41147	0.19958	0.02193
	Cultural facility (E)	0.13369	Municipal cultural facility (E1)	0.41138	Density (E11)	0.14239	0.05858	0.00783
					Location (E12)	0.19373	0.07970	0.01066
					Coverage (E13)	0.39015	0.16050	0.02146
					Accessibility (E14)	0.27373	0.11261	0.01505
			District cultural facility (E2)	0.36406	Density (E21)	0.14239	0.05184	0.00693
					Location (E22)	0.19373	0.07053	0.00943
					Coverage (E23)	0.39015	0.14204	0.01899
					Accessibility (E24)	0.27373	0.09966	0.01332
			Community cultural facility (E3)	0.22456	Density (E31)	0.22091	0.04961	0.00663
					Location (E32)	0.18358	0.04122	0.00551
					Coverage (E33)	0.18404	0.04133	0.00553
					Accessibility (E34)	0.41147	0.09240	0.01235

* *Source* Own elaboration

5.2.4.5 Level of Supply

Single Factor Evaluation of the Level of Supply

Using streets as the research unit, the four indicators (density, location, coverage, and accessibility) for the street facilities in the core area of Nanjing are calculated respectively. Next, the standardized values of the 56 basic indicators are accumulated. With 41 statistical units grouped into five categories, distribution maps of the single factors for the supply of public services and facilities in the core area of Nanjing are developed. The features of distribution are summarized in Table 5.8:

Evaluation of the Level of Supply by Category

Using streets as the research unit, the level of supply of **the five categories of facilities** in the streets of the core area of Nanjing is calculated. Next, the standardized values of the 56 basic indicators are accumulated, with 41 statistical units grouped into five categories. Finally, through visualization output, distribution maps for the level of classified supply of public services and facilities in the core area of Nanjing are developed. The features of distribution are summarized in Table 5.9:

Evaluation of Overall Level of Supply

Using streets as the research unit, the overall level of supply of public facilities in the streets of the core area of Nanjing are calculated. Through the standardized measurement approach, the 41 statistical units are grouped into five categories, and a distribution map for the overall level of supply of public services and facilities in the core area of Nanjing is developed (see Fig. 5.20). According to the map, the overall level of supply presents a general structure of **"multi-core scattered distribution that spreads out in a cross shape from the central core highland in the core area of the city to the periphery."** Specific details are as follows: (Fig. 5.19)

Level-1 streets include Xinjiekou, Yuhua, and Xinglong, forming the three core areas with the highest level of facility supply in the core area of the city;

Level-2 streets include Huaqiao Road, Chaotiangong, and Hongwu Road, forming the secondary core area in the center of the city;

Level-3 streets include Zhongyangmen, Hunan Road, Jiangdong, and Fuzimiao (Confucius Temple), spreading out southward, northward, and westward from the high-supply area in the center of the city;

Level-4 streets include Yanziji, Maigaoqiao, Hongshan, Xuanwumen, Xuanwuhu, Xiaolingwei, Yueyahu, Ruijin Road, Wulaocun, Meiyuanxincun, Jianning Road, Yuejianglou, Yijiangmen, Rehe South Road, Ninghai Road, Shuangtang, Zhonghuamen, Fenghuang, Mochouhu, Shazhou and Saihongqiao, stretching across the core area of the city center in a cross shape;

Level-5 streets include Nanyuan, Honghua, Qinhong, Guanghua Road, Daguang Road, Maqun, Mufushan, Pagodaqiao, Xiaoshi, and Suojincun, mainly scattering about the southeast and northwest sides of the core area of the city and stretching in the shape of a sector.

Table 5.8 Distribution features of single factors for facility supply

	Distribution map of the level of supply	Distribution features
Density	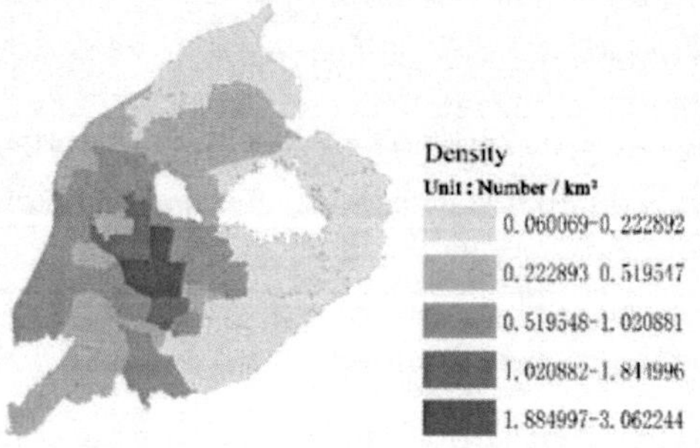	**High in the west and low in the east, with the density of facility supply progressively decreasing from the center of the old town region to peripheral circles** • Level-1 streets are primarily concentrated around Xinjiekou; • Level-2 streets extend from level-1 streets toward the southeast and northwest areas; • Level-3 streets expand and circle around streets of the first two levels; • Level-4 streets fill up the vacuum left by the first three levels while expanding north and south; • Level-5 streets are mainly concentrated on the east and north sides
Location	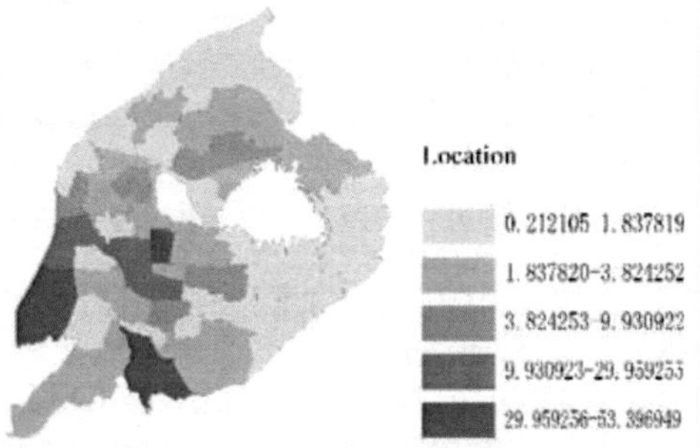	**A multi-core, multi-patch, dotted, and clustered pattern** • Level-1 streets include Xinjiekou, Yuhua, and Xinglong streets; • Level-2 streets spread between Xinglong and Xinjiekou streets, clustering along level-1 streets; • Level-3 streets scatter across the core area of the city in patches; • Level-4 streets spread out around streets of the first three levels; • Level-5 streets are dotted about the core area but most are concentrated on the east and north sides

(continued)

Table 5.8 (continued)

	Distribution map of the level of supply	Distribution features
Coverage	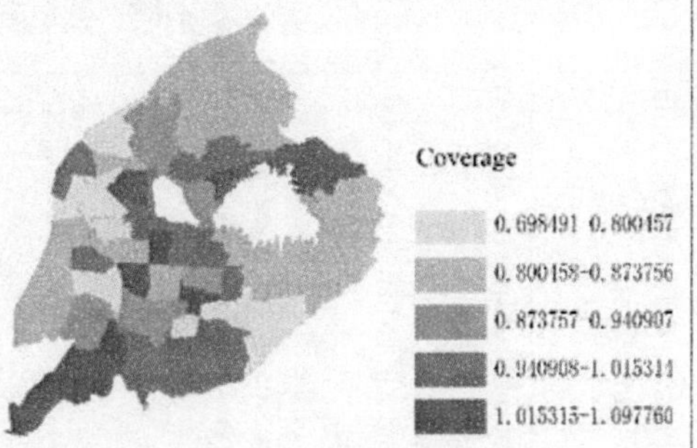	**A multi-core, sector-shaped, and dotted pattern** • Level-1 streets scatter across the central, eastern, and southern areas of the city; • Level-2 streets spread from level-1 streets to the periphery; • Combined and interconnected with streets of the first two levels, level-3 streets spread out to the periphery in a sector shape; • Level-4 streets extend from streets of the first three levels to the periphery; • Level-5 streets scatter across the core area of the city in patches
Accessibility	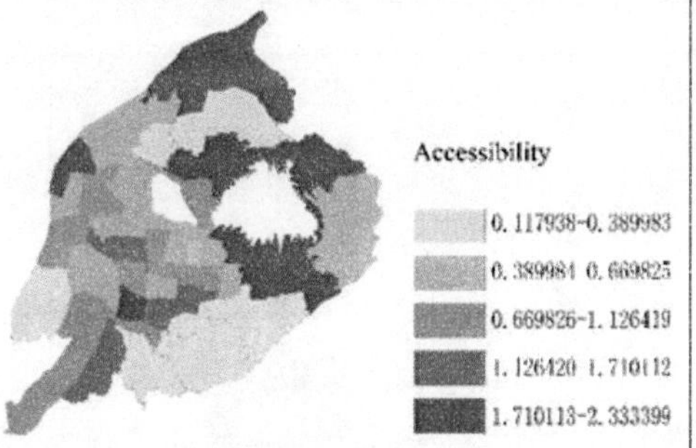	**High in the north and low in the south, forming a multi-core and sector-shaped pattern** • Level-1 streets spread around Zijin Shan (Purple Mountain), the north and west sides of the core area of the city; • Level-2 streets scatter across the central and southern regions; • Level-3 streets spread out from streets of the first two levels to the outer areas; • Connected with streets of the first three levels, level-4 streets spread out to the periphery in a sector shape; • Level-5 streets are mainly concentrated on the southeast, northeast, and southwest sides of the core area of the city

^*^*Source* Own elaboration

Table 5.9 Distribution features of the level of facility supply by category

<table>
<tr><th></th><th>Distribution map of the level of supply</th><th>Distribution features</th></tr>
<tr><td>Medical facility</td><td>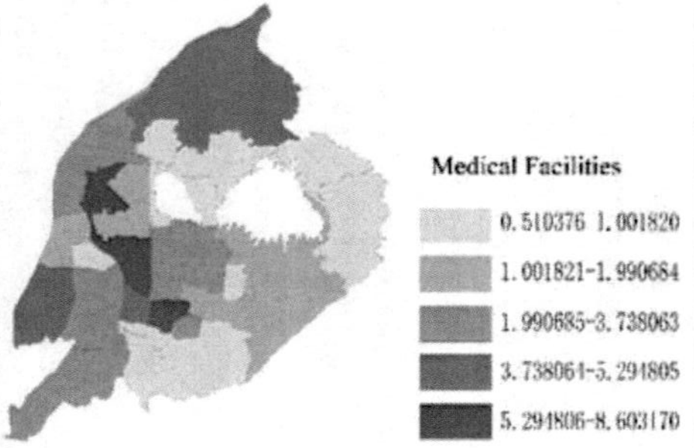
</td><td>High in the west and low in the east, clustering around streets of the first two levels
• Level-1 streets form three core patches;
• Level-2 streets cluster around three core patches on the north, southwest, and center of the core area;
• Level-3 streets spread from streets of the first two levels to the periphery;
• Level-4 streets are concentrated on the east and west sides of the central region;
• Level-5 streets are mainly concentrated on the south side of the city as well as the north side of Zijin Shan and Xuanwu Lake</td></tr>
<tr><td>Transportation facility</td><td>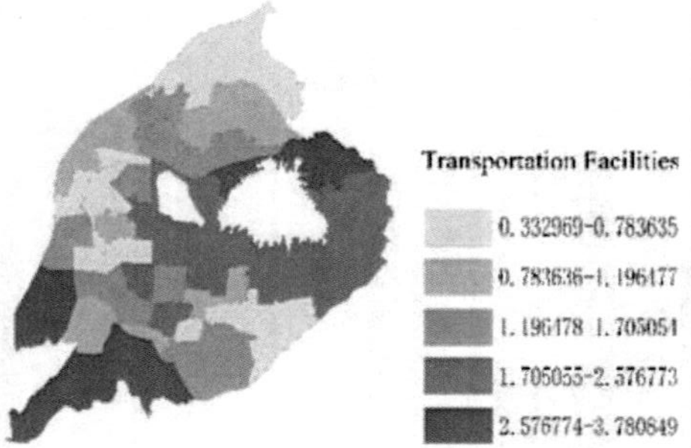
</td><td>A multi-core pattern; scattering around multiple cores to form highland-shaped concentrated areas
• Level-1 streets form three core patches;
• Level-2 streets mainly stretch around Zijin Shan and Xuanwu Lake, forming highland-shaped concentration areas;
• Level-3 streets extend from streets of the first two levels to the periphery;
• Level-4 streets spread out from streets of the first three levels to the periphery;
• Level-5 streets scatter across the core area of the city with a limited ratio</td></tr>
</table>

(continued)

Table 5.9 (continued)

	Distribution map of the level of supply	Distribution features
Commercial facility	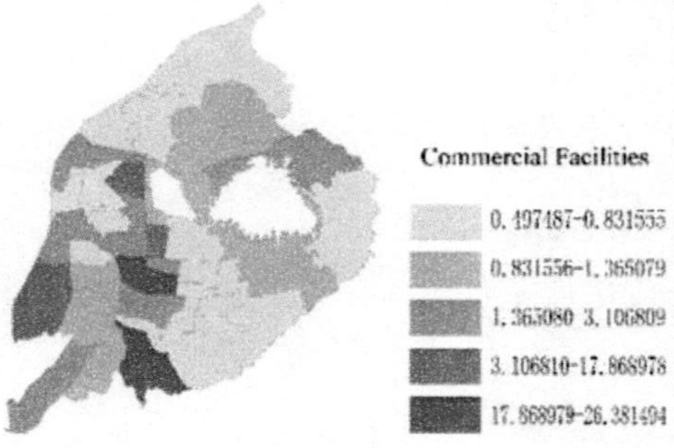	**With the center serving as the core, the overall distribution displays high levels on both sides and low levels in the middle** • Level-1 streets are concentrated in Xinjiekou and its surrounding areas; • Level-2 streets extend westward and southward from level-1 streets to form contiguous stretches of highland; • Level-3 streets extend southward and eastward from streets of the first two levels; • Level-4 streets scatter in patches within the core area of the city; • Level-5 streets mainly run through Xuanwu Lake and the core area of city in a north–south direction
Sports facility	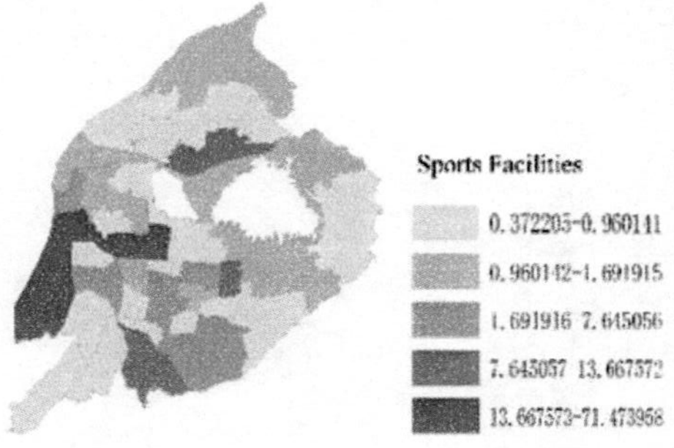	**High in the west and low in the east; spreading from west to east in a circular shape** • Level-1 streets are located in the center of the city and the streets along the river in the west; • Level-2 streets include Chaotiangong Street and its surrounding streets; • Level-3 streets scatter in patches in the east of the city and around the central region; • Level-4 streets and streets of the first three levels together form a semi-circle around Xuanwu Lake and Zijin Shan; • Level-5 streets scatter about the outer streets of the core area of the city

(continued)

Table 5.9 (continued)

	Distribution map of the level of supply	Distribution features
Cultural facility	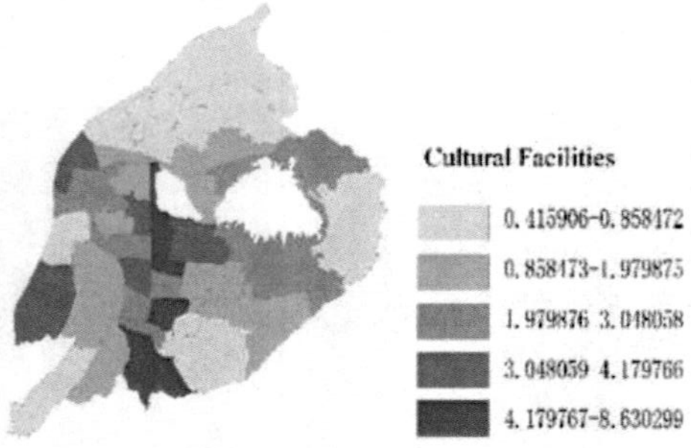	**A cross highland-shaped pattern, spreading out to the periphery in a sector shape** • Level-1streets are concentrated in Xinjiekou and its surrounding streets; • Level-2 streets spread out along level-1 streets; • Level-3 streets in combination with streets of the first two levels form a cross-axis distribution in the core area; • Level-4 streets extend along the core highland area to the periphery in the shape of a sector; • Level-5 streets are located in the outer streets of the core area of the city, with the highest concentration on the north side

**Source* Own elaboration

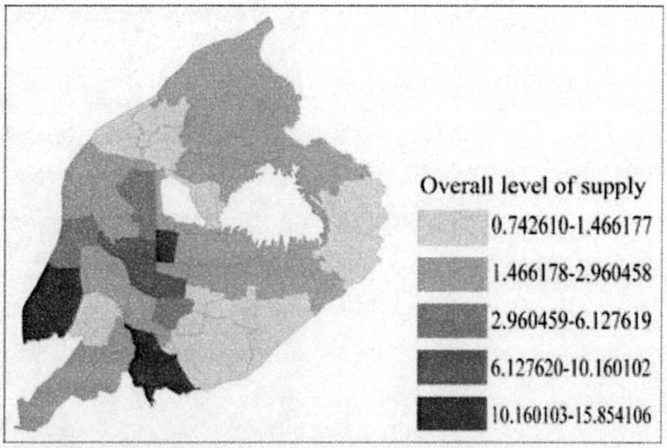

Fig. 5.20 Overall level of supply. **Source* Own elaboration

5.3 Research on the Supply of Public Services and Facilities in Migrant Worker Settlements

Based on the overview of the migrant worker group in Nanjing and the overall analysis of the public service and facility supply in its core area discussed above, this section selects typical samples of migrant worker settlements as the research subjects to establish quantitative methodology and modeling for evaluating public services and facilities. Next, following the analysis methods that combine "grade-category-ensemble" and "single factor-multiple factor," this study systematically evaluates the level and status of public service and facility supply in migrant worker settlements.

5.3.1 Definition of Research Subjects

5.3.1.1 Sample Selection

This research focuses on migrant worker settlements formed by the migrant worker group through rental housing, an active approach of residence selection. As discussed previously, the population size of the migrant worker group in Nanjing displays a "multi-core and sector-shaped" pattern, its population density manifests a "sector-shaped pattern with the southern area of the old town district as its peak," and its population ratio presents a "multi-core and block-shaped" pattern.

Given the inhomogeneous supply of public services and facilities in the core area of Nanjing, this section combines the distribution features of migrant workers and the current status of public service and facility supply in the core area of the city to offer a comprehensive study on the level of public service and facility supply in the migrant worker settlements of Nanjing. In consideration of the feasibility of actual research, a number of typical settlements are selected for in-depth empirical research. The main principles of selection are as follows:

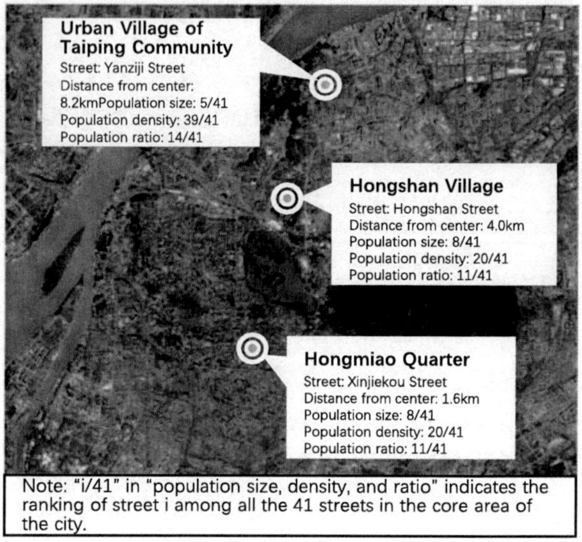

Fig. 5.21 Location and relevant information of research samples. **Source* Own elaboration

First, the selected settlement samples should be located in districts with a greater number of migrant worker settlements in Nanjing. Not only the settlement size but also the population density and ratio of the migrant worker group in each street should be considered. Moreover, the selected samples should be able to reflect the status of facility supply in different urban rings (the outer ring, inner ring, and center).

Second, the features of the current status of public service and facility supply in Nanjing should be taken into account. The selected settlement samples should cover regions with high levels of supply and those with average or below average levels of supply but high concentrations of migrant worker settlements.

Third, the feasibility of in-depth investigation and convenience of travel should be considered in sample selection. In light of this, priority should be given to appropriate settlements within the core area of the city.

Based on the above selection principles, this study selects three migrant worker settlements as the research samples: the urban village of Taiping Community in Yanziji Street (outer ring; large scale; low density; high ratio), Hongshan Village in Hongshan Street (inner ring; large scale; medium density; high ratio), and Hongmiao Quarter in Xinjiekou Street (center; medium scale; high density; medium ratio) (see Fig. 5.21 for detailed information on their indicators and locations).

5.3.1.2 Definition of Facilities

As stated above, according to the first round of 25 evaluation questionnaires on the materiality of various types of public service facilities issued to migrant workers, five categories of public service facilities closely associated with the daily lives of this disadvantaged group may be roughly established as the objects that the AHP method is applied to in the following discussion. Specifically, these five categories include medical, transportation, commercial, sports, and cultural facilities.

5.3.1.3 Overview of Samples

Based on their characteristics and distributions, the location condition and general information of the three migrant worker settlement samples are summarized as follows (see Table 5.10):

5.3.2 Design of Evaluation Methodology

5.3.2.1 Research Framework

In general, the methodology for evaluating the level of facility supply discussed in this section adheres to the analysis rationale stated above, i.e., the two dimensions of "grade-category-ensemble" and "single factor-multi-factor." First, the single factors for the level of facility supply are measured in the sequence of the three strata (grade-category-ensemble). The factors for a higher stratum are calculated with an accretion of the calculation results of the lower stratum/strata as per facility weights. Next, the single factors for the facility supply at three strata (grade-category-ensemble) are calculated with an accretion of weights, whereby the results of multi-factor comprehensive evaluation are derived (see Fig. 5.22).

5.3.2.2 Evaluation Methodology

The evaluation discussed in this section is confined to the five categories of public service facilities previously selected, with their levels of supply in different settlement samples measured respectively.

Single-factor Measurement

The methodology for single-factor evaluation on medical, commercial, sports, and cultural facilities is explained in the following. Given the travel habits and behavioral psychology of the migrant worker group, this study adopts a combination of survey sampling and calculations to measure the maximum tolerable travel circle for migrant workers in the three sample settlements. The circle derived is further used to determine the facilities to be evaluated in this section, and the single factor indicators for each category of facility are respectively calculated. Of the factors, density reflects the number and scale of facilities available within the travel circle of migrant workers; location reflects the degree of dominance shown by different categories of facility supply within the travel circle of migrant workers; coverage reflects the scope covered by various categories of facility supply within

Table 5.10 Overview of migrant worker settlement samples

Name		Urban village of Taiping community	Hongshan village	Hongmiao quarter
Distance from the city center		8.2	4.0	1.6
Street		Yanziji street	Hongshan street	Xinjiekou street
Features of migrant workers in the street	Population size	Large	Large	Medium
	Population density	Low	Medium	High
	Population ratio	High	High	Medium
Current location			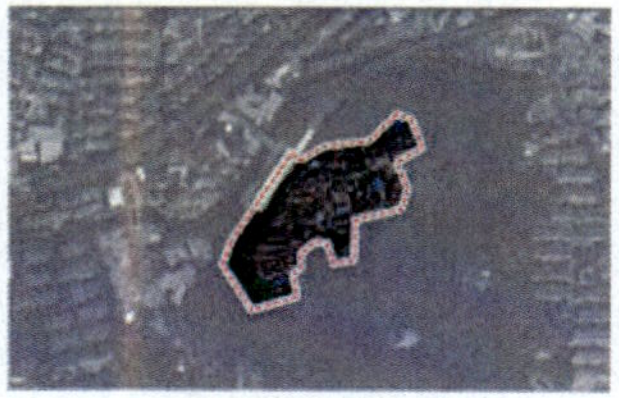	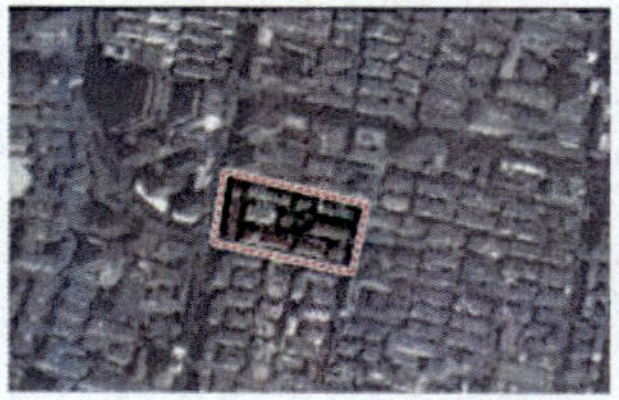
General description		The settlement is located on a slope. The constructions therein are primarily multi-storey self-built houses with three or fewer floors, spreading out along the roads in the shape of the Chinese character "feng" (丰). Its living environment is undesirable	This settlement is located at the southwest corner of Hongshan Street, Xuanwu District, adjacent to the Hongshan Zoo Subway Station. The surrounding area of Hongshan has always been a key area for migrant worker settlements. Although a large number of migrant workers have moved out due to the renovation of the zoo, some have remained	This settlement is located in Xinjiekou Street, Xuanwu District. The community therein has a high percentage of floating population, which accounts for over one-third of its permanent population. This settlement is one of the representative migrant worker settlements in the center of the core area of the city

* *Source* Own elaboration

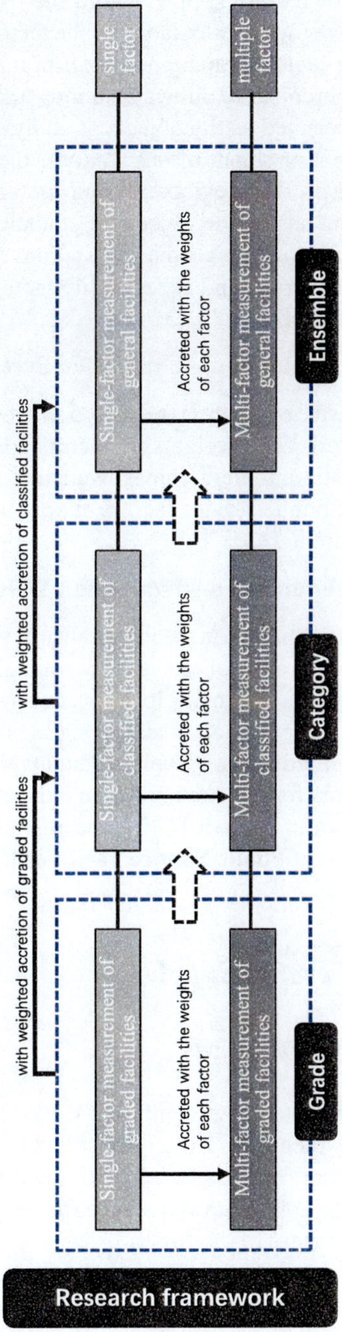

Fig. 5.22 Research framework for the level of public service facility supply in migrant worker settlements . **Source* Own elaboration

the travel circle of migrant workers; accessibility reflects the level of convenience offered by various categories of facility supply within the travel circle of migrant workers (see Sect. 5.2.2 (4) above for the calculation of the factors for each indicator).

The methodology for single-factor evaluation on transportation facilities is illustrated as follows. For the measurement of subway stations, the ones located closest to the sample settlements are selected as the objects of study, using accessibility as the evaluation factor; for the measurement of bus stations, those located within the maximum travel circle defined in the previous section are selected as the objects of study. Considering the actual usage, the objects of statistics for the density and location factors of such facilities include all the bus stations within the circle, and the objects of statistics for the coverage and accessibility factors pertain to the stops closest to the sample settlements (see Fig. 5.23).

Multi-factor Comprehensive Evaluation—Evaluation System

For the multi-factor comprehensive evaluation (see Fig. 5.24), the Analytic Hierarchy Process (AHP) is adopted herein. The weights of factors for all strata are established by constructing a multi-stratum hierarchical evaluation framework. Based on weighted accretion, the level of public service facility supply for the migrant worker group in Nanjing City is evaluated.

Multi-factor Comprehensive Evaluation—Evaluation Weights

Based on the 25 evaluation questionnaires on the materiality of various types of public service facilities completed by migrant workers and 25 evaluation questionnaires on the materiality of the single factors (i.e., density, location, coverage, and accessibility) for the supply of facilities in various grades and categories completed by urban planning specialists, judgment matrices are developed and weights of all factors are calculated. The weights of factors for the strata of basic indicators, criteria, and sub-objectives are derived and shown in Table 5.11 (for specific details on the technical route, see the section "Overview of Public Service Facilities in Nanjing" above).

5.3.3 *Measurement of Level of Supply*

5.3.3.1 Delineation of Evaluation Targets

A questionnaire survey was conducted among migrant workers in Nanjing to obtain an understanding of the travel modes and the corresponding maximum tolerable travel time for the migrant worker group to reach facilities of different categories and grades. Further, the maximum tolerable travel distance for the group to access these facilities is calculated. Next, by means of the ArcGIS Network Analyst, its maximum tolerable travel circle is established. Using this circle as a basis, the categories and grades of facilities appropriate for inclusion in the pool of evaluation targets are delineated.

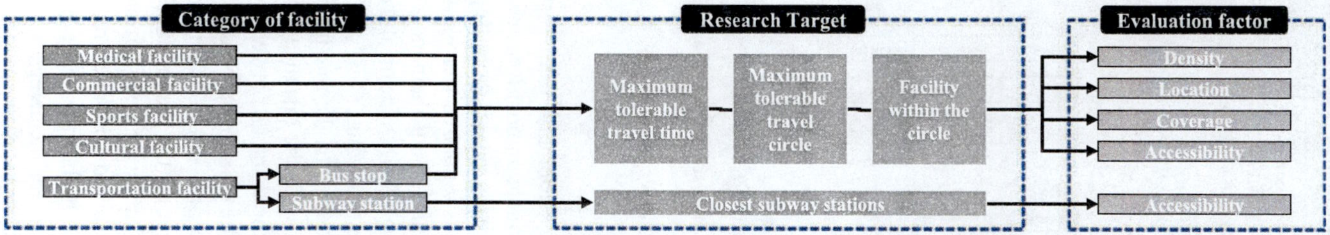

Fig. 5.23 Methodology for evaluating the level of public service facility Supply in migrant worker settlements. **Source* Own elaboration

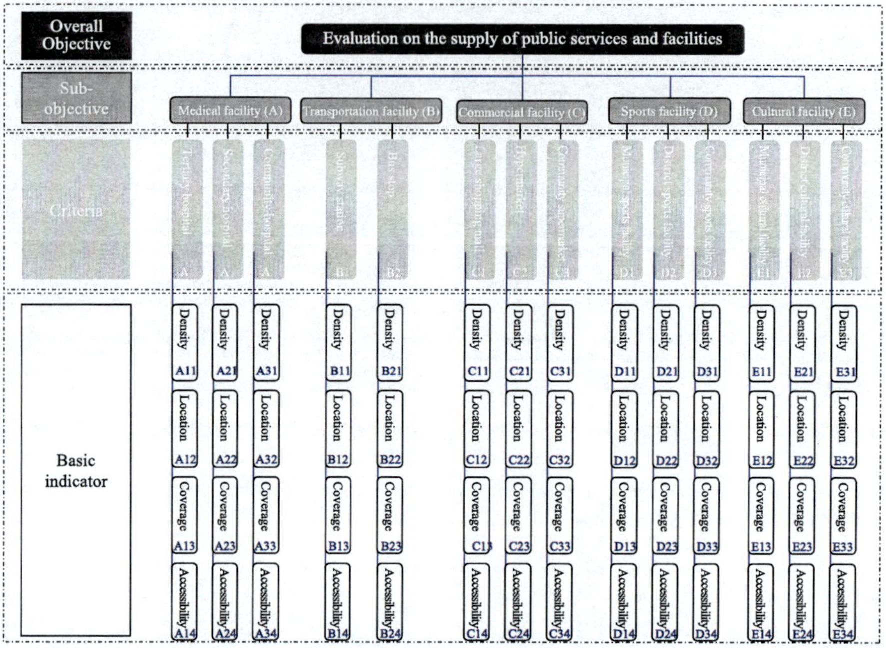

Fig. 5.24 Evaluation index system of public service facilities for migrant workers in sampled settlements. **Source* Own Elaboration

Maximum Tolerable Travel Time

Migrant workers in different sample settlements often display disparate characteristics of travel due to differences in gender composition, age distribution, income level, etc. To objectively and comprehensively reflect the differentiated characteristics of different sample settlements, 25 questionnaires were issued to migrant workers in the three sample settlements to inquire about their maximum tolerable travel time for accessing public facilities. Through data collation and calculation of mean values (see Table 5.12), this study finds the following:

For different migrant worker settlements, the maximum tolerable travel time for the migrant worker group is linked with the distance from the center of the city's core area. The greater the straight-line distance between a settlement and the center of the city's core area, the longer is the maximum tolerable travel time. However, the three settlements show minor differences in the maximum tolerable travel time for facilities of the same grade and category (the maximum difference being 14.36 min for tertiary hospitals). For different public service facilities, the maximum tolerable travel time for migrant workers is linked with the grade and category of facility. In terms of the grade of facility, the higher the grade of the facility, the longer is the maximum tolerable travel time; as for the category of facility, medical facilities boast the longest maximum tolerable travel time, whereas bus stops see the shortest maximum tolerable travel time.

Table 5.11 Index system and weights for the evaluation of public service facility supply

Goal	Sub-objective B		Criteria C		Basic indicator D		Weight to sub-goal	Weight to goal
	Name	Weight	Name	Weight	Name	Weight		
Evaluation of the supply level of public service facilities for migrant workers	Medical facility (B)	0.17335	Tertiary hospital (A1))	0.50404	Density (A11)	0.14239	0.07177	0.01244
					Location (A12)	0.19373	0.09765	0.01693
					Coverage (A13)	0.39015	0.19665	0.03409
					Accessibility (A14)	0.27373	0.13797	0.02392
			Secondary hospital (A2)	0.21010	Density (A11)	0.14239	0.02992	0.00519
					Location (A12)	0.19373	0.04070	0.00706
					Coverage (A13)	0.39015	0.08197	0.01421
					Accessibility (A14)	0.27373	0.05751	0.00997
			Community hospital (A3)	0.28586	Density (A11)	0.22091	0.06315	0.01095
					Location (A12)	0.18358	0.05248	0.00910
					Coverage (A13)	0.18404	0.05261	0.00912
					Accessibility (A14)	0.41147	0.11762	0.02039
	Transportation facility (B)	0.25301	Subway station (B1)	0.42616	Accessibility (A14)	1.00000	0.42616	0.10782
			Bus stop (B2)	0.57384	Density (A11)	0.22091	0.12677	0.03207
					Location (A12)	0.18358	0.10534	0.02665
					Coverage (A13)	0.18404	0.10561	0.02672
					Accessibility (A14)	0.41147	0.23612	0.05974
	Commercial facility (C)	0.33009	Large shopping mall (C1)	0.27203	Density (A11)	0.14239	0.03873	0.01279
					Location (A12)	0.19373	0.05270	0.01740
					Coverage (A13)	0.39015	0.10613	0.03503

(continued)

Table 5.11 (continued)

Goal	Sub-objective B		Criteria C		Basic indicator D		Weight to sub-goal	Weight to goal
	Name	Weight	Name	Weight	Name	Weight		
					Accessibility (A14)	0.27373	0.07446	0.02458
			Hypermarket (C2)	0.38902	Density (A11)	0.14239	0.05539	0.01828
					Location (A12)	0.19373	0.07537	0.02488
					Coverage (A13)	0.39015	0.15177	0.05010
					Accessibility (A14)	0.27373	0.10649	0.03515
			Community supermarket (C3)	0.33895	Density (A11)	0.22091	0.07488	0.02472
					Location (A12)	0.18358	0.06222	0.02054
					Coverage (A13)	0.18404	0.06238	0.02059
					Accessibility (A14)	0.41147	0.13947	0.04604
	Sports facility (D)	0.10986	Municipal sport facility (D1)	0.28314	Density (A11)	0.14239	0.04032	0.00443
					Location (A12)	0.19373	0.05485	0.00603
					Coverage (A13)	0.39015	0.11047	0.01214
					Accessibility (A14)	0.27373	0.07751	0.00852
			District sports facility (D2)	0.23182	Density (A11)	0.14239	0.03301	0.00363
					Location (A12)	0.19373	0.04491	0.00493
					Coverage (A13)	0.39015	0.09044	0.00994
					Accessibility (A14)	0.27373	0.06346	0.00697
			Community sports facility (D3)	0.48504	Density (A11)	0.22091	0.10715	0.01177
					Location (A12)	0.18358	0.08904	0.00978

(continued)

Table 5.11 (continued)

Goal	Sub-objective B		Criteria C		Basic indicator D		Weight to sub-goal	Weight to goal
	Name	Weight	Name	Weight	Name	Weight		
					Coverage (A13)	0.18404	0.08927	0.00981
					Accessibility (A14)	0.41147	0.19958	0.02193
	Cultural facility (E)	0.13369	Municipal cultural facility (E1)	0.41138	Density (A11)	0.14239	0.05858	0.00783
					Location (A12)	0.19373	0.07970	0.01066
					Coverage (A13)	0.39015	0.16050	0.02146
					Accessibility (A14)	0.27373	0.11261	0.01505
			District cultural facility (E2)	0.36406	Density (A11)	0.14239	0.05184	0.00693
					Location (A12)	0.19373	0.07053	0.00943
					Coverage (A13)	0.39015	0.14204	0.01899
					Accessibility (A14)	0.27373	0.09966	0.01332
			Community cultural facility (E3)	0.22456	Density (A11)	0.22091	0.04961	0.00663
					Location (A12)	0.18358	0.04122	0.00551
					Coverage (A13)	0.18404	0.04133	0.00553
					Accessibility (A14)	0.41147	0.09240	0.01235

* *Source* Own elaboration

Table 5.12 Maximum tolerable travel time for migrant workers in sample settlements

Name of sample settlement		Urban village of Taiping community (outer ring)	Hongshan village (inner ring)	Hongmiao quarter (center)
Straight-line distance to the center of the core area of the city (km)		8.2	4.0	1.6
Medical facility	Tertiary hospital	65.22	58.02	50.86
	Secondary hospital	46.96	42.57	39.04
	Community hospital	20.44	19.04	16.41
Transportation facility	Bus stop	7.03	6.47	5.33
Commercial facility	Large shopping mall	56.17	53.47	50.16
	Hypermarket	38.47	32.01	28.47
	Community supermarket	14.02	13.70	10.37
Sports facility	Municipal sports facility	48.71	46.68	47.22
	District sports facility	32.54	34.89	30.56
	Community sports facility	16.64	15.51	13.95
Cultural facility	Municipal cultural facility	49.62	52.19	45.79
	District cultural facility	38.51	36.70	32.34
	Community cultural facility	17.17	15.30	13.26

* *Source* Data from the sampling survey conducted by the research team on the level of public service facility supply in migrant worker settlements in the core area of Nanjing (2018)

Composition of Travel Modes

A data analysis of the sampling survey on the travel modes of migrant workers provides a general understanding of the travel characteristics of the three sample settlements. Specifically, higher-grade facilities involve mainly buses and subways, whereas lower-grade facilities primarily involve walking. The mode of travel varies with the category of facility, with little difference observed among the three settlements (see Figs. 5.25, 5.26, 5.27, 5.28 and 5.29).

For different categories of facilities, the factors that affect the structure of travel modes among migrant workers include the grade of facility, characteristics of usage, quantity, scale, and location distribution, among others. Municipal and district-grade

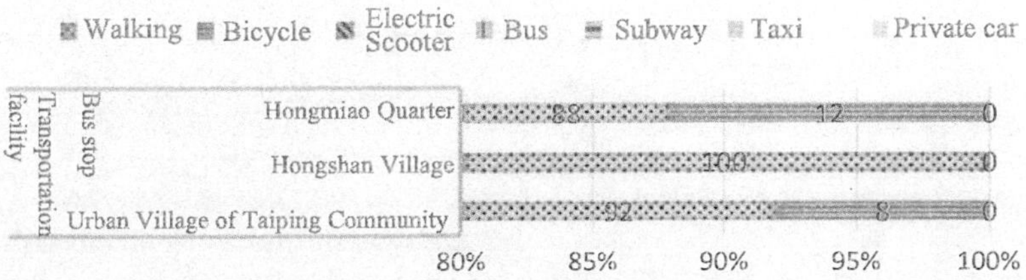

Fig. 5.25 Characteristics of travel to transportation facilities. **Source* Data from the sampling survey conducted by the research team on the level of public service facility supply in migrant worker settlements in the core area of Nanjing (2018)

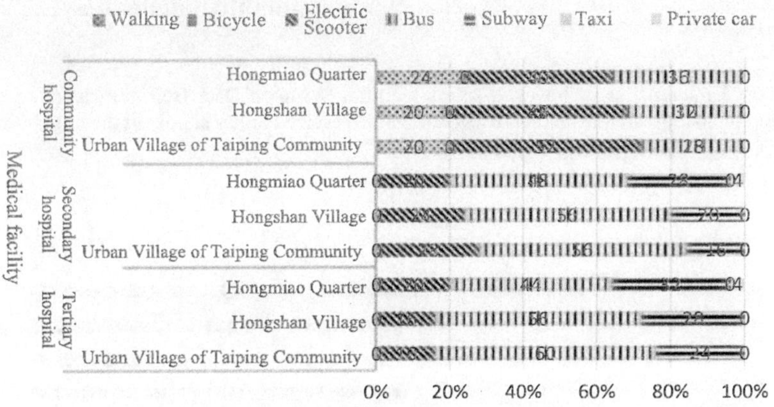

Fig. 5.26 Characteristics of travel to medical facilities. **Source* Data from the sampling survey conducted by the research team on the level of public service facility supply in the migrant worker settlements in the core area of Nanjing (2018)

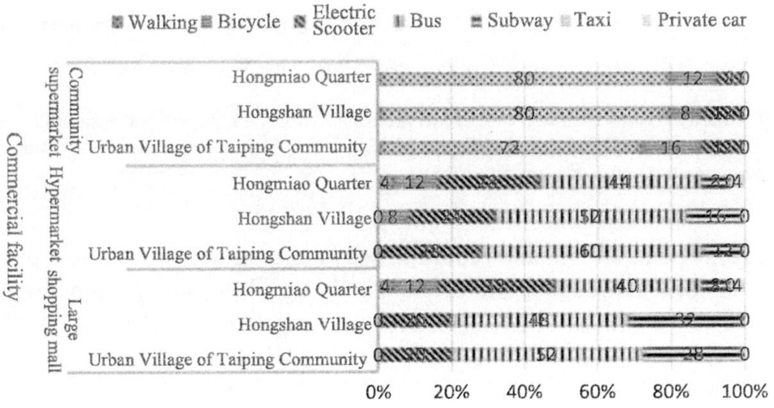

Fig. 5.27 Characteristics of travel to commercial facilities. **Source* Data from the sampling survey conducted by the research team on the level of public service facility supply in the migrant worker settlements in the core area of Nanjing (2018)

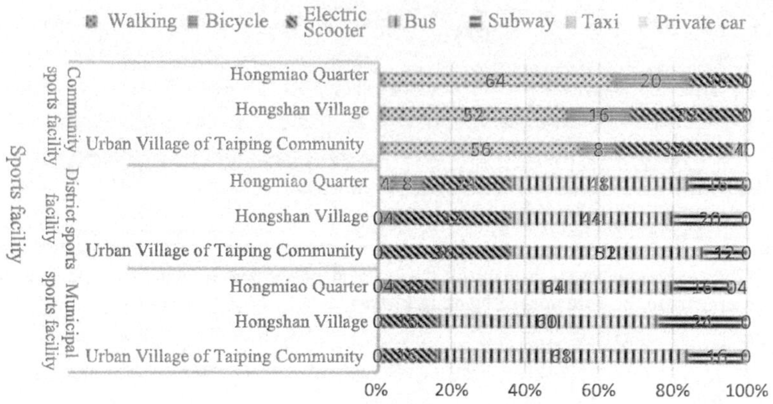

Fig. 5.28 Characteristics of travel to sports facilities. **Source* Data from the sampling survey conducted by the research team on the level of public service facility supply in the migrant worker settlements in the core area of Nanjing (2018)

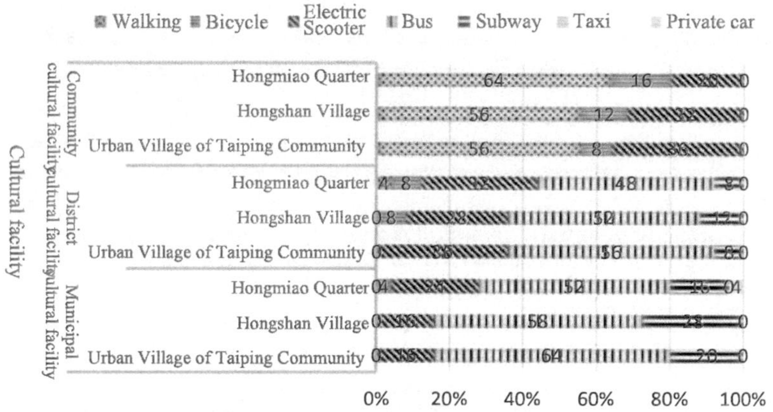

Fig. 5.29 Characteristics of travel to cultural facilities. **Source* Data from the sampling survey conducted by the research team on the level of public service facility supply in the migrant worker settlements in the core area of Nanjing (2018)

medical facilities have limited locations, mostly concentrated in the core area of the city. On this account, migrant workers typically travel to such facilities by public transportation (accounting for 44–60%). Conversely, as community-grade medical facilities have more locations, migrant workers mostly travel by electric scooter or bus (accounting for 76–80%), of which electric scooters take up a higher percentage than buses. Finally, the bus stops in the category of transportation facilities have more locations in the core area of the city within shorter distances from the settlements, and migrant workers are mostly limited to the two travel modes of walking and bicycle, with walking accounting for the vast majority (88–100%).

For different sample settlements, the structure of travel modes differs due to variances in their conditions, such as the location of settlements and traffic conditions in surrounding areas. The Hongmiao Community is located in the central block of the city and has a relatively adequate supply of facilities in the surrounding regions. For the use of municipal and district-grade facilities, buses serve as the primary travel mode (40–64%), while subway serves as the secondary travel mode. For community-grade facilities, walking is a frequently adopted travel mode (64–80%). Located beside an urban subway station, Hongshan Village has similar characteristics of travel to the Hongmiao Community, with more frequent use of subway facilities. The Urban Village of Taiping Community is farthest from the city center and has relatively few facilities in its surrounding regions. As a result, it displays the highest percentage of bus travel when accessing various municipal and district-grade facilities (52–68%). As for the use of community-grade facilities, walking accounts for the highest percentage in this settlement but is still lower than in the other two settlements (56–72%).

Maximum Tolerable Travel Time

Through the above research on the composition of travel modes and maximum tolerable travel time among migrant workers, the maximum tolerable travel distance for migrant workers in each sample settlement to access public service facilities of different categories and grades is calculated. The calculation formula is as follows:

$$\boldsymbol{Sn} = \boldsymbol{Tn} * \sum \boldsymbol{Vi} * \boldsymbol{Ain}\% \tag{5.10}$$

In the formula, ***Sn*** represents the maximum tolerable travel distance for migrant workers to access type-***n*** public service facilities, ***Tn*** represents the maximum tolerable travel time for migrant workers to access type-***n*** public service facilities derived from the sampling survey, and Vi represents the average speed of type- ***i*** travel mode (In Nanjing, the average speed of buses, subways, cars (including private cars and taxis), walking, and non-motorized vehicles (i.e., bicycles and electric scooters) is 20 km/h, (Plan and for the Reform and Development of Public Transport in Nanjing(2011–2015)[Z]. Nanjing, 2011) 36.99 km/h, (www.ditiezu.com) 21.84 km/h,(Hao et al. 2011) 4.6 km/h,(Tao and Xin 2015) and 15 km/h, respectively). Finally, ***Ain%*** represents the percentage of type-***i*** travel mode in accessing type-***n*** public service facilities derived from the sampling survey. The calculation results are as follows (see Table 5.13):

Take the inner-ring sample Hongshan Village as an example. Its maximum tolerable travel distance circles for each type of public service facility are as follows (see Fig. 5.30):

Table 5.13 Maximum tolerable distance for migrant workers in sample settlements

Name of sample settlement		Urban village of Taiping community (outer ring)	Hongshan village (inner ring)	Hongmiao quarter (center)
Straight-line distance to the center of the core area of the city (km)		8.2	4.0	1.6
Medical facility	Tertiary hospital	25.30	23.17	20.78
	Secondary hospital	16.69	15.75	15.51
	Community hospital	4.88	4.61	3.91
Transportation facility	Bus stop	0.64	0.50	0.52
Commercial facility	Large shopping mall	22.24	21.78	15.56
	Hypermarket	13.23	11.27	8.93
	Community supermarket	1.76	1.53	1.15
Sports facility	Municipal sports facility	17.79	18.11	17.31
	District sports facility	10.98	12.56	10.44
	Community sports facility	2.60	2.48	1.94
Cultural facility	Municipal cultural facility	18.69	21.19	16.33
	District cultural facility	12.55	12.38	10.10
	Community cultural facility	2.63	2.34	1.84

**Source* Own elaboration

5.3.3.2 Single-factor Measurement

As per the previously described calculation method, the indicators for all types of facilities at all grade and categories established in each maximum tolerable travel circle of migrant worker settlements are measured. Among the various types of facilities, subway stations in the category of transportation facilities undergo single-factor evaluation only in terms of the accessibility of the nearest stations to the sample settlements. All other facilities are subject to the measurement of all the four single factors: density, location, coverage, and accessibility. The calculation results are as follows (See Table 5.14):

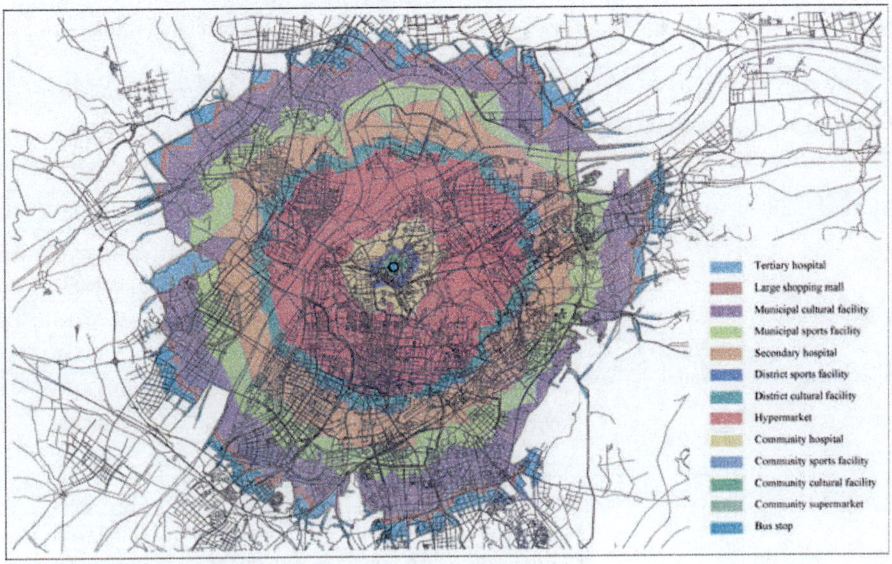

Fig. 5.30 Maximum tolerable travel distance circles for different types of public service facilities in Hongshan village. *_Source_ Own elaboration

5.3.3.3 Multi-factor Comprehensive Evaluation

To evaluate and compare the levels of supply of facilities of the same types among the three sample settlements in a comprehensive manner, all single factors are nondimensionalized by type (see Table 5.15). With the evaluation indicators standardized, the scores for the level of supply at each objective stratum are calculated in accordance with the specified weights. For detailed calculation results, please refer to the next section.

5.3.4 Evaluation of Level of Supply

In this section, a comparative analysis is provided to demonstrate the differentiated characteristics of public service facilities in terms of the quantity, scale, and spatial distribution of supply. The analysis is conducted in two steps: single-factor and multi-factor. In addition, a horizontal comparison between the three sample settlements is rendered to reveal the level of supply of public service facilities in migrant worker settlements under the influence of location differences (outer ring, inner ring, and center). The general structure of evaluation is illustrated as follows (see Fig. 5.31):

Table 5.14 Results of single-factor evaluation of public service facilities in migrant worker sample settlements

Type of facility		Single factor	Urban village of Taiping community (outer ring)	Hongshan village (inner ring)	Hongmiao quarter (center)
Category	Grade				
Medical facility	Tertiary hospital	Density	0.02646	0.02984	0.03737
		Location	0.13764	0.15523	0.19439
		Coverage	0.82196	0.91928	0.98671
		Accessibility	0.53230	4.74915	7.36185
	Secondary hospital	Density	0.03439	0.03453	5.72550
		Location	0.31071	0.31199	0.32218
		Coverage	0.86330	0.95133	0.98796
		Accessibility	0.29106	0.49374	1.60348
	Community hospital	Density	1.03158	3.09111	5.07465
		Location	0.98021	1.08530	0.47247
		Coverage	0.94818	0.99325	0.99558
		Accessibility	29.70630	78.10811	268.54345
Transportation facility	Subway station	Accessibility	0.08710	168.99762	3.37364
	Bus stop	Density	2.25662	18.13904	15.46112
		Location	1.11109	1.10002	0.74590
		Coverage	0.34874	0.63357	1.00000
		Accessibility	10.09340	264.35165	210.26695
Commercial facility	Large shopping mall	Density	0.00781	0.00799	0.01328
		Location	0.16752	0.17145	0.28491
		Coverage	0.85959	0.93707	0.98906
		Accessibility	0.06905	0.23932	4.00247
	Hypermarket	Density	0.21649	0.34216	0.64434
		Location	0.28155	0.44499	0.83800
		Coverage	0.86656	0.97339	0.99254
		Accessibility	2.21558	16.12508	74.22659
	Community supermarket	Density	1.51552	4.52778	10.03749
		Location	1.43537	1.58454	0.93148
		Coverage	0.87090	0.99185	0.99997
		Accessibility	26.59949	64.03984	168.77531
Sports facility	Municipal sports facility	Density	0.00623	0.00553	0.00452
		Location	0.26758	0.23713	0.19398
		Coverage	0.66819	0.80406	0.90240
		Accessibility	0.02726	0.08259	2.32590
	District sports facility	Density	0.00846	0.02814	0.05056

(continued)

Table 5.14 (continued)

Type of facility		Single factor	Urban village of Taiping community (outer ring)	Hongshan village (inner ring)	Hongmiao quarter (center)
Category	Grade				
		Location	0.12097	0.40259	0.72336
		Coverage	0.74632	0.88458	0.99232
		Accessibility	0.13307	0.50761	3.06341
	Community sports facility	Density	0.10754	0.30005	2.77865
		Location	0.34042	0.35097	0.86187
		Coverage	0.86696	0.98594	0.99935
		Accessibility	2.56624	2.11872	62.57223
Cultural facility	Municipal cultural facility	Density	0.01672	0.05003	0.07220
		Location	0.05980	0.17893	0.25822
		Coverage	0.83364	0.86432	0.98979
		Accessibility	0.44266	1.46206	30.61472
	District cultural facility	Density	0.04859	0.07230	0.14941
		Location	0.23172	0.34477	0.71247
		Coverage	0.77316	0.93301	0.99266
		Accessibility	0.27319	1.23747	25.26721
	Community cultural facility	Density	0.20992	0.45233	4.82793
		Location	0.52031	0.41427	1.17251
		Coverage	0.80515	0.95873	0.99967
		Accessibility	5.73747	5.14898	69.68522

**Source* Own elaboration

5.3.4.1 Comparative Analysis of the Supply Level of Graded Facilities

Single-factor Comparative Analysis

In this study, the 14 types of facilities and four types of single factors combined form 53 types of scenarios. Through data collation, the level and overall ranking of public service supply in migrant worker settlements may be categorized into the following three classes:

[Outer ring < Inner ring < Center]

Forty-three types of scenarios obtain scores that present the characteristic of "outer ring < inner ring < center" (see Fig. 5.32 and Table 5.16):

The types of facilities for which the scores of the four factors increase in accordance with this order include tertiary hospitals, secondary hospitals, large shopping malls, hypermarkets, district sports facilities, community sports facilities, municipal cultural facilities, and district cultural facilities. Of them, accessibility spans

Table 5.15 Results of nondimensionalization of levels of facility supply

Basic indicator—score of single factor					
Type of facility		Single factor	Urban village of Taiping community (outer ring)	Hongshan village (inner ring)	Hongmiao quarter (center)
Category	Grade				
Medical facility	Tertiary hospital	Density	0.00521	0.00588	0.00736
		Location	0.12682	0.14303	0.17911
		Coverage	0.99110	0.82561	0.92337
		Accessibility	0.00198	0.01768	0.02741
	Secondary hospital	Density	0.00678	0.00680	0.00703
		Location	0.28629	0.28747	0.29686
		Coverage	0.99235	0.86713	0.95556
		Accessibility	0.00108	0.00184	0.00597
	Community hospital	Density	0.20328	0.60913	1.00000
		Location	0.90317	1.00000	0.43533
		Coverage	1.00000	0.95239	0.99767
		Accessibility	0.11062	0.29086	1.00000
Transportation facility	Subway station	Accessibility	0.00052	1.00000	0.01996
	Bus stop	Density	0.14595	0.39107	1.00000
		Location	1.00000	0.99004	0.67132
		Coverage	1.00000	0.34874	0.63357
		Accessibility	0.03818	1.00000	0.79541
Commercial facility	Large shopping mall	Density	0.00078	0.00080	0.00132
		Location	0.10572	0.10820	0.17980
		Coverage	0.98909	0.85961	0.93709
		Accessibility	0.00041	0.00142	0.02371
	Hypermarket	Density	0.02157	0.03409	0.06419
		Location	0.17769	0.28083	0.52886
		Coverage	0.99256	0.86658	0.97342
		Accessibility	0.01313	0.09554	0.43980
	Community supermarket	Density	0.15099	0.45109	1.00000
		Location	0.90586	1.00000	0.58786
		Coverage	1.00000	0.87092	0.99188
		Accessibility	0.15760	0.37944	1.00000
Sports facility	Municipal sports facility	Density	0.00224	0.00199	0.00163
		Location	0.31046	0.27513	0.22507
		Coverage	0.90299	0.66862	0.80458
		Accessibility	0.00044	0.00132	0.03717

(continued)

Table 5.15 (continued)

Basic indicator—score of single factor					
Type of facility		Single factor	Urban village of Taiping community (outer ring)	Hongshan village (inner ring)	Hongmiao quarter (center)
Category	Grade				
	District sports facility	Density	0.00304	0.01013	0.01820
		Location	0.14036	0.46711	0.83930
		Coverage	0.99296	0.74680	0.88516
		Accessibility	0.00213	0.00811	0.04896
	Community sports facility	Density	0.03870	0.10798	1.00000
		Location	0.39498	0.40722	1.00000
		Coverage	1.00000	0.86753	0.98658
		Accessibility	0.04101	0.03386	1.00000
Cultural facility	Municipal cultural facility	Density	0.00346	0.01036	0.01495
		Location	0.05100	0.15260	0.22023
		Coverage	0.99011	0.83392	0.86460
		Accessibility	0.00635	0.02098	0.43933
	District cultural facility	Density	0.01006	0.01497	0.03095
		Location	0.19763	0.29404	0.60764
		Coverage	0.99298	0.77342	0.93331
		Accessibility	0.00392	0.01776	0.36259
	Community cultural facility	Density	0.04348	0.09369	1.00000
		Location	0.44375	0.35331	1.00000
		Coverage	1.00000	0.80542	0.95904
		Accessibility	0.08233	0.07389	1.00000

* *Source* Own elaboration

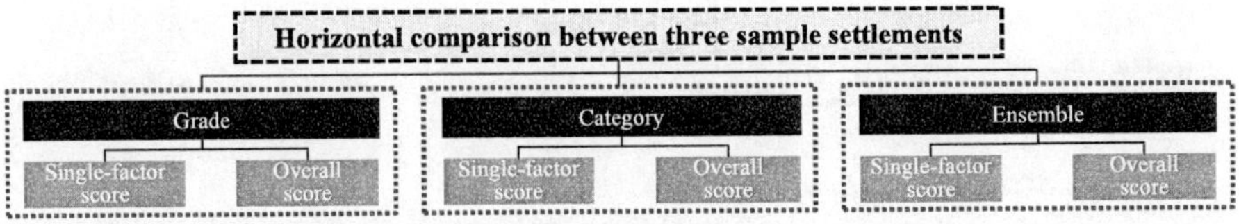

Fig. 5.31 Process and framework for evaluating level of supply. * *Source* Own elaboration

the largest range (particularly for large shopping malls, community sports facilities, municipal cultural facilities, and district cultural facilities), coverage spans the smallest range, and density spans from large to small ranges (with the largest range observed in secondary hospitals and community sports facilities and the smallest range found in tertiary hospitals).

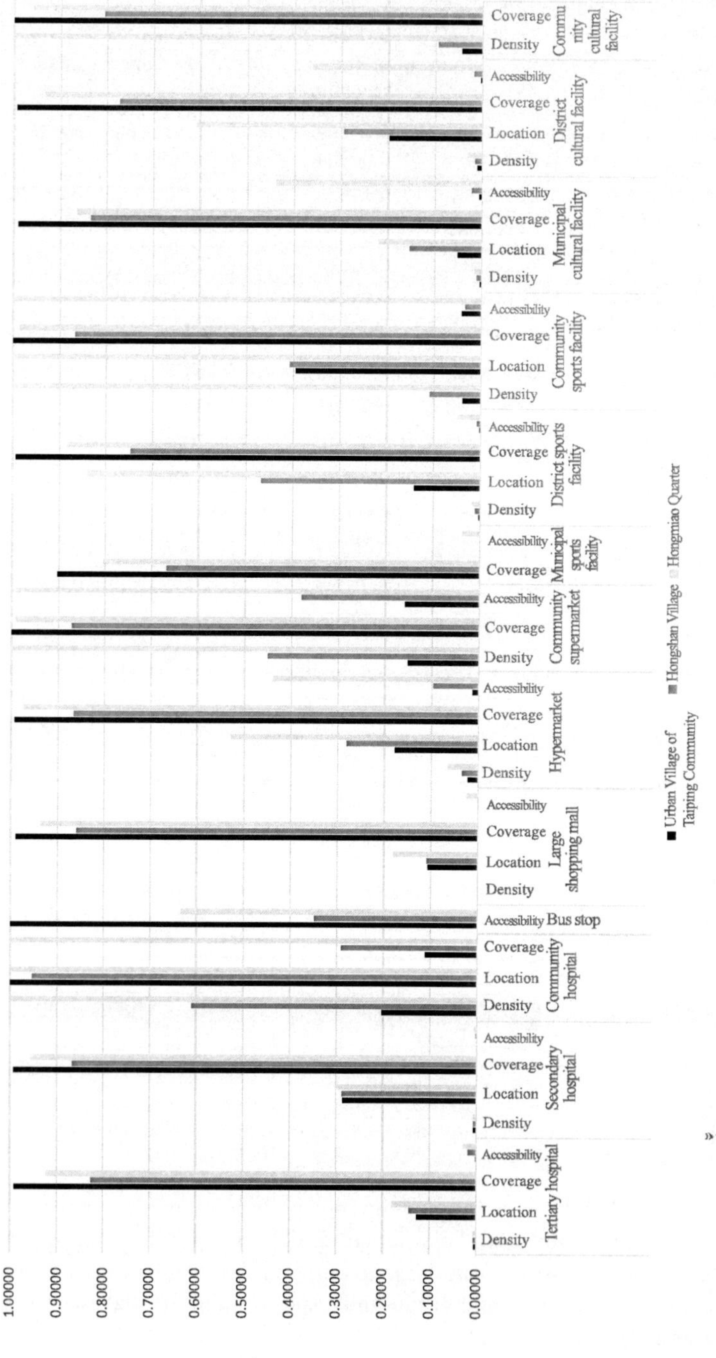

Fig. 5.32 Horizontal comparison of single-factor scores for supply level of graded facilities (I). **Source* Data from the sampling survey conducted by the research team on the level of public service facility supply in migrant worker settlements in the core area of Nanjing (2018)

Table 5.16 Comparison of single factor scores for supply level of graded facilities

Type of facility: Category	Type of facility: Grade	Single factor	Urban village of Taiping community (outer ring)	Hongshan village (inner ring)	Hongmiao quarter (center)
Medical facility	Tertiary hospital	Density	0.00521	0.00588	0.00736
		Location	0.12682	0.14303	0.17911
		Coverage	0.99110	0.82561	0.92337
		Accessibility	0.00198	0.01768	0.02741
	Secondary hospital	Density	0.00678	0.00680	0.00703
		Location	0.28629	0.28747	0.29686
		Coverage	0.99235	0.86713	0.95556
		Accessibility	0.00108	0.00184	0.00597
	Community hospital	Density	0.20328	0.60913	1.00000
		Location	0.90317	1.00000	0.43533
		Coverage	1.00000	0.95239	0.99767
		Accessibility	0.11062	0.29086	1.00000
Transportation facility	Subway station	Accessibility	0.00052	1.00000	0.01996
	Bus stop	Density	0.14595	0.39107	1.00000
		Location	1.00000	0.99004	0.67132
		Coverage	1.00000	0.34874	0.63357
		Accessibility	0.03818	1.00000	0.79541
Commercial facility	Large shopping mall	Density	0.00078	0.00080	0.00132
		Location	0.10572	0.10820	0.17980
		Coverage	0.98909	0.85961	0.93709
		Accessibility	0.00041	0.00142	0.02371
	Hypermarket	Density	0.02157	0.03409	0.06419
		Location	0.17769	0.28083	0.52886
		Coverage	0.99256	0.86658	0.97342
		Accessibility	0.01313	0.09554	0.43980
	Community supermarket	Density	0.15099	0.45109	1.00000
		Location	0.90586	1.00000	0.58786
		Coverage	1.00000	0.87092	0.99188
		Accessibility	0.15760	0.37944	1.00000
Sports facility	Municipal sports facility	Density	0.00224	0.00199	0.00163
		Location	0.31046	0.27513	0.22507
		Coverage	0.90299	0.66862	0.80458
		Accessibility	0.00044	0.00132	0.03717
	District sports facility	Density	0.00304	0.01013	0.01820

(continued)

Table 5.16 (continued)

Type of facility		Single factor	Urban village of Taiping community (outer ring)	Hongshan village (inner ring)	Hongmiao quarter (center)
Category	Grade				
		Location	0.14036	0.46711	0.83930
		Coverage	0.99296	0.74680	0.88516
		Accessibility	0.00213	0.00811	0.04896
	Community sports facility	Density	0.03870	0.10798	1.00000
		Location	0.39498	0.40722	1.00000
		Coverage	1.00000	0.86753	0.98658
		Accessibility	0.04101	0.03386	1.00000
Cultural facility	Municipal cultural facility	Density	0.00346	0.01036	0.01495
		Location	0.05100	0.15260	0.22023
		Coverage	0.99011	0.83392	0.86460
		Accessibility	0.00635	0.02098	0.43933
	District cultural facility	Density	0.01006	0.01497	0.03095
		Location	0.19763	0.29404	0.60764
		Coverage	0.99298	0.77342	0.93331
		Accessibility	0.00392	0.01776	0.36259
	Community cultural facility	Density	0.04348	0.09369	1.00000
		Location	0.44375	0.35331	1.00000
		Coverage	1.00000	0.80542	0.95904
		Accessibility	0.08233	0.07389	1.00000

**Source* Own elaboration

The types of facilities for which the scores of three factors increase in this order include community hospitals (with the exclusion of location) and community supermarkets (with the exclusion of location). As these two types of facilities have a greater number of locations, no obvious location difference is observed in the core area of the city.

The types of facilities for which the scores of two factors increase in this order include municipal sports facilities (with the exclusion of density and location) and community cultural facilities (with the exclusion of location and accessibility).

[Outer ring > Inner ring > Center]

Only three types of scenarios have scores characterized by "outer ring > inner ring > center" (see Fig. 5.33 and Table 5.16):

The type of facility with a salient increase in score from the center to the inner ring to the outer ring in terms of density and location is municipal sports facilities.

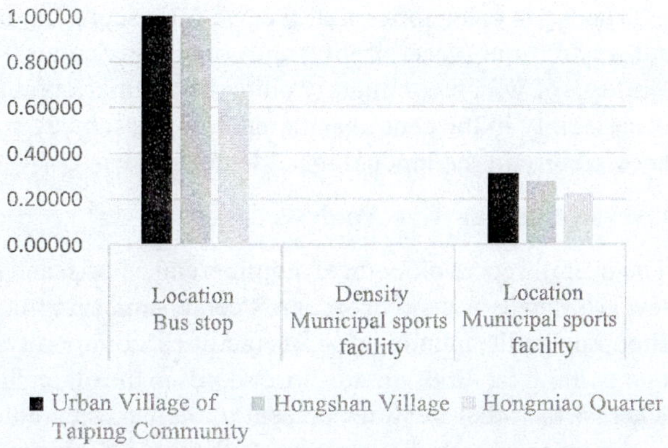

Fig. 5.33 Horizontal comparison of single-factor scores for supply level of graded facilities (II). **Source* Sample data collected by the research team on the level of public service facility supply in migrant worker settlements in the core area of Nanjing (2018)

The type of facility with a non-salient increase in score from the center to the inner ring to the outer ring in terms of location (with outer-ring and inner-ring settlements obtaining similar scores) is bus stations.

[No marked increase or decrease]

Seven types of scenarios obtain scores showing the characteristics of "no marked increase or decrease" (see Fig. 5.34 and Table 5.16):

The types of facilities that obtain high scores in two sample settlements include community hospitals (location), bus stations (density and accessibility), and community supermarkets (location). Of them, community hospitals and community supermarkets receive high scores for location in the inner-ring and outer-ring settlements, while bus stations receive high scores for density and accessibility in the center and inner-ring settlements, respectively.

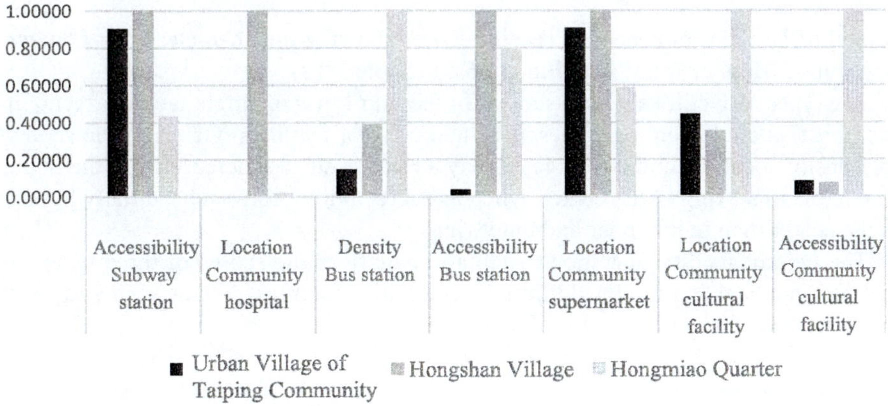

Fig. 5.34 Horizontal comparison of single-factor scores for supply level of graded facilities (III). **Source* Sample data collected by the research team on the level of public service facility supply in migrant worker settlements in the core area of Nanjing (2018)

The types of facilities that receive high scores in one sample settlement include subway stations (accessibility) and community cultural facilities (location and accessibility), of which community cultural facilities obtain high scores for location and accessibility in the central settlement, while subway stations obtain a high score for accessibility in the inner-ring settlements.[3]

Overall Comparative Analysis

The overall scores of medical, commercial, sports, and cultural facilities at all grade-levels basically ascend in order of "center-inner ring-outer ring". Among them, large shopping malls, municipal sports facilities, community sports facilities, and community cultural facilities obtain close scores in the outer-ring and inner-ring settlements, whereas the other facilities present more marked gradient increase from "the center to the inner ring to the outer ring". What is slightly unexpected is that the overall score of transportation facilities escalates in order of "inner ring-center-outer ring". This result may have to do with the fact that only one single factor of "location" is involved in the evaluation of subway stations, and that the inner-ring settlement sample is immediately adjacent to a subway station. In light of this, this finding may not necessarily represent the general situation (see Fig. 5.35 and Table 5.17).

5.3.4.2 Comparative Analysis of the Supply Level of Classified Facilities

Single-factor Comparative Analysis

In this study, five categories of facilities and four types of single factors (density, location, coverage, and accessibility) are combined to form 20 types of scenarios. The level of supply and the overall ranking of public service facilities in migrant worker settlements may be categorized into the following two classes.

[Outer ring < Inner Ring < Center]

A total of 16 types of scenarios receive scores showing the characteristic of "outer ring < inner ring < center" (see Fig. 5.36 and Table 5.18):

The types of facilities whose scores for the four types of single factors ascend in such order include commercial, sports, and cultural facilities. Of them, the scores for density, location, and coverage display a relative steady increase from the outer ring to the inner ring to the center, whereas accessibility scores substantially higher in the center than in the inner and outer rings.

The type of facility that shows such an increase in the scores of three types of factors is medical facility. In addition, transportation facilities display such an ascent in the score for the single-factor of coverage.

[3] The inner-ring resettlement sample selected for this study is located just next to Hongshan Subway Station, hence the abruptly high score it obtained in the single factor calculations of accessibility. The overall situation of subway stations in the core area of Nanjing in terms of accessibility awaits further in-depth studies.

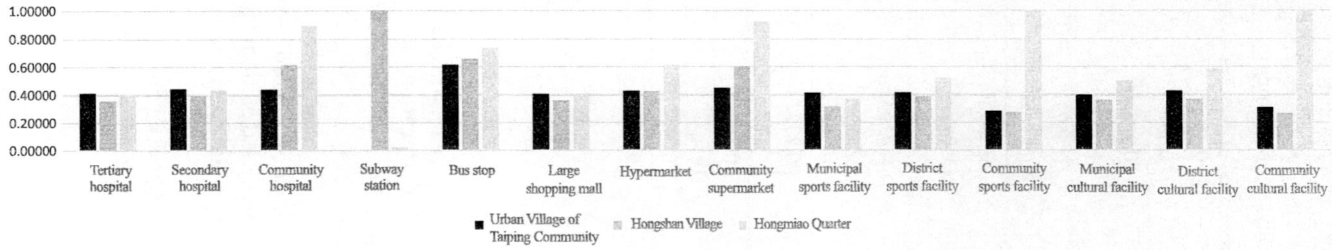

Fig. 5.35 Horizontal comparison of the overall scores of the supply level of graded facilities. **Source* Sample data from the sampling survey conducted by the research team on the level of public service facility supply in the migrant worker settlements in the core area of Nanjing (2018)

Table 5.17 Comparison of the overall scores for the supply level of graded facilities

Sub-goal—scores for the level of supply of graded facilities			
Type of facility	Urban village of Taiping community (outer ring)	Hongshan village (inner ring)	Hongmiao quarter (center)
Tertiary hospital	0.41252	0.35549	0.40350
Secondary hospital	0.44388	0.39547	0.43295
Community hospital	0.44026	0.61310	0.89591
Subway station	0.00052	1.00000	0.01996
Bus stop	0.61511	0.65728	0.73736
Large shopping mall	0.40659	0.35684	0.40712
Hypermarket	0.42833	0.42351	0.61176
Community supermarket	0.44854	0.59964	0.92284
Municipal sports facility	0.41288	0.31481	0.36791
District sports facility	0.41561	0.38552	0.52393
Community sports facility	0.28197	0.27220	0.99753
Municipal cultural facility	0.39840	0.36213	0.50237
District cultural facility	0.42820	0.36570	0.58551
Community cultural facility	0.30898	0.26419	0.99246

**Source* Own elaboration

[No marked increase or decrease]

Four types of scenarios receive scores displaying the characteristics of "no marked increase or decrease," including medical (location) and transportation facilities (density, location, and accessibility) (see Fig. 5.37 and Table 5.18):

Overall Comparative Analysis

The types of facilities whose scores ascend in the order of "outer ring-inner ring-center" include medical, commercial, sports, and cultural facilities. Of them, medical facilities manifest the steadiest increase, while sports and cultural facilities display the most marked increase in the center.

The scores of transportation facilities show an abrupt increase in the inner ring. Provided the influence of immediate adjacency to a subway station is excluded, transportation facilities also show an ascent "from the outer ring to the center" (see Fig. 5.38 and Table 5.19).

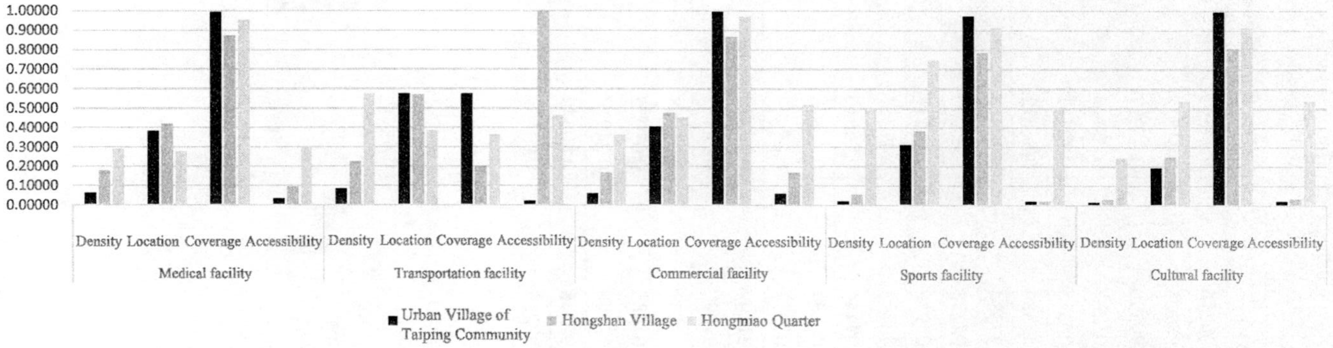

Fig. 5.36 Horizontal comparison of single-factor scores for the supply level of classified facilities (I). **Source* Sample data collected by the research team on the level of public service facility supply in the migrant worker settlements in the core area of Nanjing (2018)

Table 5.18 Comparison of single factor scores for the supply level of classified facilities

Category	Factor	Urban village of Taiping community (outer ring)	Hongshan village (inner ring)	Hongmiao quarter (center)
Medical facility	Density	0.06216	0.17852	0.29104
	Location	0.38225	0.41835	0.27709
	Coverage	0.99390	0.87057	0.95137
	Accessibility	0.03285	0.09244	0.30093
Transportation facility	Density	0.08375	0.22441	0.57384
	Location	0.57384	0.56812	0.38523
	Coverage	0.57384	0.20012	0.36357
	Accessibility	0.02213	1.00000	0.46494
Commercial facility	Density	0.05978	0.16637	0.36429
	Location	0.40493	0.47764	0.45390
	Coverage	0.99414	0.86616	0.96979
	Accessibility	0.05864	0.16616	0.51649
Sports facility	Density	0.02011	0.05529	0.48972
	Location	0.31202	0.38370	0.74333
	Coverage	0.97090	0.78322	0.91154
	Accessibility	0.02051	0.01868	0.50691
Cultural facility	Density	0.01485	0.03075	0.24198
	Location	0.19258	0.24917	0.53638
	Coverage	0.99338	0.80549	0.91082
	Accessibility	0.02253	0.03169	0.53729

**Source* Own elaboration

Fig. 5.37 Horizontal comparison of single-factor scores for the supply level of classified facilities (II). **Source* Sample data collected by the research team on the level of public service facility supply in the migrant worker settlements in the core area of Nanjing (2018)

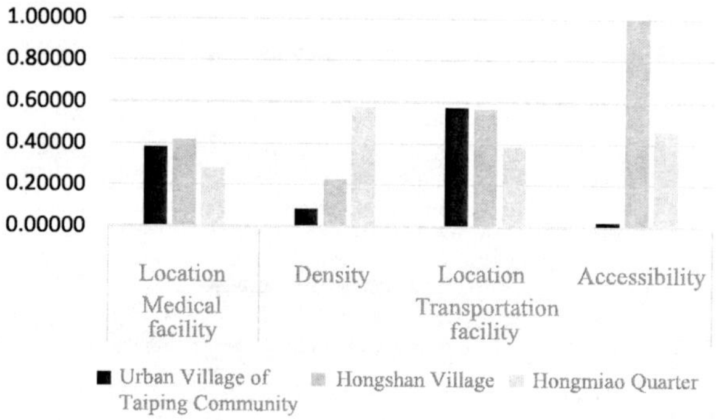

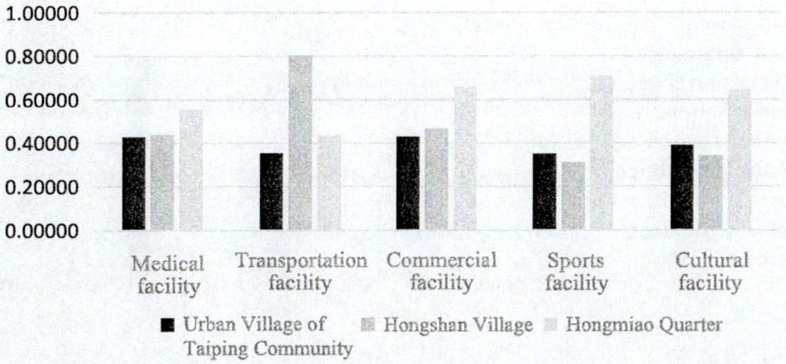

Fig. 5.38 Horizontal comparison of the overall scores of supply level of classified facilities. **Source* Sample data collected by the research team on the level of public service facility supply in migrant worker settlements in the core area of Nanjing (2018)

Table 5.19 Comparison of the overall scores for supply level of classified facilities

Category	Urban village of Taiping community (outer ring)	Hongshan village (inner ring)	Hongmiao quarter (center)
Medical facility	0.42704	0.43753	0.55045
Transportation facility	0.35319	0.80333	0.43163
Commercial facility	0.42927	0.46507	0.66153
Sports facility	0.35002	0.31053	0.70947
Cultural facility	0.38917	0.34144	0.64269

**Source* Own elaboration

5.3.4.3 Comparative Analysis of Overall Level of Supply

Single-factor Comparative Analysis

The scores for the four types of single factors ascend in the order of "outer ring-inner ring-center." Of them, the scores for the factors of location and coverage show relatively steady increase, with only minor differences in outer-ring, inner-ring, and central settlements; accessibility shows the most drastic increase in score, with greater differences observed among the three circles (see Fig. 5.39 and Table 5.20).

Overall Comparative Analysis

The overall scores for the levels of supply in the three migrant worker settlement samples ascend in the order of "outer ring-inner ring-center." In other words, the closer the settlement is to the city center, the higher is the overall score for its level of supply (see Table 5.21).

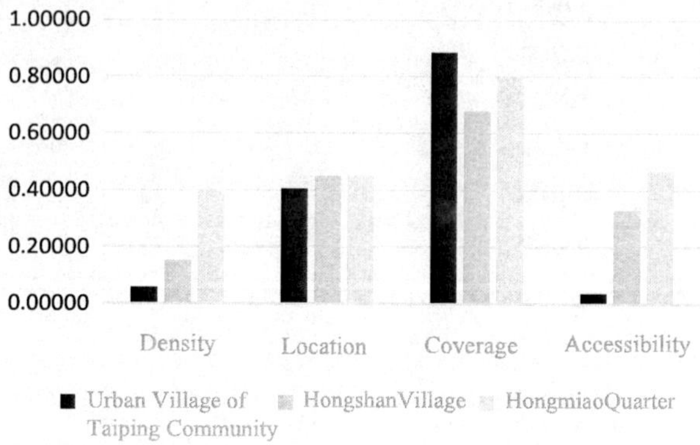

Fig. 5.39 Horizontal comparison of single factor scores for overall level of supply. **Source* Sample data collected by the research team on the level of public service facility supply in migrant worker settlements in the core area of Nanjing (2018)

Table 5.20 Comparison of single factor scores for overall level of supply

Factor	Urban village of Taiping community (outer ring)	Hongshan village (inner ring)	Hongmiao quarter (center)
Density	0.05589	0.15283	0.40203
Location	0.40514	0.44939	0.44870
Coverage	0.88510	0.68119	0.79894
Accessibility	0.03591	0.33017	0.46781

**Source* Own elaboration

Table 5.21 Comparison of overall scores for overall level of supply

Sample settlement	Urban village of Taiping community (outer ring)	Hongshan village (inner ring)	Hongmiao quarter (center)
Distance to the center of the core area	8.2	4.0	1.6
Type of settlement	Urban village	Urban village	Community
Score for the overall level of supply	0.39557	0.51237	0.58686

**Source* Own elaboration

5.3.5 *Analysis of the Factors Leading to Features of Facility Supply*

Overall, the level of supply of public service facilities in migrant worker settlements presents the characteristic of circle-based (outer ring < inner ring < center) differentiation on the three strata of "grade-category-ensemble." Such difference in supply is due to the joint effect of multiple factors, including the pattern of planning and allocation and policy orientation, among others, in the design at the top level as well as renewal and adjustment under the continuous regulation of market economy. The interplay and cross-influence between these factors may be generally summarized into the following three points (see Fig. 5.40).

5.3.5.1 Cumulative Effect of Urban Construction

Urban construction entails a process of gradual accumulation and continuous optimization, and the construction of public service facilities is no exception. As a result of the persistent advancement and cumulative effects of urban construction, facilities of all types at all grade levels will more or less manifest some circle-based differentiation.

In the process of urban expansion and construction, a clear consensus has been established among urban administrators and facility planners to provide the most consummate public service facilities possible. Even so, it is a universal fact that improvement has been achieved in a progressive manner and advancement is noted from the center to the outer circles. The old town area in the center of Nanjing has a long history. After years of development and operation, this area has established a relatively mature construction foundation manifested by the comprehensive and systematic classification, grading, distribution, and scaling of facilities of all types at all grade levels. In comparison, the inner-ring area, as an area for secondary urban development, has also undergone enhancement through several rounds of urban expansion and construction, although it does not have as solidly founded public service levels as the central area. As a follow-up circle in urban development, the outer-ring circle shows an inherently fundamental disparity in the overall level of supply compared with the center and inner-ring area, although some of its basic configurations may meet the governing standards and provide a certain degree of assurance.

5.3.5.2 Inadequate Planning Standards for Specific Strata

Public service facilities are an essential component of urban planning and construction, wherein the government serves as the main undertaker of facility supply (particularly high-grade facilities such as municipal sports and cultural facilities). On this account, inadequate institutional design and regulatory standards for the planning and construction of public service facilities will result in circle-based differences in facility supply.

Generally speaking, urban public service facilities include not only municipal and district-level facilities that require clearly defined land use indicators and boundaries in the process of overall urban planning but also community-level facilities whose land use indicators, building indicators, etc., are clarified in the process of detailed planning and specific design. However, during the actual planning process, regional authorities often formulate their own supporting standards for public service facilities on the basis of the local conditions in addition to referring to national standards. A case in point is the Supporting Planning Standard for Public Facilities in Nanjing City

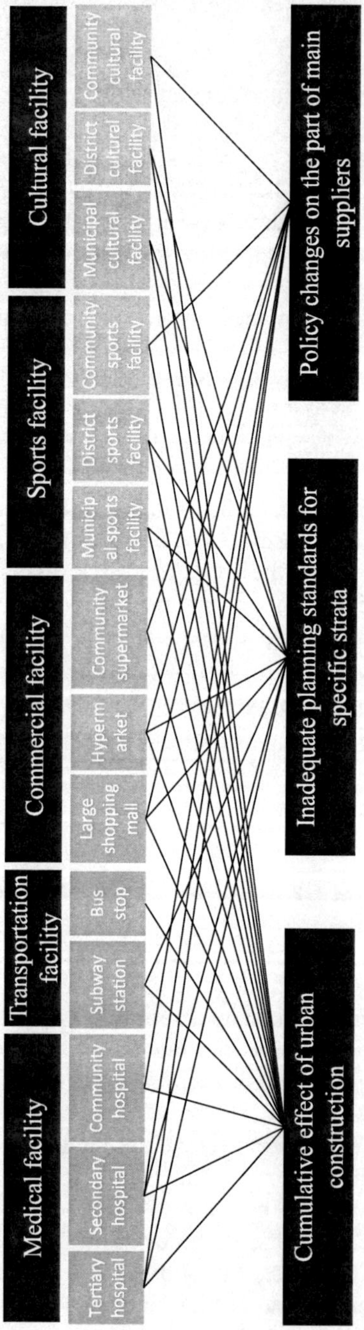

Fig. 5.40 Causes of circle-based differentiation in supply of public service facilities. **Source* Own elaboration

promulgated in Nanjing in 2015. The Standard provides no corresponding indicators to regulate the distribution and scale of municipal facilities. In terms of district-level facilities, the Standard stipulates the distribution and scale of various facilities but lacks standard settings for per capita indicators. In addition, it does not adequately specify requirements for the overall accessibility of facilities in the entire region in terms of spatial configuration. As for residential and grassroots community-level facilities (i.e., community-level facilities in this study), more detailed and comprehensive provisions are made in relation to the distribution and scale of various facilities, with the thousand-person index employed as an important reference standard. However, suggestions and guidance for more detailed classification regarding the scope of services are missing.

5.3.5.3 Policy Changes on the Part of Main Suppliers

In the general context of current economic transformation, public service facilities have been shifting from a unified supply system solely provided by the government to a coordinated supply system jointly provided by the market and the government. As a result, the spatial configuration of all types of facilities at all grade levels has come under a greater influence of market factors. Particularly under the premise that the government has limited financial capacity to undertake the construction of all facilities (especially community-level facilities such as community sports, cultural, and commercial facilities), the maturity of local market operations and degree of socioeconomic level will occasion enormous spatial differences in the supply of such products.

With the growing demand for better life and material culture among national residents, the government is unable to shoulder the total responsibility for supplying the entire spectrum of facilities as the traditional main facility supplier. Rather, it can only focus on addressing the supply of high-grade public service and public welfare facilities that are unlikely to be provided by the market. As for other facilities, market forces and capital may be introduced to compensate for deficiencies and afford further supply through government-guided multi-party cooperation platforms. However, it is evidently unlikely to keep profit-oriented capital from flowing and agglomerating into areas with larger populations, higher incomes, and better foundations. Spatially, this is reflected in the universal phenomenon that the central circle has a higher level of facility supply than the inner-ring circle and a much higher level than the outer-ring circle.

5.4 Research on Utilization of Public Service Facilities in Migrant Worker Settlements

First, we need to clarify that, as a supplement and extension to the analysis of the features of facility "supply" discussed in the preceding section, the topic of facility "utilization" to be explored herein does not reflect the subjective "demand" as is generally stated but rather the subjective selection and consumer demand of the migrant worker group under the constraint of current supply or "limited demand and consumer response under objective constraints" in a more strict and precise terminology.

Regarding the operational process of this section, the previously identified five categories of public service facilities that have relatively close association with the daily life of the migrant worker group are the research subjects of facility utilization. Based on the questionnaires on facility utilization among migrant workers distributed to the sample settlements, the general situation of the group, usage status of facilities, and contributing causes are analyzed. Further, the SPSS statistical software is employed to measure the correlation between the individual attributes and usage characteristics (including frequency of usage, duration of usage, and amount of consumption) of the migrant worker group. Finally, related items with higher degrees of relevancy are selected for a key-factor analysis to disclose the differentiated influence of the differences in the individual attributes of migrant workers on the characteristics of utilization of public service facilities.

5.4.1 Design of Research Methodology

Previous studies at home reveal different characteristics manifested by the migrant worker settlement group on the user side. Based on the migrant worker settlement samples previously determined, this study conducts further investigation and measurement and selects cross-categories that show significant correlations between individual attributes (e.g., gender, age, and education level) and characteristics of facility utilization (e.g., frequency of usage, duration of usage, and amount of consumption) among migrant workers, thus presenting the differentiated situation within the said group under the influence of these related items.

5.4.1.1 Research Framework

The research on facility utilization discussed in this section mainly involves the following three steps. First, through the collation of survey data, a general overview on the individual attributes of migrant workers (including social attributes, work income, and living conditions) and the characteristics of their utilization of public service facilities (including frequency of usage, duration of usage, amount of consumption, and type of consumption) is outlined. Next, a cross-analysis of "individual attributes-usage characteristics" is conducted using the SPSS statistical software to measure and select cross-categories that display significant correlations between the individual attributes and usage characteristics of migrant workers in the settlements. Finally, based on the selected categories, a more detailed study on intra-group differentiation is carried out (see Fig. 5.41).

5.4.1.2 Research Methodology

Target of Evaluation

Given that the grade of facilities serves as a critical factor in the evaluation and selection of facility among migrant workers, this section selects facilities of different grade levels as the objects of the study. Of them, medical facilities are further divided into tertiary hospitals, secondary hospitals, and community hospitals; transportation facilities are divided into subway stations and bus stops; commercial facilities are divided into large shopping malls, hypermarkets, and community supermarkets; sports facilities are divided into municipal sports facilities, district sports facilities, and community sports facilities; cultural facilities are divided into municipal cultural facilities, district cultural facilities, and community cultural facilities.

Evaluation Factors

A series of evaluation factors are required to be introduced for an analysis of the usage characteristics of different objects of study. Of these factors, frequency of usage reflects the frequency at which migrant workers utilize a facility; duration of usage reflects the amount of time migrant workers spend at a facility per usage; amount of consumption reflects the sum of money migrant workers spend on the use of a facility per consumption. Notably, the utilization of commercial facilities deserves special mention. Considering that the same type of commercial facility may involve different forms of consumption behaviors, and the same type of consumption behavior may also occur at commercial facilities of different grade levels, a new evaluation factor of "type of consumption" should be included in the study on the usage characteristics of commercial facilities, which reflects the type of consumption (purchase) migrant workers make when using a facility (see Fig. 5.42).

Additionally, according to the standards issued by the National Bureau of Statistics of China, household consumption can be categorized into the following types: food; clothing; household appliances and services; medical and health care; transportation and communication; culture, education, and entertainment; housing; and the products and services. Of them, housing consumption and consumption of transportation and communication have no association with commercial facilities, consumption of medical and health care is associated with medical facilities, and consumption of culture, education, and entertainment is associated with sports and cultural facilities. On this basis, the types of consumption involved in the utilization of commercial

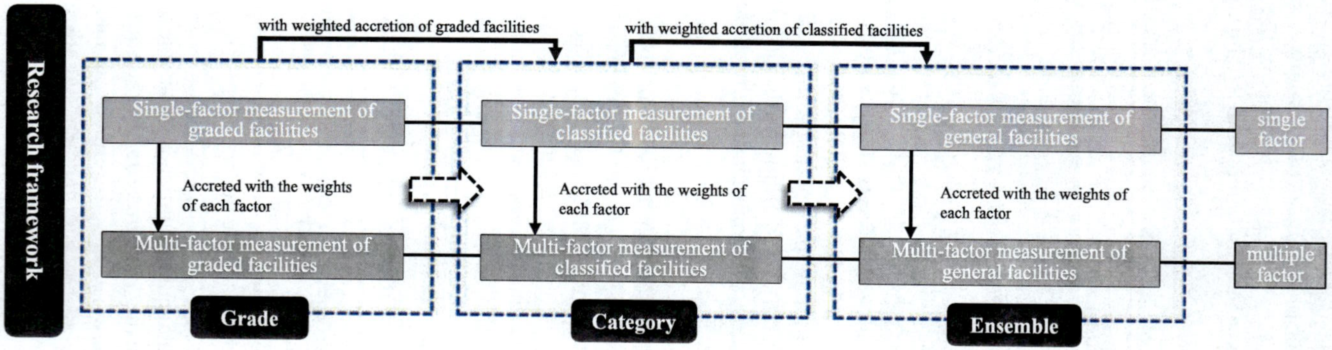

Fig. 5.41 Research framework for the level of public service facility supply in migrant worker settlements. **Source* Own elaboration

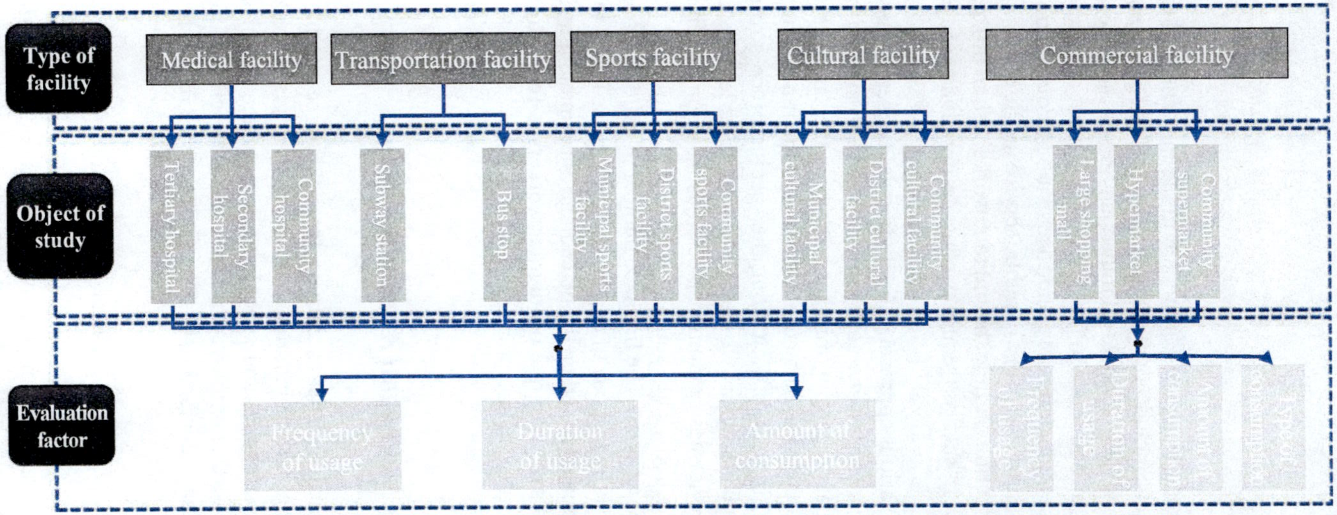

Fig. 5.42 Evaluation methodology for the utilization of public service facilities in migrant worker settlements. **Source* Own elaboration

facilities in this study include food, clothing, household appliances and services, and other products and services.

5.4.1.3 Data Collection

The questionnaires issued by the research team to the migrant worker group in the three sample settlements include two sections. The first section contains personal information questions, which facilitate an understanding of the general situation of the migrant worker group and relate to factors such as gender, age, time of residence in Nanjing, marital status, education level, and occupational composition. The second section contains questions on the usage of public service facilities by migrant workers to mainly obtain information on how the migrant worker group utilizes facilities of all types at all grade levels, such as frequency of usage, duration of usage, and amount of consumption.

For data collection, various approaches (including field observation, questionnaire statistics, and thematic interview methods) are adopted to obtain first-hand information and fundamental data regarding migrant workers in the sample settlements and their utilization of facilities. The data collected serve as the fundamental data for the analysis discussed in this section. Regarding the specific process of questionnaire distribution, this study takes migrant workers in the sample settlements as the research subjects, with 320 interviewees randomly selected using a combination of stratified and cluster sampling methods. The sampling ratio of each settlement depends on the total number of migrant workers. The finalized number of valid questionnaires totals 287 copies, reaching an effective rate of 89.7%. In addition, the actual total number of migrant workers is estimated based on the percentage of permanent residents in each settlement. The overall sampling rate of migrant workers is found to attain 6.31%, fulfilling the sampling ratio requirements (see Table 5.22).

5.4.2 *Overview of Individual Attributes*

Based on the questionnaire statistics derived from the three sample settlements, this section analyzes the current individual attributes of migrant workers from the dimensions of social, economic, and spatial attributes of the sampled group. The analysis results will be further employed as the basis for the cross-differentiation of "individual attributes-usage characteristics" in the follow-up research.

5.4.2.1 Social Attributes of Migrant Workers in Sample Settlements

In the sample settlements, male migrant workers account for a slightly higher percentage (54.4%). Majority of the migrant worker population is married (89.9%)

Table 5.22 Population size and number of sample Questionnaires in each sample settlement

Name of migrant worker settlement	Number of permanent residents and migrant workers		Number of questionnaires		Sampling rate (%)
	Permanent residents	Migrant workers	Number of questionnaires issued	Number of valid questionnaires	
Urban village of Taiping Community (outer ring)	Around 3800	Around 2200	150	133	6.04
Hongshan village (inner ring)	3106	Around 1700	110	103	6.06
Hongmiao quarter (center)	1899	Around 650	60	51	7.85

Source Own compilation and elaboration based on relevant data and actual research

and aged 30–59 (71.1%), with 6–20 years of residence in Nanjing (66.8%) and education level of junior high school (40.4%) (see Table 5.23).

5.4.2.2 Economic Attributes of Migrant Workers in Sample Settlements

Majority of the migrant workers in the sample settlements fall in the occupational category of others (including mobile vendors, contingent workers, and individuals with no fixed occupation, accounting for 29.3%), serving in individual industrial and commercial households (38.3%) and private enterprises (25.4%), and working 8–12 h per day (44.3%), 61–84 h per week (50.5%) for 6–7 days per week (61.7%), with a monthly income of CNY3,000–5,000 (47.0%) (see Table 5.24).

5.4.2.3 Spatial Attributes of Migrant Workers in Sample Settlements

Majority of the migrant workers in the sample settlements are married couples living together with their children (48.4%) in independent rental houses (77.0%) (see Table 5.25).

5.4.3 Overall Characteristics of Usage

Based on the survey statistics, an overall analysis of facility utilization among migrant workers in the sample settlements is conducted. In terms of the frequency of facility usage, the data analysis examines all questionnaire samples (287 copies in total); as

Table 5.23 Overview of social attributes of migrant workers in sample settlements

Category of attribute	Distribution graph of component attribute ratios	Major component attribute			Cumulative percentage of top three component attributes (%)
		Ranking	Category of attribute	Ratio (%)	
Gender composition	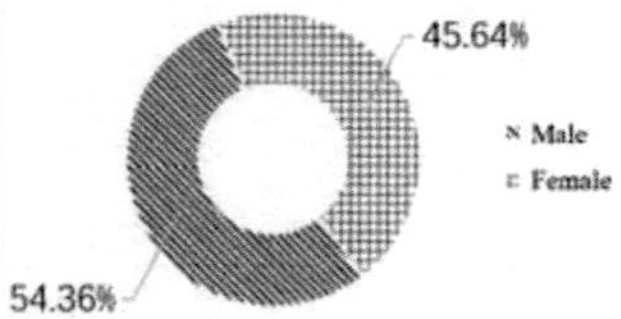	1	Male	54.4	100.0
		2	Female	45.6	
Marital status	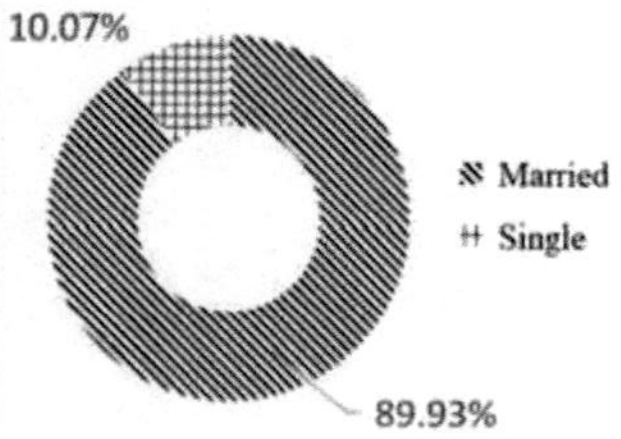	1	Married	89.9	100.0
		2	Single	10.1	
Time of residence in Nanjing	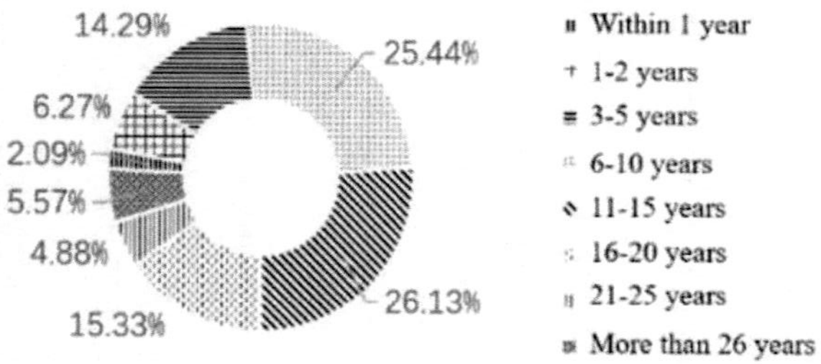	1	11–15 years	26.1	66.8
		2	6–10 years	25.4	

(continued)

Table 5.23 (continued)

Category of attribute	Distribution graph of component attribute ratios	Major component attribute			Cumulative percentage of top three component attributes (%)
		Ranking	Category of attribute	Ratio (%)	
		3	16–20 years	15.3	
Age composition	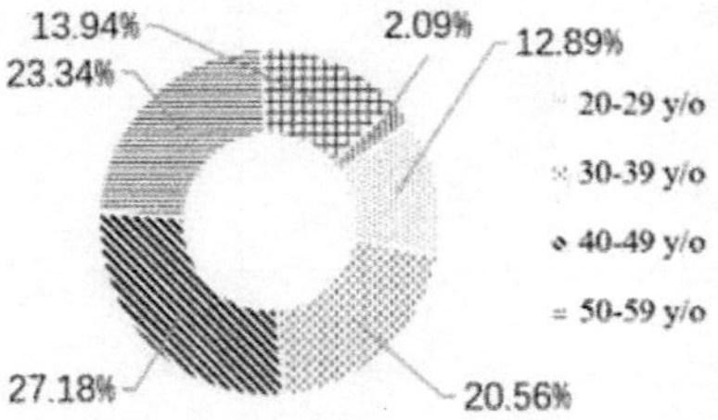	1	40–49	27.2	71.1
		2	50–59	23.3	
		3	30–39	20.6	
Education level	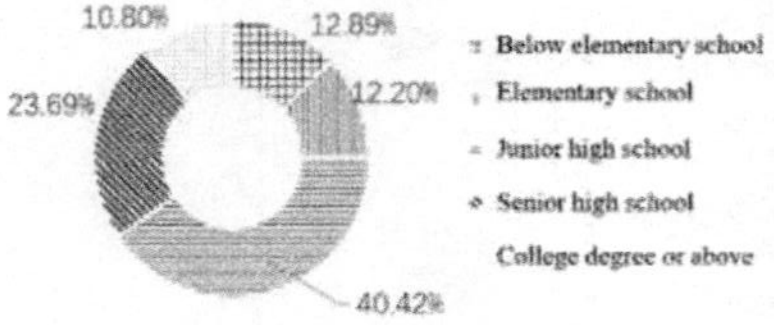	1	Junior high school	40.4	77.0
		2	Senior high school	23.7	
		3	Elementary school and below	12.9	

**Source* Sample data collected by the research team on the level of public service facility supply in migrant worker settlements in the core area of Nanjing (2018)

Table 5.24 Overview of economic attributes of migrant workers in sample settlements

Category of attribute	Distribution graph of component attribute ratios	Major component attribute			Cumulative percentage of top three component attributes (%)
		Ranking	Category of attribute	Ratio (%)	
Occupational composition		1	Others	29.3	63.0
		2	Other service industries	22.6	
		3	Wholesale and retail industry	11.1	
Nature of occupation	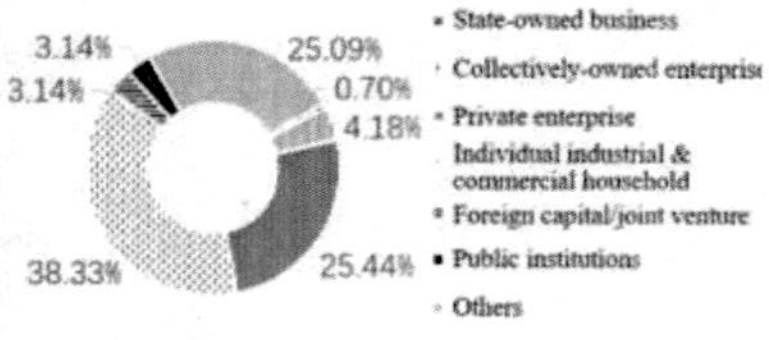	1	Individual industrial and commercial household	38.3	88.8
		2	Private enterprise	25.4	
		3	Others	25.1	
Income level	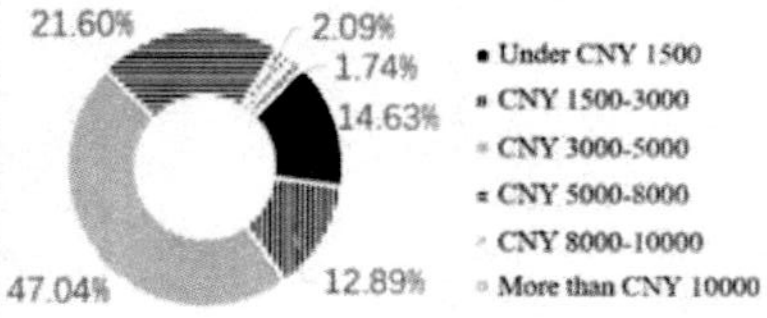	1	CNY3,000–5000	47.0	83.2
		2	CNY5,000–8000	21.6	
		3	Under CNY1,500	14.6	

(continued)

Table 5.24 (continued)

Category of attribute		Distribution graph of component attribute ratios	Major component attribute			Cumulative percentage of top three component attributes (%)
			Ranking	Category of attribute	Ratio (%)	
Working hours	Working days per week	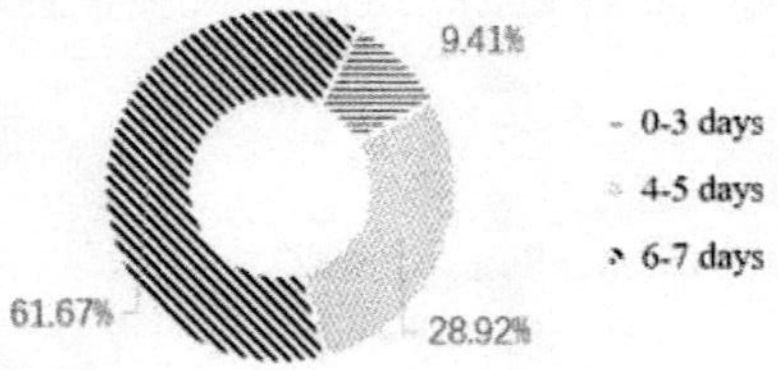	1	6–7 days	61.7	100.0
			2	4–5 days	28.9	
			3	0–3 days	9.4	
	Working hours per day	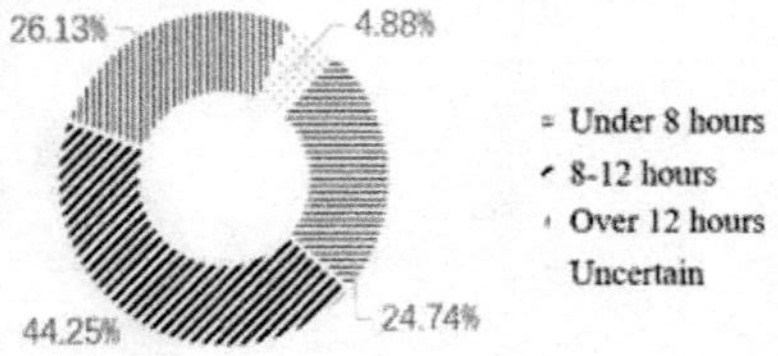	1	8–12 h	44.3	95.1
			2	Over 12 h	26.1	
			3	Under 8 h	24.	
	Working hours per week	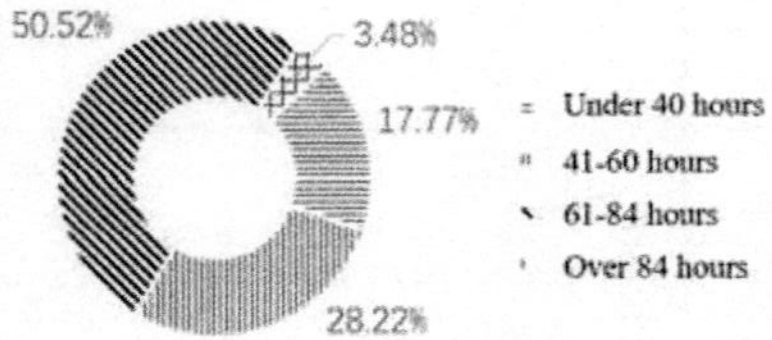	1	61–84 h	50.5	96.5
			2	41–60 h	28.2	
			3	Under 40 h	17.8	

**Source* Sample data collected by the research team on the level of public service facility supply in migrant worker settlements in the core area of Nanjing (2018)

Table 5.25 Overview of spatial attributes of migrant workers in sample settlements

Category of attribute	Distribution graph of component attribute ratios	Major component attribute			Cumulative percentage of top three component attributes (%)
		Ranking	Category of attribute	Ratio (%)	
Family structure	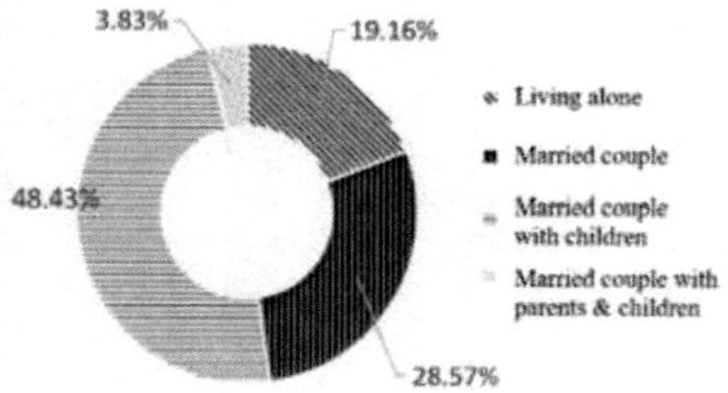	1	Married couple with children	48.4	100.0
		2	Married couple	28.6	
		3	Living alone	19.2	
Housing source	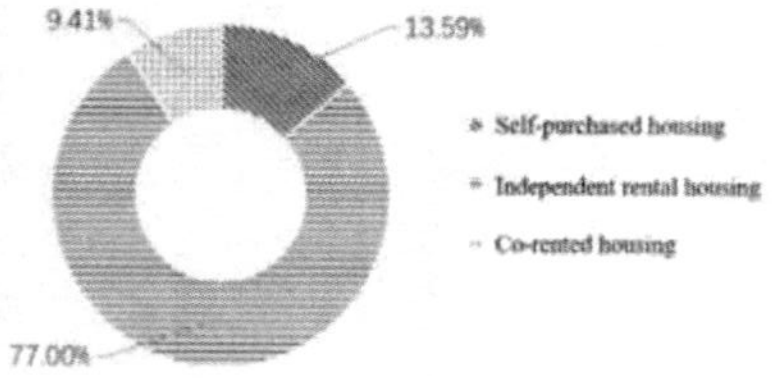	1	Independent rental housing	77.0	100.0
		2	Self-purchased housing	13.6	
		3	Co-rented housing	9.4	

*****Source* Sample data collected by the research team on the level of public service facility supply in migrant worker settlements in the core area of Nanjing (2018)

for the duration of usage and amount of consumption regarding a specific type of facility, the data analysis only focuses on the valid samples that indicate experience in using such facility.

5.4.3.1 Medical facility

The overall characteristics of medical facility usage among migrant workers in the sample settlements are as follows. The frequency of usage displays the order of "Tertiary hospitals > Community hospitals > Secondary hospitals"; the duration of usage shows the order of "Tertiary hospitals > Secondary hospitals > Community hospitals"; the amount of consumption presents the order of "Tertiary hospitals > Secondary hospitals > Community hospitals". The ratios and mean features for the characteristics of medical facility usage are shown as follows (see Table 5.26).

5.4.3.2 Transportation Facilities

The overall characteristics of transportation facility usage among migrant workers in the sample settlements are as follows. The frequency of usage displays the order of "Subway stations > Bus stops"; the duration of usage shows the order of "Bus stops > Subway stations"; the amount of consumption manifests the order of "Subway stations > Bus stops." The ratios and mean features for the characteristics of transportation facility usage are shown as follows (see Table 5.27).

5.4.3.3 Commercial Facilities

The overall characteristics of commercial facility usage among migrant workers in the sample settlements are as follows. The type of consumption displays the order of "Community supermarkets > Hypermarkets > Large shopping malls"; the frequency of usage presents the order of "Community supermarkets > Hypermarkets > Large shopping malls"; the duration of usage shows the order of "Large shopping malls > Hypermarkets > Community supermarkets"; and the amount of consumption manifests the order of "Large shopping malls > Hypermarkets > Community supermarkets". The ratios and mean features for the characteristics of commercial facility usage are shown as follows (see Fig. 5.43 and Table 5.28).

5.4.3.4 Sports Facilities

The overall characteristics of sports facility usage among migrant workers in the sample settlements are as follows. Migrant workers show average or low levels of usage frequency for sports facilities at all three grade levels, which rank in the order of "Community sports facilities > District sports facilities > Municipal sports facilities";

Table 5.26 Overview of characteristics of medical facility usage among migrant workers in sample settlements

Evaluation factor		Frequency of usage		Duration of usage		Amount of consumption	
Tertiary hospital	Summary of attribute ratios	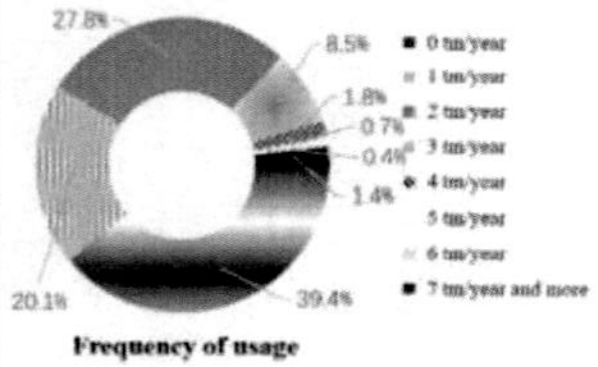		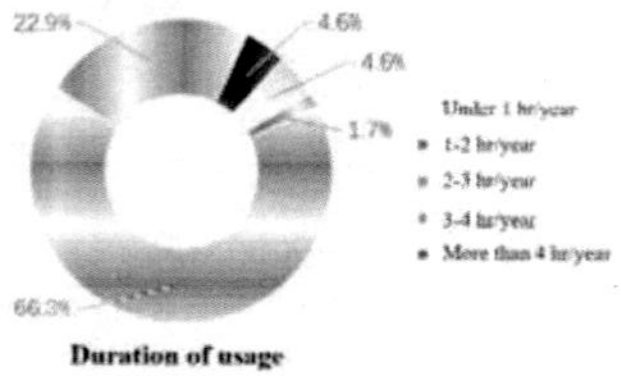		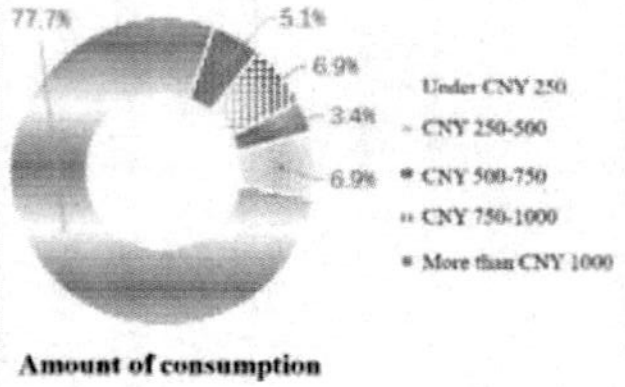	
	Top attribute	0 tm/year	39.4%	2–3 h/year	66.3%	CNY250–500	77.7%
	Mean feature	1.37 tm/year		3.21 h/year		CNY 481.4 (per usage)	
Secondary hospital	Summary of attribute ratios	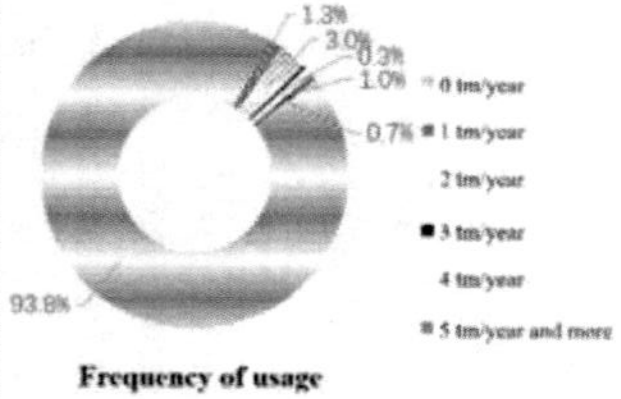		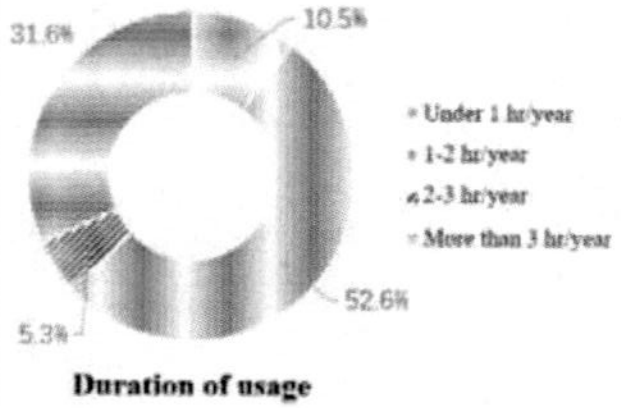		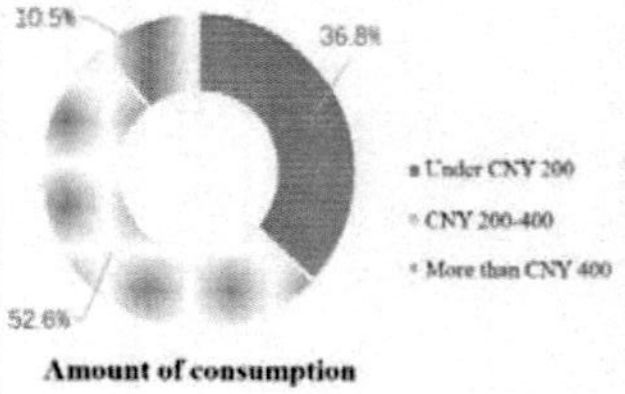	
	Top attribute	0 tm/year	93.8%	1–2 h/year	52.5%	CNY200–400	52.6%
	Mean feature	0.17 tm/year		2.26 h/year		CNY260.5 (per usage)	

(continued)

Table 5.26 (continued)

Evaluation factor		Frequency of usage		Duration of usage		Amount of consumption	
Community hospital	Summary of attribute ratios	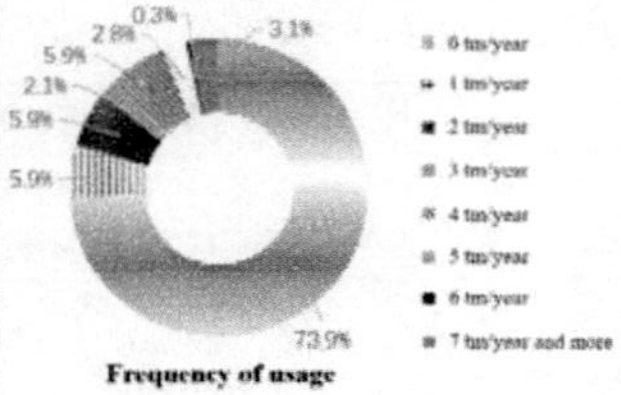		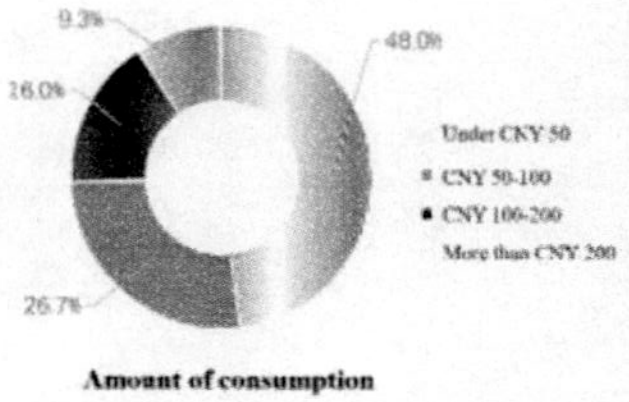		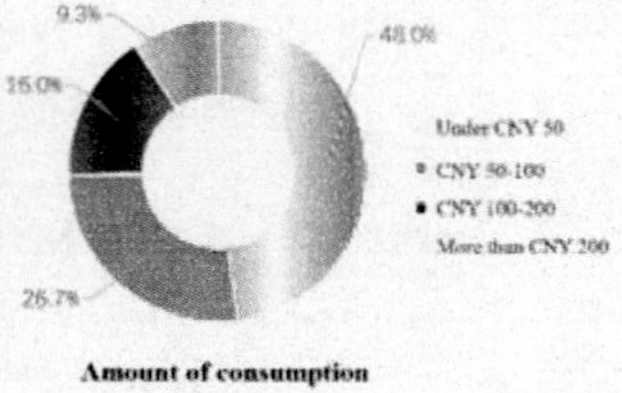	
	Top attribute	0 tm/year	73.9%	0.5–1 h/year	56.0%	Under CNY50	48.0%
	Mean feature	0.95 tm/year		1.20 h/year		CNY103.1 (per usage)	

Source Sample data collected by the research team on the level of public service facility supply in migrant worker settlements in the core area of Nanjing (2018)

Table 5.27 Overview of characteristics of transportation facility usage among migrant workers in sample settlements

Evaluation factor		Frequency of usage		Duration of usage		Amount of consumption	
Subway station	Summary of attribute ratios	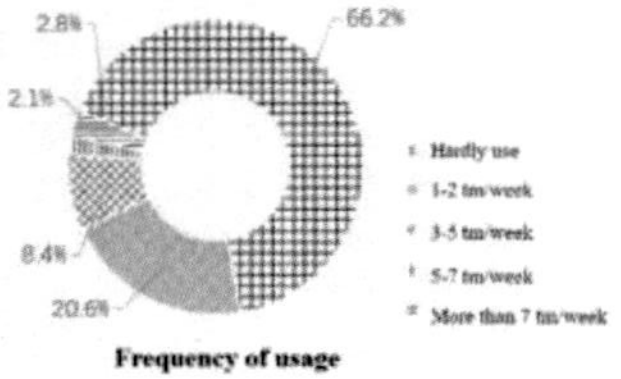		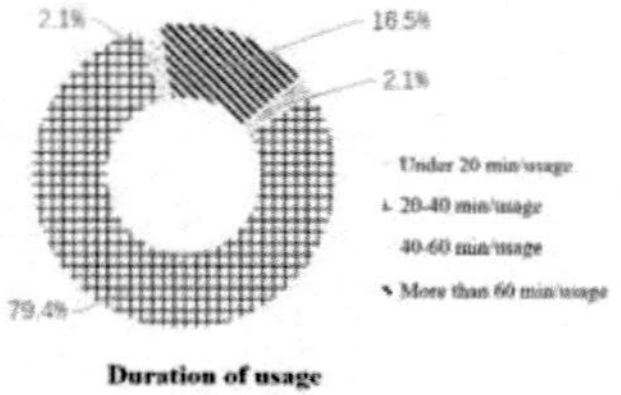		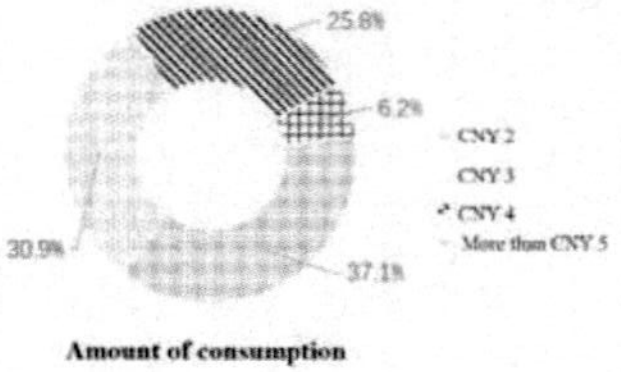	
	Top attribute	Hardly use these facilities	66.2%	20–40 min/usage	79.4%	2 CNY/usage	37.1%
	Mean feature	0.95 tm/wk		0.57 h/usage		3.02 CNY/usage	
Bus stop	Summary of attribute ratios	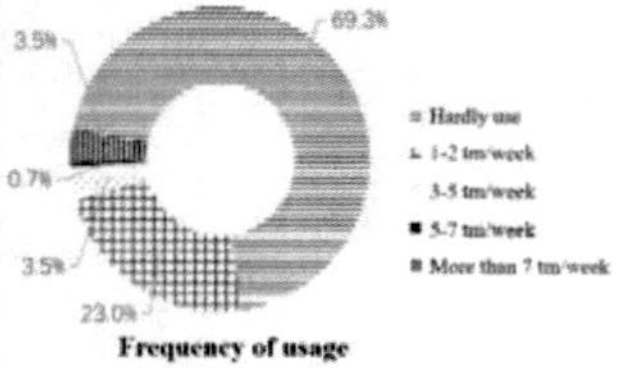		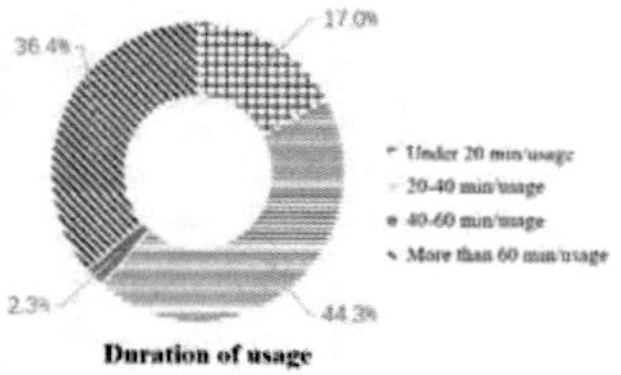		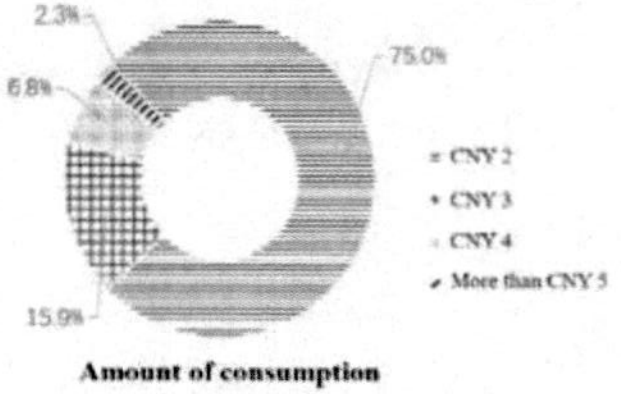	
	Top attribute	Hardly use these facilities at all	69.3%	20–40 min/usage	44.3%	2 CNY/usage	75.0%
	Mean feature	0.82 tm/wk		0.65 h/usage		2.36 CNY/usage	

* *Source* Sample data collected by the research team on the level of public service facility supply in migrant worker settlements in the core area of Nanjing (2018)

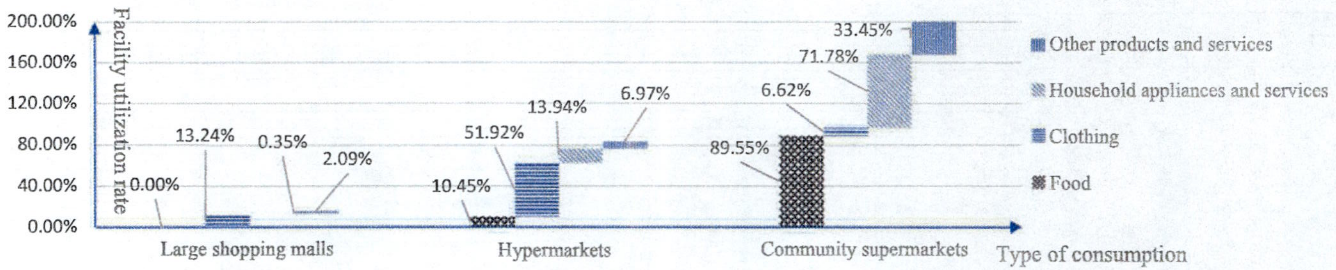

Fig. 5.43 Overview of characteristics of commercial facility usage among migrant workers in sample settlements (I)*sample data collected by the research team on the level of public service facility supply in migrant worker settlements in the core area of Nanjing (2018)

Table 5.28 Overview of characteristics of commercial facility usage among migrant workers in sample settlements (II)

Evaluation factor		Frequency of usage		Duration of usage		Amount of consumption	
Large shopping mall	Summary of attribute ratios	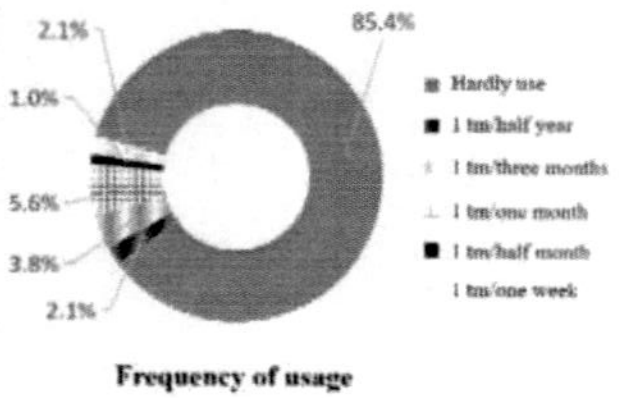		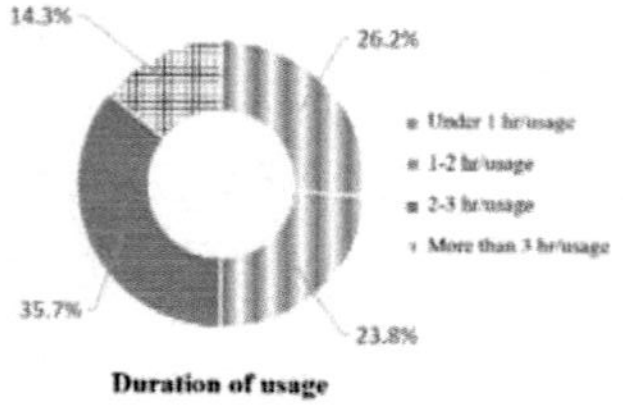		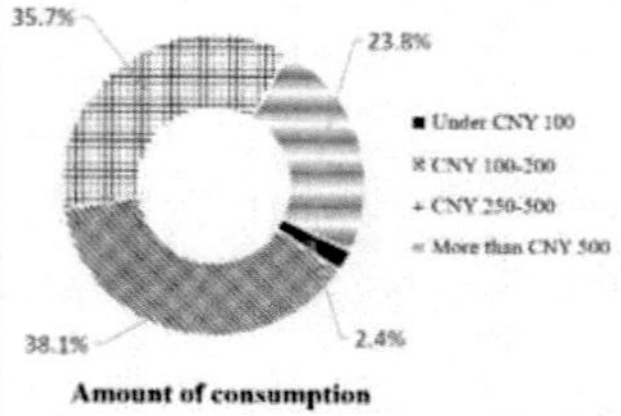	
	Top attribute	Hardly uses	85.4%	2–3 h/usage	35.7%	100–250 CNY/usage	38.1%
	Mean feature	0.04 tm/wk		2.20 h/usage		420.6 CNY/usage	
Hypermarket	Summary of attribute ratios	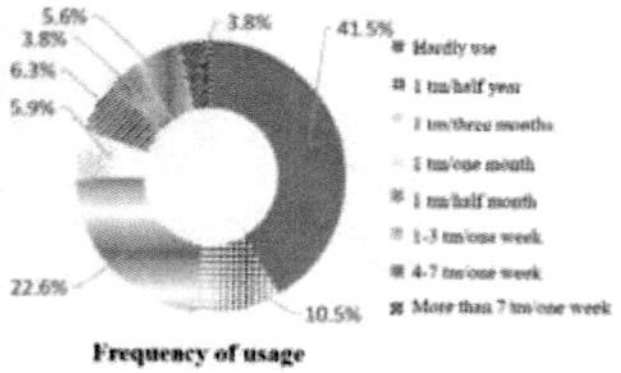		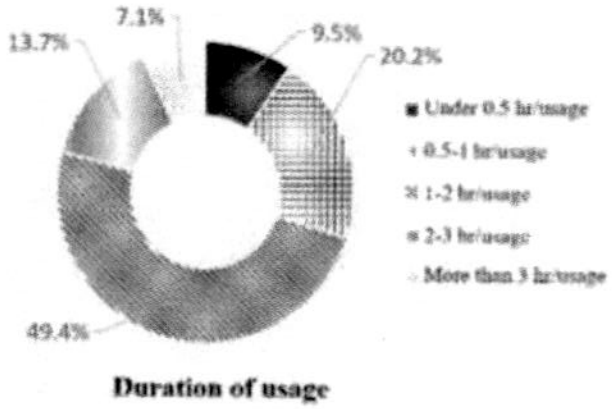		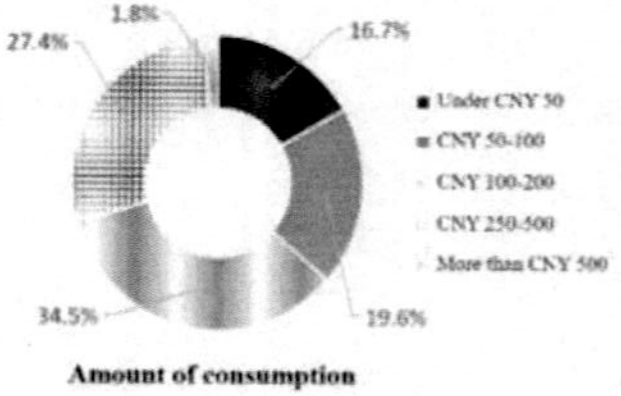	
	Top attribute	Hardly use these facilities at all	41.5%	1–2 h/usage	49.4%	100–250 CNY/usage	34.5%
	Mean feature	0.80 tm/wk		1.69 h/usage		203.6 CNY/usage	

(continued)

Table 5.28 (continued)

Evaluation factor		Frequency of usage		Duration of usage		Amount of consumption	
Community market	Summary of attribute ratios	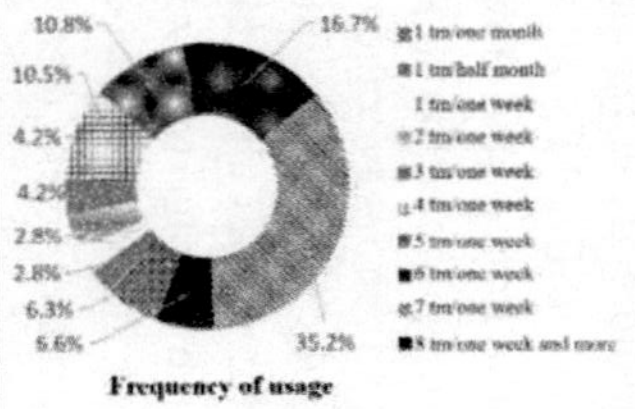		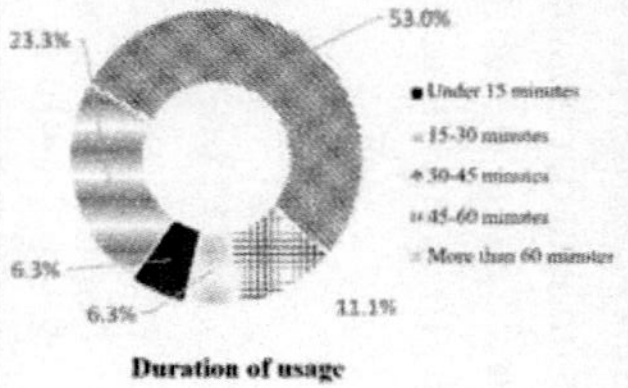		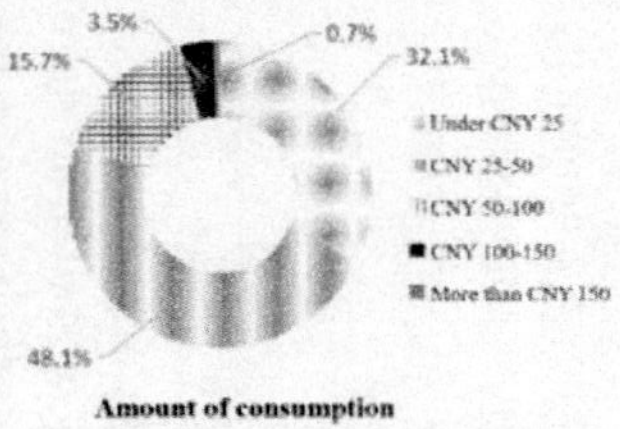	
	Top attribute	7 tm/wk	35.2%	30–45 min/usage	53.0%	25–50 CNY/usage	48.1%
	Mean feature	5.44 tm/wk		0.60 h/usage		39.0 CNY/usage	

* *Source* Sample data collected by the research team on the level of public service facility supply in migrant worker settlements in the core area of Nanjing (2018)

the duration of usage shows the order of "Municipal sports facilities > Community sports facilities > District sports facilities"; the amount of consumption displays the order of "Municipal sports facilities > District sports facilities > Community sports facilities." The ratios and mean features for the characteristics of sports facility usage are shown as follows (see Table 5.29).

5.4.3.5 Cultural Facilities

The overall characteristics of cultural facility usage among migrant workers in the sample settlements are as follows. Migrant workers also show limited levels of usage frequency for cultural facilities at all three grade levels, which rank in the order of "District cultural facilities > Community cultural facilities > Municipal cultural facilities"; the duration of usage shows the order of "Community cultural facilities > District cultural facilities > Municipal cultural facilities"; the amount of consumption manifests the order of "Municipal cultural facilities > District cultural facilities > Community cultural facilities." The ratios and mean features for the characteristics of cultural facility usage are shown as follows (see Table 5.30).

5.4.3.6 Summary

For medical facility utilization among migrant workers in the three sample settlements, tertiary hospitals achieve the highest levels in usage frequency, usage duration, and consumption amount. Regarding transportation facility utilization, subway stations show higher levels of usage frequency and consumption amount, whereas bus stops attain higher levels of usage duration. In terms of commercial facility utilization, community commercial facilities are the most commonly utilized for all forms of consumption and the most frequently utilized as well, while large-scale shopping malls manifest the highest levels in usage duration and consumption amount. As for sports facility utilization, sports facilities at all three grade levels show average to low levels of usage frequency. Of them, municipal sports facilities achieve the highest levels of usage duration and consumption amount. Finally, in terms of cultural facility utilization, cultural facilities at all three grade levels also have average to low levels of usage frequency, with community and municipal cultural facilities showing the highest levels in usage duration and consumption amount, respectively (see Table 5.31).

5.4.3.7 Analysis of the Causes of Usage Characteristics

How migrant workers in the sample settlements utilize various types of public service facilities at different grade levels is influenced by the combined effect of multiple factors. At variance with the supply of public service facilities (which is mainly affected by "external" factors such as policy systems and social economy), although

Table 5.29 Overview of characteristics of sports facility usage among migrant workers in sample settlements

	Frequency of usage			Duration of usage			Amount of consumption		
Municipal sports facility		Frequency	Percentage		Frequency	Percentage		Frequency	Percentage
	Hardly use these facilities at all	280	97.6%	1 h	4	57.1%	CNY 80	2	28.6%
	1 tm/mth	7	2.4%	2 h	3	42.9%	CNY 100	5	71.4%
				Mean duration	1.43 h		Mean amount	CNY 94.3	
District sports facility		Frequency	Percentage		Frequency	Percentage		Frequency	Percentage
	Hardly use these facilities at all	281	97.9%	30 min	4	33.3%	CNY 50	2	33.3%
	1 tm/mth	4	1.4%	60 min	2	66.7%	CNY 100	4	66.7%
	1 tm/wk	2	0.7%						
				Mean duration	40 min		Mean amount	CNY 83.3	
Community sports facility		Frequency	Percentage		Frequency	Percentage		Frequency	Percentage
	Hardly use these facilities at all	268	93.4%	Under 0.5 h	10	52.6%	CNY 10	4	21.1%
	1 tm/mth	1	0.3%	0.5–1 h	2	10.5%	CNY 20	7	36.8%
	1 tm/wk	6	2.1%	1–2 h	7	36.8%	CNY 25	1	5.3%

(continued)

Table 5.29 (continued)

	Frequency of usage			Duration of usage			Amount of consumption		
	2 tm/wk	5	1.7%				CNY 50	2	10.5%
	3 tm/wk	5	1.7%						
	4 tm/wk	2	0.7%						
				Mean duration	1.11 h		Mean amount	CNY 17.9	

* *Source* Sample data collected by the research team on the level of public service facility supply in migrant worker settlements in the core area of Nanjing (2018)

Table 5.30 Overview of characteristics of cultural facility usage among migrant workers in sample settlements

	Frequency of usage			Duration of usage			Amount of consumption		
Municipal cultural facility		Frequency	Percentage		Frequency	Percentage		Frequency	Percentage
	Hardly use these facilities at all	264	92.0%	0.5 h	1	4.3%	CNY 0	4	17.4%
	1 tm/mth	6	2.1%	1 h	3	13.0%	CNY 50	6	26.1%
	2 tm/mth	10	3.5%	1.5 h	1	4.3%	CNY	1	4.3%
	1 tm/wk	7	2.4%	2 h	16	69.6%	CNY 100	9	39.2%
				3 h	2	8.7%	CNY 120	1	4.3%
							CNY 150	2	8.7%
				Mean duration	1.43 h		Mean amount	CNY 73.5	
District cultural facility		Frequency	Percentage		Frequency	Percentage		Frequency	Percentage
	Hardly use these facilities at all	282	98.3%	1.5 h	3	60.0%	CNY 50	3	60.0%
	1 tm/mth	3	1.0%	2 h	2	40.0%	CNY 100	2	40.0%
	2 tm/mth	2	0.7%						
				Mean duration	1.50 h		Mean amount	CNY 70.0	
Community cultural facility		Frequency	Percentage		Frequency	Percentage		Frequency	Percentage
	Hardly use these facilities at all	276	96.2%	1 h	1	9.1%	CNY 40	4	36.4%
	2 tm/mth	2	0.7%	2 h	8	72.7%	CNY 60	6	54.5%
	1 tm/wk	6	2.1%	3 h	2	18.2%	CNY 100	1	9.1%
	2 tm/wk	3	1.0%						
				Mean duration	2.09 h		Mean amount	CNY 56.4	

* *Source* Sample data collected by the research team on the level of public service facility supply in migrant worker settlements in the core area of Nanjing (2018)

Table 5.31 Summary of characteristics of facility utilization among migrant workers in settlements

Type of facility		Frequency of usage			Duration of usage			Amount of consumption			Type of consumption (Usage rate)			
		Average frequency of usage	Range of highest frequency of usage		Average duration of usage	Range of longest duration of usage		Average amount of consumption	Range of greatest amount of consumption		Food	Clothing	Household appliances**appliances** and services	Other products and services
Medical facility	Tertiary hospital	1.37	0	39.4%	3.21	2–3 h	63.3%	481.4	CNY250–500	77.7%				
	Secondary hospital	0.17	0	93.8%	2.26	1–2 h	52.6%	260.5	CNY200–400	52.6%				
	Community hospital	0.95	0	73.9%	1.20	0.5–1 h	56.0%	103.1	Under CNY50	48.0%				
Transportation facility	Subway station	0.95	0	66.2%	0.57	20–40 min	79.4%	3.02	CNY2	37.1%				
	Bus stop	0.82	0	69.3%	0.65	20–40 min	44.3%	2.36	CNY2	75.0%				
Commercial facility	Large shopping mall	0.04	0	85.4%	2.20	2–3 h	35.7%	420.6	CNY100–250	38.1%	0.0%	13.2%	0.4%	2.1%
	Hypermarket	0.80	0	41.5%	1.69	1–2 h	49.4%	203.6	CNY100–250	34.5%	10.5%	51.9%	13.9%	7.0%
	Community supermarket	5.44	7 tm/wk	35.2%	0.60	30–45 min	53.0%	39.0	CNY25–50	48.1%	89.6%	6.6%	71.8%	33.5%
Sports facility	Municipal sports facility	0.00	0	97.6%	1.43	1 h	57.1%	94.3	CNY100	66.7%				
	District sports facility	0.01	0	97.9%	0.67	1 h	66.7%	83.3	CNY100	66.7%				
	Community sports facility	0.14	0	93.4%	1.11	1—2 h	36.8%	17.9	CNY20	36.8%				

(continued)

Table 5.31 (continued)

		Frequency of usage			Duration of usage			Amount of consumption			Type of consumption (Usage rate)
Cultural facility	Municipal cultural facility	0.05	0	92.0%	1.43	2 h	69.6%	82.2	CNY100	39.2%	
	District cultural facility	0.01	0	98.3%	1.50	1.5 h	60.0%	70	CNY50	60.0%	
	Community cultural facility	0.05	0	96.2%	2.09	2 h	72.7%	56.4	CNY60	54.5%	

* *Source* Sample data collected by the research team on the level of public service facility supply in migrant worker settlements in the core area of Nanjing (2018)

facility utilization is also affected by such "external" factors, the "internal" willingness of an individual to use a certain facility plays a more decisive role. For different categories and grades of facilities, distinct factors come into play in the level of willingness of migrant workers to use a specific facility, and these factors exert varying degrees of influence on the use of different types of facilities. The contributing factors for facility utilization are generally summarized into the following four aspects (see Fig. 5.44):

Inadequate Security Systems

During the period of the planned economy, China established an urban–rural dual structure. This structure has continued to exert its restraint and influence until this day when the economic system has undergone transformations and massive numbers of migrant workers are flooding into cities. The result is that migrant workers are unable to fully enjoy the same social protection as local urban residents; moreover, they have lost some protection and benefits previously obtained in rural areas. (In case of pension insurance, the Ministry of Human Resources and Social Security (MOHRSS) stipulates that "pension insurance may only be enjoyed after 15 consecutive years of insurance premium payment. For migrant workers with high levels of mobility, this requirement is difficult to fulfill. Data shows that migrant workers achieve a low coverage rate of less than 20% for the insurance policy of "five social insurance and one provident housing fund.") The social security systems have not yet been equalized among urban and rural residents, which has imperceptibly increased the levels of uncertainty and risk for migrant workers living and working in the city. In the meantime, the threshold for migrant workers to use facilities (such as medical insurance) has been passively raised and their willingness to use public service facilities has thus been reduced (e.g., they may prevent uncertainty by reducing consumption and increasing savings).

Uneven Utilization of Facilities

The overall level of facility supply has been enhanced under the impact of market orientation. As the improvement of quality has mostly been achieved through increase in prices, the issue of inequality has inevitably arisen in the utilization of public service facilities. Although market regulation plays a pivotal role in urban social and economic development, the reality of the wealth gap cannot be disregarded. Moreover, market regulation, a profit-oriented mechanism, is unlikely to ensure an equitable distribution of fundamental and public welfare facilities for disadvantaged urban groups represented by migrant workers. Evidently, there is still a long way to go before a more even distribution of facilities among various groups is achieved through rational allocation of resources, stronger government guidance, and proper market regulation, to ultimately realize greater equality in the utilization of facilities.

Disparity in Consumption Habits

Undeniably, migrant workers living in the settlements have become an inseparable part of the urban system in terms of economic sectors and living space. However, the traditional lifestyle habits and perceptions they developed previously have continued

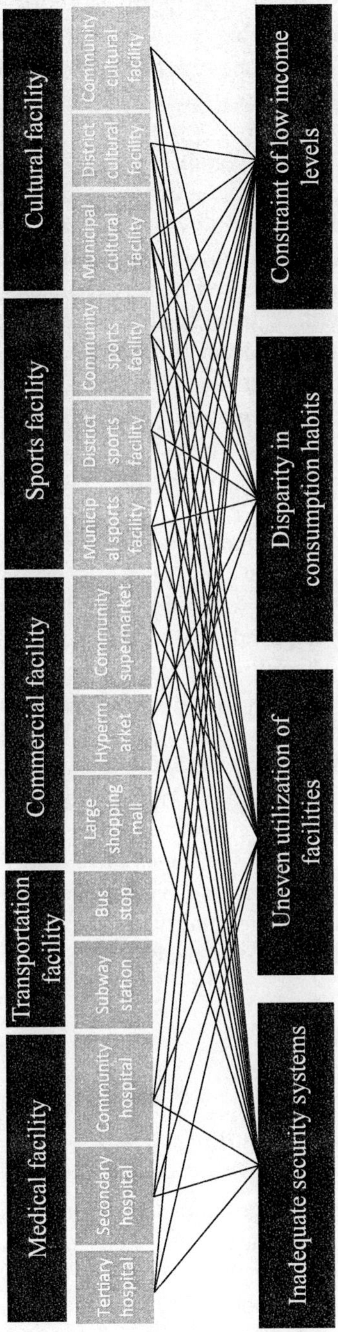

Fig. 5.44 Causes of differences in utilization of public service facilities. **Source* Own elaboration

to exert a sustained and intrinsic impact on their consumption habits (such as lack of consumer awareness about cultural and sports activities; the survey found that migrant workers in the settlements achieve low utilization rates in terms of sports and cultural facilities at 2.4%, 2.1%, and 6.6% for the three grades of sports facilities and 8.0%, 1.7%, and 3.8% for the three grades of cultural facilities). Such intrinsic differences in the consumption habits between migrant workers and urban residents make it difficult for the former to feel a genuine sense of belonging to the settlements and cities they reside in. Moreover, without active guidance from external entities, such as society, enterprises, and communities, to integrate migrant workers into urban cultural life, the barrier of duality between urban and rural consciousness is difficult to eliminate. In addition, for migrant workers in the settlements, the various consumption behaviors do not square with their original intention of controlling the cost of living and making profit in the city. This unconformity could also potentially impede their level of subjective initiative in utilizing public service facilities (particularly sports and cultural facilities) before a reasonably appropriate consumption habit is established.

Constraint of Low Income Levels

Although migrant workers have obtained relatively higher incomes in big cities compared with what they may earn in their places of origin, they still fall in the low-income class among the various groups living in the city. According to statistics, urban residents in Nanjing achieved a disposable personal income per capita of CNY54,538 in 2017, (Nanjing Municipal Bureau of Statistics 2017) with an average monthly income of approximately CNY4,545, while the minimum monthly wage in Nanjing was CNY2,020. (http://www.nanjing.gov.cn Nanjing Municipal People's Government) By contrast, the average monthly income of migrant workers in China was registered at merely CNY3,485 in 2017. While it met the minimum wage standard in Nanjing, this amount was still significantly lower than the average income of the residents therein. Such low income levels have also fundamentally limited migrant workers' ability and willingness to use public service facilities, resulting in their preference for low-priced, average-quality, mass-produced, and wide-ranging public service products to fulfill their basic needs of life.

5.4.4 Correlation Measurement

5.4.4.1 Methodology for Correlation Measurement

For an in-depth investigation of the usage characteristics regarding the five categories of facilities in migrant worker settlements, this study conducts an SPSS bivariate correlation analysis on the individual attributes of migrant workers in the settlements and the characteristics of their daily facility utilization. The analysis aims to select more pertinent attribute standards for the study of intra-group differentiation with to facility utilization.

For example, to analyze the differences in the usage of bus stations among different migrant workers in the settlements, the data is processed and summarized based on the results of thematic interviews and survey statistics into five categories: "hardly use these facilities" (69.3%), "1–2 times per week" (23.0%), "3–5 times per week" (3.5%), "5–7 times per week" (0.7%), and "7 times or more per week" (3.5%). Further, a bivariate correlation analysis is conducted using Kendall tau rank correlation coefficient, and calculations are performed using SPSS (results are shown in Table 5.32). The findings show that the indicators of gender ($p = 0.002$), marital status ($p = 0.000$), business type ($p = 0.002$), and the number of working days per week ($p = 0.001$) bear significant correlations with the frequency of bus station usage.

Table 5.32 Bivariate correlation analysis of individual attributes of migrant workers in settlements to their frequency of bus stop usage

Kendall tau rank correlation coefficient

Category of facility: transportation facility	Object of study: bus stop		Evaluation factor: frequency of usage
Individual attribute of migrant workers in the settlements	Correlation coefficient	Sig. (2-tailed)	N
Gender	−0.174**	0.002	287
Age group	−0.016	0.749	287
Marital status	0.222**	0	287
Time of residence in Nanjing	0.105*	0.025	287
Education level	0.119*	0.018	287
Type of occupation	−0.059	0.231	287
Type of business unit	−0.159**	0.002	287
Household composition	−0.032	0.544	287
Number of working days per week	0.187**	0.001	287
Number of working hours per day	−0.023	0.653	287
Total working hours per week	0.091	0.063	287
Monthly income	0.012	0.819	287
Housing situation	0.022	0.689	287

**Source* Own elaboration

***Correlation is significant at the 0.01 level of confidence (2-tailed); *Correlation is significant at the 0.05 level of confidence (2-tailed)

5.4.4.2 Results for Correlation Measurement of "Individual Attributes-Usage Characteristics"

In a similar vein, the bivariate correlation analysis based upon the Kendall tau rank correlation coefficient is conducted to explore the internal correlation between other individual attributes of migrant workers in settlements and various public service facilities at all grade levels. As such, the relevant individual attributes that have a significant impact are summarized and determined (see Table 5.33). On the basis of the results, a more specific and detailed discussion on differentiation among the user group is provided below.

5.4.5 Differentiation Among User Groups

5.4.5.1 Selection of Objects of Study

By studying the intra-group differentiation in usage characteristics under the influence of individual attributes, this section further reveals the differences in the utilization of facilities within the settlement samples based on the knowledge of the characteristics of overall utilization. To ensure that the conclusions of the intra-group differentiation analysis achieves a high level of focus and representativeness, a screening mechanism for the typicality of the cross-categories of "individual attributes-usage characteristics" should be further carried out in accordance with the following two selection principles:

Based on the results of the cross-analysis of the correlation between "individual attributes and usage characteristics" discussed above, the top five types of individual attributes with the highest cumulative total number of cross-categories that show significant correlations between "individual attributes and usage characteristics" are selected as specific objects of study (accounting for 60.3% of all cross-categories that show significant correlations). The selected attributes are listed as follows in descending order: education level, marital status, income level, time of residence in Nanjing, and occupational composition.

It is necessary to ensure an adequate coverage of facility types and adequate number of samples used (see Fig. 5.45). Take individual attribute A as an example. As shown in Fig. 4.4, if the factors of usage frequency and consumption type are taken as the objects of study, the total numbers of samples N_1 and N_2 include all survey samples (N = 287), and the number of sub-samples for each individual attribute is A_i (i = 1, 2, 3, …, n); if the factors of usage duration and consumption amount are taken as the objects of study, the total numbers of samples N3 and N4 include all survey samples that indicate use of a specific facility ($N' \leq 287$), and the number of sub-samples for each individual attribute is A'_i (i = 1, 2, 3, …, n). According to the survey results and general research experience, the actual total number of samples for each evaluation factor, N_i (i = 1, 2, 3, 4), should assuredly be no less than 30% of the total number of survey samples (i.e., $N_i \geq 86.1$), and the actual number of

Table 5.33 Results for bivariate correlation analysis of various individual attributes of migrant workers in settlements and their usage of public service facilities

Type of facility	Object of study	Evaluation factor	Gender composition	Age composition	Marital status	Time of residence in Nanjing	Education level	Occupational composition	Nature of business unit	Income level	Number of working days per week	Number of working hours per day	Total working hours per week	Family structure	Housing source
Medical facility	Tertiary hospital	Frequency of usage					■			■				■	■
		Duration of usage					■			■					■
		Amount of consumption					■				■		■	■	■
	Secondary hospital	Frequency of usage													
		Duration of usage													
		Amount of consumption													
	Community hospital	Frequency of usage			■			■							
		Duration of usage						■							
		Amount of consumption						■							

(continued)

Table 5.33 (continued)

Type of facility	Object of study	Evaluation factor	Gender composition	Age composition	Marital status	Time of residence in Nanjing	Education level	Occupational composition	Nature of business unit	Income level	Number of working days per week	Number of working hours per day	Total working hours per week	Family structure	Housing source
Transportation facility	Subway station	Frequency of usage		■	■		■	■	■	■					
		Duration of usage		■	■		■	■	■	■					■
		Amount of consumption		■	■		■	■	■	■					
	Bus stop	Frequency of usage	■		■				■		■				
		Duration of usage	■						■		■				
		Amount of consumption	■		■				■		■				
Commercial facility	Large shopping mall	Type of consumption	■	■	■		■								
		Frequency of usage	■	■	■		■								

(continued)

Table 5.33 (continued)

Type of facility	Object of study	Evaluation factor	Gender composition	Age composition	Marital status	Time of residence in Nanjing	Education level	Occupational composition	Nature of business unit	Income level	Number of working days per week	Number of working hours per day	Total working hours per week	Family structure	Housing source
		Duration of usage		■	■		■								
		Amount of consumption		■	■		■								
	Hypermarket	Type of consumption				■		■						■	
		Frequency of usage				■	■	■						■	
		Duration of usage													
		Amount of consumption													
	Community supermarket	Type of consumption			■	■									
		Frequency of usage				■									

(continued)

Table 5.33 (continued)

Type of facility	Object of study	Evaluation factor	Gender composition	Age composition	Marital status	Time of residence in Nanjing	Education level	Occupational composition	Nature of business unit	Income level	Number of working days per week	Number of working hours per day	Total working hours per week	Family structure	Housing source
		Duration of usage				■				■					■
		Amount of consumption				■				■					
Sports facility	Municipal sports facility	Frequency of usage					■			■					
		Duration of usage					■			■					
		Amount of consumption					■			■					
	District sports facility	Frequency of usage				■									
		Duration of usage				■									
		Amount of consumption				■									
	Community sports facility	Frequency of usage													

(continued)

Table 5.33 (continued)

Type of facility	Object of study	Evaluation factor	Gender composition	Age composition	Marital status	Time of residence in Nanjing	Education level	Occupational composition	Nature of business unit	Income level	Number of working days per week	Number of working hours per day	Total working hours per week	Family structure	Housing source
		Duration of usage													
		Amount of consumption							■						
Cultural facility	Municipal cultural facility	Frequency of usage		■	■	■	■	■	■	■				■	■
		Duration of usage		■	■	■	■	■	■	■				■	■
		Amount of consumption		■	■	■	■	■	■	■				■	■
	District cultural facility	Frequency of usage													
		Duration of usage													
		Amount of consumption													

(continued)

Table 5.33 (continued)

Type of facility	Object of study	Evaluation factor	Gender composition	Age composition	Marital status	Time of residence in Nanjing	Education level	Occupational composition	Nature of business unit	Income level	Number of working days per week	Number of working hours per day	Total working hours per week	Family structure	Housing source
	Community facility	Frequency of usage			■										
		Duration of usage			■										
		Amount of consumption								■				■	
Cumulative total number of cross-categories that show significant correlations between individual attributes and usage characteristics			5	10	16	12	17	11	10	14	4	0	1	8	8

**Source* Own elaboration
"■" indicates a significant correlation at the 0.01 level of confidence (2-tailed)

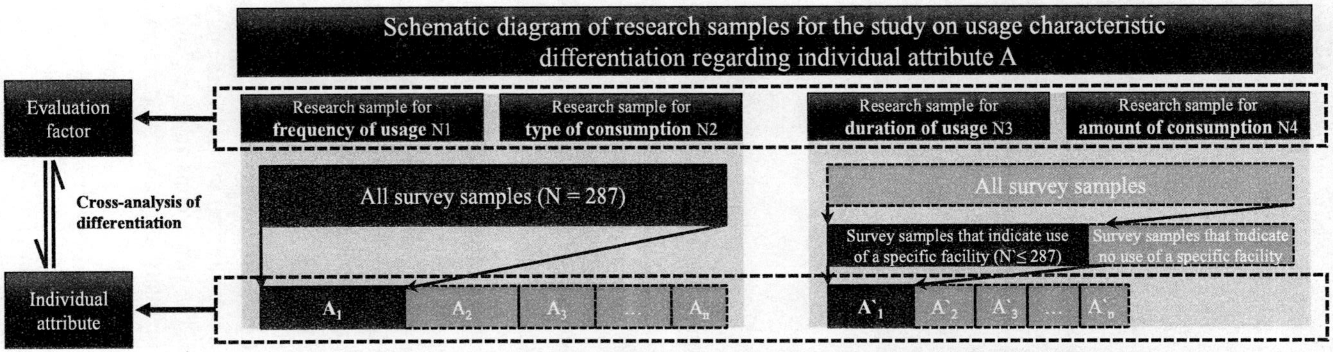

Note: Ai, A`i (i = 1, 2, 3,..., n) refer to the cumulative total number of samples that indicate use of sub-attribute i of individual attribute A.

Fig. 5.45 Schematic diagram of relevant samples for the study on differentiation within the user group. **Source* Own elaboration samples for each individual attribute, A_i and A'_i ($i = 1, 2, 3,\ldots, n$), should be no less than 10% of the total number of survey samples (i.e., A_i, and $A'_i \geq 28.7$).

Pursuant to the selection principles described above, only five types of individual attributes and the usage characteristics whose sample size N_i fulfills the requirements are selected. The cross-categories for the study of intra-group differentiation are established as follows (for specific details on the determination of A_i and A'_i, please refer to the research process for individual attributes illustrated below). As for the codes for all grades and types of facilities, the coding methodology stated previously is referred to and adopted (see Fig. 2.17). For a clearer demonstration, all evaluation factors are indicated by the codes shown in Table 5.34:

Notably, the five categories and twelve types of facilities selected in the above table involve a total of 21 evaluation factors, each of which contains two to ten sub-attributes. To facilitate the analysis and enhance information identifiability of the figures in the following discussion, the common evaluation grading legends and usage intensity information shown from Figs. 5.46, 5.47, 5.48, 5.49 and 5.50 are summarized as follows (see Table 5.35), where ········ indicates high-intensity usage, ········ indicates medium-intensity usage, and ········ indicates low-intensity usage:

5.4.5.2 Marital Status

Eleven evaluation factors show significant correlations with marital status and have adequate total number of samples for each usage characteristic (i.e., N_i fulfills the requirements stated above). Of them, the sub-attributes that meet the requirements include two types: married (89.9%) and single (10.1%). On this basis, a statistical analysis is conducted on the sub-attributes of evaluation factors for usage characteristics, excluding those whose sample sizes A_i and A'_i fall short of the requirements. Finally, the cumulative results of each sub-attribute are converted into percentages for standardized comparison (see Fig. 5.46). The findings show that the intra-group differentiation in usage characteristics under the influence of marital status involves four types of facilities: medical, transportation, commercial, and cultural facilities.

In terms of medical facilities, married migrant workers show a significantly lower frequency of overall usage of community hospitals (A3) (at least once) than single migrant workers. In addition, the former also achieve lower levels of medium- and high-intensity usage than the latter.

Regarding transportation facilities, married migrant workers display significantly lower frequency of overall usage of subway stations (B1) and bus stops (B2) (at least once) than single migrant workers. The two groups achieve close levels in medium-intensity usage; in terms of high-intensity usage, married migrant workers attain a lower level than single migrant workers.

With regard to commercial facilities, compared with single migrant workers, married migrant workers achieve significantly lower levels of usage frequency in the overall usage, medium-intensity usage, and high-intensity usage of large shopping malls (C1). However, married and single migrant workers manifest similar structures of consumption type for community supermarkets (C3), with the former achieving

Table 5.34 Selection of "individual attribute-usage characteristic" cross-categories for study on intra-group differentiation

Type of facility	Object of study	Evaluation factor	Marital status	Time of residence in Nanjing	Education level	Occupational composition	Income level
Medical facility (A)	Tertiary hospital (A1)	Frequency of usage (A1-Frequency)			■		■
		Duration of usage (A1-Duration)			■		■
		Amount of consumption (A1-Amount)			■		
	Community hospital (A3)	Frequency of usage (A3-Frequency)	■			■	
		Duration of usage (A3-Duration)				■	
		Amount of consumption (A3-Amount)				■	
Transportation facility (B)	Subway station (B1)	Frequency of usage (B1-Frequency)	■		■	■	■
		Duration of usage (B1-Duration)	■		■	■	■
		Amount of consumption (B1-Amount)	■		■	■	■

(continued)

Table 5.34 (continued)

Type of facility	Object of study	Evaluation factor	Marital status	Time of residence in Nanjing	Education level	Occupational composition	Income level
	Bus stop (B2)	Frequency of usage (B2-Frequency)	■				
		Amount of consumption (B2-Amount)	■				
Commercial facility (C)	Large shopping mall (C1)	Type of consumption (C1-type)	■		■		
		Frequency of usage (C1-Frequency)	■		■		
	Hypermarket (C2)	Type of consumption (C2-type)		■		■	
		Frequency of usage (C2-Frequency)		■	■	■	
	Community supermarket (C3)	Type of consumption (C3-type)	■	■			
		Frequency of usage (C3-Frequency)		■			

(continued)

Table 5.34 (continued)

Type of facility	Object of study	Evaluation factor	Marital status	Time of residence in Nanjing	Education level	Occupational composition	Income level
		Duration of usage (C3-Duration)		■			■
		Amount of consumption (C3-Amount)		■			■
Sports facility (D)	Municipal sports facility (D1)	Frequency of usage (D1-Frequency)			■		■
	District sports facility (D2)	Frequency of usage (D2-Frequency)		■			
Cultural facility (E)	Municipal cultural facility (E1)	Frequency of usage (E1-Frequency)	■	■	■	■	■
	Community cultural facility (E3)	Frequency of usage (E3-Frequency)	■				
Total			11	8	11	9	9
			48				

Source Own elaboration

Table 5.35 Intra-group differentiation in usage characteristics

Evaluation grading legend										
A1-Frequency	No use	1 tm/yr	2 tm/yr	3 tm/yr	4 tm/yr	5 tm/yr	6 tm/yr	7 tm/yr and above		
A1-Duration	Under 1 h	1–2 h	2–3 h	3–4 h	Over 4 h					
A3-Frequency	No use	1 tm/yr	2 tm/yr	3 tm/yr	4 tm/yr	5 tm/yr	6 tm/yr	7 tm/yr and above		
B1-Frequency	No use	1–2 tm/wk	3–5 tm/wk	6–7 tm/wk						
B1-Duration	Under 20 min	20–40 min	40–60 min	Over 60 min						
B1-Amount	CNY2	CNY3	CNY4	Over CNY5						
B1-Frequency	No use	1–2 tm/wk	3–5 tm/wk	6–7 tm/wk	7 tm/wk and above					
B1-Duration	Under 20 min	20–40 min	40–60 min	Over 60 min						
B1-Amount	CNY2	CNY3	CNY4	CNY5						
C1-Type	Food	Clothing	Household supplies, appliances, and services	Other products and services						
C1-Frequency	No use	2 tm/yr	4 tm/yr	1 tm/mth	2 tm/mth	1 tm/wk				
C2-Type	Food	Clothing	Household supplies, appliances, and services	Other products and services						

(continued)

Table 5.35 (continued)

Evaluation grading legend										
C2-Frequency	No use	2 tm/yr	4 tm/yr	1tm/mth	2 tm/mth	1–3 tm/wk	4–7 tm/wk	Over 7 tm/wk		
C3-Type	Food	Clothing	Household supplies, appliances, and services	Other products and services						
C3-Frequency	1 tm/mth	2 tm/mth	1 tm/wk	2 tm/wk	3 tm/wk	4 tm/wk	5 tm/wk	6 tm/wk	7tm/wk	8 tm/wk and above
C3-Duration	Under 15 min	15–30 min	30–45 min	45–60 min	Over 60 min					
C3-Amount	Under CNY25	CNY25–50	CNY50–100	CNY100–150	Over CNY150					
D1-Frequency	No use	1 tm/mth								
D2-Frequency	No use	1 tm/mth	1 tm/wk							
E1-Frequency	No use	1 tm/mth	2 tm/mth	1 tm/wk						
E3-Frequency	No use	2 tm/mth	1 tm/wk	2 tm/wk						

* *Source* Own elaboration

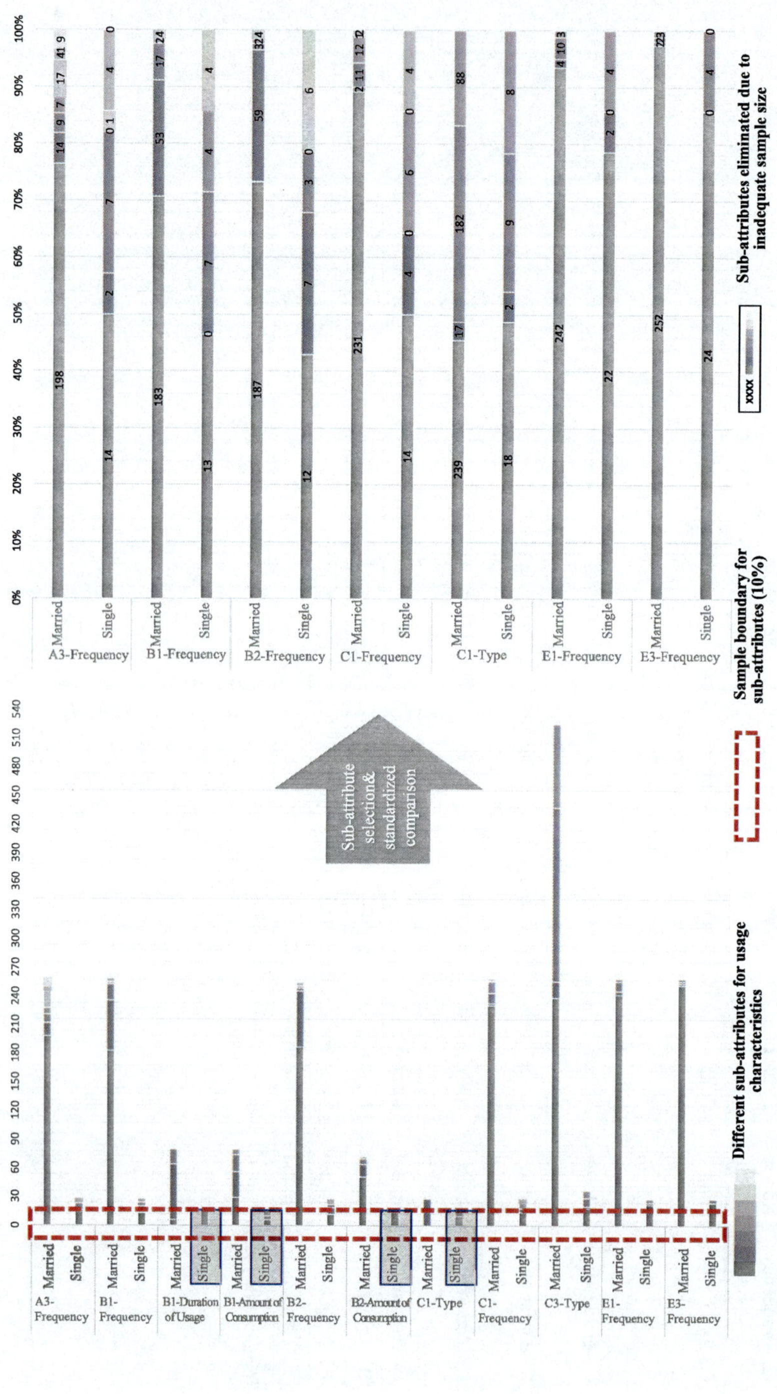

Fig. 5.46 Intra-group differentiation in usage characteristics under the influence of marital status. **Source* Own elaboration

slightly higher percentages in the consumption of household appliances, supplies, and services than the latter.

As for cultural facilities, married and single migrant workers show relatively low levels of usage frequency of municipal cultural facilities (E1) and community cultural facilities (E3). Here, married migrant workers achieve lower levels than their single counterparts.

5.4.5.3 Education Level

Eleven evaluation factors show significant correlations with education level and have adequate total number of samples for each usage characteristic (i.e., N_i fulfills the requirements stated above). Of them, the sub-attributes that meet the requirements include four types: below primary school (12.9%), primary school (12.2%), middle school (40.4%), high school (23.7%), and college and above (10.8%). On this basis, a statistical analysis is conducted on the sub-attributes of evaluation factors for usage characteristics, excluding those whose sample sizes A_i and A'_i fall short of the requirements. Finally, the cumulative results of each sub-attribute are converted into percentages for standardized comparison (see Fig. 5.47). Findings show that the intra-group differentiation in usage characteristics under the influence of education level involves five types of facilities: medical, transportation, commercial, sports, and cultural.

In terms of medical facilities, the frequency of overall usage and duration of usage of tertiary hospitals (A3) ascend with the increase in the education level of migrant workers. Those with an education level of middle or high school show the highest levels in medium-intensity usage, whereas those with an education level of college and above achieve the highest level in high-intensity usage. As for the amount of consumption, a slight declining trend is observed with the increase in the education level but to a trifling extent.

Regarding transportation facilities, migrant workers in the settlements display an ascending trend in the frequency of overall usage of subway stations (B1) with increase in their education level. In addition, migrant workers with higher education levels also account for greater percentages of medium- and high-intensity usage, with the most significant percentage achieved by those with education level of college and above.

In terms of commercial facilities, an ascending trend is observed in the frequency of overall usage of large shopping malls (C1) and hypermarkets (C2) with increase in education level, and the highest level is achieved by those with education level of college and above.

As for sports facilities, the education level of migrant workers in the settlements has limited influence on the frequency of usage of municipal sports facilities (D1), with a relatively low level achieved on the whole.

With regard to cultural facilities, an ascending trend is observed in the frequency of overall usage of municipal cultural facilities (E1) among migrant workers with

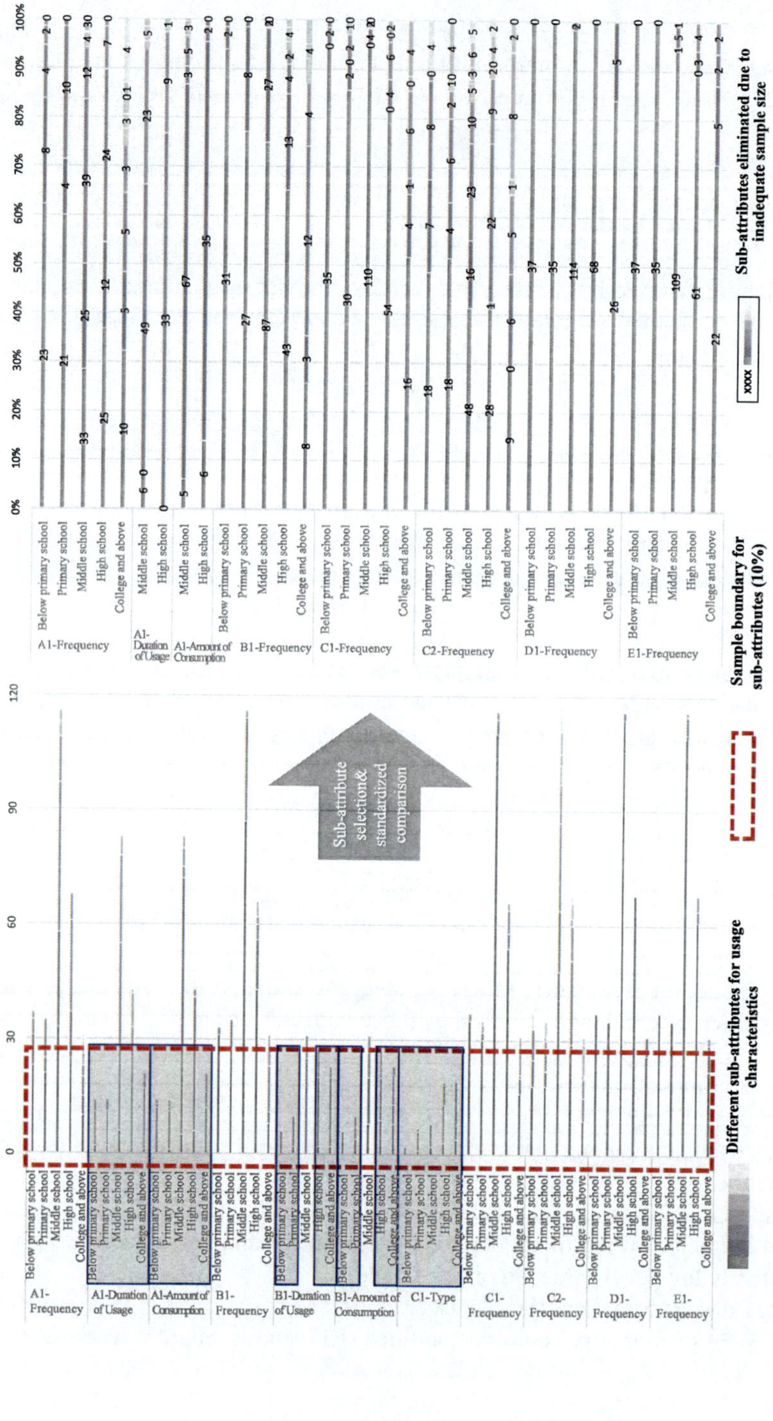

Fig. 5.47 Intra-group differentiation in usage characteristics under the influence of education level. **Source* Own elaboration

education level of middle school and above, and the highest level is achieved by those with education level of college and above.

5.4.5.4 Occupational Composition

Nine evaluation factors show significant correlations with occupational composition and have adequate total number of samples for each usage characteristic (i.e., N_i fulfills the requirements stated above). Of them, the sub-attributes that meet the requirements include five types: construction industry (10.8%), wholesale and retail industry (11.1%), hospitality industry (10.5%), industry of other service businesses (22.6%), and others (29.3%). On this basis, a statistical analysis is conducted on the sub-attributes of evaluation factors for usage characteristics, excluding those whose sample sizes A_i and A'_i fall short of the requirements. Finally, the cumulative results of each sub-attribute are converted into percentages for standardized comparison (see Fig. 5.48). The findings show that the intra-group differentiation in usage characteristics under the influence of occupational composition involves four types of facilities: medical, transportation, commercial, and cultural facilities.

In terms of medical facilities, migrant workers serving in the wholesale and retail industry display the highest frequency of overall usage of community hospitals (A3), whereas those working in the other four categories of industries achieve lower and close levels in the frequency of overall usage. In addition, those working in wholesale and retail industries, other service businesses, and others show high-intensity usage.

Regarding transportation facilities, migrant workers serving in wholesale and retail industries attain the highest frequency of overall usage of subway stations (B1), those working in the construction industry achieve an average level, while those working in the other three industry categories display low levels of usage frequency.

As for commercial facilities, migrant workers serving in these five categories of industries display similar structures in the type of consumption in hypermarkets (C2), with those working in the hospitality industry accounting for a slightly lower percentage in clothing consumption. Migrant workers serving in construction, wholesale, and retail industries and hospitality achieve higher levels in the frequency of overall usage of hypermarkets (C2), and migrant workers serving in wholesale and retail industries manifest a significantly higher level than those working in the other four industry categories in medium-intensity usage.

As regards cultural facilities, migrant workers serving in these five industry categories all display average to low levels of frequency of overall usage of municipal cultural facilities (E1). Among them, those working in other service businesses almost never use municipal cultural facilities, while those working in the hospitality industry achieve a slightly higher usage rate than those working in the construction, and wholesale and retail industries.

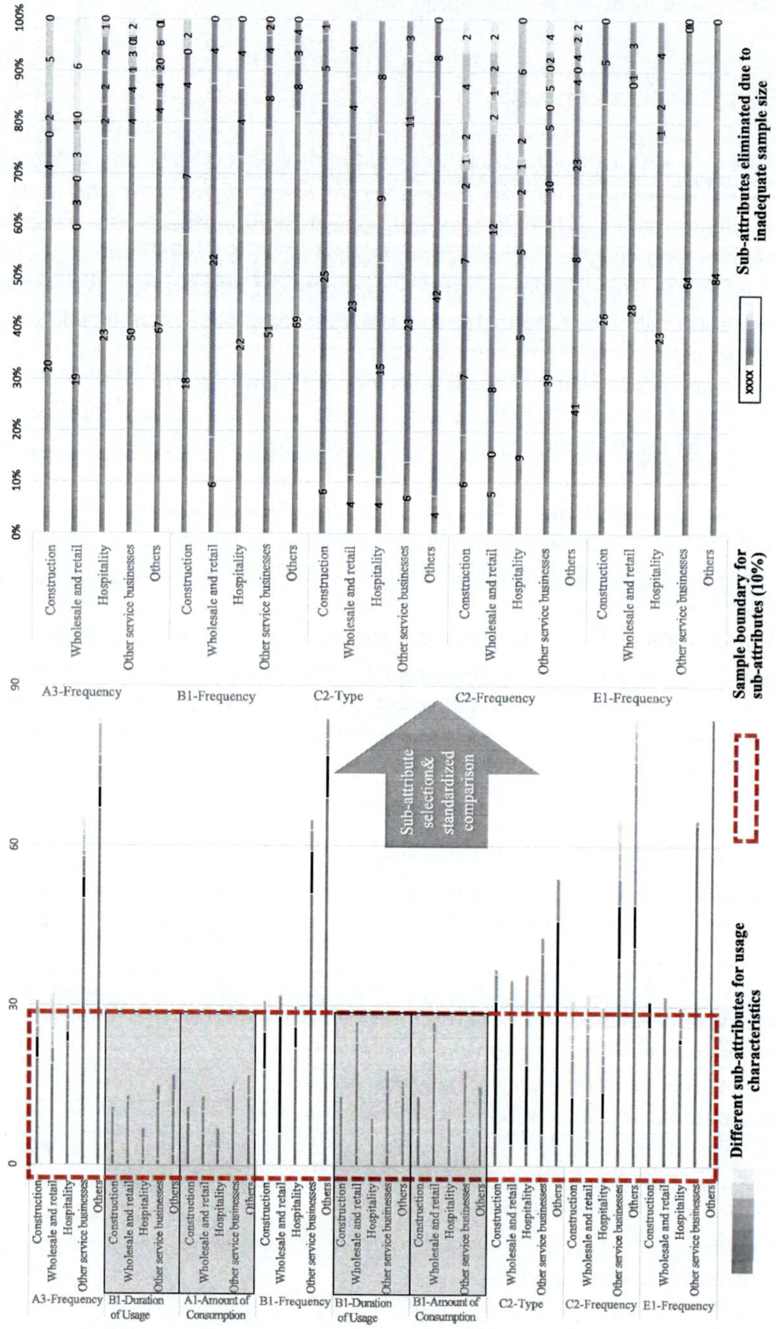

Fig. 5.48 Intra-group differentiation in usage characteristics under the influence of occupational composition. **Source* Own elaboration

5.4.5.5 Income Level

Nine evaluation factors show significant correlations with income level and have adequate total number of samples for each usage characteristic (i.e., N_i fulfills the requirements stated above). Of them, the sub-attributes that meet the requirements include four types: under CNY1,500 per month (14.6%), CNY1,500–3,000 per month (12.9%), CNY3,000–5,000 per month (47.0%), and CNY5,000–8,000 per month (21.6%). On this basis, a statistical analysis is conducted on the sub-attributes of evaluation factors for usage characteristics, excluding those whose sample sizes A_i and A'_i fall short of the requirements. Finally, the cumulative results of each sub-attribute are converted into percentages for standardized comparison (see Fig. 5.49). The findings show that the intra-group differentiation in usage characteristics under the influence of income level involves five types of facilities: medical, transportation, commercial, sports, and cultural.

In terms of medical facilities, migrant workers in the settlements display a slightly increasing trend in the frequency of overall usage of tertiary hospitals (A1) with the ascent in income levels. Migrant workers with different income levels show subtle differences in medium-intensity usage and a slightly more distinct rising trend in high-intensity usage. In addition, migrant workers display broadly similar lengths of usage of tertiary hospitals (A1), with slightly higher percentages achieved at both ends (low and high levels of usage duration) with the increase in income levels.

Regarding commercial facilities, migrant workers in the settlements manifest a slightly descending trend in the low-level duration and a slightly ascending trend in the high-level duration of usage of community supermarkets (C3). In terms of the amount of consumption, migrant workers display a descending trend in low-level amounts and a slightly ascending trend in high-level amounts with the increase in income levels.

With regard to sports facilities, migrant workers in the settlements show average to low levels of frequency of overall usage of municipal sports facilities (D1), with a slight increase observed with the ascent in income levels above CNY3,000.

In terms of cultural facilities, migrant workers in the settlements also display average to low levels of frequency of overall usage of municipal cultural facilities (E1), with an ascent observed with the rise in income levels above CNY3,000.

5.4.5.6 Time of Residence in Nanjing

Eight evaluation factors show significant correlations with the time of residence in Nanjing and have adequate total number of samples for each usage characteristic (i.e., N_i fulfills the requirements stated above). Of them, the sub-attributes that meet the requirements include four types: 3–5 years (14.3%), 6–10 years (25.4%), 11–15 years (26.1%), and 16–20 years (15.3%). On this basis, a statistical analysis is conducted on the sub-attributes of evaluation factors for usage characteristics, excluding those whose sample sizes A_i and A'_i fall short of the requirements. Finally, the cumulative

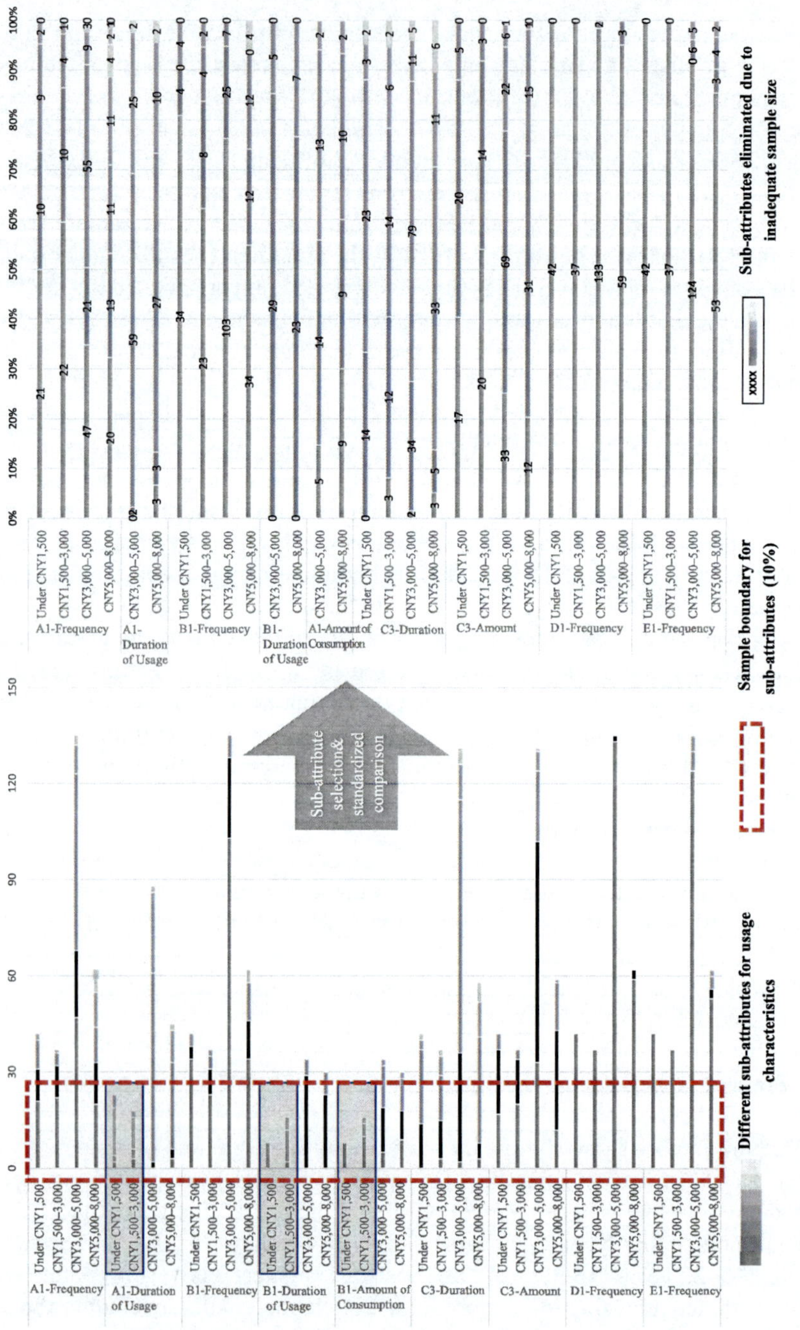

Fig. 5.49 Intra-group differentiation in usage characteristics under the influence of income level. **Source* Own elaboration

results of each sub-attribute are converted into percentages for standardized comparison (see Fig. 5.50). The findings show that the intra-group differentiation in usage characteristics under the influence of income level involves three types of facilities: commercial, sports, and cultural.

With regard to commercial facilities, migrant workers in the settlements show no significant difference in the composition of the type of consumption in hypermarkets (C2) and community supermarkets (C3) with the increase in their time of residence in Nanjing, and an ascending trend is observed in the percentages of medium- and high-intensity usage for the former and a descending trend is noted in the same for the latter. In addition, migrant workers display a slightly ascending trend in medium- and high-level duration of usage and high-level consumption amounts in community supermarkets.

In terms of sports facilities, all migrant workers show average to low frequency of usage of district sports facilities (D2) regardless of their time of residence in Nanjing, except that a small number of those with 11–15 years of residence in Nanjing show low-level frequency of usage.

Regarding cultural facilities, migrant workers also display relatively low levels of frequency of usage of municipal cultural facilities (E1), except those with 3–5 years of residence in Nanjing.

5.5 Match Patterns of Facility "Supply-Utilization" in Migrant Worker Settlements and Corresponding Strategies for Enhancing Facility Supply

In Section 5.3, the level of public service facility supply in migrant worker settlements is evaluated through the statistical analysis and measurement of public service facilities in and around the settlements. In Sect. 5.4, based on the survey of migrant workers in the settlements, the characteristics of facility utilization among this disadvantaged group (i.e., limited demand under objective constraints) and the intra-group differentiation under the influence of individual attributes are revealed. In this section, the aforementioned research conclusions on facility supply and utilization in migrant settlements will be combined to derive match patterns between the two. On the basis of the match classification, corresponding strategies for enhancing different types of public service facilities in migrant worker settlements will be proposed.

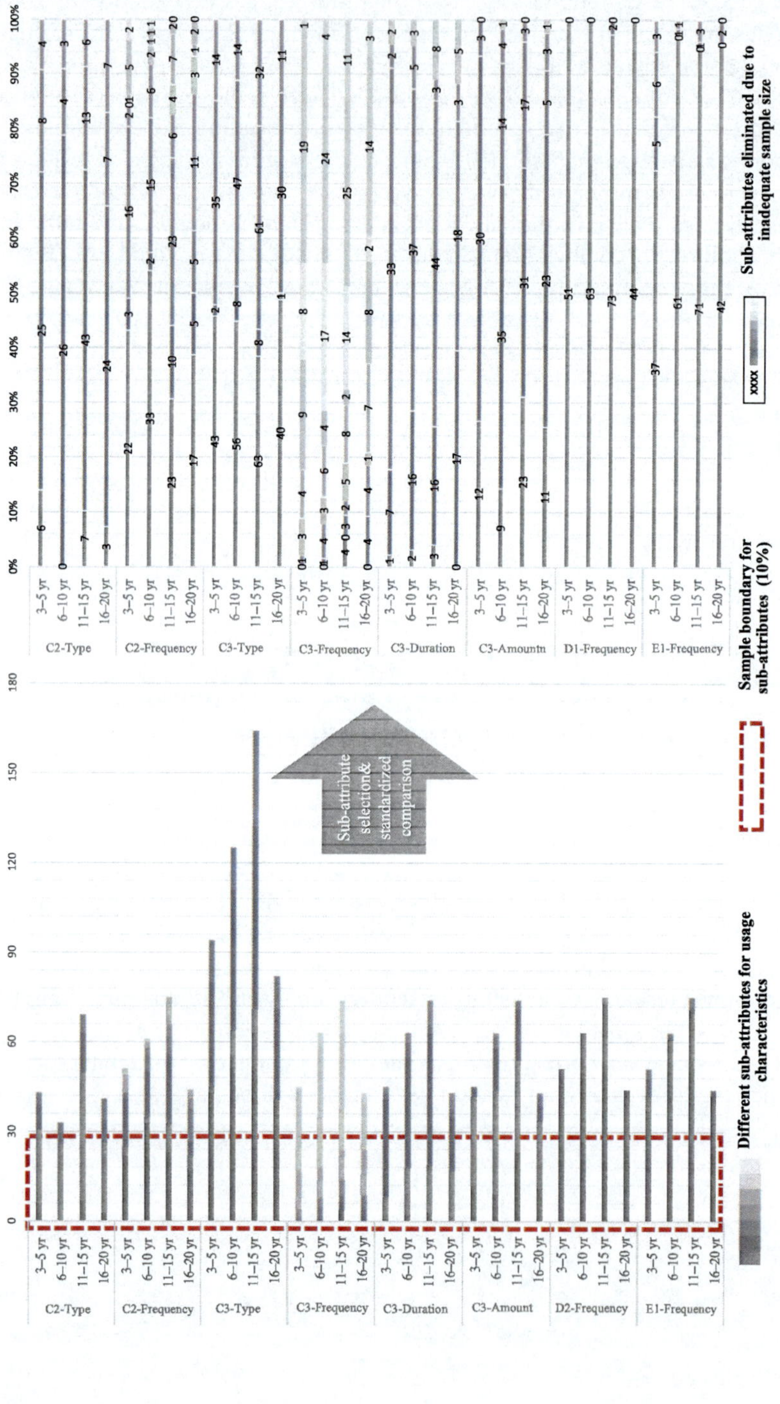

Fig. 5.50 Intra-group differentiation in usage characteristics under the influence of time of residence in Nanjing. **Source* Own elaboration

5.5.1 Research on Match Patterns of Facility "Supply-Utilization"

5.5.1.1 Research Methodology

Research Framework

To shed light on the matching characteristics regarding the supply and utilization of public service facilities in migrant worker settlements, a BCG Matrix is constructed for the match patterns of "classified and graded" facilities on the basis of the research results previously derived, specifically, the average scores for various grades and types of facilities in different circles.[4] Four "supply-utilization" match types are developed for the four matrix quadrants. They are "high supply-high utilization, high supply-low utilization, low supply-high utilization, and low supply-low utilization" (see Fig. 5.51). These four patterns further serve as the foundation for subsequent strategy formulation.

Sources of Data

Regarding the sources of data, the evaluation results of the supply level of facilities by category are illustrated in Table 5.18 and those of facilities by grade are shown in Table 5.16.

To evaluate the level of utilization of facilities by category and grade, a similar methodology is adopted. First, 25 copies of the weighted scoring questionnaire for the four types of factors (i.e., frequency of usage, duration of usage, amount of consumption, and type of consumption) are distributed to urban planning specialists. Subsequently, in pursuance of the 1–9 ratio scale in the analytic hierarchy process (AHP), the judgment matrix and weight of each factor are established. Finally, the nondimensionalized evaluation factors are measured, with the scores of overall facility usage calculated using the weighted accumulation mechanism. The weight of each factor and scores of facility utilization thus derived are shown as follows (see Table 5.36).

Matrix Construction

First, the average level of supply of public service facilities in migrant worker settlements is taken as the boundary between high and low levels of supply (i.e., $x = 0$ in the match matrix). In the matrix, the average supply level defined for facilities by category is the average score for all five categories of facilities in all three circles, and the average supply level defined for facilities by grade is the average score for different grades of facilities in the three circles. Furthermore, the average

---

[4] The BCG Matrix was created in 1970 by Bruce Henderson, a well-known American management expert and founder of the Boston Consulting Group. It is also known as the Growth Market Share Matrix, the Boston Consulting Group method, the four quadrants analysis approach, product structure management method, etc. The purpose of the matrix is to provide different definitions and corresponding strategic countermeasures for the four quadrants where corporate products are located.

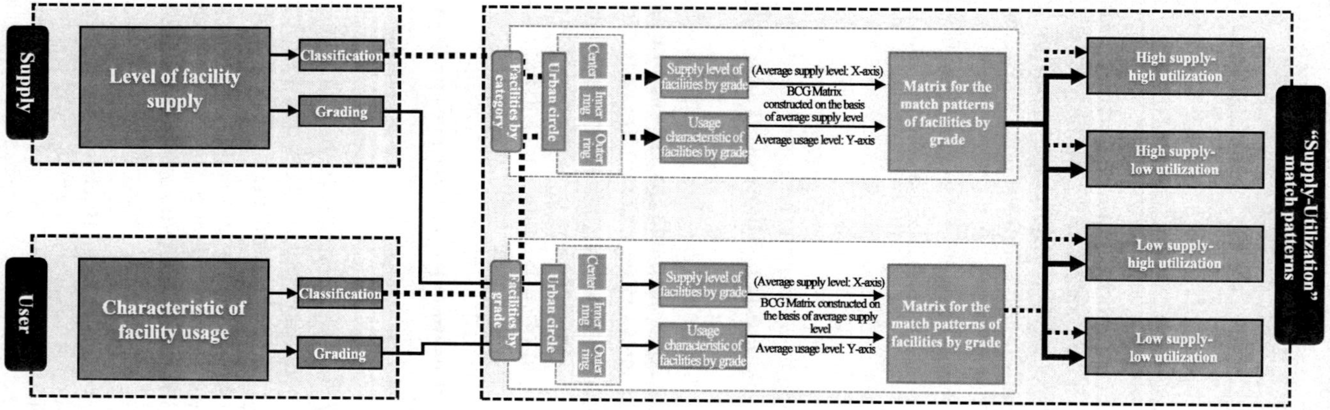

Fig. 5.51 Evaluation framework for "supply-utilization" match patterns of public service facilities in migrant worker settlements. **Source* Own elaboration

Table 5.36 Scores for level of facility utilization among migrant workers in settlements

Facility by category		Facility by grade			Single factor for utilization evaluation			
Facility	Score	Facility	Weight	Score	Type of single factor	Weight of single factor	Nondimensionalized single factor	Weighted score of single factor
Medical facility	0.51248	Tertiary hospital	0.50404	0.50263	Frequency of usage	0.51158	0.02776	0.01420
					Duration of usage	0.33888	1.00000	0.33888
					Amount of consumption	0.14954	1.00000	0.14955
		Secondary hospital	0.21011	0.32151	Frequency of usage	0.51158	0.00346	0.00177
					Duration of usage	0.33888	0.70472	0.23882
					Amount of consumption	0.14954	0.54115	0.08093
		Community hospital	0.28586	0.67022	Frequency of usage	0.51158	1.00000	0.51158
					Duration of usage	0.33888	0.37367	0.12663
					Amount of consumption	0.14954	0.21411	0.03202
Transportation facility	0.92623	Subway station	0.42616	0.95632	Frequency of usage	0.51989	1.00000	0.51989
					Duration of usage	0.36407	0.88002	0.32039

(continued)

Table 5.36 (continued)

Facility by category		Facility by grade			Single factor for utilization evaluation			
Facility	Score	Facility	Weight	Score	Type of single factor	Weight of single factor	Nondimensionalized single factor	Weighted score of single factor
					Amount of consumption	0.11604	1.00000	0.11604
		Bus stop	0.57384	0.90388	Frequency of usage	0.51989	0.86439	0.44939
					Duration of usage	0.36407	1.00000	0.36407
					Amount of consumption	0.11604	0.77929	0.09043
Commercial facility	0.43036	Large shopping mall	0.27203	0.33361	Type of consumption	0.21996	0.00149	0.00034
					Frequency of usage	0.45044	0.00816	0.00367
					Duration of usage	0.08000	1.00000	0.08000
					Amount of consumption	0.24961	1.00000	0.24961
		Hypermarket	0.38902	0.24971	Type of consumption	0.21996	0.00793	0.00174
					Frequency of usage	0.45044	0.14615	0.06583

(continued)

Table 5.36 (continued)

Facility by category		Facility by grade			Single factor for utilization evaluation			
Facility	Score	Facility	Weight	Score	Type of single factor	Weight of single factor	Nondimensionalized single factor	Weighted score of single factor
					Duration of usage	0.08000	0.76719	0.06138
					Amount of consumption	0.24961	0.48380	0.12076
		Community supermarket	0.33895	0.71534	Type of consumption	0.21996	1.00000	0.21996
					Frequency of usage	0.45044	1.00000	0.45044
					Duration of usage	0.08000	0.27213	0.02177
					Amount of consumption	0.24961	0.09285	0.02318
Sports facility	0.59992	Municipal sports facility	0.28314	0.51885	Frequency of usage	0.48156	0.00085	0.00041
					Duration of usage	0.35002	1.00000	0.35002
					Amount of consumption	0.16842	1.00000	0.16842
		District sports facility	0.23182	0.31290	Frequency of usage	0.48156	0.00146	0.00070

(continued)

Table 5.36 (continued)

Facility by category		Facility by grade			Single factor for utilization evaluation			
Facility	Score	Facility	Weight	Score	Type of single factor	Weight of single factor	Nondimensionalized single factor	Weighted score of single factor
					Duration of usage	0.35002	0.46667	0.16334
					Amount of consumption	0.16842	0.88384	0.14885
		Community sports facility	0.48504	0.78442	Frequency of usage	0.48156	1.00000	0.48156
					Duration of usage	0.35002	0.77368	0.27081
					Amount of consumption	0.16842	0.19029	0.03205
Cultural facility	0.56648	Municipal cultural facility	0.41138	0.49898	Frequency of usage	0.48615	0.02027	0.00985
					Duration of usage	0.37899	0.93478	0.35427
					Amount of consumption	0.13486	1.00000	0.13486
		District cultural facility	0.36406	0.43827	Frequency of usage	0.48615	0.00258	0.00126
					Duration of usage	0.37899	0.85000	0.32214

(continued)

Table 5.36 (continued)

Facility by category		Facility by grade			Single factor for utilization evaluation			
Facility	Score	Facility	Weight	Score	Type of single factor	Weight of single factor	Nondimensionalized single factor	Weighted score of single factor
					Amount of consumption	0.13486	0.85185	0.11488
		Community cultural facility	0.22456	0.89797	Frequency of usage	0.48615	1.00000	0.48615
					Duration of usage	0.37899	1.00000	0.37899
					Amount of consumption	0.13486	0.24339	0.03282

* *Source* Own elaboration

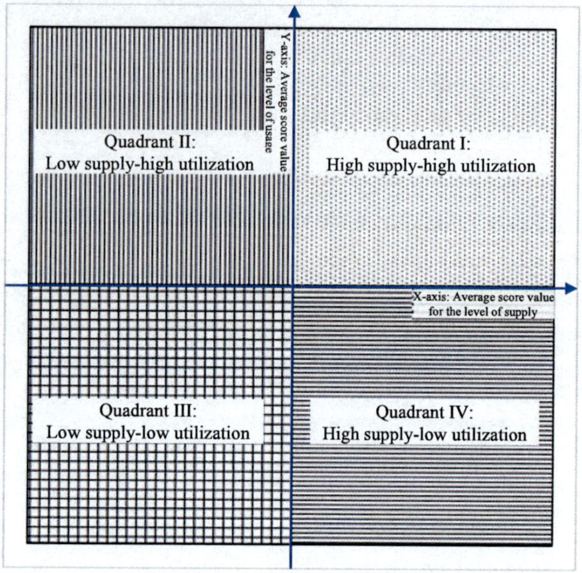

Fig. 5.52 Matrix for the "supply-utilization" match model regarding public service facilities in migrant worker settlements. **Source* Own elaboration

value for facility usage characteristics is used as the boundary between high and low levels of usage (i.e., y = 0 in the match matrix). In the matrix, the average usage level defined for facilities by category is the average score for the usage of the five categories of facilities, and the average usage level defined for facilities by grade is the average score for the usage of different grades of facilities. On this basis, the matrix for the "supply-utilization" match is constructed. The quadrant distribution and corresponding match types are shown in Fig. 5.52, where in.

Quadrant I: This quadrant represents above-average levels of facility supply and usage characteristic;

Quadrant II: This quadrant represents below-average levels of facility supply but an above-average level of usage characteristic;

Quadrant III: This quadrant represents below-average levels of facility supply and usage characteristic;

Quadrant IV: This quadrant represents above-average levels of facility supply but below-average levels of usage characteristic.

Analysis of Characteristics of the "Supply-Utilization" Match

The characteristics of the "supply-utilization" match regarding various types of facilities at all grade levels are summarized below. In addition, the match patterns for public facilities in each circle are analyzed. The results may serve as a guiding foundation for the research on corresponding enhancement strategies to be discussed below.

Characteristics of the "Supply-Utilization" Match for Facilities by Category.

Based on the match patterns between the supply and utilization of facilities by category in different urban circles (see Table 5.37 and Fig. 5.53), this study finds the following:

Overall, zero facilities land in the quadrant that indicates the "supply-utilization" match type of "high supply-high utilization." Majority of the facilities fall in the quadrants of "low supply-low utilization" (six types) and "high supply-low utilization" (six types); a small number of facilities fall in the quadrant of "low supply-high utilization" (three types, all of which are transportation facilities). The distribution

Table 5.37 Characteristics of the "supply-utilization" match for facilities by category

Type of facility		Outer ring		Inner Ring		Center	
		Supply level	Usage characteristic	Supply level	Usage characteristic	Supply level	Usage characteristic
Facility by category	Medical facility	■	□	■	□	■	□
	Transportation facility	□	■	□	■	□	■
	Commercial facility	□	□	□	□	■	□
	Sports facility	□	□	□	□	■	□
	Cultural facility	□	□	□	□	■	□

**Source* Own elaboration

Note "■" indicates high supply or utilization

"□" indicates low supply or utilization

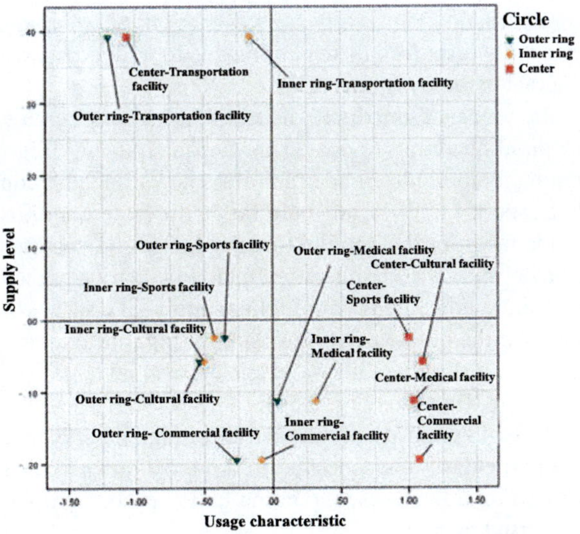

Fig. 5.53 Characteristics of "supply-utilization" match for facilities by category. **Source* Own elaboration

shows that migrant workers in the settlements achieve average or low levels in overall facility utilization and that the overall level of facility supply is higher than that of facility utilization but with similar high-low limits.

In terms of urban circles, all facilities in the center fall in the fourth quadrant of "high supply-low utilization," except that transportation facilities land in "low supply-high utilization"; all facilities in the inner ring fall in the two quadrants under the average level of utilization (y = 0) that indicate "low-utilization", except for transportation facilities, which are located in the fourth quadrant of "high supply-low utilization"; the outer ring displays the same match patterns as the inner ring.

Characteristics of "Supply-Utilization" Match for Facilities by Grade

To further explore and elucidate the "supply-utilization" match patterns, the 14 types of facilities at three grade levels in three urban circles are examined and their "supply-utilization" match patterns are summarized in Table 5.38 and Fig. 5.54. The results show the following:

Overall, a relatively small number of facilities land in the quadrant that indicates the "supply-utilization" match type of "high supply-high utilization" (seven types), whereas similar numbers of facilities fall in the three quadrants of "low supply-high utilization" (eleven types), "low supply-low utilization" (twelve types), and "high supply-low utilization" (twelve types). The distribution shows that there are limited types of facilities in the settlements that achieve high levels of supply and utilization and that the three circles display similar patterns in the types of facilities of low supply and low utilization.

A circle-based analysis of "supply-utilization" match patterns for facilities by grade is detailed in the following:

Center (see Fig. 5.55):

On the whole, majority of the facilities land in the first and fourth quadrants near the axis of average supply level (y = 0) and are particularly concentrated in the fourth quadrant. Only subway stations and community cultural facilities present relatively dispersed distributions due to wider disparities in supply levels and usage characteristics, respectively.

In terms of quadrant distribution, public service facilities in the center fall in three quadrants. Located in the quadrant of "high supply-high utilization" are mainly community-grade facilities, which include community hospitals, supermarkets, sports facilities, and cultural facilities; tertiary hospitals are the only municipal-grade facilities that land in this quadrant. The quadrant of "high supply-low utilization" primarily contains municipal- and district-grade facilities, which are relatively concentrated in areas with higher supply levels (x = 0.20). Only one type of facility lands in the quadrant of "low supply-high utilization," i.e., subway stations, which has only a slightly above-average (y = 0) level of usage characteristic.

Inner Ring (see Fig. 5.56):

Overall, majority of the facilities in the inner ring are located near the origin and are particularly concentrated in the third quadrant. Only subway stations, community cultural facilities, and community sports facilities are distributed in a relatively dispersed manner.

Table 5.38 Characteristics of "supply-utilization" match for facilities by grade

Type of facility			Outer ring		Inner ring		Center	
			Supply level	Usage characteristic	Supply level	Usage characteristic	Supply level	Usage characteristic
Facility by grade	Medical facility	Tertiary hospital	□	■	■	■	■	■
		Secondary hospital	■	□	□	□	■	□
		Community hospital	□	■	□	■	■	■
	Transportation facility	Subway station	□	■	■	■		■
		Bus stop	■	□	■	□	■	□
	Commercial facility	Large shopping mall	□	□	□	□	■	□
		Hypermarket	□	□	□	□	■	□
		Community supermarket	□	■	□	■	■	■
	Sports facility	Municipal sports facility	■	□	□	□	■	□
		District sports facility	□	□	□	□	■	□
		Community sports facility	□	■	□	■	■	■
	Cultural facility	Municipal cultural facility	□	□	□	□	■	□
		District cultural facility	□	□	□	□	■	□

(continued)

Table 5.38 (continued)

Type of facility			Outer ring		Inner ring		Center	
			Supply level	Usage characteristic	Supply level	Usage characteristic	Supply level	Usage characteristic
		Community cultural facility	□	■	□	■	■	■

* *Source* Own elaboration
Note "■" indicates high supply or utilization
"□" indicates low supply or utilization

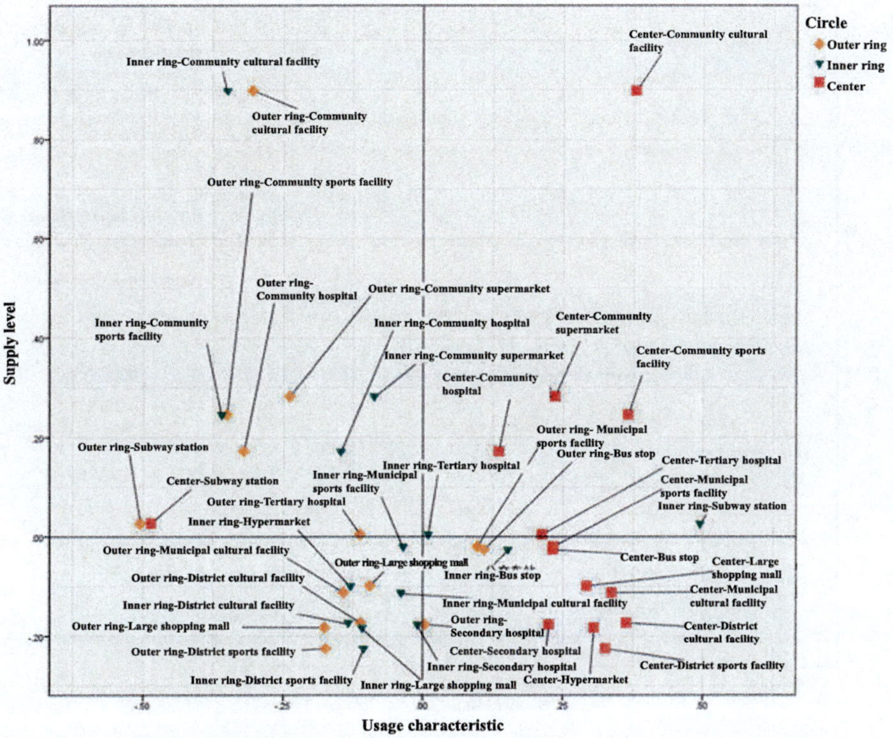

Fig. 5.54 Characteristics of “supply-utilization” match for facilities by grade under the influence of circle differences. **Source* Own elaboration

In terms of quadrant distribution, public service facilities in the inner ring are found in all four quadrants. Only subway stations land in the quadrant of “high supply-high utilization”, with the level of utilization being slightly above the average (y = 0). The quadrant of “low supply-high-utilization” contains four types of community-grade facilities, which include community hospitals, supermarkets, sports facilities, and cultural facilities. The quadrant of “low supply-low utilization” contains the most types of facilities, which include medical, commercial, sports, and cultural facilities of municipal and district grades. Only one type of facility lands in the quadrant of “high supply-low utilization”, i.e., bus stops, which has only a slightly above-average (y = 0) level of usage characteristic.

Outer Ring (see Fig. 5.57):

On the whole, public service facilities in the outer ring display a more dispersed distribution than those in the center and inner ring, with a relatively higher concentration near the negative semi-axis of the average supply level (y = 0), particularly in the third quadrant.

In terms of quadrant distribution, public service facilities in the outer ring are distributed in three quadrants. There are mainly community-grade facilities in the

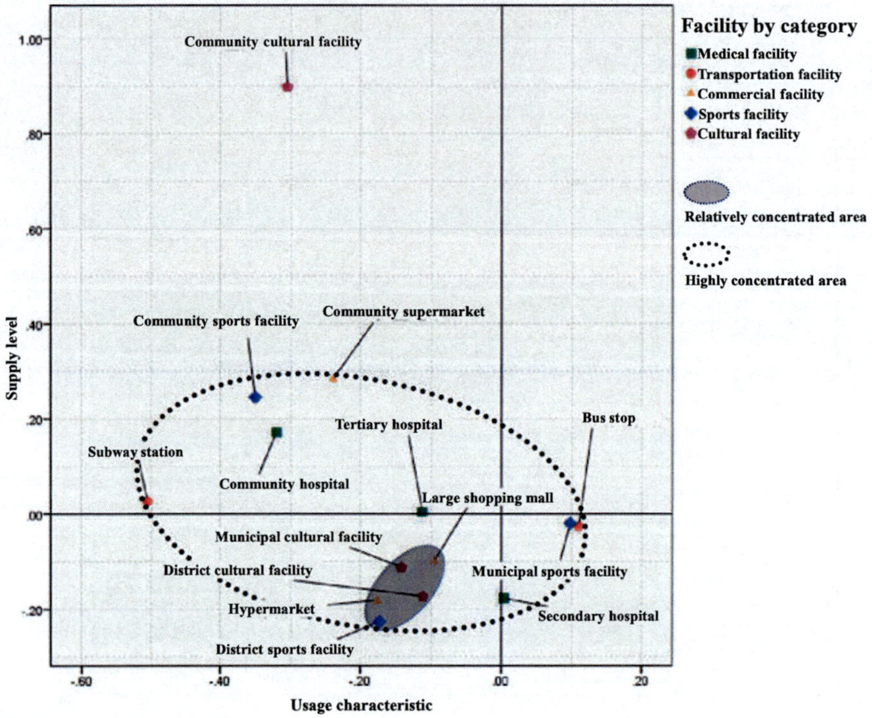

Fig. 5.55 Characteristics of "supply-utilization" match for facilities by grade in the center. **Source* Own elaboration

quadrant of "low supply-high-utilization that include community hospitals, supermarkets, sports facilities, and cultural facilities. In addition, subway stations and tertiary hospitals are located in this quadrant, with slightly above-average (y = 0) levels of supply. Located in the quadrant of "low supply-low utilization" are purely municipal- and district-grade facilities, which are highly concentrated due to their similarities in match patterns. The quadrant of "high supply-low utilization" contains three types of facilities, of which bus stops and municipal sports facilities have a level of usage characteristic close to the average (y = 0), and secondary hospitals have a level of supply close to the average (x = 0).

5.5.1.2 Summary of "Supply-Utilization" Match Issues

Based on the above discussion on the "supply-utilization" match patterns, various issues associated with the supply and utilization of public service facilities in migrant worker settlements are summarized as follows (see Table 5.39).

The above table indicates that only a small number of facilities (e.g., community supermarkets in the center, subway stations in the inner ring) display the "high

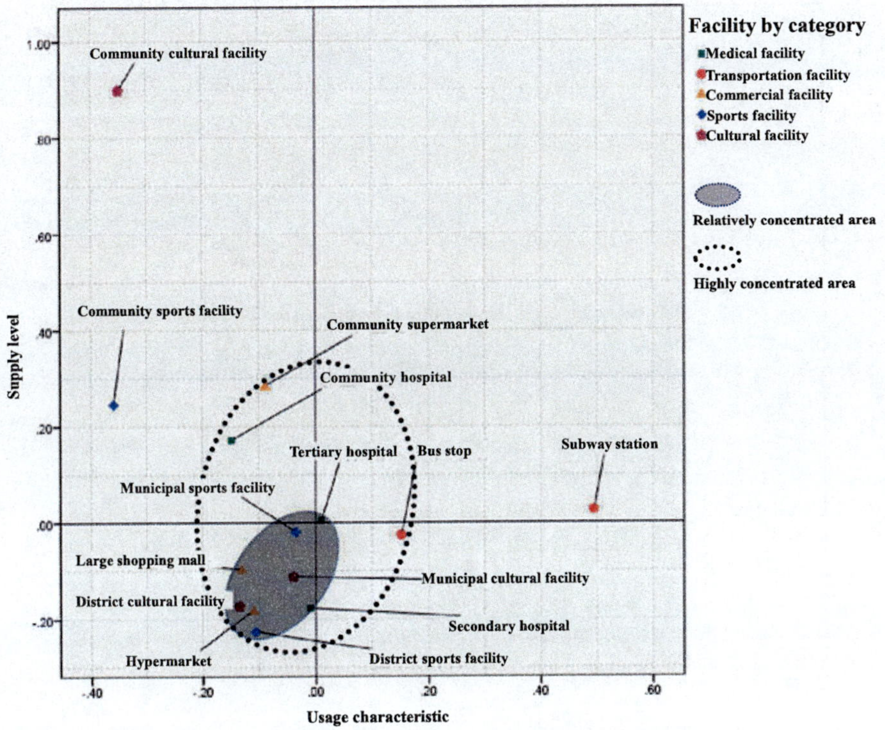

Fig. 5.56 Characteristics of "supply-utilization" match for facilities by grade in the inner ring. **Source* Own elaboration

supply-high utilization" match type. The overwhelming majority of facilities manifest inadequate levels in at least one side. In addition, facilities in different quadrants show different types and patterns of concentrated distribution due to variances in grade. Based on the various "supply-utilization" match types, differentiated strategies for the enhancement of different types of facilities are proposed in the following.

5.5.2 *Research on Strategies for Public Facility Enhancement*

5.5.2.1 General Concept

For the citizenization of "migrant agricultural population (represented by migrant workers)", promoting equalization of public services is of critical importance. How can the entire permanent population - inclusive of migrant workers - truly enjoy the same basic public services as the registered population? Unquestionably, the

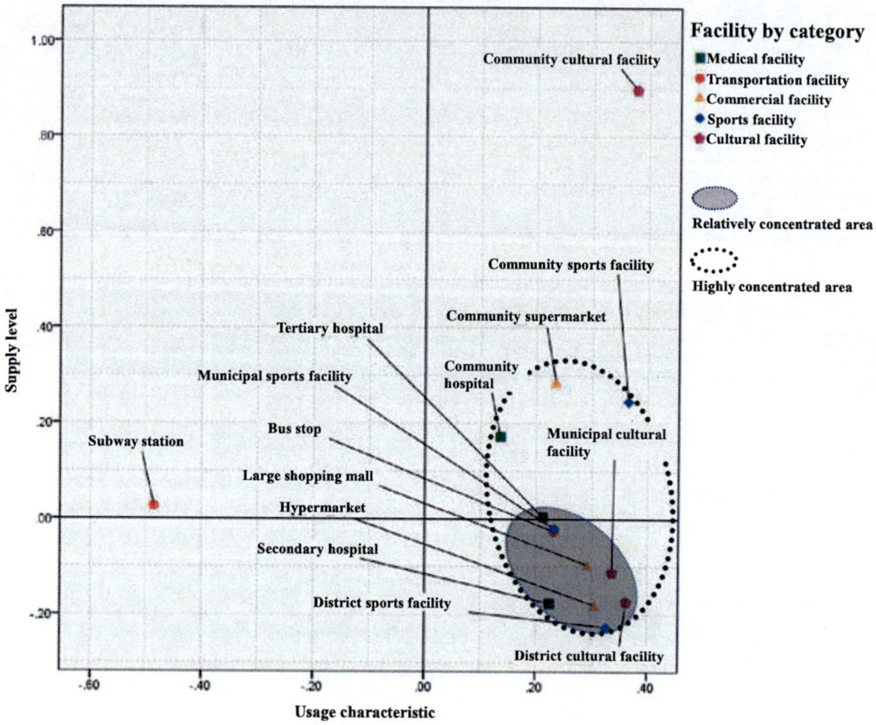

Fig. 5.57 Characteristics of "supply-utilization" match for facilities by grade in the outer ring. **Source* Own elaboration

reform and liberalization on the household registration system is an essential prerequisite. The Opinions on Promoting the System and Mechanism Reform for Social Mobility of Labor Force and Talented People issued by the state in 2019 clearly stated the following: the restrictions on urban household registration shall be completely removed in cities with a permanent population of less than three million; the requirements for urban household registration shall be comprehensively relaxed in large cities with a permanent population of three to five million; the point-based household registration policy shall be optimized for super-large cities and megacities with a permanent population of more than five million, wherein the point program shall be streamlined while ensuring that the scores for social insurance payment period and residence period account for a major proportion.

Following this significant step forward, we may further enhance the level of public services in migrant worker settlements by adopting strategies that are "developed using relevant supporting standards for public services as the baseline and specifically tailored for various types of facility on the foundation of the match analysis of "supply-utilization" in each quadrant." The "relevant supporting standards for public services" stated herein refer to the baseline standards for the public services in migrant worker settlements constructed in accordance of the Supporting Planning Standard

Table 5.39 Overview of the issues relating to the public service facilities in migrant worker settlements

"Supply-utilization" match pattern

Urban circle where a migrant worker settlement is located			Center			Inner ring			Outer ring			
Facility grade			Municipal	District	Community	Municipal	District	Community	Municipal	District	Community	Total
High supply-high utilization	Quadrant I	No	1	0	4	2	0	0	–			7
		Issue	–			–						
High supply-low utilization & Low supply-high utilization	Quadrant II	No	1	0	0	0	0	4	2	0	4	11
		Issue	Subway stations display a broadly average level of usage characteristic but low level of supply			Except transportation facilities, majority of the community-grade facilities display inadequate levels of supply			Community-grade facilities and subway stations display excessively low levels of supply; community cultural facilities also manifest a level of supply far below the willingness to use them; tertiary hospitals show a slightly inadequate level of supply			
	Quadrant III	No	–			3	4	0	2	3	0	12
		Issue				Municipal and district facilities display inadequate levels of supply and inadequate levels of willingness to use is observed among migrant workers			Municipal and district facilities show limited disparities from the average levels in supply and utilization			
Low supply-low utilization	Quadrant IV	No	3	5	0	0	1	0	1	2	0	12

(continued)

Table 5.39 (continued)

"Supply-utilization" match pattern						
Urban circle where a migrant worker settlement is located			Center	Inner ring	Outer ring	
		Issue	Municipal and district facilities show inadequate levels of willingness to use among migrant workers despite having adequate levels of supply	Bus stops have a slightly below-average level of utilization	Migrant workers display close-to-average levels of willingness to use in regard of bus stops and municipal sports facilities but a relatively inadequate level of willingness to use in terms of secondary hospitals	

* *Source* Own elaboration

for Public Facilities in Nanjing City (2015), on the basis of Nanjing City, the empirical region of this study. The standards also serve as a fundamental prerequisite for ensuring the daily needs of migrant workers. The items of the Standard associated with the five categories of facilities examined in this study are summarized as follows (see Table 5.40).

As for "on the foundation of the match analysis of 'supply-utilization' in each quadrant," this principle entails that differentiated strengthening strategies should be adopted for different match types indicated by each quadrant. In other words, for the first quadrant, which represents "high supply-high utilization," an "improving steadily"-based strategy should be adopted; for the second quadrant, which represents "low supply-high utilization," a "strengthening supply"-based strategy should be adopted; for the third quadrant, which represents "low supply-low utilization," a two-pronged strategy that combines "strengthening supply" and "activating utilization" should be adopted; and for the fourth quadrant, which represents "high supply-low utilization," an "activating utilization"-based strategy should be adopted.

Admittedly, for different types and grades of public service facilities located in the same quadrant, although the same general strategy is applied, specific measures will be proposed to provide "idiosyncratically" critical service content of different facility types based on relevant results and special studies (see Fig. 5.58).

5.5.2.2 Medical Facilities

As shown in Fig. 5.59, medical facilities located in Quadrant I include tertiary and community hospitals in the center and tertiary hospitals in the inner ring, for which future development should focus on "improving steadily"; those located in Quadrant II include community hospitals in the center and outer ring and tertiary hospitals in the outer ring, for which the future development should focus on "strengthening supply"; located in Quadrant III are secondary hospitals in the inner ring, for which the future development should focus on "strengthening supply" and "activating utilization"; finally, those located in Quadrant IV include secondary hospitals in the center and outer ring, for which the future development should focus on "activating utilization."

Strengthening Supply

For medical facilities, strengthening work should center on the supply of facilities at all grade levels in the inner and outer rings. In the Standard, no specific regulatory indicators are stipulated for the supply of municipal medical facilities; at the district level, only the scale of land use and number of hospital beds are regulated; at the community level, a more detailed scale and 1000-person indices are provided.

To strengthen the level of medical facility supply, the following measures are required. For outer ring-tertiary hospitals and inner ring-secondary hospital, regulatory indices shall be added for municipal and district medical facilities, including the number of hospital beds per 1000 people, the number of doctors per 1000 people, construction land per bed, and total land use. As for inner ring-community hospitals and outer ring-community hospitals, construction and arrangement shall be executed

Table 5.40 Summary of related key items of supporting planning standard for public facilities in Nanjing City

			Medical facility	Transportation facility	Commercial facility	Sports facility	Cultural facility
Municipal grade	Municipal	Scope of service	·Municipal public facilities refer to public facilities that serve the entire city and larger areas				
		Content of regulation	·Public facilities shall be reasonably arranged and coordinated in accordance with relevant standards, using the city and district (area) as the unit and accommodating the positioning of planned functions, the objectives of socioeconomic development, and the requirements of society ·The content and scale of municipal facilities shall be determined according to the stage objectives, general arrangement, and construction order of urban development				
District grade	District	Scope of service	·District public facilities mainly serve regions with a population of 200,000 to 300,000, providing residents with a comprehensive and assorted range of life services				
		Content of regulation	General hospitals with 300–500 beds, various specialized hospitals, health and epidemic prevention facilities, maternal and child health centers, rehabilitation centers for the disabled, nursing homes, etc., with a land use scale of 7–9 hectares	A hub of district public facilities is formed in the central area of the region with high levels of accessibility by combining junction stations of rail and bus transit	Large hypermarkets, shopping centers, banks, professional shopping malls, hotels and restaurants, telecommunications branches, etc	Standard stadiums, gymnasiums, national fitness centers, fitness trails, etc., with a land use scale of 3–5 hectares	Public libraries, cultural centers, museums, art galleries, etc., with a land use scale of 2–3 hectares
Community grade	Residential community	Scope of service	·Residential community-grade public facilities mainly serve regions within a service radius of 500–600 m and with a population of 30,000–50,000, providing residents with relatively comprehensive and wide-ranging daily life services ·Residential community centers should be arranged in a centralized manner to form a central land, with the scale of land use controlled at 4–5 hectares, of which 2.2–4 hectares are allocated for public facilities				

(continued)

Table 5.40 (continued)

				Medical facility	Transportation facility	Commercial facility	Sports facility	Cultural facility
		Content of regulation	Arrangement	In general, one community health service center is set up in each street or for every 30,000–100,000 permanent residents. Health service stations are set up in areas which community health centers cannot cover; generally, one community health service station is set up for every 10,000 residents	Community comprehensive service centers are centrally arranged in the form of a complex, with a land use scale of 1.4–2.8 hectares, providing a wide range of services, which include public culture, sports, administrative management and community services; community commercial services; and a variety of facilities such as traditional markets, post offices, motor vehicle parking lots, and public toilets			
					Bus terminals	Including traditional markets, post offices, and community commercial service facilities	Sports activities centers	Cultural activities centers
			Regulatory index	Content, construction scale, land use scale, 1000 person index of construction scale, and 1000 person index for land use scale	Land use scale, and 1000-person index for land use scale	Content, construction scale, and 1000 person index for construction scale	Content, construction scale, and 1000-person index for construction scale	Content, construction scale, land use scale, 1000 person index for construction scale, and 1000 person index for land use scale

(continued)

Table 5.40 (continued)

				Medical facility	Transportation facility	Commercial facility	Sports facility	Cultural facility
	Grassroots community	Scope of service		·Grassroots community-grade public facilities mainly serve regions within a service radius of 200–300 m and with a population of 5000–10,000, providing residents the most basic daily life services ·Grassroots community centers should be arranged in a centralized manner, with the scale of land use controlled at 7000–8000 square meters, of which 2000–3000 square meters are allocated for public facilities				
		Content of regulation	Arrangement	Community health service stations	Motor vehicle parking lots; arranged and constructed in accordance with the Standards and Guidelines Governing the Construction of Parking Facilities for Buildings in Nanjing City (Revised in 2012) and supplementary regulations	Small-sized business and financial service facilities	Sports activities station/court	Cultural activities room
			Regulatory index	Content, construction scale, and 1000 person index of construction scale		Content, land use scale, and 1000 person index for land use scale	Content, construction scale, land use scale, 1000 person index of construction scale, and 1000 person index for land use scale	Content, land use scale, and 1000 person index for land use scale

**Source* Nanjing Municipal People's Government; the Supporting Planning Standard for Public Facilities in Nanjing City (2015) [Z]. Nanjing, 2015
Note The "content" stated in the indices refers to the service content the said facility should undertake

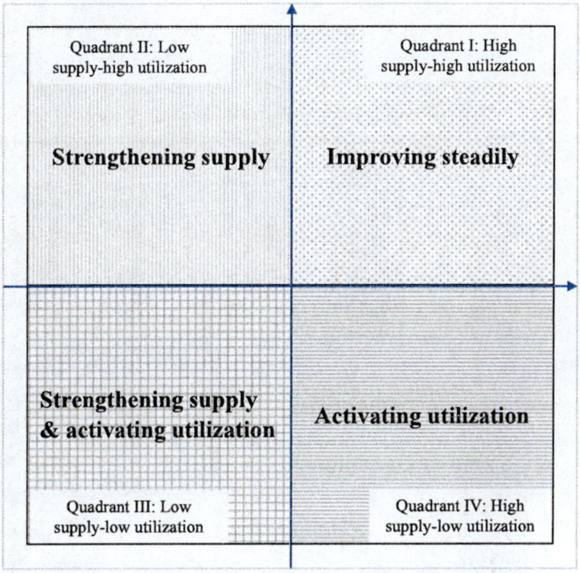

Fig. 5.58 "Differentiated" strategies for enhancing public service facilities in migrant worker settlements. **Source* Own elaboration

in strict accordance with relevant standards, and additional facilities shall be provided for areas with inadequate facility supply (Wei-peng and Zheng 2006).

Activating Utilization

Among the three grades of medical facilities, migrant workers in the three circles show inadequate levels of willingness to use secondary hospitals. This is mainly due to the common practice of "going to the hospital only for serious conditions to cut down on the cost of living" among migrant workers. This phenomenon also reflects the three major challenges confronting this group. First, under the urban–rural dual structure and the current household registration system, migrant workers are unable to fully enjoy urban medical protection; in the meantime, they have also lost some protection and benefits previously obtained in rural areas. Second, in terms of improving the medical insurance system for migrant workers, the government has not given full play to its corresponding functions. Third, enterprises have an inadequate sense of corporate social responsibility, while migrant workers have insufficient awareness of the medical insurance system, resulting in reduced rates of migrant workers taking out medical insurance.

In light of this, to improve the utilization of medical facilities among migrant workers, response measures should be introduced in three dimensions. First, the household registration system shall be reformed, and a unified (rather than urban–rural divided), labor relations-based medical support system shall be established. Second, the government shall give full play to its supervisory and advocacy role in enhancing medical support for migrant workers. Third, enterprises and migrant workers shall receive more rigorous education on corporate social responsibility and medical insurance awareness, respectively, whereby the coverage rate of the

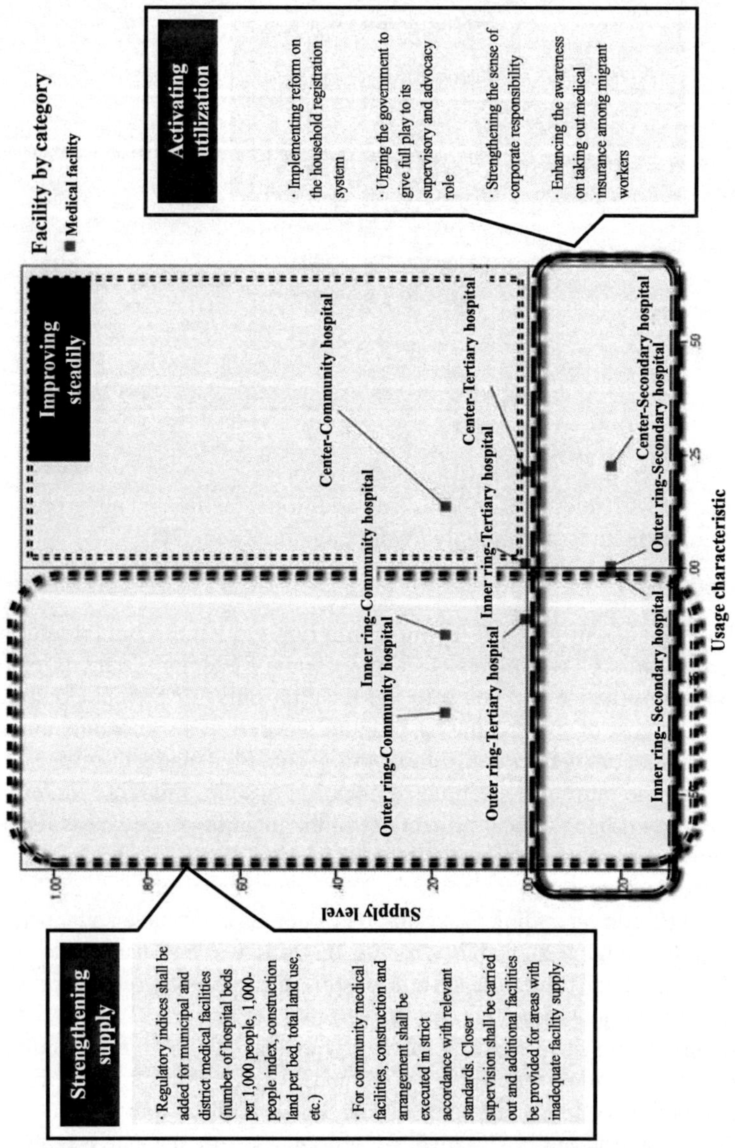

Fig. 5.59 Strategies for enhancing medical facilities developed on the basis of the "supply-utilization" match patterns. **Source* Own elaboration

medical insurance system may be raised among migrant workers. Through these efforts, migrant workers may establish reliable and conscious medical practices and self-insurance awareness (Ying-ying and Li-qing 2016).

5.5.2.3 Transportation Facilities

As shown in Fig. 5.60, transportation facilities located in Quadrant I are subway stations in the inner ring; those located in Quadrant II include subway stations in the center and outer ring; and those located in Quadrant IV include bus stops in the center, inner ring, and outer ring.

In terms of utilizing transportation facilities at all grade levels, all three circles display levels close to the overall average (showcasing flattened and uniform distributions). On this account, special strengthening measures need not be introduced for usage characteristics. Regarding the level of facility supply, the three circles display adequate supply levels of bus stops. As far as subway stations are concerned, the center achieves a slightly higher level of supply than the outer ring (the inner-ring sample shows an inadequate level of comparability for discussion due to its particularity[5]). This gap is mainly associated with the overall planning and gradual construction of subway lines and stations. The circle-based differentiated spatial distribution of subway stations not only conforms to the normal pattern of urban planning and construction but also ubiquitously exists and will continue to exist in cities. In light of this, the strategy of "strengthening supply" pursues by no means an instant, absolute equalization of supply but rather an overall equilibrium achieved in combination with the development of other transportation means.

5.5.2.4 Commercial Facilities

As shown in Fig. 5.61, commercial facilities located in Quadrant I are community supermarkets in the center, for which future development should focus on "improving steadily"; those located in Quadrant II include community supermarkets in the inner and outer rings, for which the future development should focus on "strengthening supply"; located in Quadrant III are large shopping malls in the inner and outer rings, and hypermarkets in the inner and outer rings, for which the future development should focus on "strengthening supply" and "activating utilization"; finally, those located in Quadrant IV include large shopping malls and hypermarkets in the center, for which the future development should focus on "activating utilization."

Strengthening Supply

For commercial facilities, strengthening work is required for the supply of facilities at all grade levels in the inner and outer rings. The Standard does not provide specific

[5] As stated previously, the inner-city sample settlement is located right outside the Hongshan subway station. On this account, it is not representative in the research of such facilities.

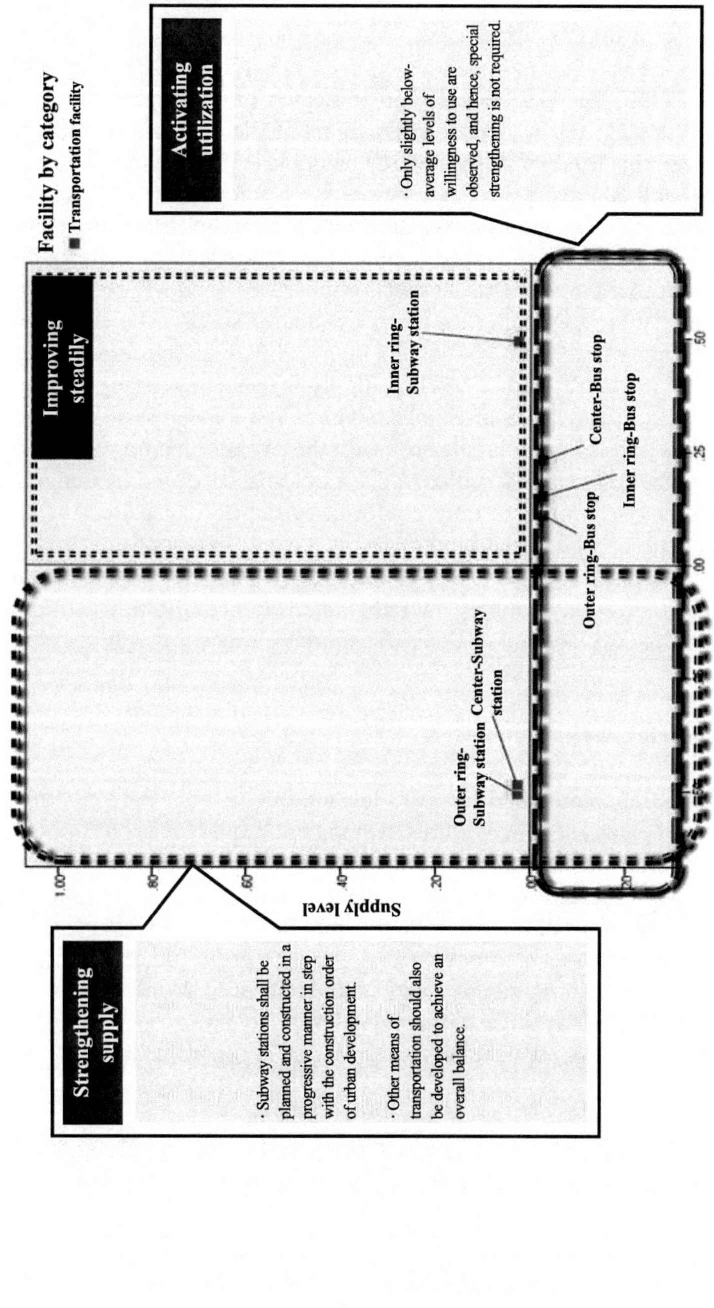

Fig. 5.60 Strategies for enhancing transportation facilities developed on the basis of "supply-utilization" match patterns. **Source* Own elaboration

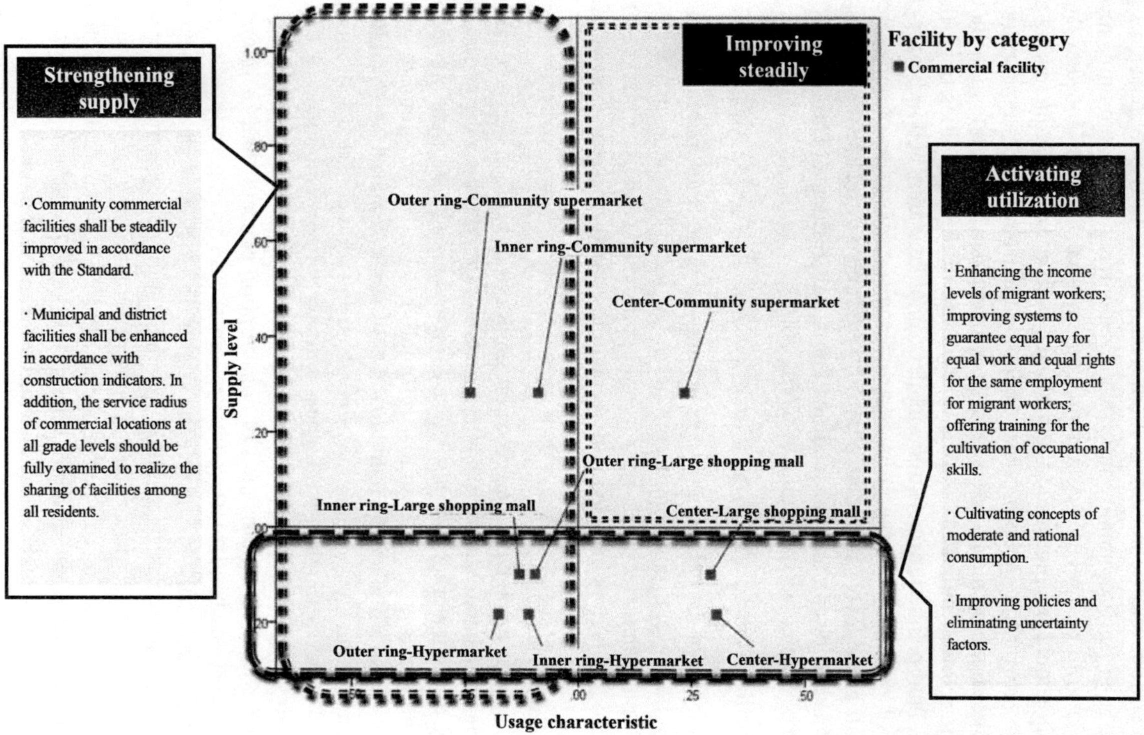

Fig. 5.61 Strategies for enhancing commercial facilities developed on the basis of "supply-utilization" match patterns. **Source* Own elaboration

regulatory indicators for the supply of municipal commercial facilities; regarding district commercial facilities, only the type of facility is regulated. The Standard merely provides regulatory indices for community commercial facilities, regulating areas such as the scale of land use, construction scale, and 1000 person indicators.

To enhance the level of commercial facility supply, the following measures are required. For outer ring-community supermarkets and inner ring-community supermarkets, the overall coverage of community commercial facilities and density of facility supply should be improved in compliance with construction indicators and related content. Regarding inner ring-large shopping malls, inner ring-hypermarkets, outer ring-large shopping malls, and outer ring-hypermarkets, close attention should be paid to the relevant indicators for the distribution of these facilities (e.g., their proportion in land planned for construction, per capita land area, and indicators for the land use of commercial centers). In addition, the service radius of commercial locations at all grade levels should also be fully examined to realize the sharing of municipal and district commercial facilities among all residents.(The Ministry of Construction of the People's Republic of China 2008).

Activating Utilization

In general, in terms of the willingness to use commercial facilities, migrant workers display the highest level in community commercial facilities but inadequate levels in higher-grade commercial facilities, i.e., large shopping malls and hypermarkets. The primary influencing factors include family economic conditions, consumption references, and monthly income levels. The better the economic condition of their family, the less constrained are migrant workers by the burden of their family when making purchases. They may engage in diversified consumptions as per their personal wishes while ensuring an adequate standard of living for their family. During their consumption practices, migrant workers will also make comparisons with surrounding people; as a result, the increase in the overall consumption level of the group to which they belong will also stimulate the individual's willingness to purchase. In addition, the increase in income levels will stimulate the consumption confidence of migrant workers, thereby increasing their willingness and level of consumption.

On this basis, to improve the utilization of commercial facilities among migrant workers, income levels and cultivation of consumption concept shall be improved simultaneously. On one hand, more sound systems should be implemented to guarantee equal pay for equal work and equal rights for the same employment for migrant workers. This will help maintain income stability for the migrant worker group. Meanwhile, a government-enterprise cooperation training mechanism should be established and cultivation of occupational skills should be strengthened. On the other hand, the consumption concept of migrant workers should be transformed. In addition to improving the level of social security for migrant workers and reducing uncertainty factors that pose hindrances to consumption practices (e.g., medical insurance, work-related injury insurance), various measures, such as aid, advocacy, and education, should be employed to enhance the ability of migrant workers to reasonably manage their incomes. Moderate and rational consumption practices should be encouraged (Liu Yi-feng and Meng-ran 2019).

5.5.2.5 Sports Facilities

As shown in Fig. 5.62, sports facilities located in Quadrant I are community sports facilities in the center, for which future development should focus on "improving steadily"; those located in Quadrant II include community sports facilities in the inner and outer rings, for which future development should focus on "strengthening supply"; located in Quadrant III are municipal sports facilities in the inner ring and district sports facilities in the inner and outer rings, for which future development should focus on "strengthening supply" and "activating utilization"; finally, those located in Quadrant IV include municipal sports facilities in the center and outer ring and district sports facilities in the center, for which future development should focus on "activating utilization."

Strengthening Supply

For sports facilities, strengthening work should be focused on the supply of facilities at all grade levels in the inner and outer rings. The Standard does not offer specific regulatory indicators for the supply of municipal sports facilities. Moreover, the Standard stipulates the provision of facilities and the scale of land use for district sports facilities and recommends a concentrated arrangement that combines the community comprehensive service center with other public service facilities as community sports facilities. The indicators involved therein include construction scale, land use scale, and 1000 person indicators. In China, special plans for public sports facilities are often dependent on the general plan or formulated for specific sports events. The main purpose of this planning model is to ensure the organization of large-scale events or construction of professional sports stadiums. However, the development of national fitness-oriented public sports facilities is either lagging or lacking. Due to such deficiencies, current public sports facilities are unable to accommodate the general and basic sports needs of urban residents today.

To strengthen the level of sports facility supply, the following measures are required. The main service objects in planning sports facilities should be shifted from large-scale sports events to daily sports activities for residents. In addition, more equalized distributions of public service facilities should be achieved and a comprehensive network of sports facilities at all grade levels should be constructed. By adopting the distribution pattern of "dispersion on the whole and concentration in part" for sports facilities at all grade levels, a supply system with multi-stratum indicators may be established. In particular, the supply level of community sports facilities should be strengthened to realize a rational structural transformation in the supply of sports facilities at all grade levels from the "inverted triangle" structure to the "regular triangle" structure (Yan Yong-tao and Zhi-dong & Li Zi-ming 2015).

Activating Utilization

Overall, in terms of the willingness to use sports facilities, migrant workers show the highest level in community sports facilities but relatively inadequate levels in municipal and district sports facilities. In 2006, against the backdrop of the integration of urban and rural areas, the General Administration of Sport of China specifically

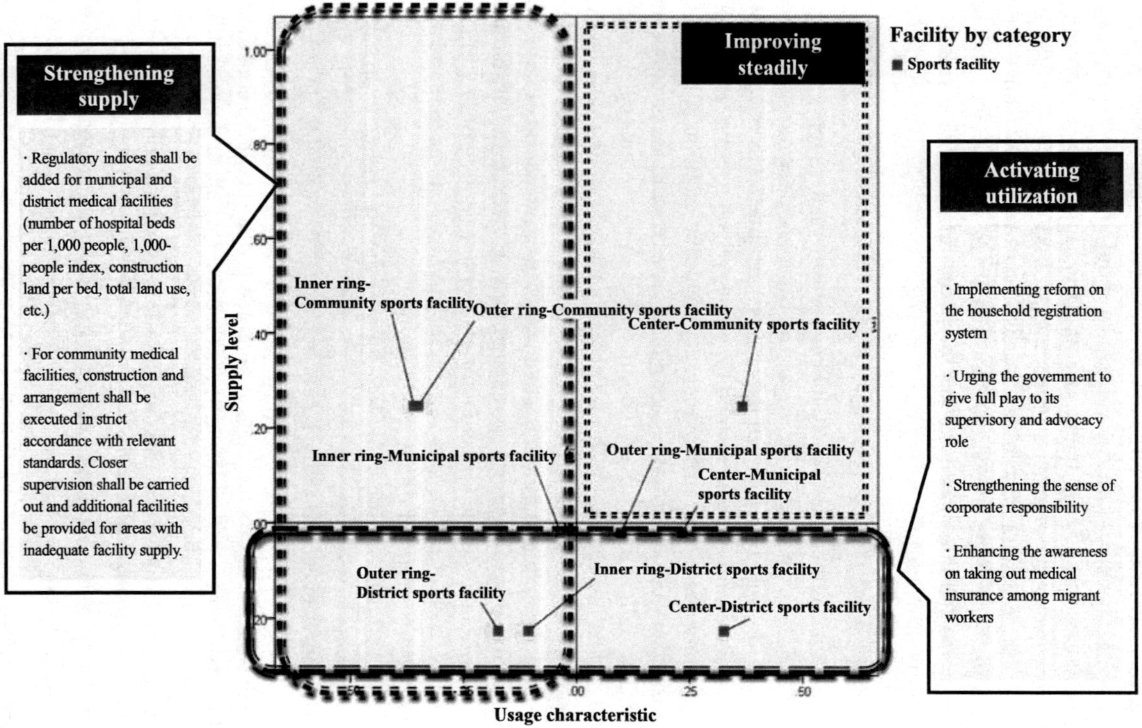

Fig. 5.62 Strategies for enhancing sports facilities developed on the basis of "supply-utilization" match patterns. **Source* Own elaboration

requested that the Migrant Workers Fitness Program be taken as an important item for sports work today and for a long time in the future. In academia, extensive research has also been conducted on the protection of sports-related rights and interests of migrant workers, characteristics of the utilization of sports facilities, and various other topics. On the whole, there are two main reasons for the low willingness of migrant workers to use public sports facilities. First, migrant workers lack awareness about sports rights protection. Currently, few can break free from the shackles of traditional rural thinking to integrate into modern urban sports culture. Second, migrant workers are subject to tight economic constraints. This group may be unable to engage in sports activities due to lack of funds or limited incomes. However, even when their income levels improve, they may find it difficult to effectively increase their willingness to use sports facilities since participating in sports and fitness activities appears to contradict their original intention of coming to the city to earn a living.

On this account, to enhance the utilization of sports facilities among migrant workers, the cultivation of sports habits and construction of a sports support system should be considered the key missions. On one hand, through the effect of cultural penetration, the following goals shall be achieved: enhancing the integration of migrant workers into modern urban sports culture, fostering their sports awareness, fitness values, etc., and further increasing their willingness to use sports facilities. On the other hand, the management system should be improved and a sports activity support system should be established for migrant workers to ensure that the spirit of mass sports can be fully realized in each and every individual while promoting the physical and mental health of migrant workers through sports activities (Ming-xing and Qing-ding 2014).

5.5.2.6 Cultural Facilities

As shown in Fig. 5.63, cultural facilities located in Quadrant I are community cultural facilities in the center, for which future development should focus on "improving steadily"; those located in Quadrant II include community cultural facilities in the inner and outer rings, for which future development should focus on "strengthening supply"; located in Quadrant III are municipal cultural facilities in the inner and outer rings and district cultural facilities in the inner and outer rings, for which future development should focus on both "strengthening supply" and "activating utilization"; finally, those located in Quadrant IV include municipal and district sports facilities in the center, for which future development should focus on "activating utilization."

Strengthening Supply

For cultural facilities, strengthening work should be focused on the supply of facilities at all grade levels in the inner and outer rings. The Standard does not offer specific regulatory indicators for the supply of municipal cultural facilities. For district cultural facilities, the Standard stipulates the provision of facility supply

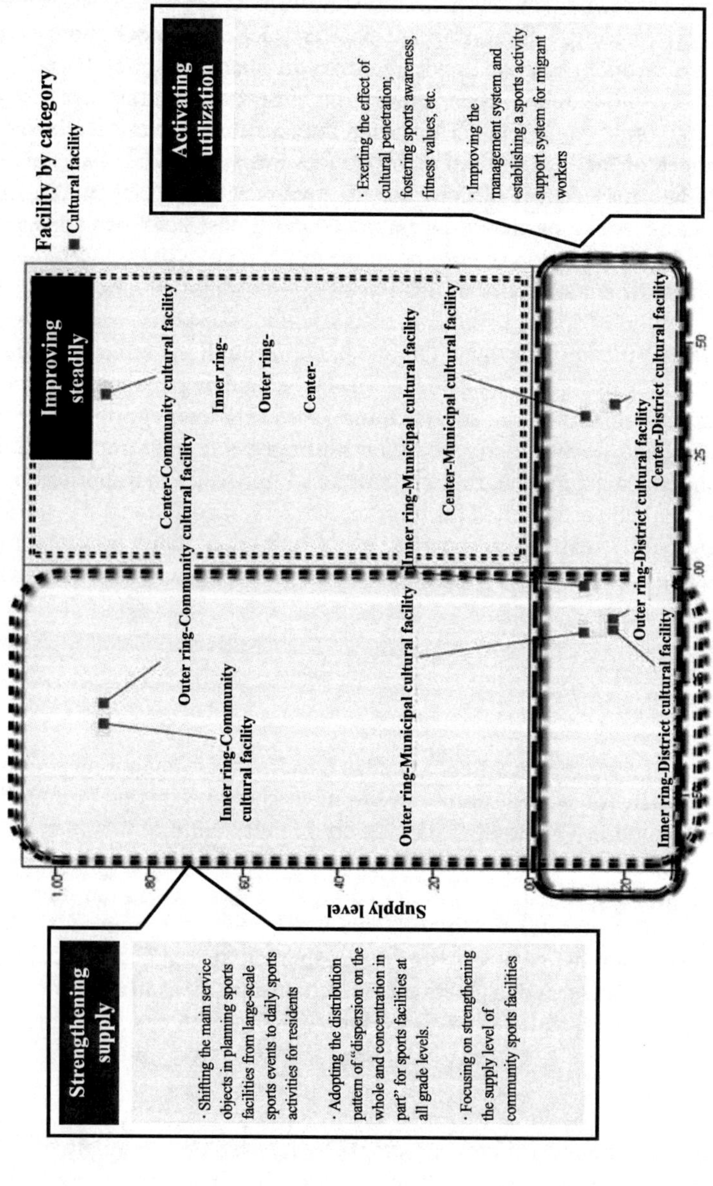

Fig. 5.63 Strategies for enhancing cultural facilities developed on the basis of "supply-utilization" match patterns. **Source* Own elaboration

and the scale of land use; for community cultural facilities, the Standard recommends a concentrated arrangement that integrates cultural activities centers, cultural activities rooms, etc., into community comprehensive service centers in combination with other public service facilities. The regulatory indicators provided therein include construction scale, land use scale, and 1000-person indicators. The economic development in China has brought about the fulfillment of material life and, in turn, given rise to a strong demand for spiritual life among residents. Admittedly, the supply of cultural facilities has become inseparable from the market economy; hence, more market economy-related factors should be taken into account.

To improve the level of cultural facility supply, the following measures are required. First, a "full-range"-oriented strategy for planning and construction should be devised and recommended indicators for the supply of cultural facilities at all grade levels should be developed to strengthen guidance for planning and achieve higher levels of equalization and scientificity of facility supply on the basis of consumer needs and spatial distribution. In addition, financing channels should be broadened to encourage the involvement of social funds in public cultural investment and increase the engagement of overall society in the supply of cultural facilities. In the meantime, more efforts should be dedicated to supervision and management work to organize and integrate the network of cultural facilities, ensure improvement in the overall utilization efficiency of cultural facilities, and optimize the overall supply level of urban cultural facilities (Yue 2014).

Activating Utilization

In terms of the willingness to use cultural facilities, in general, migrant workers display the highest level in community cultural facilities but inadequate levels in municipal and district cultural facilities. In the process of citizenization of migrant workers, "cultural integration" has become one of the significant measurement indicators. The efficient use of cultural facilities among migrant workers may help increase their recognition of urban cultural life and further eliminate the urban–rural barrier in their consciousness. However, migrant workers are confronted with manifold obstacles in using cultural facilities. First, high-intensity workload and long working hours may curtail the time they have available for cultural activities. The low income levels also limit their cultural consumption ability. Moreover, the low degree of recognition of their urban resident status presents a psychological obstacle for them in obtaining urban information and engaging in urban cultural activities. In the meantime, although the government has organized specific cultural activities for migrant workers, the overall inadequacy of public cultural services for migrant workers has not fundamentally improved.

On this ground, to improve the utilization of cultural facilities among migrant workers, close collaboration and joint efforts by companies, the government, and society are required. First, the government should establish a comprehensive service system of cultural facilities in migrant worker settlements and provide migrant workers sufficient, convenient, and inexpensive access to cultural activities. Companies should attach greater significance to establishing cultural awareness among migrant workers while combining corporate culture with the cultural education for migrant workers and nourishing their enthusiasm for engagement in cultural activities. Furthermore, the society should strengthen efforts to organize and advocate cultural activities for migrant workers, increase their awareness and participation in cultural activities, and gradually eliminate their "psychological thresholds" (Hai-shan and Sheng 2016).

5.6 Conclusion

This chapter provides a specialized study on "the inflow of the migrant agricultural population into cities" regarding the topic of "integration-response." First, clear definitions are offered for various related concepts, including migrant workers, migrant worker settlements, and public service facilities. The formation mechanism of migrant worker settlements and the overall conditions of migrant workers and public service facilities are summarized. Next, with a special focus on migrant worker settlements formed through active selection, empirical research is conducted using a combination of qualitative and quantitative approaches on the level of supply and usage characteristics (limited demand and consumption response under supply constraints). In terms of the level of supply, factor indicators and evaluation methods are first established, followed by an analysis of the overall characteristics of the level of facility supply and contributory causes conducted using the two dimensions of "single factor and multi-factor" and "grade-category-ensemble." In the study of usage characteristics, the research framework and evaluation methodology are first established and data are collected. Following this, the individual attributes and overall usage characteristics of migrant workers in the settlements are outlined. Finally, through an SPSS bivariate correlation analysis and cross-analysis of "individual attributes-facility utilization," this study sheds light on intra-group differentiation regarding the utilization of public service facilities among migrant workers in the settlements. In the last section of this chapter, based on the analysis of facility "supply-utilization" match patterns, different strengthening strategies are proposed for different types of public service facilities in and around migrant worker settlements to effectively promote the "citizenization of the migrant agricultural population."

The main conclusions of this chapter include:

(1) In terms of the supply level of public service facilities at different grade levels in migrant worker settlements, single factor scores display the following features. Regarding density, coverage, and accessibility, the factor scores for the 14 types of facilities generally rank as "Outer ring < Inner ring < Center" (35 out of 40 types of scenarios match this order). For the factor of location, 13 types of facilities (except subway stations) present more diversified score rankings, i.e., eight types of scenarios show the order of "Outer ring < Inner ring < Center", two types of scenarios manifest the order of "Outer ring > Inner ring > Center", and three types of scenarios display no marked ascending or

descending order for different circles. The overall factor scores showcase the following characteristics. In terms of the overall level of supply, twelve types of facilities show the order of "Outer ring < Inner ring < Center, whereas the other two types of facilities display the order of "Outer ring < Center < Inner ring."

(2) In terms of the supply level of public service facilities of different categories in migrant worker settlements, single factor scores manifest the following features. Regarding density, location, coverage, and accessibility, the factor scores for the five categories of facilities generally rank as "Outer ring < Inner ring < Center" (16 out of 20 types of scenarios match this order, whereas the remaining four types of scenarios show no marked ascending or descending order for different circles). In terms of the overall level of supply, four of the five categories of facilities are characterized by the order "Outer ring < Inner ring < Center" and one category of facilities shows the order of "Outer ring < Inner ring < Center."

(3) In terms of the overall supply level of public service facilities in migrant worker settlements, the scores for the four single factors of density, location, coverage, and accessibility rank in order of "Outer ring < Inner ring < Center, while the overall score also presents the order of "Outer ring < Inner ring < Center.

(4) The overall usage characteristics of public service facilities in migrant worker settlements are as follows. In terms of medical facility utilization among migrant workers, tertiary hospitals achieve the highest levels in usage frequency, usage duration, and consumption. Regarding transportation facility utilization, subway stations show higher levels of usage frequency and consumption amount, whereas bus stops attain higher levels of usage duration. In terms of commercial facility utilization, community commercial facilities are the most commonly utilized for all types of consumption and the most frequently utilized category as well, while large-scale shopping malls manifest the highest levels in usage duration and consumption amount. As for sports facility utilization, sports facilities at all three grade levels show average to low levels of usage frequency. Of them, municipal sports facilities achieve the highest levels of usage duration and consumption amount. Finally, in terms of cultural facility utilization, cultural facilities at all three grade levels also have average to low levels of usage frequency, with community and municipal cultural facilities showing the highest levels in usage duration and consumption amount, respectively.

In terms of the differentiated effects that individual attributes have on the characteristics of usage of public service facilities among migrant workers in the settlements, education level bears the closest correlation with usage characteristics, followed by marital status, income level, time of residence in Nanjing, and occupational composition. These individual attributes exert varying degrees of influence on different types and grades of facilities.

(6) Regarding the "supply-utilization" match patterns of public service facilities in migrant worker settlements, the discussion on facilities by category shows average to low levels of overall usage of various types of facilities; the overall

Table 5.41 Strategies for enhancing various types of facilities developed on the basis of "supply-utilization" match patterns

Category	"Strengthening supply" strategy for quadrants II and III		"Activating utilization" strategy for quadrants III and IV	
Medical facility	Type	Inner ring-Community hospital; Outer ring-Tertiary hospital; Outer ring-Community hospital; Inner ring-Secondary hospital	Type	Inner ring-Secondary hospital; Outer ring-Secondary hospital; Center-Secondary hospital
	Strategy	·Regulatory indices shall be added for municipal and district medical facilities (number of hospital beds per 1000 people, 1000-people index, construction land per bed, total land use, etc.) ·For community medical facilities, construction and arrangement shall be executed in strict accordance with relevant standards. Closer supervision shall be carried out and additional facilities be provided for areas with inadequate facility supply	Strategy (5)	• Implementing reform on the household registration system • Urging the government to give full play to its supervisory and advocacy role • Strengthening the sense of corporate responsibility • Enhancing the awareness on taking out medical insurance among migrant workers
Transportation facility	Type	Outer ring-Subway station; Center-Subway station	Type	Outer ring-Bus stop; Center-Bus stop; Inner ring-Bus stop
	Strategy	• Subway stations shall be planned and constructed in a progressive manner in step with the construction order of urban development • Other means of transportation should also be developed to achieve an overall balance	Strategy	• Only slightly below-average levels of willingness to use are observed; hence, special strengthening is not required
Commercial facility	Type	Inner ring-Community supermarket; Outer ring-Community supermarket; Inner ring-Large shopping mall; Outer ring-Large shopping mall; Inner ring-Hypermarket; Outer ring-Hypermarket;	Type	Inner ring-Large shopping mall; Outer ring-Large shopping mall; Inner ring-Hypermarket; Outer ring-Hypermarket; Center-Large shopping mall; Center-Hypermarket
	Strategy	• Community commercial facilities shall be steadily improved in accordance with the Standard • Municipal and district facilities shall be enhanced in accordance with construction indicators. In addition, the service radius of commercial locations at all grade levels should also be fully examined to realize the sharing of facilities among all residents	Strategy	• Enhancing the income levels of migrant workers; improving systems to guarantee migrant workers equal pay for equal work and equal rights for the same employment; offering training for the cultivation of occupational skills • Cultivating concepts of moderate and rational consumption • Improving policies and eliminating uncertainty factor
Sports facility	Type	Inner ring-Community sports facility; Outer ring-Community sports facility; Inner ring-Municipal sports facility; Inner ring-District sports facility; Outer ring-District sports facility	Type	Inner ring-Municipal sports facility; Inner ring-District sports facility; Outer ring-District sports facility; Outer ring-Municipal sports facility; Center ring-Municipal sports facility; Center-District sports facility

(continued)

Table 5.41 (continued)

Category	"Strengthening supply" strategy for quadrants II and III		"Activating utilization" strategy for quadrants III and IV	
	Strategy	• Shifting the main service objects in planning sports facilities from large-scale sports events to daily sports activities for residents • Adopting the distribution pattern of "dispersion on the whole and concentration in part" • Focusing on strengthening the supply level of community sports facilities	Strategy	• Exerting the effect of cultural penetration; fostering sports awareness, fitness values, etc • Improving the management system and establishing a sports activity support system for migrant workers
Cultural facility	Type	Inner ring-Community cultural facility; Outer ring-Community cultural facility; Inner ring-Municipal cultural facility; Outer ring-Municipal cultural facility; Inner ring-District cultural facility; Outer ring-District cultural facility	Type	Inner ring-Municipal cultural facility; Outer ring-Municipal cultural facility; Inner ring-District cultural facility; Outer ring-District cultural facility; Center-Municipal cultural facility; Center-District cultural facility
	Strategy	• Formulating a "full-range"-oriented strategy for planning and construction • Strengthening guidance for planning • Broadening financing channels • Enhancing supervision and management	Strategy	• Establishing a comprehensive service system of cultural facilities • Working with companies to cultivate awareness of cultural participation among migrant workers and increase their willingness to participate in cultural activities • Enhancing efforts to organize and advocate cultural activities for migrant workers; eliminating their "psychological thresholds"

* *Source* Own elaboration

level of supply is higher than that of utilization but with close high-low limits. The discussion on facilities by grade reveals that, on the whole, a limited number of facility types are characterized by "high supply-high utilization," whereas similar facility types manifest "low supply-low utilization."

(7) Following the reform of the household registration system and equalization of basic public services, we may further enhance the level of public services in migrant worker settlements by adopting strategies that are "developed using relevant supporting standards for public services as the baseline and specifically tailored for various types of facilities on the foundation of the match analysis of "supply-utilization" in each quadrant." The specifics are detailed as follows (see Table 5.41).

References

Chengrong D. Some issues on the floating population and the floating population studies in China. Popul Res. 1999;2:48–54.

Dawei Z, Weidong C, Xueping L, et al. Research on the code of urban community public service facilities plan and its implementation unit: a case of Wuhan. Urban Plann Forum. 2006;3:99–105.

Deyu J, et al. On Chinese farmers in the transformation of modernization. Nanjing: Nanjing University Press; 1998. p. 165.

Huan-chen Z, Shu-bai X, Sheng J. In: Analytic hierarchy process—a simple decision making method. Beijing: Science Press; 1996.

Hai-shan J, Sheng L. the status of public cultural services for migrant workers in Beijing from the perspective of urban integration. Popul Soc. 2016;32(2):48–55.

Hao Y, Ming Y, Xiao-zhou H, et al. 2014 Annual report on Nanjing transportation development. Nanjing: Nanjing Transportation Planning & Design Research Institute Co., Ltd.; 2011.

Jianmin W, Qi H. The floating population in China. Shanghai: Shanghai University of Finance and Economics Press; 1995.

Junbo G. In: Research on the supply mode and formation mechanism of public services and facilities in Chinese cities during the transition period: based on the case of Guangzhou. (Doctoral Dissertation). Guangzhou: Sun Yat-sen University, 2010.

Junbo G, Chunshan Z, Haiyan J, Changdong Y. The research on the spatial differentiation of the urban public service facilities distribution in Guangzhou. Hum Geogr. 2010;25(3):78–83.

Jun-bo G. Research on the supply mode and formation mechanism of public service facilities in Chinese cities during the transition period: based on the example of Guangzhou. [Doctoral Dissertation]. Guangzhou: Sun Yat-sen University; 2010.

Liu Yi-feng W, Meng-ran P-W. Analysis of the factors influencing the consumption behavior of migrant workers—based on the case of Beijing City. J Commer Econ. 2019;6:57–9.

Ming-xing Z, Qing-ding Y. Research on sports rights protection of migrant workers against the background of urban-rural integration. J Jilin Sport Univer. 2014;30(2):40–3.

Nanjing Municipal Bureau of Statistics. Statistical Bulletin on the National Economic and Social Development of Nanjing, 2017.

Qiang L. Urban migrant workers and social stratification in China. Beijing: Social Science Academic Press; 2012.

Rongshi L. Floating population in contemporary china: some considerations. Popul Res. 1996;1:10–5.

Ruijun W. An exploration on the meaning of the floating population. Popul Econ.s. 1990;3:53–5.

Samuelson P. The pure theory of public expenditure. Rev Econ Stat. 1954;6:387–9.

Shanyu Z. Population geography. Shanghai: East China Normal University Press; 1999.
Shu-bai X. Principle of analytic hierarchy process: using decision making methods. Tianjin: Tianjin University Press; 1995.
Talen E, Anselin L. Assessing spatial equity: an evaluation of measures of accessibility to public playgrounds. Environ. Plann. 1998; 30(4): 595–613.
Tao Z, Xin C. Speed investigation and signal design for special pedestrians crossing the street. Modern Transport Technol. 2015;12(5):75–8.
The Five Year Action Plan for the Reform and Development of Public Transport in Nanjing (2011–2015). Nanjing, 2011.
The Ministry of Construction of the People's Republic of China. GB50442–2008. Code for urban public facilities planning. Beijing: China Construction Industry Press; 2008.
Weilue H. Completion of population planning and grimness of population situation. In: Excerpted from an analysis of current chinese society and forecast of its future: 1993–1994. Beijing: China Social Sciences Press; 1994: 73–77.
Xiao W. In: A study on the settlement patterns of the floating population against the background of urbanization in China: based on Beijing, Nanjing and Shenzhen. (Doctoral Dissertation). Nanjing: Southeast University; 2001.
Xiao W. Analysis of the employment space for metropolitan floating population: an empirical study for migrant workers. Nanjing: Southeast University Press; 2010. pp. 144–55.
Xiao W, et al. An analysis of the living space of the floating population in China's major cities: an empirical study for urban migrant workers. Nanjing: Southeast University Press; 2010. p. 138.
Xiao-hong Z. The impact of mobility and urban experience on the modernity of Chinese peasants. J Sociol. 1998;5:60–1.
Yude J et al (ed). In: Farmers in the process of China's modernization. Nanjing: Nanjing University Press; 1998:393.
Wei-peng L, Zheng Y. Reform of sanitation system and planning of urban sanitation facilities. Urban Plann Rev. 2006;4:47–50.
Ying-ying C, Li-qing M. Research on the medical security problems for Chinese migrant workers and countermeasures. Rural Econ Sci Technol. 2016;27(14):188–9.
Yong-tao Y, Zhi-dong X, Zi-ming L. Exploration on the special planning of public sports facilities for national fitness—based on the example of Guangzhou. Planners 2015; 31(7): 11—16
Yue Y. Research on the "Full-range" strategy for the planning and construction of public cultural facilities in Guangdong. In: The urban planning society of china and shenyang municipal people's government. 60 years of planning: achievements and challenges——the proceedings of 2014 annual meeting of china urban planning. (12th Five-Year Plan Implementation and Management). The Urban Planning Society of China and Shenyang Municipal People's Government: the Urban Planning Society of China, 2016:11.
http://www.nanjing.gov.cn Nanjing Municipal People's Government
www.ditiezu.com.

Chapter 6
Conclusion: Integration-Response: Survival and Evolution of Settlement Space for Special Communities

Abstract The preceding chapters covered a wide-ranging discussion from the large-scale agglomeration of international students who come to China alongside the wave of globalization to land-lost peasant resettlements formed under the guidance of relevant policies and from minority settlements encountering the invasion of new cultures in the process of modernization to the migrant worker boom arising in the wake of China's economic development and system completion. Through the analysis of four typical cases that are highly unique yet consistent with the development of times, this research outlines a differentiated picture of the survival and evolution of special communities in large and medium-sized cities in China. In the process, the settlement spaces have not only brought multiple changes to urban areas but also responded to drastic social changes. Their marginal yet universal life structure has become one of the typical variables and key samples used to analyze while influencing the social spatial structure of Chinese cities. In light of this, drawing on Western migration theories and related research progress, this chapter builds on the two-way perspective of "integration-response" (The two corresponding concepts of "migrant settlements and the place of population immigration" have given rise to the two-way interactive process of integration-response. The effects and influences of the place of population immigration on the formation and evolution of migrant settlements are manifested as the "integration" process of migrants, whereas the impact and changes brought about by migrant settlements on the place of population immigration are manifested as the "response" process of the place of population immigration.) between the settlement space of special communities and the place of population inflow and proposes a two-way linkage framework for general interpretation that involves "two ends-five dimensions." Further, following the notion of "community type-model scenario-evolution result" hypotheses and explanations are rendered for the evolution model. On this basis, multi-scenario model interpretations are provided for three types of "settlement spaces for special communities" that bear the most typical significance in China. Through the interpretations, this research sheds light on the actual condition and underlying mechanism of their survival and evolution and reveals some of the fundamental universal patterns of contemporary Chinese urban social space.

X. Wu and L. Wang, *Settlement Spaces: Urban Survival Prospects of China's Special Communities*, Springer Geography,
https://doi.org/10.1007/978-981-16-4892-2_6

Keywords Special settlement space · "Integration-response" model · Multi-scenario interpretations

6.1 Interpretation of Classic Theories

The survival and evolution of the "settlement space of special communities" are driven by diverse and complex motivations. In essence, they involve a continuous interaction process of "integration-response" between such space and the place of population inflow. On this ground, this section draws on the classic theories related to the evolution and integration mechanisms for the settlement space of special communities (which involve perspectives and dimensions of integration) to build up solid support for the establishment of theoretical and analytical frameworks below.

6.1.1 Results of Relevant Research on Evolution of Settlement Space of Special Communities

Western research on the settlement space of special communities mainly focuses on marginal groups, such as floating population (or migrants) and racial (or ethnic) groups, and their settlements. Since the beginning of the twentieth century, the settlement space of special communities has gradually become a popular research topic in Western sociology, economics, and geography. Among such studies, the theoretical research on the floating population (or migrants) started from E. G. Ravenstein's definition of the concept of "migration" in his book *The Laws of Migration.* On this foundation, a succession of theoretical models have been developed, including the push and pull theory, new economics theory of migration, dual labor market hypothesis, and world-systems theory. Special efforts have been devoted to explaining the phenomenon of migration, with focus on issues such as "reasons for migration," "influencing factors for migration," and "how migration persists." In contrast, the research on racial (or ethnic) group theories can be traced back to the U.S. "Chicago school of sociology" in the 1920s, whose central viewpoints include the theories of assimilation, multiculturalism, segmented assimilation, etc.

Regarding the theoretical research on the floating population (or migrants), Western research began with the push and pull theory. The theory holds that population migration is triggered by a series of "forces," including "push forces" that induce a person to leave a place and "pull forces" that draw the person to move to another place (Ya-nan 1979). On this basis, a number of theoretical models have been developed successively. In particular, the new economics theory of migration introduces various dimensions for analysis, such as society, population, and family (Stark and Bloom 1985). The dual labor market theory argues that the labor market under modern capitalism contains dual segments, i.e., the high-wage, high-benefit primary segment, and the low-wage, low-benefit secondary segment. As employers recruit international immigrants to fill the vacancies in the secondary segment, population migration comes about accordingly (Piore 1970). The world-systems theory holds that with the transnational transfer of production factors and the adjustment and upgrading to the industrial structure of countries around the world, cross-border migration inevitably occurs Massey et al. (1993). In comparison with Western research, Chinese scholars primarily worked on qualitative research before 2000, devoting special attention to the characteristics and formation mechanisms of social space (Han-sheng et al. 1997; De and Chang-chun 2009). In 2000, Chinese researchers started utilizing quantitative research approaches to investigate the following research topics: reconstruction of the geographic and physical space of settlements (Pei-lin 2002; Xiao et al. 2003; Shuang et al. 2009), cooperative governance and social integration in settlements (Zhi-gang et al. 2011; Jing-xiang 2014), and economic characteristics and formation mechanism of settlements (Dan and Jing-xiang 2015; Xiao-ying 2010). Generally speaking, Western researchers have established relatively complete systems for migration theories, whereas Chinese researchers draw on Western theories and mainly focus on the impact China's political and economic systems exert on the settlements; however, inadequate discussions have been provided regarding the interaction between economy and settlements, and the overall analytical framework lacks the guidance of multiple theories.

As regards the study on racial (or ethnic) or other marginal groups and their settlements, the most renowned Western research is the assimilation theory advanced by Rorber Park in 1930. The theory posits that the evolution of the racial relations of migrants includes four stages: contact, competition, accommodation, and assimilation (Park 1928). Migrant settlements represent only a transitional stage. They will decline and wither away with the removal of migrants due to economic, cultural, and spatial assimilation brought about by the improvement of their economic and social status. The spatial assimilation model may well reflect the experience of most early European migrant groups. After World War II, multiculturalism was born. The theory contents that while migrants adapt themselves to new social and cultural environments, their original ethnic cultural characteristics do not necessarily disappear in the process and that migrants tend to rebuild their cultural traditions and relationship networks in the places of population inflow, exhibiting the diversity of migrant settlements in space. Representative cases include poverty-stricken black neighborhoods, Chinatowns and "ethnoburbs" (Gordon 1964; Hirschman 1983; Perlman and Waldinger 1997). The theory of segmented assimilation was proposed in the 1980s. Portes argues that racial migrant groups will produce three different integration results due to their different socioeconomic backgrounds in place of population inflow, namely, integration into mainstream society, integration into the urban underclass, and selective integration. This theoretical model serves as an important supplement to the theories of assimilationism and multiculturalism. With the development of globalization, transnationalism became a significant perspective for research on the social space of immigrants by the end of the twentieth century. Portes and Zhou (1993) indicated that migrants establish and maintain multiple social relationships between the country of origin and the country of immigration by crossing national boundaries. The reconstruction of global economy and changes in migration policies stimulate professionals and entrepreneurs to dive into the new wave of migration. Instead of assuming the identity of permanent residents in the country of immigration, these migrants travel back and forth two countries. With continuously increasing transnational activities, many migrant communities are no longer subject to the confines of racial and national borders and have thus become transnational communities (Myers et al. 2009; Entzinger and Biezeveld 2003) while displaying more distinct transnational networks and characteristics of social spaceDongmei et al 2016. In comparison, Chinese scholars are primarily concerned with the formation and evolution patterns of minority settlements. In particular, Hong-yan and You-tao (2004) argued that the Hui communities in large cities undertake social and cultural functions, providing urban cultural capital for urban modernization and internationalization and furthering the development and progress of the Hui culture in the meantime, thereby functioning as a "bridge" that connects Chinese Muslims with people around the world Hong-yan and You-tao (2004). Conversely, Jing-yu (2003) held that research on the development of urban minority settlements should focus on the following aspects: social distance and cultural integration; ethnic cultural identity; interaction of culture and economy; influence of urban ethnic distribution patterns, geography, and other factors; social and historical background of settlement development; influence of ethnic policies and characteristics of the relationship between the state and society (Jing-yu 2003). In general, under the guidance of social space dialectics, in-depth discussions have been conducted on the formation mechanism, reaction, and spatial effects of transnational migrant settlements. The relevant theoretical trends and research frameworks are significant for this research.

6.1.2 *Results of Relevant Research on Integration Mechanisms*

Migrant integration as a public policy originated in Western developed countries. This notion involves multiple disciplines, primarily sociology, demography, management science, political science, and urban planning. Researchers at home and abroad mainly focus on integration-related theoretical research as well as the content, dimensions, and influencing factors of integration.

The integration of the settlement space of special communities signifies an evolutionary process resulting from mutual influences between the settlement space and the place of population inflow. Current research perspectives mostly focus on the process of "**one-way integration,**" emphasizing that social integration is the process of transition from integrators to urban residents, which is manifested as the influences that the place of population inflow has on the settlements in terms of residence, employment, values, etc. In addition, some studies focus on the impact that the settlement space of special communities has on the place of population inflow to explore the "**reverse response**" effect of urban migration. Regarding research conducted from the **two-way perspective** of "integration-response," the discussion chiefly emphasizes the joint action and mutual integration of the integrators and the integrated, through which the two groups build a positive relationship of "integration-response" and closely interact with each other until mutual recognition is reached (Alba and Nee 2005; Yuan and Min-le 2006).

In terms of the analytical dimension of "integration" and "response,"[1] Gordon (1964) first proposed the "structural (institutional)-cultural" dual basic model for social integration. On this basis, seven dimensions for measuring the social integration of ethnic relationships are proposed, including cultural integration, structural integration, marriage integration, identity integration, attitude acceptance, behavior acceptance, and public affairs integration. Based on this foundation, Alba and Nee (2005) introduced other essential dimensions, including socioeconomic integration and spatial (or residential) integration. Further, foreign scholars have successively constructed similar frameworks for integration measurement and refined the measurement indicators under each dimension, such as education level, occupation, and income level for the dimension of socioeconomic integration. In comparison, majority of the Chinese scholars draw on the social integration theories and empirical measurement indicators proposed by European and U.S. scholars and take account of the actual situation in China to establish their own measurement dimensions for integration or response. In terms of the dimensions for integration analysis, Tian (1995) pointed out that the social integration of the floating population includes three levels: economic, social, and psychological or cultural levels. Zhu (2002) argued that these three different aspects develop in a progressive manner. Regarding the dimensions for response analysis, economic and social dimensions are generally employed to characterize the impact of special communities on urban areas. The summarization of measurement dimensions in existing research reveals the following (see Table 6.1): The analysis of the degree of "integration-response" is multi-dimensional, complex, and sequential, mainly including three progressive basic dimensions, i.e., cultural integration, socioeconomic integration, and psychological integration, while allowing for expansion and supplement in research where appropriate based on actual situations.

This research primarily focuses on the settlement space of special communities in China. Drawing on the basic viewpoints of Western migration theories (e.g., multiculturalism, transnationalism) and research framework of measurement dimensions, this research positions the two-way integration process of "integration-response" between the settlement space and place of population inflow under the same measurement dimension framework. Through this, the research interprets the evolution model of the settlement space of special communities in China, sheds light on the actual situation of their survival and evolution, and provides supplement and improvement on the fundamental framework for Chinese research on the settlement space of special communities at the community scale.

[1] The term "response" stated in this study refers to the one-way impact of migrants on the place of immigration. Keyword search can be conducted in combination with other terms such as "influence" and "effect" of population migration.

Table 6.1 Analytical dimensions for social integration of migrants in existing research

End		Researcher	Analytical dimension
Two-way		Gordon (1964)	"Structure-culture" integration model: culture, structure, marriage, identity, attitude, behavior and public affairs
		Alba and Nee (2005)	On the foundation of Gordon's theory, dimensions such as socioeconomic integration and spatial (or residential) integration are added
		Yuan and Min-le (2006)	Social capital, management system (household registration, public welfare, etc.), education and training, labor market
		Zhou (2012)	Settlement/economic integration, cultural adaptation, social adaptation, structural integration, assimilation/identity
One-way	Integration	Junger-Tas (2001)	"Three-dimensional" model: structural integration, social subdivision integration, and political legitimacy integration
		Entzinger and Biezeveld (2003)	"Four-dimensional" model: socioeconomic integration, political integration, cultural integration, social identity
		Tian (1995), Zhu (2002)	Economic, social, and psychological (or cultural) integration
		Feng (2004)	Economy, psychology, environment, life
		Tong and Ma (2008)	"Dual communities", "harmonious relationship" and "homogeneous identity"
		Zhang and Lei (2008)	Culture, psychology, identity (or institution), economy
		Yu et al. (2012)	Economic adaptation, social acceptance, cultural and psychological integration
		Wei and Lu (2012)	Economic capital, human capital, social capital, life experience
		Liu (2012)	Public services, economic status, social security, community engagement, and identity recognition

(continued)

Table 6.1 (continued)

End		Researcher	Analytical dimension
		Zhao and Li (2013)	Institutional, housing, socioeconomic, and personal characteristics factors
		Tian and Peng (2014)	Economic, social, psychological, institutional, and spatial integration
		Xiao et al. (2019)	Political, economic, public service, psychological and cultural integration
	Response	Ji et al. (2020)	Economic dimension: foreign investment R&D embedding, human capital accumulation, etc
		Xu et al. (2019)	Economic dimension: from labor mobility to industrial structure upgrading
		Du et al. (2012)	Economic dimension: consumer demand and production capacity
		Li et al. (2009)	Social dimension: the degree of acceptance of migrant groups among residents in the place of population inflow

6.2 Construction of "Integration-Response" Model of Evolution

6.2.1 Proposal of Logical Framework

Using social integration theories (including the theories of assimilation, multiculturalism, and segmented assimilation) as reference sources and based on Gordon's "structural (institutional)-cultural" dual basic model, this research proposes a general logical framework with "**two ends and five dimensions**":

6.2.1.1 Two Ends: "Integration End" and "Response End"

The survival and evolution process of the settlement space of special communities is a dynamic and progressive interactive process. As stated previously, in various research studies on marginal groups, such as floating population (or migrants) and racial (or ethnic) groups and their settlement spaces, both topics of "integration" and "response" are involved. However, most studies are limited to or focused on a one-way interpretation of a single end, with a vast majority of research dedicated to the analysis of "integration" and further details provided for the basic dimensions and essential components regarding the evaluation of special community integration. By contrast, relatively limited research has been conducted on "response," and far less research simultaneously introduces a two-way perspective of "integration" and "response." At least no existing research connects the two and places them under the same theoretical framework to offer a unified, two-way linkage interpretation. As a result, this research seeks to coordinate and examine the "integration end" (i.e., the integration characteristics of the settlement space of special communities under the influence of the place of population inflow) and the "response end" (i.e., the impact of the emergence of special community settlement space on the place of population inflow). Thereby, a basic framework may be constructed for interpreting the interaction between the settlement space of special communities and the place of population inflow, and the evolution of the former.

6.2.1.2 Five Dimensions: "Spatial, Economic, Social, Institutional, and Psychological" Dimensions

This study draws on and synthesizes the analytical dimensions of "integration" and "response" in existing research. Notably, classic integration theories have keenly incorporated a two-way perspective, especially the "structural (institutional)-cultural" dual basic model of social integration proposed by Gordon and the multi-dimensional model of "structure (institution)-culture-social economy-space" derived therefrom. These models lay a theoretical foundation for the construction of the theoretical model in this research. With reference to the influencing factors and analytical dimensions for special community integration (Table 6.1), this research combines the different objects the "integration end" and "response end" act on respectively and the characteristics of the process to construct a unified explanatory framework for "integration-response" evolution. In line with the theories, the framework contains five basic dimensions, i.e., spatial, economic, social, institutional, and psychological dimensions. However, it should be stressed that even within the same basic dimensions, the "two ends" shall contain distinct specific content and constituent elements to ensure the overall systemicity and inherent distinctiveness of the research system.

On this basis, the research breaks down the discrete and one-way nature of the actions by the "integration end" and the "response end" and transcends the finiteness of analytical dimensions to establish the theoretical explanatory framework. In other words, five dimensions (i.e., economic, institutional, spatial, social, and psychological) are selected as the common dimensions that simultaneously characterize the interaction and evolution analysis of the settlement space of special communities (i.e., "integration end") and place of population inflow (i.e., "response end"). However, even in the same dimension, the analysis of different cases will hold different connotations, and cross-dimensional effects may occur under specific conditions (Fig. 6.1).

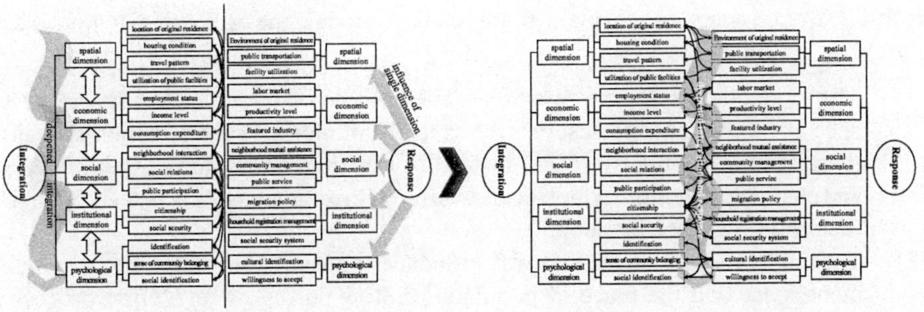

Fig. 6.1 "Integration End-Response End" connected explanatory framework for the evolution mechanism of the settlement space of special communities in China

Consider migrant worker settlements in China (that have the largest size and widest distribution among all settlement types) as an example. The five dimensions will have effects of varying degrees on the "integration-response" process of this economic-type special community. Of them, production and livelihood in the economic dimension has the most profound impact. In terms of the "integration end," the place of population inflow (the level of economic development, industry type, labor market demand, etc.) exerts a differentiation effect on the employment status, income level, and consumption expenditure of migrant workers. At the opposite "response end," due to the gradual "integration" of migrant workers (in terms of employment status, income level, housing conditions, etc.), "response" and structural effects will be generated in different dimensions of the place of population inflow. For example, with a large number of migrant workers engaging in self-employed catering and commercial service industries, not only is the labor market in the economic dimension of the "response end" affected but also the level of public services in the social dimension is altered. This series of influences will, in turn, produce a reverse effect on the economic dimension (e.g., income level) and spatial dimension (e.g., utilization of public facilities) of migrant worker settlements at the "integration end." In general, the "integration end" and "response end" of the settlement space of special communities exhibit progressive deepening in the degree of integration through interactive changes in the five dimensions.

6.2.2 Key Points for Model Construction

Should the "integration-response" explanatory framework be employed as a theoretical tool to study the survival and evolution of the settlement space of special communities, the "integration end" and "response end" may be juxtaposed in a common five-dimensional system for comparative analysis and further exploration and elucidation be attempted following the notion of "**community type-model scenario-evolution**

result." To construct and explain the evolution model, the following hypotheses should be established:

(1) The model delineates the research scope based on the "settlement space" of a special community and observes the special community at the community scale (determining people by space) instead of tracking "individuals" that migrate and change constantly, which leads to arbitrary expansion of the research scope (determining space by people).
(2) The model focuses on the two-way influence and interaction between the settlement space and the place of population inflow during the evolution process. The interference and influence of internal factors on the two in the evolution process should be reduced. For example, in terms of land-lost peasant resettlements, more attention should be devoted to the influence of external factors (e.g., supply of public service facilities, degree of acceptance for the community settlement space among urban residents) on the degree of integration in different dimensions (facilitating or impeding integration) without considering the individual differences and intra-group differentiation of the community per se.
(3) The ultimate state of the model is that the "integration-response" ends can maintain a relatively stable dynamic equilibrium in all dimensions for a long time, whereby the future state of the settlement space may be inferred through the analysis of the current situation and development trend under this theoretical framework.

First, in view of the complexity and difference of population sources and causes, this research classifies the settlement space of special communities in China into sub-categories for model construction. According to the main reasons for the migration of special communities and their internal relations with the existing structural system of the city, three largest and most typical types of special community settlement spaces are selected as community representatives for the next step of scenario simulation. The settlement types selected include settlement spaces of economic-type special communities (e.g., migrant worker settlements, multinational enterprise labor communities), social-type special communities (e.g., international student settlements, minority settlements), and engineering/construction-type special communities (e.g., land-lost peasant resettlements, ecological migrant resettlements). On this basis, through the five basic dimensions (i.e., economic, institutional, spatial, social, and psychological) and their specific constituent elements, the dynamic evolution process of the "integration end" and "response end" of settlement space of the three scenario types is simulated. Finally, under the interaction of the "integration end" and the "response end" jointly influenced by the five dimensions, two types of results (assimilation or coexistence)[2] for the evolution of special community settlements are derived, and the actual situation of Chinese special community settlements is presented in a point-to-plane manner.

---

[2] With reference to the fundamental viewpoints of the classical assimilation theory, plural assimilation theory, and segmented assimilation theory in classic social integration theories, two possible

Following the construction rationale of the model, two key issues in scenario simulation need to be clarified:

(1) **Time gradient between the five analytical dimensions, i.e., the time sequence of the evolution of "integration-response":** Based on the agreed conclusions on the measurement dimensions in research on the integration mechanism described above, the level of difficulty and fundamentality of integration in each dimension is comprehensively considered. The spatial dimension refers to the spatial carrier of the special community settlement space embedded in the place of population inflow; the economic dimension indicates the substantial foundation and driving force for integration; the social dimension pertains to the humanistic connection between the settlement space and the place of population inflow; the institutional dimension involves the structural elements of social integration; and the psychological dimension represents in-depth integration on the spiritual level. Different dimensions may act synchronously or alternately. However, the final realization of integration and response differs in the level of difficulty and sequence of time, usually ascending progressively from "spatial to economic to social to institutional to psychological" dimensions. Different types of settlement space for special communities may have slight differences, default or even crossover. For example, in terms of migrant worker settlements, two-way influences are present throughout the five dimensions. As for minority and land-lost peasant settlements, the two-way impact of the institutional dimension may be disregarded (it is not that these groups do not pay subjective attention to this dimension but that objective institutional policy protection has reduced the significant obstacles to their integration-response). In terms of international student settlements, the influence of the economic dimension is missing at the integration end (economic needs such as earning a living and making a profit are the main reasons for the transnational migration of this group) and so is the two-way influence of the institutional dimension. However, for all types of settlement space, two-way influences are assuredly observed throughout spatial, social, and psychological dimensions.

(2) **Linkage relationship between the two ends on the time gradient, i.e., the association of "integration-response" evolution:** The motive power (trigger point) for the evolution process is driven by the pulling force (positive benefits in economic, policy, social, and other aspects) from the place of population inflow to the settlement space of special communities. On this account, the evolution process often starts from the "integration end," manifested as a multi-dimensional, continuously acting two-way process between the "integration

evolution results may be derived for the evolution of migrant settlements: assimilation or coexistence. First, based on the evolution result of migrant settlements advocated by the classical assimilation theory, the settlements will eventually give up their unique identity and ethnic characteristics and integrate into the mainstream society alongside their economic, cultural, and spatial assimilation. Second, based on the basic viewpoint of pluralism, migrants do not lose their original ethnic cultural characteristics in the process of adapting to the new social and cultural environment. Instead, they rebuild their cultural traditions and relationship networks in the place of immigration.

and response ends." In addition, the two ends display a certain consistency in the influence gradient of each dimension, i.e., the easier the "integration" is in a dimension, the more positive and rapid the "response" will be. For instance, the land-lost peasant group, which emerges as a result of land development and construction, will often go through the steps of organized site selection, planning and design, development and construction, and collective resettlement arranged by the resettlement area under the powerful protection and promotion of relevant policies and systems. This group may thereby complete the process of "passive integration" in the urban spatial dimension in a relatively swift manner. Subsequently, with new demands and markets brought about by a large number of new residents, the spatial environment of the resettlement and its surrounding areas may experience a reverse effect and improve (e.g., addition to some service facilities) or decrease (e.g., destruction to some public space) in quality. This, in turn, induces a local "response" in the spatial dimension. It can be concluded that the linkage between the "integration end" and the "response end" often occurs first in the same dimension, and the "two ends" display the closest two-way interaction in the same dimension.

6.2.3 Interpretation of Evolution Model

From the perspective of the general logic of evolution, the interaction between the settlement space of special communities and the place of population inflow in the five dimensions of "space, economy, society, institution, and psychology," exhibit the two-way characteristics and patterns of "integration end" and "response end" in a progressive manner (see Fig. 6.2).

The spatial dimension reflects the embedded relationship between the settlement space of special communities and the place of population inflow in physical space. The characteristics of evolution on the "integration end" are mainly manifested in the following: the actual state of community locations, housing conditions, transportation and travel, service facilities, public spaces, etc., that the special communities should be embedded in and adapted to after (active or passive) the selection of residence under the constraints of various conditions, including capital markets, community ties, and land policies. On the "response end," the characteristics of evolution are shown as follows: new changes and influences brought about by new users, demands, and markets embedded into the settlements and surrounding areas. Apart from community locations and housing conditions, which are difficult to change instantly, evolution tends to be more prominently reflected in service facilities, public spaces, and even transportation and travel (in the form of improvement or decline).

The economic dimension refers to the fundamental driving force and restrictive conditions for the survival and evolution of the special community settlement space in the place of population inflow. On the "integration end," evolution is mainly reflected in the following: under the combined effect of the industry structure and the labor market at the place of population inflow, whether the special communities can

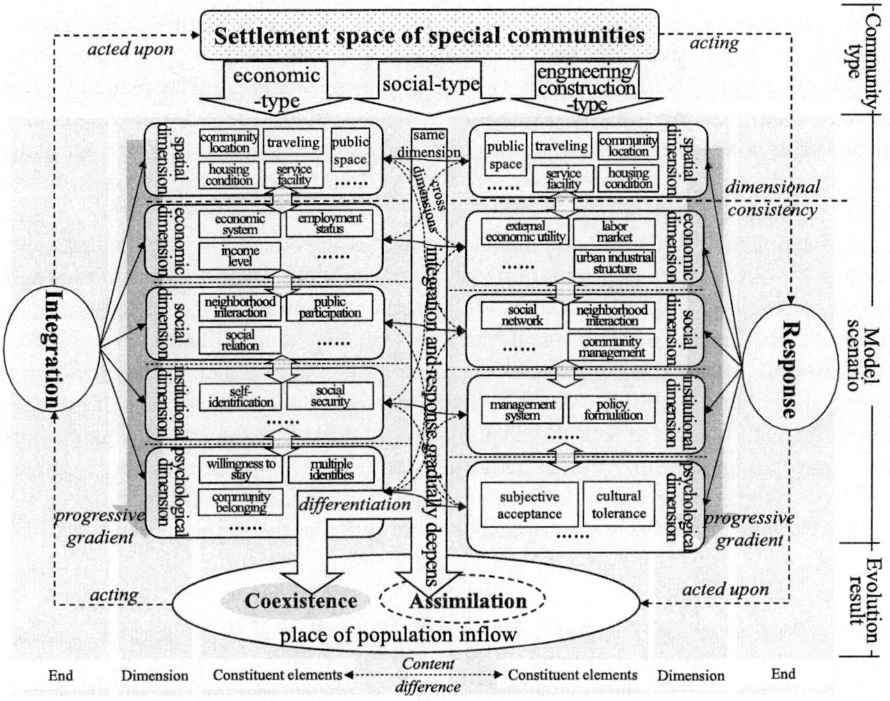

Fig. 6.2 Schematic diagram of the integration-response evolution model

successfully integrate into the urban economic system through marketized employment while ensuring and enhancing the employment status and income level in the settlements per se. On the "response end," evolution is manifested in the external economic utility the special communities produce due to large-scale agglomeration as the supplier of labor as well as the different changes (optimization or deterioration) they bring to the urban industrial structure and labor market.

The social dimension pertains to the process in which special communities and the mainstream society in the place of population inflow gradually identify with each other in terms of social relations. Primary manifestations on the "integration end" include the following: the social barriers that special communities have to overcome and the various types of characteristics they exhibit in their attempt to integrate into local social relations, neighborhood interactions, and public participation in the face of the unfamiliar social environment (cultural differences, policy neglect, social network isolation, etc.) in the place of population inflow. On the "response end," the process is manifested as follows: the changes occurring in the social network, community management, and neighborhood interactions among others of the corresponding mainstream society in the face of special communities (inclusiveness or inertia of resistance to changes brought about by external forces).

The institutional dimension denotes the process by which special groups gradually become "regularized" through policy implementation and feedback while being incorporated into the management system of the place of population inflow. The primary manifestations on the "integration end" include the following: whether special communities can be included into the existing regular institution regarding aspects such as identity (household registration) and social security (citizenship treatment such as pension and medical care). On the "response end," the process is manifested as follows: the response and changes introduced by the management department of the place of population inflow in formulating various institutions and policies (removing institutional thresholds or maintaining the original status) when faced with the large-scale influx and settlement of special communities.

The psychological dimension refers to the degree of acceptance of the other party among special communities and the mainstream community in the place of population inflow. All the above four dimensions have an effect on the level of integration in the psychological dimension. In addition, the degree of cultural difference also, to a large extent, determines the limits of mutual acceptance. On the "integration end," acceptance is embodied by special communities in the willingness to stay, the sense of community belonging, social identification, self-identification, and identification on other dimensions, among others. On the "response end," acceptance is manifested in the degree of subjective acceptance, cultural tolerance, etc., of the special communities among the mainstream community.

The survival and evolution of the "settlement space of special communities" involves a process of continuous "integration-response" between the settlement space and the place of population inflow. The result is essentially determined by the interaction between external influences such as repulsion and organization by the mainstream society and internal cohesion of the special communities (Knox and Pinch 2005). If a certain special community does not diverge greatly from the mainstream society in terms of cultural background, ideological concepts, lifestyles and customs, etc. (displaying a weaker "repulsion" force), the community will be more easily incorporated into the organizational system by the mainstream society. If the community does not have strong internal cohesion, it may easily achieve a relatively stable equilibrium between the "integration end" and the "response end" in the above dimensions. This indicates that the spatial settlement formed by this community is highly likely to be a transitional stage toward being organized and assimilated by the extensive urban social space. If the settlement does not have continuous and large-scale input of new members from the special community subsequently, it will eventually become "assimilated" by the mainstream society. The opposite scenario would be the following: when equilibrium is unlikely to be achieved between the "integration end" and the "response end," this community will "coexist" within the mainstream society in the spatial form and life structure of a special community settlement on a long-term basis due to its strong internal cohesion.

6.3 Multi-Scenario Interpretation of "Integration-Response" Evolution Model

According to the "integration-response" evolution model developed on the basis of the notion of "community type-model scenario-evolution result," evolution scenarios are respectively derived and interpreted for settlement spaces of the three representative yet different types of special communities described previously, i.e., economic, social, and engineering/construction types of special communities.

6.3.1 Scenario a: Settlement Space of Economic-Type Special Communities

Economic-type special community settlement space refers to the form of settlement established by special communities that migrate to and make profits in the place of population inflow driven by economic factors such as enhancing employment or income levels (e.g., migrant workers, transnational laborers). Due to economic reasons such as housing, rent and commuting, these special communities spontaneously reside (become embedded) in the existing communities of the city and establish close associations with the industrial economy, job market, and other economic systems of the city they have moved to. This type of settlement space contains two major sub-types: mixed settlements and relationship-related settlements. The former are typically characterized by "individual or family migration, spontaneous embeddiment, and internal heterogeneity" due to lack of dominant traditional bonds, economic connections, and social communication. One representative example is the mixed settlements of migrant workers that exist in large numbers and with wide distributions in urban villages throughout China. Regarding relationship-related settlements, due to the continuation of traditional bonds and interweaving of multiple connections among residents, settlement members maintain a certain degree of continuity with the place of population outflow in terms of cultural background, conceptual awareness, lifestyles and habits, as well as the cohesion generated therefrom while they are establishing various economic ties with the place of population inflow. Classic examples include geography-based settlements such as the "Zhejiang Village" and "Henan Village" in Beijing and transnational ethnic communities such as the "Guangzhou Xiaobei Black Neighborhood." Owing to the diversity of the group composition, the "integration-response" evolution process of the economic-type special community settlement space is manifested by the complexity in the five dimensions and the interactivity between the two ends. Based on Chinese migrant worker settlement samples, the evolution model is analyzed from the five dimensions in accordance with the gradient (see Table 6.2):

Table 6.2 Interpretation for evolution model of scenario A

Dimension	End		Schematic diagram of the evolution model
	"Integration end"	"Response end"	
Spatial	[Community location] Spontaneously residing in and scattering across urban fringe areas; [Transportation and travel] [Housing condition] Actively adapting to the current status; [Service facility] Meeting the rigid demand of daily life	[Urban social space] Heterogeneity; [Public service facilities] [Public space] An increase in the usage rate or the phenomenon of expropriation	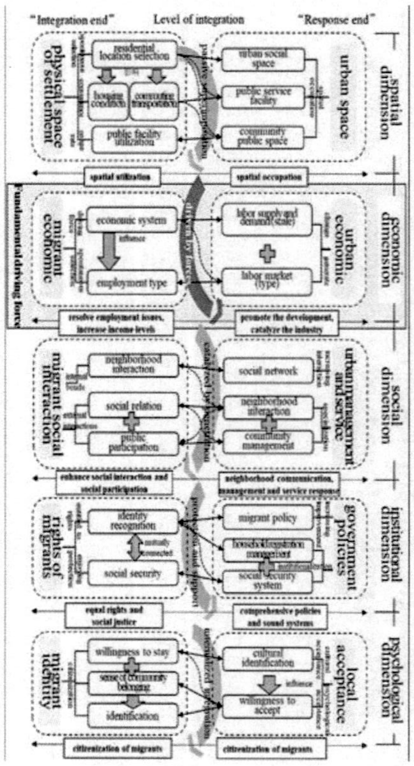
Economic	[Economic system] Relationship-related settlement or settlement equipped with an industrial system; mixed settlements cannot form a large-scale economy [Employment type] Individual employment; lower-end employment	[Labor supply and demand] Job supply and demand fluctuate; [Labor market] Monopolize middle and low-end markets	
Social	[Neighborhood interaction] No social bonds; [Social relation] [Public participation] Barriers exist	[Social Network] Heterogeneity; [Community Management] [Neighborhood interaction] No evident exclusion	
Institutional	[Identity recognition] [Social security] Increasing urbanization and regularization;	[Migration policy] [Household registration Management] [Social Security System] Continuous feedback, improvement and systematization	
Psychological	[Willingness to stay] Relatively strong; [Identity recognition] Desire for citizenship; [Sense of community belonging] Polarization	[Cultural identity] [Willing to accept] Positively correlated to social influence	

6.3.1.1 Spatial Dimension

Alongside the occurrence of consumption and trading behaviors in rental-based space, the corresponding settlement space for economic-type special communities is generated therefrom. The overall manifestation involves a process of superficial integration under "passive spatial embeddiment." As the largest economic-type special community in China, in terms of the evolution of the spatial dimension, the migrant worker group usually actively choose to reside and agglomerate in urban fringe areas (which are characterized by relatively low rents, dilapidated spatial environments, and inadequate supply of public services) based on a comprehensive consideration of factors such as cost, housing, transportation, and management. Migrant workers become embedded into the spatial system of local communities and even cities in an isolated state and present the typical characteristic of "marginalized location selection under multiple constraints (time, rent, housing, etc.) as well as active adaptation and embeddiment into the current state in terms of housing conditions, transportation and travel, service facilities, etc. On the corresponding "response end," due to the new demands and markets brought about by this type of settlement, phenomena such as "increases in the supply and demand of service facilities, disorderly occupation in public space, and heterogenization of social space" appeared in the communities and surrounding areas. These phenomena have also posed higher requirements for the level of urban management and public security (which involves the response in the institutional dimension).

6.3.1.2 Economic Dimension

Economic-type special communities generally migrate to large cities to increase income levels and improve living standards. On this ground, integration in the economic dimension becomes the fundamental driving force for their survival and evolution and directly determines whether they can gain a foothold in the place of population inflow. The overall manifestation involves a solid process of integration "positively driven by economic forces." Among such communities, majority of the relationship-related settlements establish their own effective forms of industrial organization while providing residents with necessary residential places and employment opportunities (the few relationship-related settlements without a self-built industrial system are akin to mixed settlements). The migrant workers sampled in this study are mainly sourced from typical mixed settlements. Residents basically live in a discrete state, caring about their own affairs and, hence, lacking essential economic connections and social interactions with each other. However, as individuals in the labor force, these migrant workers may still achieve non-agricultural employment through the urban job market but mostly flow to middle and low-end commercial service economic activities. As such, a large number of cheap labor resources have been injected into the mainstream market. However, they are also likely to compress the local middle and low-end labor markets. In sum, the "integration end" shows the characteristics that "the settlements are unable to establish their own industrial system and that community members tend to enter low-end employment," while the "response end" shows the characteristics of "fluctuations in the structure of urban labor supply and demand and marked involution of migrant workers in middle and low-wage job markets."

6.3.1.3 Social Dimension

Regarding the social dimension, while relationship-related settlements still retain traditional community ties and display stronger community cohesion and integration, the more widely distributed mixed settlements become makeshift and single places of residence for lack of their own industrial system and unified community bonds, manifested as a process of limited integration with "barriers to social connections." As far as migrant workers are concerned, their settlements are formed mainly through the common location selection to make profits. Residents contribute to urban economy yet encounter exclusion and alienation in social and cultural life. Despite their prolonged residence in and familiarity with the city, they experience profound difficulty surpassing the multiple thresholds for citizenization. Due to their dispersed residential pattern, these individuals may easily become "invisible" in an obscure corner of the city and turn into the so-called "marginal community." To sum up, the "integration end" presents the characteristics of "barriers to external social connections of the community and lack of common bonds within the community." As a result of the direct impact of the spatial and economic dimensions on the mainstream society, the "response end" exhibits the characteristics of "heterogenization of the mainstream social network and a certain extent of repulsion among urban residents." Admittedly, owing to the lack of solid social ties, such special community settlements are more likely to dissolve as a result of internal differentiation (some members are assimilated in the mainstream society due to upward mobility).

6.3.1.4 Institutional Dimension

Due to the high relocation rates among individuals and families, economic-type communities often do not receive timely and effective support from local policies when they first move into the place of population inflow. There may even be institutional barriers to identity recognition, mostly manifested as a process of structural integration with "regularization of special identities." Migrant workers mostly come from rural surplus labor force suppressed by the urban—rural dual structure and migrate spontaneously. Serious institutional imbalances exist between migrant peasants and urban residents in terms of household registration status and social security, among others. On the other hand, the "chain migration" and "marginal agglomeration" characteristics of migrant workers also pose challenges and pressures to urban population and public security management. This can be described as a process of citizenization for the migrant agricultural population. In sum, the "integration end" displays the evolutionary characteristics of "active familiarization with the rules of the mainstream society and regularization and citizenization of special identities," while the "response end" is manifested as the process of "self-improvement and systematization of the original management system and social security under the impact and challenge of non-regular communities."

6.3.1.5 Psychological Dimension

Mobility per se is often a typical manifestation of market-oriented employment competition. This is especially true for economic-type special communities. While being absorbed into the grand urban economic system, they often display a subjective willingness to stay in the place of population inflow. The overall manifestation involves a process of deep integration featuring "internal two-way identification." As far as the case of migrant workers is concerned, profound differences exist in the various norms in urban and rural cultures. If migrant workers actively adapt to and accept urban cultures and codes of conduct from the heart, an upward mobility process of "citizenization of special communities" occurs. On the contrary, deviant behaviors may arise due to a sense of incompatibility with the urban norms, which, in turn, will produce a negative impact of "isolation and repulsion" on the psychology of the original residents in the place and community of population inflow. As such, the "integration" in the psychological dimension shows subjective initiative and, hence, the "integration end" manifests the following characteristics: "a stronger subjective willingness to stay, consistent psychological tendency toward desiring citizenship, and polarization between adaptation and repulsion to the culture and codes of conduct (sense of belonging) in the place of population inflow." On the corresponding "response end," the following characteristic is observed: "positive correlations exist between the cultural identification and psychological acceptance of the place of population inflow, and the social influence of special communities."

Based on the evolution scenarios of economic-type special community settlements and analysis of typical cases, differentiated forecasts may be provided regarding the evolution results from the ideal state and actual situation. In terms of the ideal state, this type of special community appears more likely to achieve the ultimate integration with the place of population inflow due to a lack of multiple bonds and rooted identities. If demolition and relocation in the spatial dimension are disregarded, such settlements are likely to lose their heterogeneity and become "assimilated to the place of population inflow." On the other hand, as far as reality is concerned, economic-type special communities tend to spontaneously reside in existing communities located in the urban–rural fringe, an area with low economic costs and limited conditions, in view of multiple constraints and considerations. This unanimous consensus also results in the anchorage nature of the settlement space. Whenever a member moves out because of upward mobility, a steady stream of new members enters the community, thereby maintaining the survival of the space in a dynamic equilibrium. On this ground, the following inference may be derived: this type of special community settlement space is more likely to "coexist" within the place of population inflow for a long time in the real world.

6.3.2 *Scenario B: Settlement Space of Social-Type Special Communities*

The settlement space of social-type special communities largely refers to the form of settlement space established on the foundation of kinship, geography, or industry-based relationships. This type of settlement is formed by special communities that migrate into cities for social and cultural affairs (e.g., pursuing higher education), historical reasons (e.g., the overall migration of ethnic minorities based on traditional bonds), etc. These special communities gather as groups from the same village, industry, or ethnicity (race or nationality). Such settlement space reflects an extension of traditional bonds under the conditions of modern cities while displaying the characteristics of "group migration, rapid scaling-up, and internal homogeneity." Classic examples include international student settlements comprising foreign students in China and minority settlements characterized by geographic relationship and national character. Influenced by strong internal ties (mostly ethnicity, kinship, or geographical relationship) and rooted identities, the evolution process of social-type special community settlement space manifests relative closedness in dimensions and imbalances between the two ends. Based on the samples of international student settlements and Hui communities, this research analyzes their evolution models from four dimensions (without involving the institutional dimension due to the "regularity" of their identity) in accordance with the gradient (see Table 6.3).

6.3.2.1 Spatial Dimension

Social-type special communities often "form clustered settlements that are mixed with dwellings of original urban residents" in the place of population inflow due to their own cultural ties. The overall manifestation involves a process of active agglomeration characterized by "spatial isolation and juxtaposition." Ethnic minorities (represented by the Hui people) were among the earliest Chinese special communities that migrated out from their places of origin and gathered in other places. Affected by the wave of modernization, their settlements have undergone a transformation from local agglomeration to dispersion across the city, gradually shifting from ethnic aggregation to pluralism in terms of personnel composition, land use, and spatial distribution. However, a considerable number of Huis still live in this type of settlement and are termed as "Place Name-Hui District," which indicates an agreed-upon spatial separation from the city. By contrast, as emerging products in the context of globalization, majority of the international students in China reside in areas around colleges and universities in large cities with advanced education levels (exclusive of the dormitories provided by colleges and universities for international students). They spontaneously embed themselves in existing communities as students taking up short-term relocation, living together with the original urban residents at the community level, but forming a separate residential pattern at the unit level. As such, the settlement space of social-type special communities mostly

Table 6.3 Interpretation for evolution model of scenario B

Dimension	End		Schematic diagram of the evolution model
	"Integration end"	"Response end"	
Spatial dimension	[Residential location] Dispersed pattern with small-range isolation [Housing condition] [Commuting transport] [Public facility] Active adaptation	[Urban social space] Heterogenization; [Community Space] Labelized [Public transportation] [Public service facilities] Expropriation	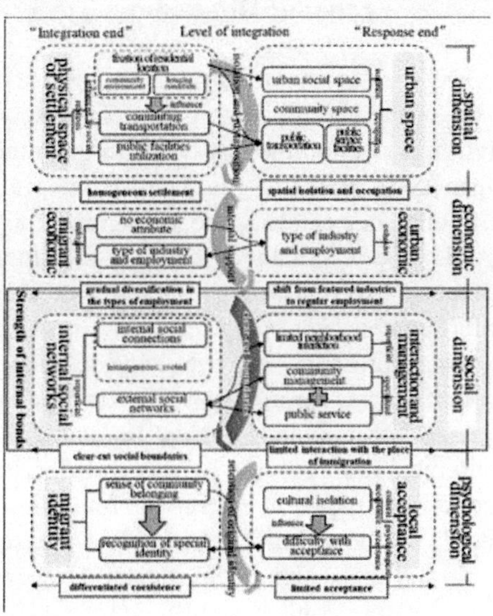
Economic dimension	[Type of industry employment] Transition from traditional featured industries to diversified employment	[Type of industry and employment] Transition from special types of employment to regular labor employment	
Social dimension	[Internal social connections] Maintaining internal cohesion [External social networks] Slowly establishing limited connections	[Neighborhood Interaction] Gradually deepened [Community Management] [Public Service] Systematization and Specialization	

present the following characteristics on the "integration end": dependent settlements at the city level, separate units mixed with original urban residences at the community level, and seeking active adaptation and embeddiment into the current state in housing conditions, transportation and travel, and service facilities, among others. On the "response end," due to the embeddiment of new residents and demands, the following characteristics are manifested: "the heterogenization of urban social space and labeling of community space" (e.g., the emergence of featured industrial spaces catalyzed by international students finds response in the economic dimension).

6.3.2.2 Economic Dimension

This type of special community exhibits two differentiated evolutionary characteristics in the economic dimension. The first type of special community does not display significant economic attributes (e.g., international students who do not engage in business activities to generate incomes); hence, evolution in the economic dimension does not exist on the "integration end." However, the domestic market finds in this type of special community a tremendous potential as a consumer group and has accordingly developed a series of exotic-style featured businesses in catering, entertainment, shopping, etc. As a result, the feedback of "constant emergence of featured industries" has been formed on the "response end." In comparison, the other type of special community relies on the featured industries that have been created on the basis of its unique culture to maintain its own identity in the city. The minority settlements represented by the Hui districts have experienced such changes in the economic dimension. Before the founding of the People's Republic of China, the Hui people owned a Hui-featured industry based on the community and supported by the cattle and sheep slaughter industry. However, with the sweeping wave of modernization, the Hui people have broken through the limitations of ethnic attributes, striving to seek and expand more employment opportunities outside the community. For this reason, the "integration end" shows the characteristics of "transition from traditional ethnic industries to diversified employments," while the "response end" has gone through the evolution process "from providing special industries and services to providing diversified industries (including ethnic industries) and services for cities."

6.3.2.3 Social Dimension

The social dimension serves as the effective bond and strong rooting force for the survival and evolution of the social-type special community settlement space, which is generally manifested as a process of root integration and resistance with "distinct community boundaries." International students have two major types of social networks: profession-based connections with colleges and universities and ethnic social networks among international students. However, with their long-lasting establishment in cities, minority settlements have already formed social relations characterized by "a continuation of the internal ties of the community and gradual strengthening of external social connections." Conversely, to guide, manage, and help the above two types of special communities, the place of population inflow has also successively established organizations and institutions as well as a series of policies and systems. Although the two community types are in different stages regarding urban connections in the social dimension, they display a similar evolutionary trajectory: the "integration end" shows "a slow transition from homogeneous to heterogeneous social connections while maintaining its internal cohesion"; the "response end" displays "increasing acceptance among urban residents, and gradual systematization and specialization of urban management and community governance."

6.3.2.4 Psychological Dimension

Having long relied on strong cultural roots and community bonds, this type of special community tends to maintain a certain degree of "independence and alienation" psychologically. Such closedness and isolation derived from a subconsciousness of self-protection is reflected in the aforementioned spatial, economic, and social dimensions. The general manifestation involves a process of limited integration with "retention of special identities." As transnational migrants, international students and the like are removed from the mainstream society in terms of cultural background, ideological concepts, and other aspects. Therefore, mutual acceptance becomes rather difficult on a psychological level. Despite their prolonged coexistence, minority groups remain "isolated" from the mainstream society and come under the accumulated influence of traditional bonds. Due to such influence, the "integration end" often presents "an intense sense of belonging within the settlement, tendency to retain the special identity, and weak willingness to become citizenized," while the "response end" exhibits "a slow process of cultural acceptance and difficulty in bridging psychological distance."

Based on the evolution scenarios of social-type special community settlements and analysis of typical cases, the following observations are derived. The integration process of this type of settlement shows endogenous and cohesive inertia of resistance to changes brought about by external forces. In other words, while the strong connections among community members (mostly ethnicity, kinship, geography-based relationships) and their rooted identities may ensure their rapid adaptation in the economic and institutional dimensions within a relatively short period of time, such communities always maintain a clear-cut boundary in social integration. This also leads to an evolution result of an incomplete integration in the psychological dimension, and the possibility for such special communities to achieve long-term "coexistence" with the place of population inflow by maintaining their own cultural characteristics and unique social networks.

6.3.3 Scenario C: Settlement Space of Engineering/Construction-Type Special Communities

Engineering/construction-type special community settlement space generally refers to the form of settlement established by passive migration groups who have lost their land due to urban and rural engineering development and infrastructure construction and migrated to designated residential areas on a large scale and in an organized manner under the guidance and assistance of relevant supporting policies.

In essence, this type of settlement falls in the category of passively selected residential communities led by the government or enterprises and organized through unified planning and construction. On this basis, such settlement space is often characterized by "overall migration, fixed scale, and controllable personnel," primarily including land-lost peasant resettlements, ecological migrant resettlements, and relocated corporate employee communities. The survival and evolution of the settlement space of this type of special community are often motivated by the guidance along the institutional dimension, which is manifested in the "integration-response" interaction process under compulsory constraints. In other words, the "institutional dimension" is a prerequisite for the consensus attainment between migrants and the place of population inflow as well as the formation of settlements (hence dispensable from the dimensional analysis of the evolution process). Based on the sample of land-lost peasant resettlements, this research analyzes the evolution model from four dimensions in accordance with the gradient (See Table 6.4).

6.3.3.1 Spatial Dimension

In the spatial dimension, engineering/construction-type special community settlement space displays passive adaptability triggered by passive selection of residence under policy guidance. The general manifestation involves a process of passive embeddiment with "overall segmentation." For example, the land-lost peasant resettlements based on government investment and administrative allocation of land are often located in the urban fringe areas characterized by relatively low land prices. This location selection may increase the spatial agglomeration of low-income groups and spatial differentiation among different classes. In addition, a series of constraints and problems have emerged for land-lost peasants, including inconvenient commuting between work and residence, inadequate public services, and unfamiliar spatial arrangement. The "integration end" exhibits the following characteristics: "agglomeration of poverty neighborhoods in urban fringe areas and passive adaptation under multiple constraints such as inconvenient transportation and travel, inadequate service facilities, and unfamiliar spatial arrangement." On the other hand, the "response end" shows the characteristic of "relatively independent settlement space, increased isolation and differentiation from urban residents, and limited influence on the urban community system."

6.3.3.2 Economic Dimension

Integration in the economic dimension also provides a material basis for the survival and evolution of this type of special community (except for a small number of corporate employee communities that have been relocated in their entirety). The general manifestation involves a process of sudden isolation featuring "forced severance of the economic source." For example, land-lost peasants become potential new urban

Table 6.4 Interpretation for evolution model of scenario C

Dimension	End		Schematic diagram of the evolution model
	"Integration end"	"Response end"	
Spatial dimension	[Community location] Large-scale agglomeration on urban fringe areas; [Housing condition] Changes in the mode of housing cause difficulty for group adaptation; [Transportation and travel] [Public facility] Fulfillment of rigid needs	[Spatial isolation] Relatively independent and self-contained [Urban System] Minor impact	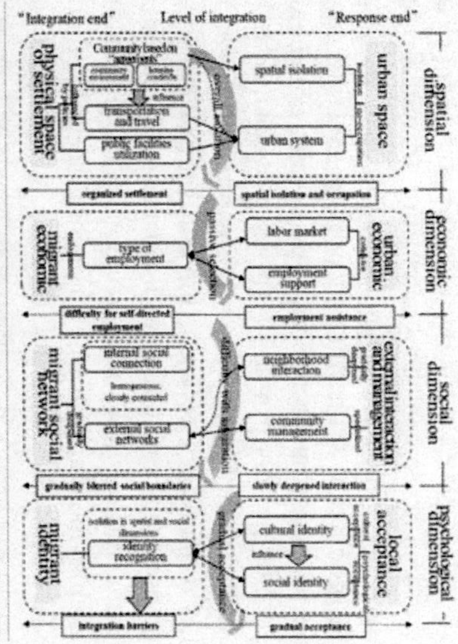
Economic dimension	[Type of employment] Great difficulty with self-directed employment; unstable employment	[Labor market] Providing opportunities [Employment assistance] Increasingly improved training and education systems	
Social dimension	[Internal social connection] Closed connected with geography and industry-based relationships; [External social network] Slowly deepening degrees of connection	[Neighborhood Interaction] Alienated relationships [Community Management] Difficulty with public security management	
Psychological dimension	[Identity] Changes in lifestyle and living environment lead to self-identity confusion	[Social Identity] [Cultural Identity] Slowly deepening	

labor resources after being cut off from the previous sources of income from agricultural activities as a result of land acquisition (originally self-sufficient in food supply). For lack of competitive advantages over urban residents in terms of education level and production skills, they face more exacting employment thresholds and economic pressure than economic-type special communities. In response to employment difficulties confronting land-lost peasants in real life, pertinent local departments in the place of population inflow often promote and implement corresponding skills training, entrepreneurship cultivation, and self-employment policy support among others in conjunction with the urban job market. As such, this type of settlement space tends to show the characteristics of "difficulties with self-directed re-entry into employment and insufficient income security" on the "integration end," while the place of population inflow displays the characteristics of "creating re-employment opportunities for new urban immigrants and providing targeted employment assistance and increasingly consummate training systems for the immigrants" in cooperation with the employment market" on the "response end."

6.3.3.3 Social Dimension

This type of special group tends to maintain the internal bonds of the original community (before relocation) and rarely seeks major social breakthroughs despite relocation to a new place and change of settlement space. In general, the social dimension shows a process of limited integration with "difficulties in transcending social boundaries." There are extremely limited boundary-crossing external exchanges. For instance, although the new settlement space and modern community form have changed the mode of social interaction among land-lost peasants, from the traditional "door-to-door," face-to-face social interaction to telephone contact, no substantial change has been observed regarding the objects they interact with. Confined by psychological barriers and remote locations, land-lost peasants rarely take the initiative to establish new social connections with the residents in the place of population inflow. In sum, the "integration end" mainly presents the characteristics of "continuing the previous homogeneous mode of social interactions internally and maintaining apparent social isolation externally," while the "response end" manifests "social alienation from the place of population inflow and relatively serious difficulty with urban management and community governance."

6.3.3.4 Psychological Dimension

For this type of special community, the key to integration in the psychological dimension lies in whether they can adapt to and accept the "new" environment and the "new" lifestyle. The overall manifestation involves a process of in-depth integration with "slow two-way acceptance." Land-lost peasants generally consider that although they have become registered permanent urban residents and moved into storied buildings, they still feel different from original urban residents. They are still surrounded by peasants who have relocated from the same villages; only their living environment has improved. On this basis, the "integration end" always exhibits the characteristic of "self-identity confusion," while the corresponding "response end" presents an overall characteristic of "a slow process of urban social and cultural identity."

The integration of the engineering/construction-type special community settlement space starts from the phenomenon of overall migration brought about by the institutional "agreement" between the special community and the city due to urban and rural engineering development and infrastructure construction. Certain connections or bonds exist among the internal individuals (mostly kinship, geography, industry-based relationships). Based on the evolution patterns, two possible evolution results are inferred: assimilation or coexistence. First, if the effect of the institutional dimension is discontinued, and the constituent members become accustomed to the place of population inflow, this type of settlement will eventually forsake its unique identity and gradually integrate into the mainstream society. However, the nature of its inherent identity will lengthen the process of "assimilation." By contrast, if special policies and management regulations formulated specifically for this community continue to function, or community members remain unable to give up their original identities and characteristics over a long span of time, the settlement may passively maintain a "coexisting" relationship with the place of population inflow. Nevertheless, such relationship may be forcibly terminated due to its negative impact or changes in external policy.

6.4 Conclusion

As a quintessential special community type in the Chinese social structure, special communities tend to present unfamiliar yet rich and diverse features in the survival and evolution of their settlement space. On the foundation of reviews on and references from existing theoretical research, this study establishes a general explanatory framework for the settlement space of special communities using the perspective of "integration-response" and provides multi-scenario interpretations for different types of settlement spaces. Through the analytical perspective of the "integration-response" two-way linkage, we can perceive the survival and evolution of "settlement space of special communities" as a process of interactions between two relative variables, i.e., special communities and urban residents or settlement space and the place of population inflow. On one hand, the process exhibits a series of complex situations (e.g., tolerance and resistance; acceptance and separation) experienced by special communities in the process of embedding themselves into the city. On the other hand, it reflects the dramatic changes that have occurred in the urban social space as a result. Against the backdrop of globalization and new-type urbanization in China, the discussion on the survival and evolution of Chinese special community settlement space provides the following inspirations:

First, the "two ends-five-dimensions" explanatory framework has revealed specific characteristics and patterns regarding the evolution of the special community settlement space and the place of population inflow in multiple dimensions as well as their constituent elements. These include changes in dominant characteristics, such as spatial and facility utilization, supply—demand relations of the labor market, and social security systems, as well as fluctuations in internal mechanisms, such as social network connections, psychological identification, and acceptance. On this ground, to promote "people-oriented" positive interactions between the settlement space of special communities and the place of population inflow, the engagement of all sectors of society is required and so is the guidance on all "visible" and "invisible" levels.

Second, the analysis of the "integration-response" evolution model shows that while the "integration end" and the "response end" maintain "progressive temporality in dimensions and consistency between two ends," multiple contradictory interaction effects have arisen in each dimension. For example, the fact that the settlement space of special communities occupies the urban space may produce negative impact such as traffic congestion and decline in the quality of the public environment. In terms of social management, the place of population inflow has been unable to achieve equality and unity in the treatment of special communities/their settlement spaces and urban residents/communities, which has imperceptibly engendered an outward "push force" against the special community. In fact, the mutual integration between the settlement space and the place of population is not "without frictions or obstructions." As such, seeking a path to deepen integration amidst the conflict of interest between the "push and pull" factors requires long-term exploration.

Third, the scenario interpretations for the three types of typical special community settlement spaces demonstrate that these settlement spaces have rather similar basic dimensions and model paths for their evolution. However, the same dimension bears varying degrees of significance for the evolution of different settlements, exhibits distinct characteristics and implications, and presents different levels of difficulty for "integration-response" (see Table 6.5). Fundamentally speaking, such variance is closely associated with the causes of settlement formation. For migrant worker settlements, a quintessential economic-type special community settlement space, the "integration" and "response" in the economic dimension (e.g., employment type, income level) serve as the fundamental driving force for their evolution and exert a certain degree of impact on other dimensions. As for engineering/construction-type special community settlement space (e.g., land-lost peasant resettlements), relevant policy preferences, management regulations, and other institutional factors have existed as fixed "agreements" since the establishment of the settlements; therefore, in their subsequent evolution process, the institutional dimension can no longer be perceived as an effective dimension to characterize the evolution of "integration" and "response." In general, in-depth investigations should be conducted to explore the specific scenarios, mechanisms, and problems for the evolution of each type of special community settlement space as well as corresponding sub-types and typical cases on the basis of the common characteristics and universal patterns for their evolution. Furthermore, hierarchical, departmentalized, and humanized response measures should be put forward in a targeted manner.

Table 6.5 Comparison of difficulty levels for "Integration-Response" of three types of scenarios in different dimensions

	Economic-type special community settlement space		Social-type special community settlement space		Engineering/construction-type special community settlement space	
	Level of difficulty for integration	Level of difficulty for response	Level of difficulty for integration	Level of difficulty for response	Level of difficulty for integration	Level of difficulty for response
Spatial dimension						
Economic dimension						
Social dimension						
Institutional dimension	(N/A)	(N/A)	—	—	—	—
Psychological dimension						

**Note* This table offers a horizontal comparison of the three types of settlement spaces in the same dimension and does not involve comparisons between dimensions. (The darker the color, the higher is the level of difficulty for integration/response.)

References

Alba R, Nee V. Remaking the American mainstream: assimilation and contemporary immigration. Boston: Harvard University Press; 2005.
Dan Y, Jing-xiang Z. Research on production of informal space from the perspective of practice of everyday life: a case study of Kongpu District in Ningbo. Hum Geogr. 2015;30(5):57–64.
De T, Chang-chun F. Review and progress of the research on Urban Villages. Hum Geogr. 2009;24(6):29–35.
Dongmei M, Xiaojian C, Ningna Y. Distribution characteristics and influencing factors of muslim settlements in Xihaigu Region of Ningxia. J Arid Land Res Environ. 2016;30(4):1–7.
Du J, Hou J, Wei X. Study on the transfer of Surplus Labor force in China's countryside. Learning Explo. 2012;11:112–115.
Entzinger H, Biezeveld R. Benchmarking in immigrant integration. Rotterdam: Erasmus University Eotterdam; 2003.
Feng X. "Ground roots"? Social adaptation of rural migrants in the Three Gorges. Sociological Res. 2004;5:21–29.
Gordon M. Assimilation in American life. Chicago: University of Chicago Press; 1964.
Han-sheng W, Shi-ding L, Li-ping S, et al. "Zhejiang Village": A unique way for Chinese farmers to enter the city. Soc Stud. 1997;1:58–69.
Hirschman C. America's melting pot reconsidered. Annu Rev Sociol. 1983;9(1):397–423.
Hong-yan Z, You-tao B. Study on the culture change of Nanjing Qijiawan Traditional Hui Community. Ethno-National Stud. 2004;4:38–47.
Jing-xiang Z, Yi H, Dong-qi S. The physical space change and social variation in urban village from the perspective of space production: a case study of Jiangdong Village in Nnajing. Hum Geogr. 2014;29(2):1–6.
Ji S, Lin X, Cui X. The embedment of foreign R&D and human Capital Formation: The action of population migration. J Central Univ Finance Econ. 2020;2:79–95.
Jing-yu L. From Closedness to Openness: The Pattern of Transformation in Urban Hui Communities. J Minzu Univer China (Philos Soc Sci Edn) 2003;20(1):7378
Junger-Tas J. Ethnic minorities, Social integration and crime. European J. Criminal Policy Res. 2001;9(1):5–29.
Knox P, Pinch S. Urban social geography: an introduction. Shanghai: The Commercial Press; 2005.
Li Z, Xue D, Du F, et al. The local response of "Social space of Transnational Immigrants" under globalization: A Case study of Xiaobei Black District in Guangzhou. Geographical Res. 2009;28(4):920–932.
Liu X. Translation of Chinese and American websites into English. China Translation. 2012;33(001):92–97.
Massey DS, et al. Theories of international migration: a review and appraisal. Popul Dev Rev. 1993;19:43166.
Myers D, Gao X, Emeka A. The gradient of immigrant age-at-arrival effects on socioeconomic outcomes in the U.S. Int Migr Rev. 2009;43(1):205–29.
Park R. Human migration and the marginal man. Am J Sociol. 1928;33(6):881–93.
Pei-lin L. Tremendous changes: the end of villages—a study of villages in the center of Guangzhou City. Soc Sci China. 2002;1:168–79.
Perlman J, Waldinger R. Second generation decline? Children of immigration, past and present- a reconsideration. Int Migr Rev. 1997;31(4):893–922.
Piore MJ. The dual labour market, theory and application. In: Barringer R, Beer SH, editors. The state and the poor. Cambridge: Cambridge Mass Winthrop; 1970.
Portes A, Zhou M. The New Second Generation: segmented assimilation and its variants. Annu Am Acad Polit Soc Sci. 1993;530(11):74–96.
Shuang C, Wan-min Z, Si-Run H. Urban Village renovation from the perspective of human settlement environment theory: a case study of Wuhan. City Plan Rev. 2009;33(8):37–42.
Stark O, Bloom OE. The new economics of labor migration. Am Econ Rev. 1985;75:173–8.

Tian K. Investigation, analysis and reflection on urban adaptability of migrant workers. Social Science Res. 1995;5:90–95.

Tian M, Peng Y. Spatial differences of urban integration of floating population: A case study of six cities in the east Coast. Urban Planning Rev. 2014;000(006):9–16,31.

Tong X, Ma X. "Harmonizer" and "Integration"—Community integration of new immigrants in cities. Social Science Res. 2008;001:77–83.

Wei W, Lu S. Endowment characteristics and Opportunity Structure: An Analysis of intergenerational differences in social Integration of urban migrant Population. China Rural Survey. 2012;000(001):13–21.

Xiao W. An investigation into marginal community. City Plan Rev. 2003;27(7):40–45; Urban Plan. 2003;27(7):4045.

Xiao Z, Xu S, Liu J. Evaluation of the social integration of Urban Floating Population in China: A Case study of 50 major cities. Population Res. 2019;5:96–112.

Xiao-ying F. Transformation of "Urban Village" and Cooperative Governance of the Floating Population Concentrated Communities in Beijing. Popul Res. 2010;34(6):55–66.

Xu Q, Fan T, Yuan Q. Research on the impact of China's population migration policy on industrial structure upgrading—An empirical test based on panel Data of 31 Provinces in China from 2000 to 2016. Macro Quality Res. 2019;4.

Ya-nan W. Anthology of capitalism and classical political economy. Beijing: The Commercial Press; 1979. p. 74.

Yu Y, Gao X, Guo Q. Research on the social integration of the New Generation of rural-urban floating population—Based on the investigation and analysis of Shanghai. Population and Economy. 2012;1:57–64.

Yuan R, Min-le W. Social integration of floating population in Urban China: a literature review. Popul Res. 2006;30(3):87–94.

Zhang W, Lei K. Analysis on the structure, status and influencing factors of social integration of urban new immigrants. Sociological Res. 2008;005:117–141.

Zhao X, Li Z. Local identity and integration of new immigrants in China's big cities. Urban Planning Rev. 2013;(12):22–29.

Zhi-gang L, Ye L, Hong-sheng C. Characteristics, mechanism, and spatiality of neo-migrants' homeplace-based communities in China: a case study of Hubei Village, Guangzhou. J Geogr Res. 2011;30(10):1910–20.

Zhou H. Measurement and theoretical thinking of social integration of floating Population. Population Res. 2012;3:27–37.

Zhu L. Introduction to social problems. Social Sciences Academic Press; 2002.

Appendix

Appendix 1: Questionnaire for the Community of International Students

1. **Gender: ___Age: ___ University: _______ Nationality: _____ Marital Status:**☐ Single ☐ Married ☐ Children **Religion:** ☐ Buddhism ☐ Christianity ☐ Catholicism ☐ Others ___ ☐ None
2. **Your Residence:** ☐ Dormitory of university ☐ Outside residential area _______ (Community name)**, if living outside of school, the reason:** ☐ More facilities for living ☐ More freedom to work and rest ☐ Choose your roommate freely ☐ Other reasons ______
3. **The degree you are studying for:** ☐ Communication ☐ Bachelor ☐ Master ☐ PhD ☐ Postdoctoral
4. **Your purpose of studying in China (Multiple Choice):** ☐ Learning Chinese ☐ Learning Chinese Culture ☐ Expertise ☐ Other ______
5. **Time to Nanjing:** _________
6. **Would you like to stay in Nanjing after graduation?** ☐ Yes ☐ No. The reason: ________
7. **Rent for accommodation / dormitory (RMB / month) ______; apart from rents, your monthly expenses (RMB) ___; do you have part-time employment ___, if have, income___**
   (a) ¥1000 (b) ¥1000–2000 (c) ¥2000–4000 (d) ¥4000 or more
   【Living situation】
8. **Your housing area: _____ m², apartment layout: ________, whether shared:** ☐ Yes ☐ No, **how many people living together:** ______
9. **What are the internal facilities (multiple choice):** ☐ kitchen ☐ toilet ☐ water ☐ Internet ☐ TV ☐ water heater ☐ air-conditioning ☐ washing machine ☐ other ______
   【Learning Activities】

X. Wu and L. Wang, *Settlement Spaces: Urban Survival Prospects of China's Special Communities*, Springer Geography,
https://doi.org/10.1007/978-981-16-4892-2

10. **Where do you usually study (multiple choice):** ☐ Classroom ☐ Study Room ☐ Studio ☐ Library ☐ Residence ☐ Other _________
11. **How do you usually learn (multiple choice):** ☐ class ☐ self-study ☐ discussion and communication ☐ other _________, **how many people study together:** ____

【Shopping Activities】

12. **(multiple choice) Where do you buy daily necessities (food, soap, laundry detergent, etc.):** ____ **Where to buy the optional products (clothing, cosmetics, books, etc.):** ____, **Where to buy the special products (home appliance, electronics, etc.):**_____ (a) supermarkets of university (b) Convenience stores or small business near your house (c) Commercial streets near your house (d) Large supermarkets or shopping centers slightly further (e) Commercial streets slightly further (f) City shopping centers/shopping mall (g) Other ____
13. **How often do you buy daily necessities (food, soap, laundry detergent, etc.):** ____, **optional products (clothing, cosmetics, books, etc.):** ____, **special products (home appliance, electronics, etc.):** ____ (a) once a day (b) 4 to 6 times a week (c) 1 to 3 times a week (d) 2 to 3 times a month (e) once a month or more (f) never

【Leisure activities】

14. **Which are your main leisure activities (multiple choice, "☑"), and choose the frequency of the activities (fill the right letter in "__"):** ☐ Cultural and entertainment (sing and dance, film, Internet, radio, etc.), **frequency** ____; ☐ Communication (chat, parties, etc.), **frequency** ____; ☐ Sports (ball games, walking, fitness, etc.), **frequency** ____; ☐ Go sightseeing (go to parks, hiking, excursions, etc.),**frequency** ____; ☐ Interests (poetry, painting, photography, Writing, invention, etc.), **frequency** ____; ☐ Public welfare (volunteers, community management, special services, etc.), **frequency** ____. (a) once a day (b) 4 to 6 times a week (c) 1 to 3 times a week (d) 2 to 3 times a month (e) once a month or more (f) never

【Communication and recognition】

15. **How much do you know about Chinese culture and customs?** ☐ Know well ☐ Understand some ☐ Do not know well ☐ Do not know, **if you celebrate Chinese Festival:** ☐ Yes ☐ No
16. **What are your main contacts in Nanjing (multiple choice):** ☐ People in the same country ☐ International students in other countries ☐ Chinese schoolmates or friends ☐ Others _____
17. **How often do you communicate with the Chinese:** ☐ Very frequent ☐ In general, you may meet or chat on the road ☐ Occasionally, only know a few ☐ Not familiar at all
18. **What problems do you think exist in the current living environment (multiple choice):** ☐ The environment is not good enough ☐ The medical facilities are not available ☐ The sports and entertainment facilities are not full ☐ Fewer specialty stores of your country ☐ Not friendly enough around ☐ Others ________

19. **Please remember the track of your life yesterday** (Tick "√"in the right blank of activities at different times, and select corresponding letters of specific activities, trip distance and transportation)

Time/o'clock	**Type of Activities** (Please choose only one type of activities at every time period)							**distance**	**Transportation**
	Sleep	**Housework**	**Study**	**Shopping**	**(a)** daily necessities (food, soap, laundry detergent, etc.) **(b)** optional products (clothing, cosmetics, books, etc.) **(c)** special products (home appliance, electronics, etc.)	**Leisure**	**(a)** Cultural and entertainment (eg. sing /dance, film, Internet, radio) **(b)** Communication (eg. chat, parties) **(c)** Sports (eg. ball games, walk, fitness) **(d)** Sightseeing (eg. visit parks, hike, excursions) **(e)** Interests (eg. painting, photography, writing, invention) **(f)** Public welfare (eg. volunteers, community service)	**(a)** at home **(b)** $\leq$ 0.5 km **(c)** 0.5–1 km **(d)** 1–2 km **(e)** 2–3 km **(f)** > 3 km	**(a)** walk **(b)** bike **(c)** motorcycle **(d)** bus **(e)** subway **(f)** taxi **(g)** private car
Example				√	*c)*			*f)*	*e)*
0–2 a.m									
2–6 a.m									
6–8 a.m									
8–10 a.m									
10–12 a.m									
12–14 p.m									
14–16 p.m									
16–18 p.m									

(continued)

(continued)

Time/o'clock	Type of Activities (Please choose only one type of activities at every time period)							distance	Transportation
	Sleep	**Housework**	**Study**	**Shopping**	**(a)** daily necessities (food, soap, laundry detergent, etc.) **(b)** optional products (clothing, cosmetics, books, etc.) **(c)** special products (home appliance, electronics, etc.)	**Leisure**	**(a)** Cultural and entertainment (eg. sing /dance, film, Internet, radio) **(b)** Communication (eg. chat, parties) **(c)** Sports (eg. ball games, walk, fitness) **(d)** Sightseeing (eg. visit parks, hike, excursions) **(e)** Interests (eg. painting, photography, writing, invention) **(f)** Public welfare (eg. volunteers, community service)	**(a)** at home **(b)** $\leq$ 0.5 km **(c)** 0.5–1 km **(d)** 1–2 km **(e)** 2–3 km **(f)** > 3 km	**(a)** walk **(b)** bike **(c)** motorcycle **(d)** bus **(e)** subway **(f)** taxi **(g)** private car
18–20 p.m									
20–22 p.m									
22–24 p.m									

Appendix 2
Questionnaire for Resettlement Areas of Land-Lost Farmers

1. **Basic personal information:**
   Sex: ☐ male ☐ female Age: ______ Domicile: ☐ urban or rural
   Marital Status: ☐ Single ☐ Married ☐ Divorced ☐ Widowed
2. **Your degree of education:** ☐ did not go to school ☐ primary school ☐ junior high school ☐ technical secondary school ☐ college or junior college or above
3. **Your current professional identity:** ☐ management ☐ professional and technical personnel ☐ service personnel
   ☐ commercial service personnel ☐ farmers, forestry, animal husbandry and fishermen Staff ☐ Production and transport equipment operators ☐ Construction ☐ Other _____
   Nature of work: ☐ temporary ☐ contract ☐ unit worker **Your working intensity:** _____ days/week and h/day
4. **When you move into the community: ________ years. After you move into this community for ______ times work. Your average duration per job ________**
5. **Your family monthly income:** ☐ less than 500 yuan ☐ 500–1500 yuan ☐ 1500–2500 yuan ☐ 2500–5000 yuan ☐ 5000 yuan or more
6. **Your source of income (multiple choice):** ☐ social security ☐ rent ☐ salary ☐ independent operation income ☐ financial management income ☐ other _____
7. **Your average monthly balance:** ☐ Less than 300 yuan ☐ 300–500 yuan ☐ 500–1000 yuan ☐ 1000–1500 yuan ☐ 1500 yuan or more
8. **Your daily expenses in order (from highest to lowest, the most is 1):** Children's education___ food___daily necessities___medical care___ entertainment and culture___transportation___ correspondence _____ others___
9. **Your original place of residence is _______**
   How you are currently living (multiple options): ☐ Living alone ☐ With parents ☐ With spouse ☐ With children
10. **Your basic information to live**: family number _____ room number _____ total area _____ _____

X. Wu and L. Wang, *Settlement Spaces: Urban Survival Prospects of China's Special Communities*, Springer Geography,
https://doi.org/10.1007/978-981-16-4892-2

11. **Your housing source:** ☐ purchase of commercial housing ☐ purchase of affordable housing ☐ purchase of demolition and resettlement housing ☐ rental of public housing ☐ rental of commercial housing ☐ other __
12. **Your working location:** _____area_____street. **Time from home to work:** _____minutes
13. **Which mode of transportation do you usually choose (multiple options)?** walk/bike/electric car/bus/subway/rent/drive/other ___________________
14. **Rank your satisfaction with the facilities in the community (from high to low, the most satisfied is 1):**
    Medical____business___education____traffic______cultural_____sports____
    What do you think needs improvement?
    ____
15. **Your Social Security Participation (multiple choice):**
    ☐ Endowment insurance ☐ Medical insurance ☐ Unemployment insurance ☐ Work injury insurance ☐ Maternity insurance ☐ Housing accumulation fund
16. **Your own identity:** ☐ urban ☐ rural people
    How do you like this neighborhood? ☐ very much ☐ like ☐ general ☐ dislike ☐ very much
17. **What is the gap between you and the citizens (both inside and outside the community)?**
    ☐ There is no gap at all ☐ There is a small gap ☐ There is a big gap ☐ There is a big gap
18. **Your main contact objects:** ☐ accompanied by the original village residents ☐ other citizens (inside and outside the community)
    The number of friends you usually have with the public (inside and outside the community): ☐ No ☐ 1–3 ☐ 4–6 ☐ more than 6
19. **Do you have frequent contacts with citizens (both inside and outside the community)?**
    ☐ Very frequent ☐ Frequent ☐ Few ☐ No acquaintance, limited to greetings ☐ Almost no communication
20. **In case of difficulty, please prioritize the objects you ask for help (the order of the first object for help is 1):** Relatives _______ friends _______ neighbours _______ _______ leadership colleagues _______ other _______
21. **Do you agree with the way of life and behavior of citizens (in and out of the community)?**
    ☐ Strongly agree ☐ Generally disagree ☐ Strongly disagree
22. **Do you feel that your relationship with the citizens (inside and outside the community) is harmonious?**
    ☐ Very good ☐ good ☐ fair ☐ not good ☐ very bad
23. **Do you adapt to the new way of living and working in the city?**
    ☐ Very good ☐ Good ☐ Normal ☐ Not good ☐ Very good
24. **Are you willing to settle down in the city (as opposed to the countryside)?**
    ☐ Very much ☐ Yes ☐ Generally not ☐ Very much not

25. **Are you willing to work in the city (relative to the countryside)?**
☐ Very much ☐ Yes ☐ Generally not ☐ Very much not
26. **Are you willing to let their children in the city (relatively rural) development?**
☐ Very much ☐ Yes ☐ Generally not ☐ Very much not

Appendix 3
Survey Questionnaire of Hui Communities

1. **Your occupation is (please help the investigator to record the working place):**
   In the early years of the founding of China: □ religion related □ catering service □ handicraft □ industry □ non-staple food processing industry □ institutions □ retired or unemployed
   1958–78: □ Religion Related □ Catering Services □ Handicrafts □ Industry □ Non-staple Food Processing Industry □ Institutions □ retired or unemployed
   1978–90: □ Religion related □ Catering service □ Handicrafts □ Industry □ Non-staple food processing □ Institutions □ retired or unemployed
   1990–98: Religion related □ Catering service □ Handicrafts □ Industry □ Non-staple food processing □ Institutions □ retired or unemployed
   1998-present: □ Religion related □ Catering service □ Handicraft □ Industry □ Non-staple food processing industry □ Institution □ retired or unemployed
2. **Your monthly income is:**
   The early days of the founding of China: □ 20 yuan or less □ 20–50 yuan □ 50–100 yuan □ 100 yuan or more
   1958–78: □ less than 20 yuan □ 20–50 yuan □ 50–100 yuan □ more than 100 yuan
   1978–90: □ less than 60 yuan □ 60–100 yuan □ 100–200 yuan □ more than 200 yuan
   1990–98: □ less than 200 yuan □ 200–400 yuan □ 400–600 yuan □ more than 600 yuan
   1998 to date: □ 600 yuan or less □ 600–1000 yuan □ 1000–3000 yuan □ 3000 yuan or more
3. **Your commuting time is:**
   In the early days of New China: □ 15 min or less □ 15–30 min □ 30–60 min □ more than 60 min
   1958–78: □ Less than 15 min □ 15–30 min □ 30–60 min □ more than 60 min
   1978–90: □ less than 15 min □ 15–30 min □ 30–60 min □ more than 60 min

X. Wu and L. Wang, *Settlement Spaces: Urban Survival Prospects of China's Special Communities*, Springer Geography,
https://doi.org/10.1007/978-981-16-4892-2

1990–98: ☐ Less than 15 min ☐ 15–30 min ☐ 30–60 min ☐ more than 60 min
1998-present: ☐ less than 15 min ☐ 15–30 min ☐ 30–60 min ☐ more than 60 min

4. **Your commuting vehicle is (please ask the investigator to help record the commuting costs):**
In the early days of China: ☐ walking ☐ bicycle ☐ public transportation ☐ others ___
1958–78: ☐ walking ☐ bicycle ☐ bus ☐ taxi ☐ private car ☐ others ___
1978–90: ☐ walking ☐ bicycle ☐ bus ☐ taxi ☐ private car ☐ others ___
1990–98: ☐ walking ☐ bicycle ☐ bus ☐ subway ☐ taxi ☐ private car ☐ others ___
1998-present: ☐ walking ☐ bike ☐ bus ☐ subway ☐ taxi ☐ private car ☐ other ___

5. **The source of your housing is (please help the investigator to record the residence location):**
The early days of China: ☐ self-built housing ☐ to buy/rent public housing units
1958–78: ☐ Self-built housing ☐ Purchase/lease of public housing
1978–90: ☐ Self-built housing ☐ Purchase/lease of public housing ☐ Purchase/lease of commercial housing
1990–98: ☐ Self-built housing ☐ Buying/leasing public housing ☐ Buying/leasing commercial housing ☐ Buying/leasing affordable housing
From 1998 to now: ☐ Self-built housing ☐ Buy/rent public housing ☐ Buy/rent commercial housing ☐ Buy/rent affordable housing

6. **Your living status is:**
Early days: ☐ collective dormitory ☐ single room ☐ one room one hall ☐ two rooms one hall ☐ two rooms one hall or more (Building condition: one story/multiple story/high story)
1958–78: ☐ collective dormitory ☐ single room ☐ one room one living room ☐ two rooms one living room ☐ two rooms one living room or more (Building condition: one story/multiple story/high story)
1978–90: ☐ collective dormitory ☐ single room ☐ one room and one living room ☐ two rooms and one living room ☐ two rooms and one living room or more (Building condition: one story/multiple story/high story)
1990–98: ☐ collective dormitory ☐ single room ☐ one room and one living room ☐ two rooms and one living room ☐ two rooms and one living room or more (Building condition: one story/multiple story/high story)
1998 to now: ☐ collective dormitory ☐ single room ☐ one room one hall ☐ two rooms one hall ☐ two rooms one hall or more (Building condition: one story/multiple story/high story)

7. **Community public service facilities that you and your family often use include (three options):**
In the early years of China: ☐ Kindergarten ☐ Primary School ☐ Vegemets and meat shops ☐ Restaurations restaurants ☐ Hardware grocery stores ☐ Tailors ☐ barber shops ☐ Medical facilities ☐ Indoor/outdoor activities facilities ☐ Almost not used

1958–78: ☐ Kindergarten ☐ Primary School ☐ Vegetables and meat shops ☐ Restaurants restaurants ☐ Hardware grocery stores ☐ Tailors shop ☐ barber shop ☐ Medical facilities ☐ indoor/outdoor activities facilities ☐ Almost not used
1978–90: ☐ Kindergarten ☐ Primary School ☐ Vegetables and meat shops ☐ Restaurants restaurants ☐ Hardware grocery stores ☐ Tailors shop ☐ barber shop ☐ Medical facilities ☐ indoor/outdoor activities facilities ☐ Almost not used
1990–98: ☐ Kindergarten ☐ Primary School ☐ Vegetables and meat shops ☐ Restaurants restaurants ☐ Hardware grocery stores ☐ Tailors shop ☐ barber shop ☐ Medical facilities ☐ indoor/outdoor facilities ☐ Almost not used
1998-present: ☐ Kindergarten ☐ Primary School ☐ Vegetables and meat shops ☐ Restaurant restaurants ☐ Hardware grocery stores ☐ Tailors shop ☐ barber shop ☐ Medical facilities ☐ indoor/outdoor activities facilities ☐ Almost not in use

8. **Your family structure is:**
   In the early days of China: ☐ Living with spouse ☐ Living with children ☐ Living with more than two pairs of children or siblings ☐ Other
   1958–78: ☐ Living with spouse ☐ Living with children ☐ Living with more than two pairs of children or siblings ☐ Other
   1978–90: ☐ Living with spouse ☐ Living with children ☐ Living with more than two pairs of children or siblings ☐ Other
   1990–98: ☐ Living with spouse ☐ Living with children ☐ Living with more than two pairs of children or siblings ☐ Other
   1998 to present: ☐ Living with spouse ☐ Living with children ☐ Living with more than two pairs of children or siblings ☐ Other
9. **Whether you and your family strictly abide by religious taboos (diet, clothing, hygiene, marriage, funeral, business, interpersonal communication, spiritual life):**
   Early days of China: ☐ Yes ☐ No **1958–78:** ☐ Yes ☐ No **1978–90:** ☐ Yes ☐ No **1990–98:** ☐ Yes ☐ No **1998-present:** ☐ Yes ☐ No
10. **How often do you attend mosque services?**
    In the early days of China: ☐ 1–5 times a day ☐ 1–5 times a week ☐ 1–3 times a month ☐ less than 1 times a month ☐ almost not
    1958–78: ☐ 1–5 times a day ☐ 1–5 times a week ☐ 1–3 times a month ☐ less than 1 times a month ☐ hardly
    1978–90: ☐ 1–5 times a day ☐ 1–5 times a week ☐ 1–3 times a month ☐ less than 1 times a month ☐ hardly
    1990–98: ☐ 1–5 times a day ☐ 1–5 times a week ☐ 1–3 times a month ☐ less than 1 times a month ☐ hardly
    1998-present: ☐ 1–5 times a day ☐ 1–5 times a week ☐ 1–3 times a month ☐ less than 1 times a month ☐ hardly
11. **Your daily contacts include:**
    In the early days of China: ☐ relatives ☐ neighbors ☐ colleagues ☐ others (mostly Hui/Han)

1958–78: ☐ Relatives ☐ Neighbors ☐ Colleagues ☐ Others (mostly Hui/Han)
1978–90: ☐ Relatives ☐ Neighbors ☐ Colleagues ☐ Others (mostly Hui/Han)
1990–98: ☐ Relatives ☐ Neighbors ☐ Colleagues ☐ Others (mostly Hui/Han).
1998-present: ☐ Relatives ☐ Neighbors ☐ Colleagues ☐ Others (mostly Hui/Han)

12. **Your neighborhood communication frequency is:**
In the early days of China: ☐ daily ☐ 1–5 times a week ☐ 1–3 times a month ☐ Almost no contact
1958–78: ☐ Daily ☐ 1–5 times a week ☐ 1–3 times a month ☐ Almost no contact
1978–90: ☐ Daily ☐ 1–5 times a week ☐ 1–3 times a month ☐ Almost no contact
1990–98: ☐ Daily ☐ 1–5 times a week ☐ 1–3 times a month ☐ Almost no contact
1998-present: ☐ Daily ☐ 1–5 times a week ☐ 1–3 times a month ☐ Almost no contact

Appendix 4
Questionnaire on the Use of Public Service Facilities in Migrant Workers' Communities

Gender: Age: Marital Status: Lining time:

1. **Your level of education is:** ☐ primary school ☐ primary school ☐ junior high school ☐ junior college and above
2. **You are mainly engaged in:** ☐ Manufacturing ☐ Service ☐ Transportation ☐ Construction Machinery ☐ Others
3. **Your current company belongs to:** ☐ state-owned enterprises ☐ collective enterprises ☐ private enterprises ☐ foreign investment/joint ventures ☐ individual businesses ☐ others
4. **The composition of your family in Nanjing:** ☐ living alone ☐ husband and wife two people ☐ husband and wife with children ☐ parents
5. **Your working days per week:** ☐ 1–3 days ☐ 4–5 days ☐ 6–7 days
6. **Your working hours per day:** ☐ within 8 h ☐ 8–12 h ☐ more than 12 h ☐ uncertain
7. **Your average monthly salary is:** ☐ 1500 yuan or less ☐ 1500–3000 yuan ☐ 3000–5000 yuan ☐ 5000–8000 yuan ☐ 8000 yuan or more
8. **Your local housing situation:** ☐ self-purchase ☐ independent rent ☐ shared rent ☐ company dormitory ☐ others
9. **How you use the following facilities:**

category	level	instructions	Frequency/times per week	Duration/each time · hour	Consumption quota/yuan
sports	The city level	City gymnasium, swimming pool, etc			
	regional	District gymnasium, swimming pool, etc			

(continued)

X. Wu and L. Wang, *Settlement Spaces: Urban Survival Prospects of China's Special Communities*, Springer Geography,
https://doi.org/10.1007/978-981-16-4892-2

(continued)

category	level	instructions	Frequency/times per week	Duration/each time · hour	Consumption quota/yuan
	Community level	Community Room			
culture	The city level	Municipal Cultural Center, Art Museum, etc			
	regional	District Cultural Center, Art Museum, etc			
	Community level	Community Cultural Center			
The traffic	The city level	The subway station			
	regional	The bus stop			

10. **How you use the following facilities:**

category	level	instructions	Frequency/times per year	Duration/each time · hour	Consumption quota/yuan
medical	The city level	Tertiary hospitals			
	regional	The secondary hospital			
	Community level	Community health posts			

11. **Your shopping situation is about**

category	instructions	Grade (√)			frequency/Once a week	The length/hour at a time	Consumption quota RMB/
		The city level	regional	Community level			
Daily provisions	Commodities, consumables, etc						
Clothing shoes package	Clothes, trousers, shoes, socks, etc						
Health food	Medicine, raw food, cooked food, etc						
Home appliance digital	Home appliances, mobile phones, cameras, etc						
Office hardware	Books, pens, stationery, etc						
The film and television entertainment	KTV, movies, etc						
other							

Epilogue

Compared with the long-established productive community research framework developed in the West, China's planning sector is currently undergoing a paradigm shift and practical sea change from being a "work-unit-orientated nation" to becoming a "community-orientated nation." When inventory planning and infrastructure construction gradually replace incremental planning as the "new norm" in the development of Chinese cities and the "people-centered" ideas strongly advocated by the central government since the 18th National People's Congress lead to a new national development model and momentous change in political direction, community research (theoretical and practical) will play a more comprehensive social role, ushering in more new opportunities and challenges. For China's cities undergoing complex, overlapping social changes, "special community settlement spaces" undoubtedly constitute a special window in urban social spaces.

In this light, this study focuses on representative spaces that typically give rise to, or are currently undergoing, dramatic changes against the backdrop of social shifts in the landscape of China's cities. From the perspective of "communities," we hope to connect and analyze various representative "special community settlement spaces," launching an interdisciplinary line of research that incorporates theoretical interpretation with empirical analysis. The empirical section is conducted around four types of representative samples, namely, international student settlements, resettlement areas for land-lost peasants, minority settlements, and migrant worker settlements. We conduct an in-depth quantitative and qualitative analysis from several angles. The theoretical section uses the bidirectional "integration-response" theoretical framework to interpret the survival and evolutionary logic of special community settlements in Chinese cities in a wide variety of contexts. The four special communities and their settlement spaces selected for this study can be considered to manifest the common logic of "integration-response" in the different areas of "international migrants (moving into a new country)," "land-lost peasants (requisitioning land in the city)," "ethnic minorities (modernization)," and "migrant workers (moving into the

X. Wu and L. Wang, *Settlement Spaces: Urban Survival Prospects of China's Special Communities*, Springer Geography,
https://doi.org/10.1007/978-981-16-4892-2

city)." We hope to use this framework to adumbrate and examine the macro picture of contemporary Chinese urban social spaces.

With "special community settlement spaces from a community perspective" as one of the long-term prioritized research directions, our research group conducted standardized and specialized studies in a continuous manner for the last decade. By undertaking relevant research projects, publishing papers and journals, winning prizes with our results, fostering talent, and innovating and raising methodological standards, we accumulated a large volume of data and established a solid research foundation, gradually building a highly individual and academically influential research framework and body of research results. This success must first be attributed to my academic mentor Mr. Wu Mingwei. Mr. Wu's erudition and keenness of mind, strict academic standards, and long-term ardent support have been a constant influence and encouragement for me to continue on this journey. At the same time, I could not have done it without the hard work of my senior colleagues in the group, Wang Chenghui, Sun Shiji, Hu Mingxing, and several cohorts of doctoral and masters students, all of whom contributed to our research foundations and technical prowess. Their scholarly entrepreneurship, professional attitude, and team support gave me hope and strength, ensuring that I never suffered the hardships of academic work alone or was in poor spirits. I offer them my most heartfelt thanks and respect.

During this long journey of discovery and investigation, I also benefited from the constructive professional guidance and enlightening, authoritative instruction of fellow scholars and senior colleagues. They helped me extensively in building the current research framework, applying technology in this research, advancing this research as a specialized study, and evaluating the postgraduate theses. Here, I would first like to thank academician Wang Jianguo for his unfailing support and guidance and making time in his busy schedule to write the preface for this book. I would also like to thank Prof. Luo Xiaolong (Nanjing University), Prof. Yao Shimou, Prof. Shen Daoqi (Nanjing Institute of Geography and Limnology at the Chinese Academy of Sciences), Prof. Luo Renchao (Nanjing Tech University), Prof. Li Zhiming (Nanjing Forestry University), Senior Planner Xu Haixian (Jiangsu Province Institute for Urban Planning Design), Cui Shuping (Jiangsu Province Urban–Rural Development Center), and seasoned researchers and experts Prof. Kong Linglong, Associate Prof. Guo Di, Associate Prof. Chen Xiaodong, Associate Prof. Yin Ming, and Associate Researcher Ma Xiaosu (Southeast University).

Due to management problems or simply differences in the statistical scope, the empirical line of this study is often faced with various obstacles and challenges posed by restriction of access to data. In these difficult circumstances, my research group integrated techniques such as fieldwork, statistical questionnaires, and interviews in our independent process of relevant data collection. We also enjoyed the selfless help and warmhearted support of relevant governmental departments, administrative staff, volunteers, residents, and colleagues. Here, I would like to thank the following:

Nanjing Urban Planning and Natural Resources Bureau, Nanjing Bureau of Statistics, Nanjing Urban–Rural Construction Council, Jiangsu Islam Association, Nanjing University Office of International Cooperation and Exchanges, Hopkins-Nanjing Center, Nanjing Normal University Office of International Cooperation and

Exchanges, Hunan Road Sub-district Office, Nanxiu Village Residents Committee, Nanxiu Village Community Service Station for Foreigners, Nanjing Jingjue Temple, Caoqiao Mosque, Jizhaoying Mosque, Tianjing shan Residents Committee, Mufu shanzhuang Residents Committee, Yaolinxianju Residents Committee, Xishan Gardens Residents Committee, Hongmiao Estate Residents Committee, Hongshan Village Residents Committee, Taiping Village Residents Committee.

This book was composed alongside the long-term investigations of my research group and the continual fostering of graduate students. It undoubtedly represents a long and difficult journey of exploration, improvement, and supplementation. Now that I am able to declare an end to this journey, with all its interruptions and detours, and submit this book for publication, I am sincerely indebted to Springer Press for their patience and assistance. In addition, thanks to the master's students Shao Yuntong and Zhang Ruiqi for their assistance in basic work such as graphic editing. Finally, this book, which is the summary and review of my group's research of many years, is not entirely free of subjective views, fragmentary knowledge, and omissions. I respectfully entreat my senior colleagues and fellow researchers to be forthcoming in their criticism.

Xiao Wu

January 2021, Wenchangqiao

Index

X. Wu and L. Wang, *Settlement Spaces: Urban Survival Prospects of China's Special Communities*, Springer Geography,
https://doi.org/10.1007/978-981-16-4892-2

D

R

S

T

U